Transcending Humanitarian Engineering Strategies for Sustainable Futures

Yiannis Koumpouros
University of West Attica, Greece

Angelos Georgoulas
University of West Attica, Greece

Georgia Kremmyda
University of Warwick, UK

A volume in the Advances in Environmental Engineering and Green Technologies (AEEGT) Book Series

IGI Global
PUBLISHER of TIMELY KNOWLEDGE

Published in the United States of America by
IGI Global
Engineering Science Reference (an imprint of IGI Global)
701 E. Chocolate Avenue
Hershey PA, USA 17033
Tel: 717-533-8845
Fax: 717-533-8661
E-mail: cust@igi-global.com
Web site: http://www.igi-global.com

Library of Congress Cataloging-in-Publication Data

Names: Koumpouros, Yiannis, editor. | Georgoulas, Angelos, 1975- editor. | Kremmyda, Georgia, 1979- editor.
Title: Transcending humanitarian engineering strategies for sustainable futures / Yiannis Koumpouros, Angelos Georgoulas, Georgia Kremmyda, editors.
Description: Hershey, PA : Engineering Science Reference, [2022] | Includes bibliographical references and index. | Summary: "This book discusses the most recent challenges and approaches in the field of humanitarian engineering and presents research, case studies, and innovative models"-- Provided by publisher.
Identifiers: LCCN 2022006383 (print) | LCCN 2022006384 (ebook) | ISBN 9781668456194 (h/c) | ISBN 9781668456200 (s/c) | ISBN 9781668456217 (eISBN)
Subjects: LCSH: Sustainable engineering. | Humanitarian assistance.
Classification: LCC TA7 .T73 2022 (print) | LCC TA7 (ebook) | DDC 620.00285--dc23/eng/20220503
LC record available at https://lccn.loc.gov/2022006383
LC ebook record available at https://lccn.loc.gov/2022006384

This book is published in the IGI Global book series Advances in Environmental Engineering and Green Technologies (AEEGT) (ISSN: 2326-9162; eISSN: 2326-9170)

British Cataloguing in Publication Data
A Cataloguing in Publication record for this book is available from the British Library.

For electronic access to this publication, please contact: eresources@igi-global.com.

Advances in Environmental Engineering and Green Technologies (AEEGT) Book Series

Sang-Bing Tsai
Zhongshan Institute, University of Electronic Science and Technology of China, China & Wuyi University, China
Ming-Lang Tseng
Lunghwa University of Science and Technology, Taiwan
Yuchi Wang
University of Electronic Science and Technology of China Zhongshan Institute, China

ISSN:2326-9162
EISSN:2326-9170

Mission

Growing awareness and an increased focus on environmental issues such as climate change, energy use, and loss of non-renewable resources have brought about a greater need for research that provides potential solutions to these problems. Research in environmental science and engineering continues to play a vital role in uncovering new opportunities for a "green" future.

The **Advances in Environmental Engineering and Green Technologies (AEEGT)** book series is a mouthpiece for research in all aspects of environmental science, earth science, and green initiatives. This series supports the ongoing research in this field through publishing books that discuss topics within environmental engineering or that deal with the interdisciplinary field of green technologies.

Coverage

- Alternative Power Sources
- Green Transportation
- Pollution Management
- Industrial Waste Management and Minimization
- Green Technology
- Biofilters and Biofiltration
- Policies Involving Green Technologies and Environmental Engineering
- Renewable Energy
- Radioactive Waste Treatment
- Contaminated Site Remediation

IGI Global is currently accepting manuscripts for publication within this series. To submit a proposal for a volume in this series, please contact our Acquisition Editors at Acquisitions@igi-global.com or visit: http://www.igi-global.com/publish/.

The Advances in Environmental Engineering and Green Technologies (AEEGT) Book Series (ISSN 2326-9162) is published by IGI Global, 701 E. Chocolate Avenue, Hershey, PA 17033-1240, USA, www.igi-global.com. This series is composed of titles available for purchase individually; each title is edited to be contextually exclusive from any other title within the series. For pricing and ordering information please visit http://www.igi-global.com/book-series/advances-environmental-engineering-green-technologies/73679. Postmaster: Send all address changes to above address.

Titles in this Series

For a list of additional titles in this series, please visit: www.igi-global.com/book-series/advances-environmental-engineering-green-technologies/73679.

Food Sustainability, Environmental Awareness, and Adaptation and Mitigation Strategies for Developing Countries
Ahmad Ni'matullah Al-Baarri (Universitas Diponegoro, Indonesia) and Diana Nur Afifah (Universitas Diponegoro, Indonesia)
Engineering Science Reference • © 2023 • 293pp • H/C (ISBN: 9781668456293) • US $110.00

Perspectives on Ecological Degradation and Technological Progress
Veli Yilanci (Faculty of Political Sciences, Canakkale Onsekiz Mart University, Turkey)
Engineering Science Reference • © 2023 • 300pp • H/C (ISBN: 9781668467275) • US $225.00

Climate Change and the Economic Importance and Damages of Insects
Ahmed Karmaoui (Moulay Ismail University of Meknes, Morocco & Moroccan Center for Culture and Science, Morocco)
Engineering Science Reference • © 2023 • 279pp • H/C (ISBN: 9781668448243) • US $215.00

Implications of Nanoecotoxicology on Environmental Sustainability
Rafiq Lone (Central University of Kashmir, India) and Javid Ahmad Malik (Guru Ghasidas Vishwavidyalaya, India)
Engineering Science Reference • © 2023 • 360pp • H/C (ISBN: 9781668455333) • US $250.00

Human Agro-Energy Optimization for Business and Industry
Pandian Vasant (MERLIN, Faculty of Electrical and Electronics Engineering, Ton Duc Thang University, Vietnam) Roman Rodríguez-Aguilar (Panamerican University, Mexico) Igor Litvinchev (Universidad Autónoma de Nuevo León, Mexico) and Jose Antonio Marmolejo-Saucedo (Panamerican University, Mexico)
Engineering Science Reference • © 2023 • 322pp • H/C (ISBN: 9781668441183) • US $240.00

Applying Drone Technologies and Robotics for Agricultural Sustainability
Pethuru Raj (Reliance Jio Platforms Ltd., Bangalore, India) Kavita Saini (Galgotias University, India) and Vinicius Pacheco (Office of the Prosecutor General, Brazil)
Engineering Science Reference • © 2023 • 318pp • H/C (ISBN: 9781668464137) • US $250.00

Intelligent Solutions for Optimizing Agriculture and Tackling Climate Change Current and Future Dimensions
Ahmed Karmaoui (Moulay Ismail University of Meknes, Morocco & Moroccan Center for Culture and Science, Morocco)
Engineering Science Reference • © 2023 • 269pp • H/C (ISBN: 9781668446492) • US $215.00

IGI Global
PUBLISHER of TIMELY KNOWLEDGE

701 East Chocolate Avenue, Hershey, PA 17033, USA
Tel: 717-533-8845 x100 • Fax: 717-533-8661
E-Mail: cust@igi-global.com • www.igi-global.com

Table of Contents

Detailed Table of Contents

the suitability of the TRM initiative for the polder in East Beel Kuksia. The research was carried out using coupled 1D and 2D hydrodynamic models, built with the software tool MIKE11 and coupled with MIKE21. Erosion was calculated based on information on bed shear stress from the hydrodynamic model and erosional critical shear stress. With the implementation of TRM, the drainage capacity of the rivers downstream of the studied polder consequently increased.

Land subsidence and its consequences are at an alarming level in the Southern and low-lying areas of the Mekong Delta. This chapter will discuss the situation of groundwater uses and urban development activities that caused the negative impacts of land subsidence and sea level rise in Ho Chi Minh City and the Mekong Delta region of Vietnam. The PSInSAR technique was applied for land subsidence monitoring and showed that an average subsidence rate at 15 mm/year which is 5 times greater than the effects resulting from the rise of sea level. This implies that land subsidence coupled with sea level rise have been as a major factor contributing to increasing the depth of annual flood and causing inconveniences to the daily lives of residents that suffered waterlogging of between 0.4 – 0.5m. Apart from the restriction of groundwater abstraction, rainwater harvesting – an alternative water sources – is a humanitarian engineering solution to minimize groundwater extraction which in turn to prevent from flood risks caused by land subsidence and the rise of sea level.

Cultural heritage sites are fundamental elements of national and international heritage and are regarded as sensitive and critical structures of high and of national importance, exposed to a range of natural hazards, leading often structural failures. Cultural heritage sites include structures and monuments with effectively limitless lifetime, while admitting only limited intervention options. Therefore, the efficient assessment of their structural risk forms a very demanding task in comparison to similar analyses applied to modern structures. The current study is involved with natural hazards and natural disasters that have affected cultural heritage sites in the past and their consequences regarding their structural integrity. After a description of the aforementioned, specific scientific tools and advanced methods for the investigation of the geophysical/hydrological and climatological/meteorological natural hazards and their consequences at cultural heritage sites, regarding their structural integrity are presented and discussed.

Antibiotics are certainly miracle drugs and have comprehensive usages in humans and livestock for the treatment of various bacterial diseases. However, inappropriate or misuse of antibiotics can lead to the development of antimicrobial-resistant bacteria. Antibiotic resistance happens when bacteria change in responding to the usage of these medicines. These bacteria can also be disposed of from animals to humans via contact between humans and animals directly, or through the food chain and the environment. Antibiotic resistance is a great threat to global humanity as it can prolong treatment in hospitals, increase medical costs, increase mortality, and burden the family. Residual antibiotics above the maximum residue limit (MRL) are harmful to human health and animals and need to be monitored. By increasing national policies, international declarations, agreements for enough chemical and analytical instrumental facilities, and above all, creating awareness about the harmful effect of resistance, the situation of antibiotics resistance can be improved.

The purpose of this study was to examine neuroimaging technologies for quantifying cognitive load in intelligent interactive multimedia systems for experimental applications by conducting a systematic review of all relevant papers published up to April 2020. The study's most striking finding is that electroencephalography, functional magnetic resonance imaging, functional near-infrared spectroscopy, and transcranial doppler ultrasonography are the most frequently used neuroimaging equipment in cognitive load research in multimedia learning. Forty papers were selected depending on the equipment that should be understood in the field of neuroimaging in examining cognitive load in multimedia learning, the benefits and drawbacks of neuroimaging devices, and the experimental protocol for cognitive load in multimedia research. The study's findings were analyzed, and numerous discrepancies in the research on cognitive load and multimedia learning were discovered.

Pesticides are chemical compounds that are toxic and are used to control different types of pests and uncontrolled growth of weeds for plant protection in crop production and food storage. A large number of

This chapter presents results and analysis from the ENHANCE project survey for the views of employers and educators towards the impactful improvement of engineering degree programs in Vietnam. This building capacity project aims to find ways of strengthening the employability of engineering graduates by improving the quality of engineering programs. The survey was carried out by inviting respondents from industry (employers) and from higher education (educators) to answer online questionnaires. The number of respondents was 116, of which 69 are educators and 47 are employers. This chapter presents results that encompass the following three topics: university graduate attributes, capabilities, knowledge, and understanding; the teaching and learning methods used in universities; and the future of engineering programs in Vietnam.

Over the years humanitarian issues have increased in many parts of the world including Indonesia. Such issues drive the need to evaluate Indonesia's engineering higher education curriculum. Aiming to provide a constructive contribution to inherent humanitarian aspects in the engineering education curriculum, this study revisits the teaching and learning programs in each curriculum of three public universities in Indonesia (i.e., ITB, UB, and UGM). The survey was conducted on 112 respondents from higher education providers and quality assurance agency and 43 respondents from the industry. Results show that about half of postgraduate and a quarter of undergraduate programmes have already run humanitarian engineering programs. According to the industry, undergraduates have better average scores in creativity and leadership, while postgraduates have higher average scores in other parameters such as analytical skills, experience in real-life applications, practical skills, and teamwork skills.

Indonesia is an earthquake-prone country. In many earthquakes, the housing sector suffers the most damage. Repairing and rebuilding houses is often the most expensive part of post-disaster recovery, placing a burden on the government. Standards and guidelines for earthquake-resilient houses exist, yet houses continue to be damaged in earthquake events. Although Indonesia has numerous engineers, both experienced and newly qualified, only a few are interested in pursuing and identifying affordable solutions to the housing problems caused by earthquakes that are sustainable and resilient. Moreover, people are unaware of and unable to implement the proposed and promoted solutions. With many people living in earthquake-prone areas with little access to affordable and appropriate technology, the challenge facing Indonesia today is to produce more engineers who have the level of sensitivity, empathy, and passion required to address this great humanitarian need. The promotion of humanitarian engineering to provide appropriate solutions must start early on in higher engineering education.

Quality assurance in higher education is a global practice. However, the quality assurance aspect in the engineering educational institutions of Bangladesh is still at the inception level. It is crucial to monitor, assess, and improve the system to make the system effective in the context of Bangladesh. This study highlights the present practice and the perception of the professionals involved with the agencies as well as the way forward to further improve the quality assurance aspects. A questionnaire survey was conducted to assess the current status of quality assurance practices in the educational institutions of Bangladesh. It is found that there is a limited scope of quality assurance in engineering education in Bangladesh. It is also revealed that there appears to be a lack of direction as there are a number of ideas for quality assurance that have a similar level of support among the respondents. There is a long way to go for the quality assurance set-up in Bangladesh to mature and be more effective.

Developing countries like Bangladesh are facing multidimensional challenges due to climate change. The traditional engineering sector needs to evolve to contribute to the solution of the emerging problems by incorporating social and technical solutions. Integration of humanitarian and soft engineering and proper project formulation and implementation can go a long way in addressing the challenges. Harmonised and clear policies to this end are of high importance, along with political will and accountability in the public investments. Efficient utilisation of resources are needed through objective and proper prioritisation of projects – free from external influence or interference. Geopolitical cooperation between countries, coordination between the government and non-government sectors, as well as within the departments of the government is crucial for the success of humanitarian and soft engineering in resource-constrained countries.

Preface

Engineering disciplines have a pivotal role to play in the solution of global humanitarian challenges, enabling our society to take steps towards sustainable human development. Engineering can be used as the catalyst for the change that the world needs; from water supply to renewable energy provision, engineering knowledge and application underpin the responses needed for us all to pursue a sustainable future. Because the issue of humanitarianism is not just engineering problems, there is a need to engage with professionals, breakdown previously siloed approaches and obdurate practices, and introduce interdisciplinary education and training to enhance combinational expertise. The need for a more human-centered approach to engineering has relatively recently led to what we call humanitarian engineering.

This book aims at providing relevant frameworks and, the latest research and education outcomes and impact in the theme of humanitarian engineering. It is written from researchers and professionals who want to improve the strategic role of humanitarian engineering at different levels of the modern world and the various challenges facing our society.

Transcending Humanitarian Engineering Strategies for Sustainable Futures covers research, practice, and applications associated to humanitarian engineering and, offers insights to educational frameworks as a means for future-proofing our communities. Covering topics such as disaster mitigation, natural hazards, land use change and, education for training future leaders in engineering and sustainable development, this premier highly interdisciplinary book is an excellent resource for engineers, scientists, environmentalists, sociologists, government officials, students and educators of higher education, non-profit organizations, researchers, and academicians.

The book investigates transcending challenges and approaches in the field of humanitarian engineering, presenting research, real-life case studies, innovative models and approaches to education, and other work that lies in the presented theme. The interdisciplinarity of the authors combined with the chapters that analyze and focus on real events and conditions in the most vulnerable countries (e.g., Indonesia, Bangladesh, Vietnam, etc.) makes it a valuable tool for academics, researchers, and students in the field.

The first part of the book brings together a series of chapters showcasing recent research outcomes and/or the impact of research, practice and field experience at the intersection of engineering and science, technology, design and community development across a range of topics such as land use and ecosystems, land subsidence, natural hazards, health and neuroimaging, and environmental impact of human activity.

The second discusses the delivery of engineering education globally by introducing modern, innovative pedagogical approaches, identifying, and tackling inherent barriers in quality, knowledge organisation, content, and hierarchy to support the advancement of emerging skills. Reference is made to needs and requirements imposed to education approaches from Quality Assurance, Industry and Higher Education providers.

Yiannis Koumpouros
University of West Attica, Greece

Angelos Georgoulas
University of West Attica, Greece

Georgia Kremmyda
University of Warwick, UK

Acknowledgment

This publication was supported by the project ENHANCE - ENabling Humanitarian Attributes for Nurturing Community-based Engineering (project No: 598502-EPP-1-2018-1-UK-EPPKA2-CBHE-JP (2018-2582 / 001-001), funded by the Erasmus + KA2 Cooperation for innovation and the exchange of good practices – Capacity building in the field of Higher Education. The publication reflects the views only of the authors, and the Commission cannot be held responsible for any use which may be made of the information contained therein.

Chapter 1
Impact of Land Use Change on Ecosystem Services in a Coastal Town of Bangladesh

Md. Al-Amin
Local Government Engineering Department, Bangladesh

G M Tarekul Islam
Bangladesh University of Engineering and Technology, Bangladesh

ABSTRACT

There is an interplay between land use change and ecosystem services. The relationship between land use change and ecosystem services was explored in Barisal Sadar Upazila, a small coastal town in Bangladesh. Land use changes in this municipality were analyzed using Landsat satellite images from 1980 to 2010 at 10 years intervals. It was found that the town experienced a rapid change in its ecosystem services in proportion to land use change, especially the conversion of agricultural land to urban settlement. In the town, the agricultural land, wetlands, and water bodies decreased by 30%, 47%, and 6% respectively from 1980 to 2010. On the contrary, urban land and rural land area increased by 201% and 23% respectively during the same period. Even though the land use change had brought economic development to the people of the town, there had been a negative impact on ecosystem services which calls for a balancing trade-off between economic development and conservation of ecosystem services for the greater benefit of the society, especially for the sustenance of the ecosystem.

INTRODUCTION

Land use has experienced massive changes worldwide over the past half-century i.e. from forest, grassland & wetland to arable & developed land; due to the increase in population, commercial advancement, urban stretch, hasty expansion of industrial and agricultural activities, etc (Chen et al. 2014). Sometimes these changes seemed advantageous but disrupt the various ecosystem services so gravely which is past repairable. Appreciating the necessity of balanced ecosystem services; this study quantifies the land use

DOI: 10.4018/978-1-6684-5619-4.ch001

changes of Barisal Sadar Upazila, a small coastal municipality in Bangladesh using Landsat satellite images from 1980 to 2010 at 10 years intervals. The impact of those changes on ecosystem services has also been explored here.

Ecosystem services are the features of ecosystems that are enthusiastically or submissively consumed to yield mankind's well-being (Fisher et al. 2009). As per the 2006 Millennium Ecosystem Assessment (MA), ecosystem services are "the benefits people obtain from ecosystems" (MA, 2005). These services include provisioning services such as food and water; regulating services such as flood and disease control; cultural services such as spiritual, recreational and cultural benefits which affect the people directly and finally supporting services such as nutrient cycling, that sustain the surroundings for life on earth (Mooney et al. 1996). The poor and marginalized people of society depend on ecosystem services especially the provisioning type of ecosystem services. The demolition of ecosystem services may cause societal collapse. Since the early 1990s, people become more concerned about rapid land use change and understand that the provision of ecosystem services is directly affected by land use changes (DeGroot et al. 2002). Many studies were accomplished in the last couple of decades worldwide to quantify these land use changes and figured out the effect of those changes on ecosystem services. Some of them only computed the area to find the change while others linked the land use changes with ecosystem services.

The world's net cultivated area has grown by 12 percent over the last 50 years, mostly at the cost of the forest, wetland and grassland habitats (Dubois, 2011). From 1975 to 2000, Africa had expanded 57% of its agricultural land at the expense of 16% of its forest and 5% of its woodlands and grasslands; where almost 50000 sq. km of natural vegetation was lost per year (Eva et al. 2006). In China urbanization increased from 17% in 1978 to 41% in 2006 (Kabba et al. 2011) and from 1990 to 2000 China's cropland increased by 27900 sq. km (Liu et al. 2005). Vihervaara et al. (2012) studied the land use changes in Uruguay from 1987 to 2007 and found a significant loss of grassland area to plantations, which affects the provision of ecosystem services seriously. The land use changes in the Small Sanjiang Plain in China experienced 17300 sq. km of cropland increases at the expense of wetland, forest, grassland, and wasteland from 1980 to 2010 (Chen et al. 2014). Kodaikanal Taluk, a hilly area in India had lost 395 sq. km (almost 50%) of its forest area for the expansion of agricultural, harvested and built-up areas (Prakasam, 2010). Bangladesh lost more than 10000 sq. km of cropland in the period from 1976 to 2010 (Hasan et al. 2013).

Kumar and Yashiro (2014) studied the linkage between ecosystem services and aspects of extreme poverty. Their findings suggest that the poor are often more vulnerable to the loss of ecosystem function that restricts the supply of natural goods and services. From 1995 to 2006, the loss of ecosystem services value along the coast of the Gulf of Mexico was found approximately $0.20 million/sq. km/year (Mendoza-González et al. 2012). Due to coastal erosion in Europe ecosystem services value decreased approximately from $23.10 billion in 1975 to $22.30 billion in 2006 and for the year 2050, the predicted loss stuck between $20.80 billion to $20.1 billion per year (Roebeling et al. 2013). In the Small Sanjiang Plain, China excluding food production all other ecosystem services were reduced and the estimated loss of ecosystem services was $2.37 billion from 1980 to 2000 and $1.67 billion from 2000 to 2010 (Chen et al. 2014). Islam et al. (2015) identified the agricultural land use changes in the Ganges delta and their implications on ecosystem services. They found an almost 50% reduction of agricultural land and over 500% expansion of wetlands from 1980 to 2008 because of a significantly increasing trend of saltwater shrimp farming. The reduction of agricultural lands and the rise of wetlands are endorsed by the fact that the agricultural lands have been transformed into shrimp farms in the study area. During this period settlement & other land use types also increased to nearly 5% but the area of Sundarbans remains steady.

Hossain et al. (2016) studied the recent changes in ecosystem services and human well-being of the Bangladesh coastal zone and found many ecosystem services had gone through noteworthy changes from 1975 to 1980 period. Contrary to these studies, ecosystem services value in the Huairou reservoir basin, China showed an increasing trend. Compared with the ecosystem services value in 1990; it increased by 0.57, 1.58, 2.46 and 2.88% in 1995, 2000, 2005 and 2008 respectively (Wang et al. 2012). Researchers preferred remote sensing and GIS tools to determine the land use change and these tools not only provide better results but also gave a good map to visualize the land use change. An open-source GIS and Remote Sensing software "Integrated Land and Water Information System (ILWIS version 3.7.2)" has been used to process the Landsat images and conduct spatial analysis in this study.

Land Use Change in Bangladesh

The total land area of Bangladesh increased from 142800 sq. km to 148400 sq. km during 1976-77 to 2010-11 due to the reclamation of char lands but during the same period net cultivable land decreased from 93900 sq. km to 85200 sq. km at a rate of 0.27 percent annually (Agricultural Statistics Yearbook of Bangladesh, 2011). Soil Resource Development Institute (SRDI) studied the land use changes in Bangladesh from 1976 to 2010 and the average loss of agricultural land was found at 331.4 sq. km yearly. However, the agricultural land decreased from 133000 sq. km in 1976 to 127400 sq. km in 2000 with an annual loss of 233.91 sq. km, and with an annual loss of 565.37 sq. km, it reached 121700 sq. km in 2010. These findings indicate that the agricultural land was decreasing much faster from 2000 to 2010 than from 1976 to 2000 (Hasan et al. 2013). In this study, agricultural land consists of cropland, forest, mangrove forest, river, lake, wetlands, tea estate, salt pan and non-agricultural land consists of rural settlements, accreted areas, urban and industrial areas. It has been estimated that more than 809 sq. km of agricultural land are converted into cities, roads and infrastructure annually in Bangladesh (Sayed and Haruyama, 2015). Dewan et al. (2009) studied the land use in the Dhaka Metropolitan area from 1960 to 2005 and identified that the built-up areas increased by 159.24 sq. km, agricultural land decreased by 76.14 sq. km, vegetation decreased by 23.36 sq. km, wetland/lowland decreased by 63.85 sq. km, water bodies decreased about 8.64 sq. km during this period. Overall, the urban areas increased from 11% in 1960 to 344% in 2005 and the landfill/bare soil category increased by about 256%. In Manikganj Sadar Upazila, built-up areas increased by approximately 12%, while agricultural land decreased by 7%, water bodies decreased by about 2% and bare land decreased by about 2% between 1989 and 2009 (Sayed and Haruyama, 2015). Chakoria Sundarban, a mangrove forest area in Bangladesh, experienced remarkable land use change from 1989 to 1999, when the shrimp farm and salt bed are taking place inside the mangrove forest area (Musa, 2008). For sustainable development, it is very essential to quantify this land use change and determine its impact on society. This study attempted to quantify the land use changes and their impacts on ecosystem services in Barisal Sadar Upazila, a small coastal town in Bangladesh.

MATERIALS AND METHODS

Study Area

Barisal is one of the most important cities in the southern part of Bangladesh and the center of this city is Barisal Sadar Upazila. It is the entry point of Jhalokhati, Patuakhali, Barguna, Bhola, Bagerhat and

Figure 1. Location of Barisal Sadar Upazila

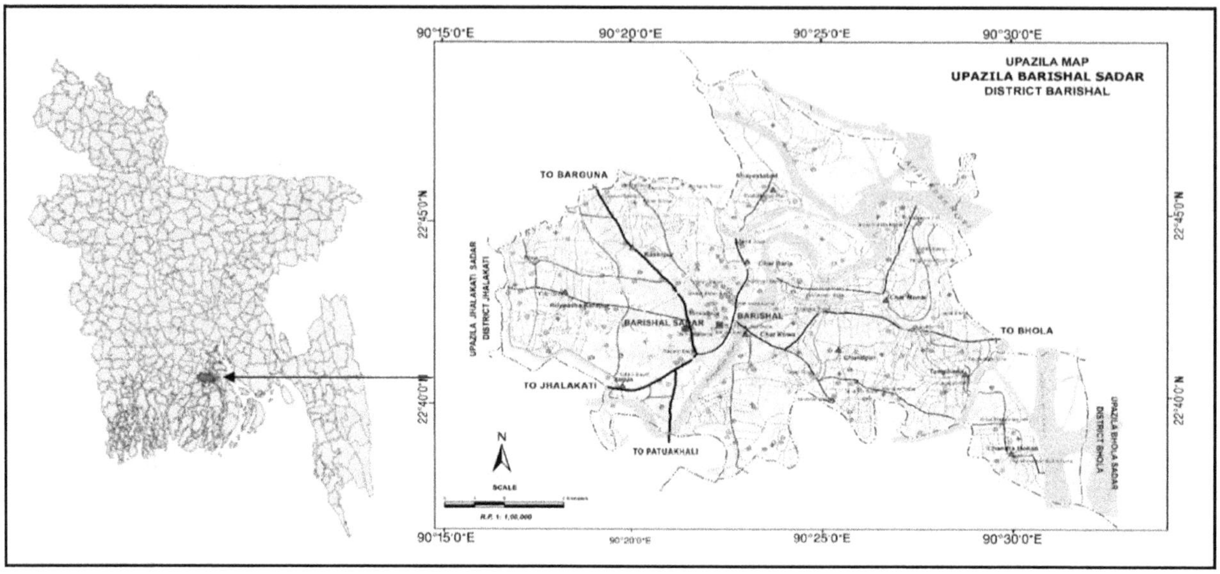

Pirojpur districts; all of them are part of the southwest coastal zone of Bangladesh and exposed to the sea. It is also the administrative and judicial headquarter of both the Barisal Division and Barisal

District. The area of Barisal Sadar Upazila increased from 308 sq. km in 1981 to 318 sq. km in 2001 and further increased to 324 sq. km in 2011 including 292 sq. km of land area and 32 sq. km area of the water body. Figure 1 shows the location of the study area between latitudes 22°37'N to 22°43'N and longitudes 90°16'E and 90°32'E.

The total population of Barisal Sadar Upazila in 1991, 2001 & 2011 was 414281, 463032 & 527017 with a decadal growth rate of 8, 12 & 14% respectively. Literacy rate (for 7 years and above) was found to be 54, 65 and 69% in 1991, 2001 and 2011 respectively. The literacy rate of females was found less than males, though the difference is minimal in recent years. The drinking water supply system of Barisal Sadar Upazila was vastly improved during the last two decades. In 1991, 46% of households used tube well, 27% used supplied water, 1% used dug well, 19% used ponds and 7% used canals or rivers as the main source of drinking water. In 2011, the scenario improved significantly; about 92% of general households used deep tube-well, 6% used to supply water and the remaining 2% used other sources for drinking water. Sanitation Facilities of Barisal Sadar Upazila showed a huge improvement from 1981 to 2011 period. The use of sanitary latrines increased from 39% of households in 1981 to 80% of households in 2011 and households with no toilet facility dropped from 11% to 1% during this time (BBS Population Census 1981, 1991 & 2011).

Historically, Barisal Sadar Upazila is an agriculture-dependent area. The net cultivated area of this Upazila was 150.07 sq. km, 152.06 sq. km and 150.21 sq. km in 1983-84, 1996 and 2008 respectively. Different species of rice namely Aus, Aman, Boro & IRI, Pulses, Vegetables, Jute and Spices are the main crops of this area. Besides rice; wheat, corn and different kind of oilseeds are also cultivated here. Both local and hybrid Aman, Aus and Boro are cultivated here. Cow, goat, chicken and duck are the main livestock and poultry in Barisal Sadar Upazila. Recently, a significant number of pigeons nurtured here. A few buffalo and sheep were also raised here. Barisal is renowned for different types of fish production. Due to a large number of canals, ponds, semi-closed water bodies and a significant area of floodplain and river; a huge amount of fish is captured here. Hilsa, catfish, carp and shrimp are the major types of fish here. The total number of the population engaged in agriculture work was 52770 including 33366 males and 19404 females in Barisal Sadar Upazila. In 1983-84, 1996 and 2008 total agri-labor households were 5470, 11554 and 15031 respectively; which is almost 17, 26 and 15% of all holdings of Barisal Sadar Upazila respectively. This significant decrease in the number of agri-labor households from 1996 to 2008 increased the percentage of the population engaged in other services and businesses (BBS Census of Agriculture 1983-84, 1996 and 2006).

A tropical environment exists in Barisal Sadar Upazila. The average annual temperature is 25.9°C. May to August is the hottest time and January is the coldest time of the year. The average annual rainfall is 2171 mm. Most rainfall occurs in July averaging 416 mm and January is the driest month with an average of less than 10 mm. After the statistical trend test was done; no significant trend was found for the annual maximum temperature and annual mean temperature in Barisal Sadar Upazila. But a decreasing trend was found for annual mean precipitation at both 5% and 10% significance levels during the 1949 to 2010 period.

Methodology

In this study, land use changing patterns in Barisal Sadar Upazila have been quantified by digital interpretation of Landsat images and the impact of land use change on ecosystem services has been studied using both primary and secondary data. For the accuracy assessment, ground-truthing data have been collected on selected locations. For this purpose, the research framework illustrated in Figure 2 has been prepared.

Digital Interpretation of the Land Use Change

Four Landsat images have been used in this study to determine the land use change in Barisal Sadar Upazila. All of them were collected from the webpage of the United States Geological Survey (www.earthexplorer.usgs.gov). The images were taken from November and December as shown in Table 1 when the sky remains free from the cloud and temporal changes in water bodies are insignificant during this period. The path and row numbers are a unique way to indicate the location of the scene whereas the resolution expresses the minimum dimension of the aspect which can be recognized from the image.

In this study, land use patterns were classified into five categories as described in Table 2 including agricultural land, rural area, urban area, water body and wetlands.

To prepare land use maps for 1990, 2000 and 2010, band combinations of 7-4-2 have been used. These three bands with three different wavelength properties when combined can infiltrate atmospheric elements and deliver a “natural-like” interpretation. After importing each band individually a sample set was made using the above-mentioned combination. To train for selecting each land use category a significant number of pixels were selected. Using this sample set maximum likelihood technique was used for supervised classification. Four land use maps for 1980, 1990, 2000 and 2010 have been obtained after completing this process.

Figure 2. Research framework

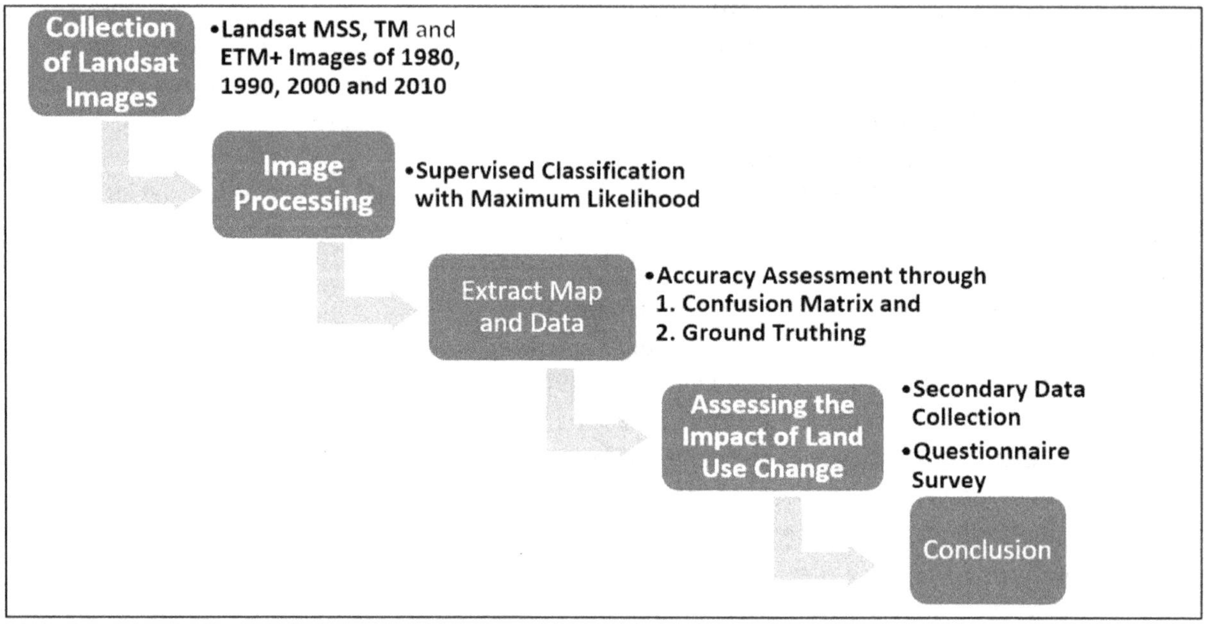

Confusion Matrix and Ground Truthing

To ensure the accuracy of land use classification, a confusion matrix has been created using ground truthing as mentioned in Figure 2. In the confusion matrix, both correct and incorrect numbers of predictions are summarized with count values and broken down by each class. Ground truthing is a method, where

Table 1. Details of landsat satellite images

Satellite Name	Path and Row	Acquisition Date	Resolution (m)
Landsat Multispectral Scanner (MSS)	147 and 44	Dec 4, 1980	60
Landsat Thematic Mapper (TM)	137 and 44	Dec 9, 1990	30
Landsat Enhanced Thematic Mapper Plus (ETM+)	137 and 44	Nov 26, 2000	30
Landsat Thematic Mapper (TM)	137 and 44	Nov 30, 2010	30

Table 2. Land use classification scheme of Barisal Sadar Upazila

Agricultural land	Crop fields, pasture, and bare fields
Rural area	Homestead, forest land, orchards, groves, vineyards, and nurseries, village road
Urban area	Residential, industrial and commercial, transportation, communications, mixed urban or built-up land, other urban or built-up lands
Waterbody	Permanent open water, lakes, reservoirs, streams, bays and estuaries
Wetland	Non-forested wetlands including the sand bar

a pixel on a Landsat image is compared to the prevailing land use. After a supervised classification was done by the remote sensing software, the ground-truthing helps to find out the percentage of accuracy in the land use classification.

Determination of the Impact of Land Use Change on Ecosystem Services

Table 3 illustrates the various source of secondary data which were collected from the Bangladesh Bureau of Statistics, Ministry of Fisheries And Livestock, Bangladesh and Bangladesh Meteorological Department to determine the impact of land use change. After collecting these data, land use information extracted from Landsat images has been analyzed to infer the impact of land use change on ecosystem services.

Table 3. Details about the main sources of secondary data

Department	Publication
Bangladesh Bureau of Statistics	Population Census 1981, 1991, 2001, 2011
	Census of Agriculture 1983-1984, 1996, 2006
Department of Fisheries, Bangladesh	Fisheries Statistical Yearbook 1983-1984, 1990-1991, 2000-2001, 2010-2011
Bangladesh Meteorological Department	Temperature and Rainfall Data

QUESTIONNAIRE SURVEY

Land use changes have social and environmental impacts as well. There is a wide range of unstable subtleties between ecosystem services and human well-being (Hossain et al. 2016). To assess the social impact of land use change, a questionnaire survey was conducted. Non-probability sampling technique was used for selecting interviewees to ensure diversity and avoid sampling errors containing biases (Pereira et al. 2005). Sixty persons were interviewed and asked about their educational level, occupation, future plan, their perception of land use change, agricultural work in their area, gardening, water supply and sanitation facilities in their house, occupation and education of other family members, and the overall situation of the society. The interviewees were farmers, landless farmers, housewives, fishermen, poultry farmers, woodcutters, sawmill workers, carpenters, shopkeepers, fish farmers, auto-rickshaw pullers, businessmen, politicians, city council members, teachers, students, etc. Both men and women were interviewed for this survey. Special attention was given to crop production, livestock and poultry, fish production, lifestyle and livelihood status, women's involvement and other social elements.

ASSESSMENT OF THE LAND USE CHANGE

Four land use maps as revealed in Figure 3 for the years 1980, 1990, 2000 and 2010 were obtained by analyzing the Landsat images.

The extracted data has been presented in Figure 4, which showed a significant decrease in agricultural land and wetlands from 1980 to 2010. In this period, agricultural land, wetlands and water bodies

Figure 3. Land use maps of Barisal Sadar Upazila in 1980, 1990, 2000 and 2010

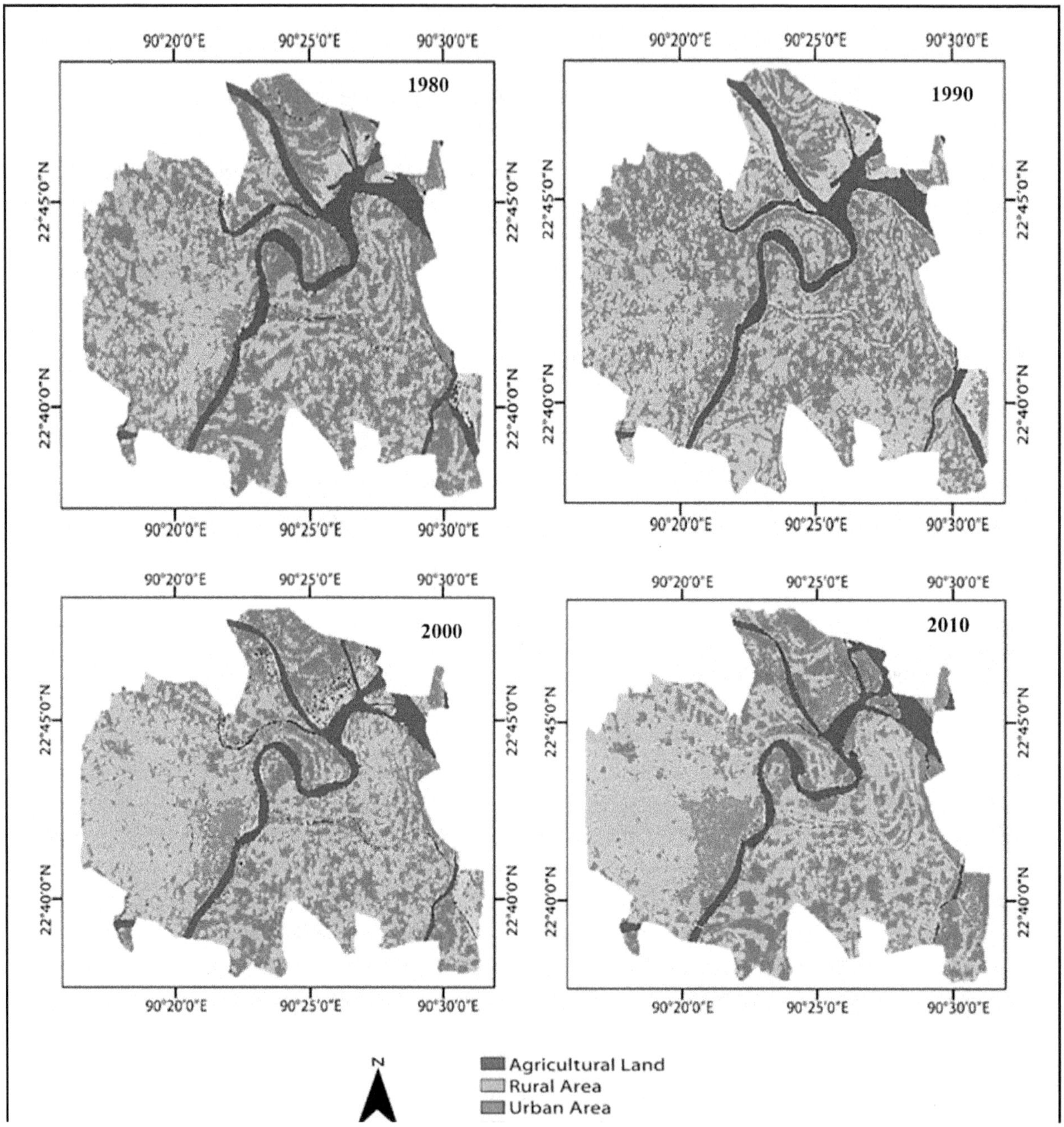

decreased by about 30, 47 and 6% respectively. The agricultural area increased from 125.08 sq. km to 133.26 sq. km from 1980 to 1990, after that it showed a decreasing trend and in 2010 reduced to 87.86 sq. km (Figure 4a). Many wetlands i.e. char areas were transformed into agricultural areas from 1980 to 1990 period (Figure 4b). As a result, the agricultural area showed an increasing trend during that time. Wetlands decreased from 24.69 sq. km to 14.92 sq. km from 1980 to 1990 period because of the transformation to the agricultural area, after that, it remained almost stable and in 2010 it became 13.02 sq. km. Waterbody decreased from 34.19 sq. km to 32.05 sq. km from 1980 to 2010 (Figure 4c).

On the contrary, urban land and rural land increased by 201% and 23% respectively during the same period. The rural area decreased from 95.6 sq. km to 85.50 sq. km from 1980 to 1990, after that it showed an increasing trend and in 2010 increased to 117.80 sq. km (Figure 4d). Barisal Sadar Upazila was established in 1984. As a result, urbanization showed an increasing pattern at that time at the expense of rural settlement (Figure 4e). Moreover, from 1990 to 2000 rural areas showed an increasing trend at the expense of agricultural areas (Figure 4a). The urban area showed an increasing trend and from 1980 to 2010 period it increased from 14.33 sq. km to 43.15 sq. km which is more than three times its area in 1980 (Figure 4e). From Figures 4a and 4e, it seems that this transformation mainly happened at the expense of agricultural areas from 2000 to 2010 period but further study is essential to confirm this linear trend.

Assessing the Accuracy of Land Use Classification

For this study, overall land use classification accuracy was found to be 97, 97, 95 & 98% for the years 1980, 1990, 2000 & 2010 respectively from ILWIS auto-generated confusion matrix technique. Moreover, ground-truthing was done after preparing the land use map for the accuracy assessment in this study. In this research, 36 sites were randomly selected throughout the study area for ground-truthing including 9 sites in agricultural areas, 3 sites in the water body, 1 site in wetland, 7 sites in the rural area and 16 sites in the urban area. Their location was identified with a GPS instrument. After that field information on the observed site was compared with the land use map for the year 2010. Out of 36 sites, 29 sites matched with the land use map of the year 2010 and the accuracy was found 81% for the land use map of 2010 in this study. Barisal Sadar Upazila has a very dynamic land area. Every day the land use pattern is changing. Many lowland areas are continuously filled with sand for settlement. As a result, ground-truthing was challenging in 2018 for the land use map of 2010 and the accuracy seemed low.

IMPACT OF LAND USE CHANGE ON ECOSYSTEM SERVICES IN BARISAL SADAR UPAZILA

For a specific area, the presence of all the ecosystem services might not possible. In this research, the impact of Land-use changes on agricultural areas and the livelihood of the community have been scrutinized. Moreover, some provisioning and cultural services such as food and fiber (including livestock, poultry, and fish production), transportation facilities and educational facilities were also studied.

These services were chosen because they are directly connected to the agricultural area and livelihood of the community. As a consequence, the social and economic environment of the study area depends on these services. To understand the impact of land use change and local people's perception of this matter clearly, a questionnaire survey was conducted in this study as well as many secondary data were used. To link with the land use change; the area used for a certain purpose and the product received from that area was compared. Sixty persons were interviewed and asked about their educational level, occupation, future plan, perception about land use change, agricultural work in their area, gardening, water supply and sanitation facilities in their house, occupation and education of other family members and the overall situation of society. The interviewee was selected by using the non-probability sampling technique and included farmers, landless farmers, housewives, fisherman, poultry farmer, woodcutter, sawmill worker,

Figure 4. Temporal change of agricultural area use in Barisal Sadar Upazila from 1980 to 2010

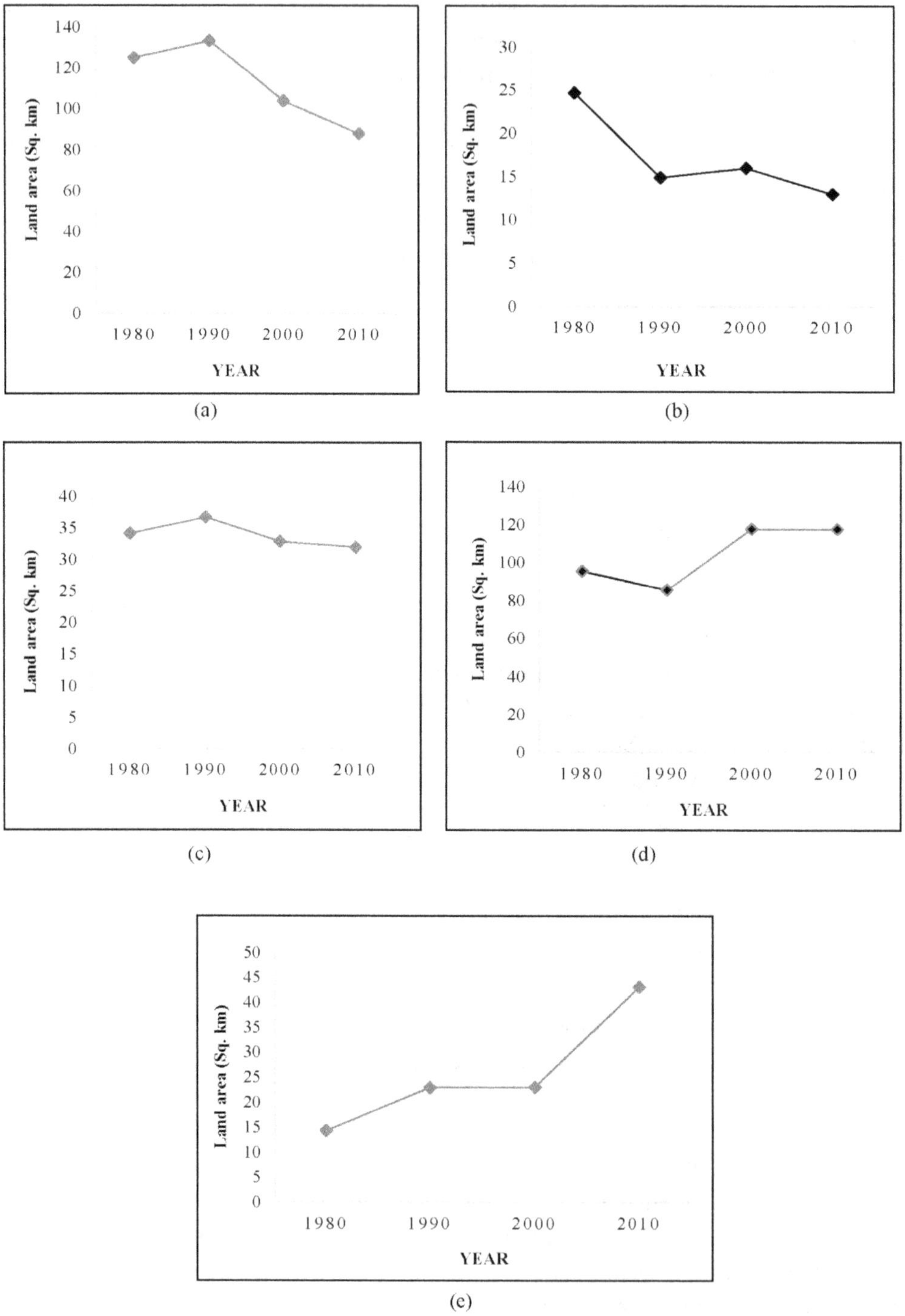

carpenter, shopkeeper, fish farmer, auto-rickshaw puller, businessman, politician, Union Porisod member, teacher, student, etc. Both men and women were interviewed for this survey. Special attention was given to crop production, livestock and poultry, fish production, lifestyle and livelihood status, women's

Figure 5. Temporal change of wetlands use in Barisal-Sadar Upazila from 1980 to 2010

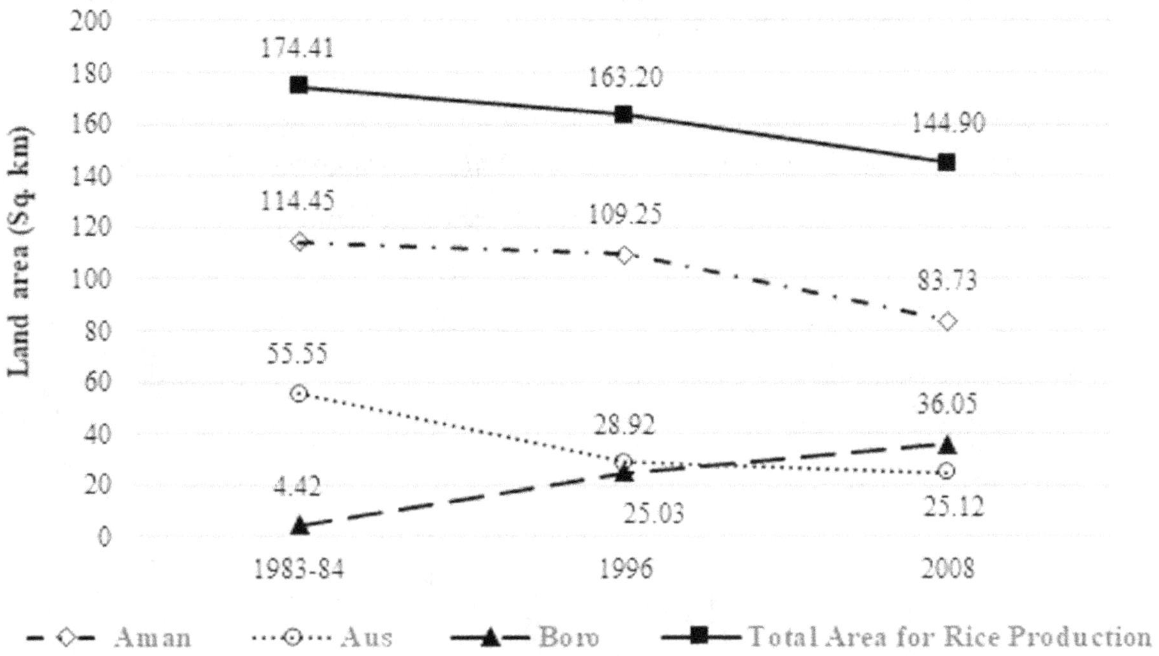

Figure 6. Temporal change of waterbody use in Barisal-Sadar Upazila from 1980 to 2010

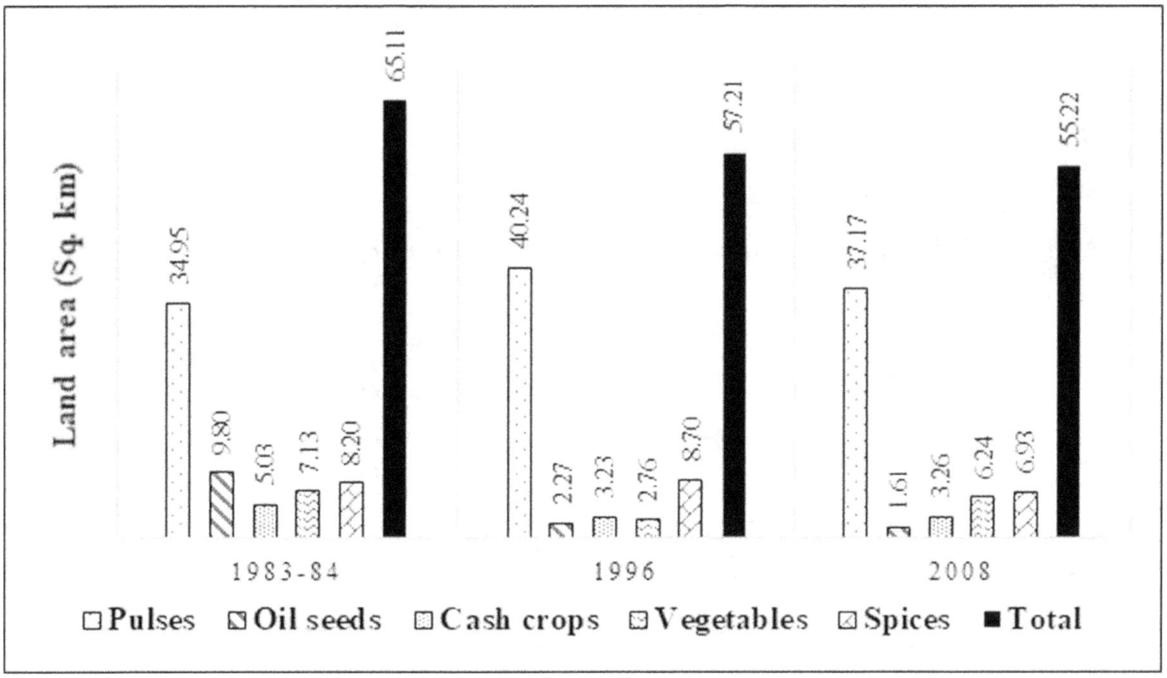

involvement and other social elements. A significant impact on different ecosystem services was found due to land-use changes in Barisal Sadar Upazila. After analyzing the various secondary data and results of the questionnaire survey, the following things have been noticed.

Figure 7. Temporal change of rural area use in Barisal-Sadar Upazila from 1980 to 2010

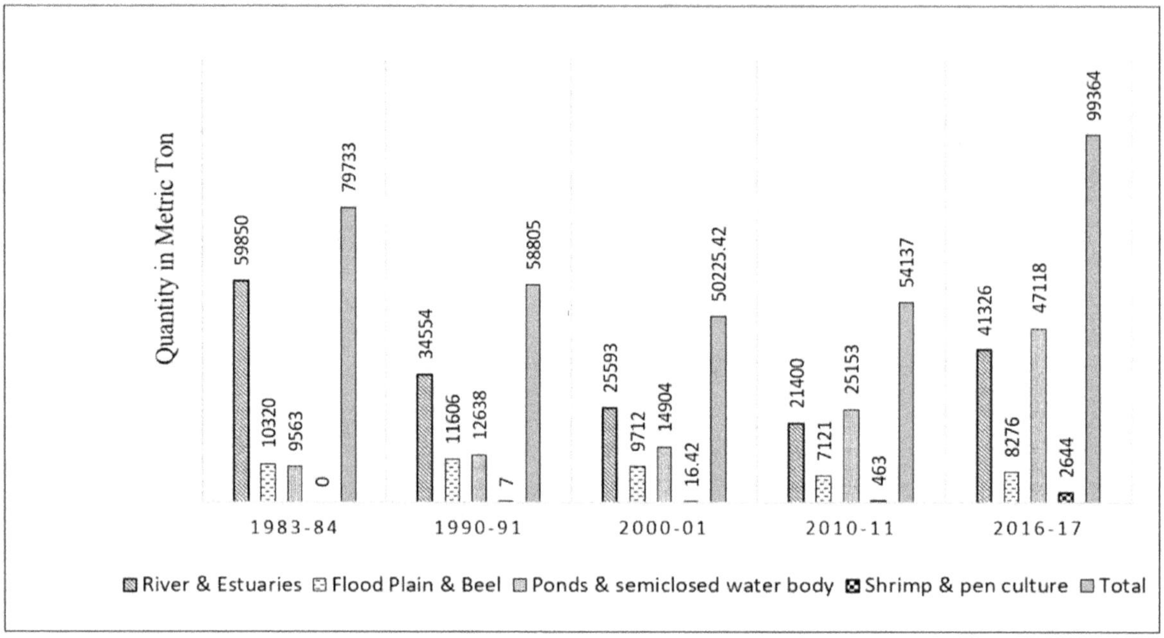

Figure 8. Temporal change of urban area use in Barisal-Sadar Upazila from 1980 to 2010

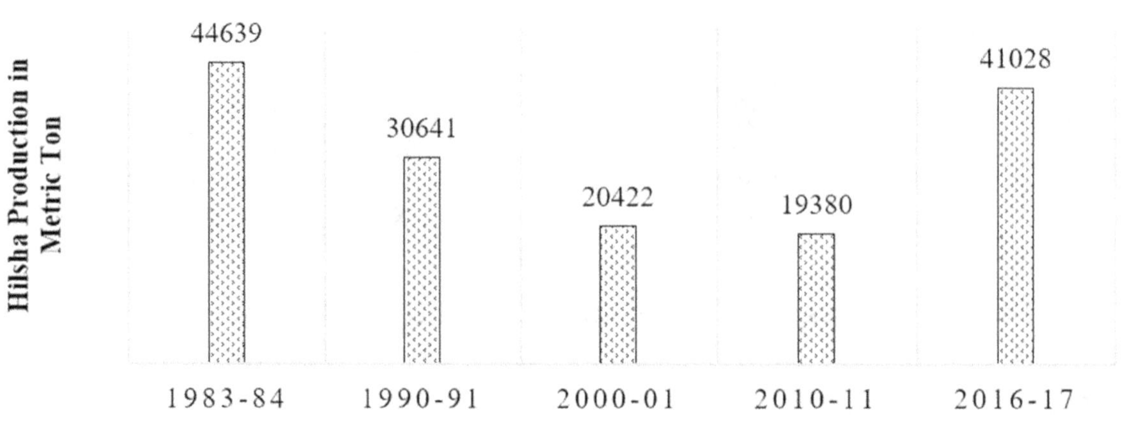

Figure 9. Temporal changes in the area of rice production in Barisal Sadar Upazila

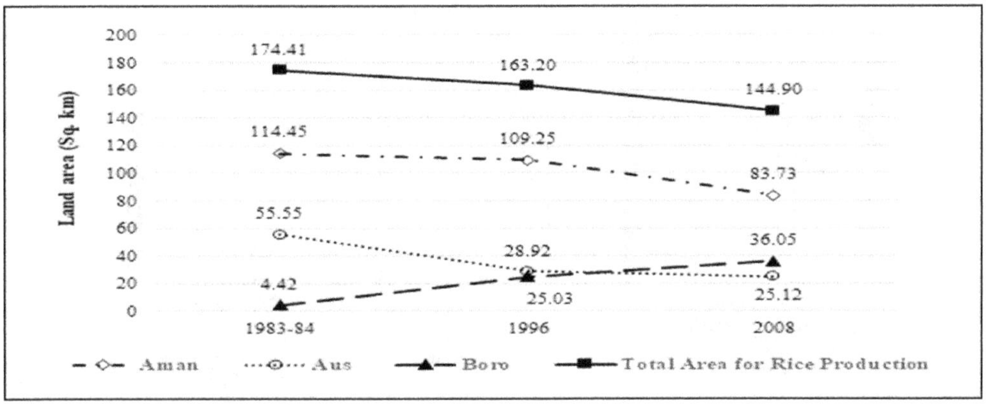

Food and Fiber (Including Livestock, Poultry, and Fish Production)

Rice is the main crop of Barisal Sadar Upazila. The four major types of rice in Bangladesh namely Aman, Aus, IRI and Boro are cultivated here. During 1983-1984 to 2008, the area of Boro production rises significantly, but the loss of area for Aman and Aus production was also noteworthy. Figure 5 signifies that the Barisal Sadar Upazila lost 29.51 sq. km of rice production area within 24 years, that is, almost 1.22 sq. km of rice production area were lost annually. As a result, the production of rice showed a decreasing trend, though productivity almost doubled during this time. Recently, IRI became the favorite type of production in this area.

To overcome the problem of continuous reduction of the agricultural area; high yielding varieties (HIV) of rice and other products were introduced with consistent backing to provide contemporary inputs including modern techniques, chemical fertilizers, pesticides, irrigation equipment and infrastructures, institutional credit, product procurement, storage and marketing facilities (Rahman, 2010). During the 1979-80 to 2016-17 period; Aus productivity increased from 88.71 tons/sq. km to 162.1 tons/ sq. km, Aman productivity increased from 107.49 tons/sq. km to 200.40 tons/sq. km, Boro productivity increased from 228.82 tons/sq. km to 383.26 tons/sq. km in this area. But continuous use of hybrid seeds, insecticides, pesticides, fly spray, etc. to increase productivity has reduced the fertility of land very much. Because of using excess pesticides, the roes are destroying alarmingly. The taste of the vegetables also changed.

Besides rice; pulses, oilseeds, cash crops, vegetables and spices are cultivated here and the general situation of these crops is visible in Figure 6 and shows a decreasing trend. The area of pulses production increased from 34.95 sq. km in 1983-84 to 40.24 sq. km in 1996 but dropped to 37.17 sq. km in 2008. The area for vegetable production decreases from 7.13 sq. km in 1983-84 to 2.76 sq. km in 1996 but significantly increased to 6.24 sq. km in 2008. The area of oilseeds, cash crops and spices production dropped from 9.80 sq. km, 5.03 sq. km and 8.20 sq. km to 1.61 sq. km, 3.26 sq. km and 6.93 sq. km respectively during the period of 1983-84 to 2008. Wheat and some other minor cereals are cultivated in a little amount of this area also; though they showed an increasing pattern, it is insignificant considering the study area. Overall, the area of production for pulses, oilseeds, cash crops, vegetables, spices, wheat and other minor cereals decreased from 65.38 sq. km in 1983-84 to 58.45 sq. km in 1996 and further dropped to 56.16 sq. km in 2008.

Figure 10. Temporal change in the area of crop production (excluding rice) in Barisal Sadar Upazila

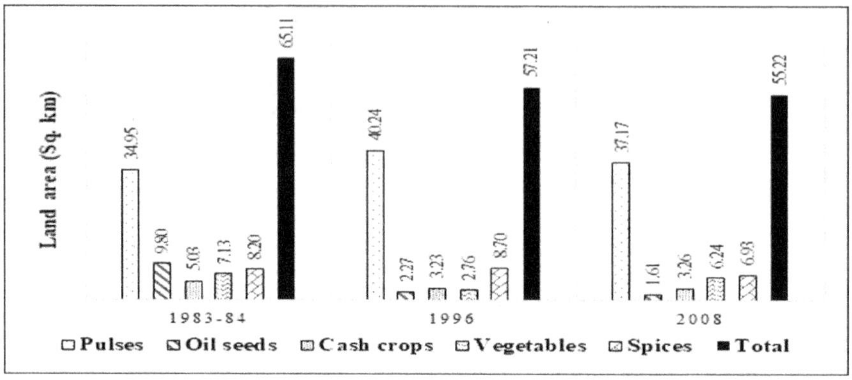

Barisal is also famous for the production of guava, hog plum, coconut, lemon and banana. These fruits are full of vitamins and show a diversity of food production. Though dependable data related to the production of these fruits are not available, their presence is noteworthy. Almost every households in the rural area produce some of these fruits.

In the Barisal district, fish production dropped from 79733 metric tons to 54129 metric tons during the 1983-84 to 2010-11 period. But after that, a significant increase to 99364 metric tons till 2016-17 was found because the fish production from rivers and estuaries as well as the fish production from ponds & semi-closed water bodies increased remarkably in recent years. To improve fish production and protect biodiversity more than 500 fish sanctuaries were established all over the country. To conserve and develop the production of hilsa which is the national fish of Bangladesh, five fish sanctuaries were established in the selected river system. Many new techniques were introduced to increase the cultured fisheries. Figure 7 describes the overall condition of fish production in the Barisal district, where the fish production from the rivers and estuaries decreased from 59850 metric tons to 41326 metric tons, fish production from floodplains and wetlands decreased from 10320 metric tons to 8276 metric tons, and fish production from ponds and semi-closed water bodies increased from 9563 metric tons to 47118 metric tons, fish production from shrimp and pen culture increased from zero metric ton to 2644 metric ton; overall fish production increased from 79733 metric ton to 99364 metric ton in Barisal district during 1983-1984 to 2016-1917 period.

Barisal is famous for hilsa fish and Figure 8 represents the scenario of hilsa production. The production of Hilsa decreased from 44639 Metric tons to 19380 Metric tons from 1983-84 to 2010-11 period but increase to 41028 Metric tons in 2016-17. The district fishery officer mentioned the success of the anti-hilsa fry campaign, which contributed enormously to increasing hilsa production.

Though this analysis was done for Barisal district instead of Barisal Sadar Upazila due to insufficient data, it is easy to predict that the outcome for Barisal Sadar Upazila would have been similar to this and due to land-use change captured fisheries production decreasing but cultured fisheries production increasing.

Cow and goats are the main livestock in this study area. Besides these, some buffalo and sheep are also found in this area. From 1981 to 2011 number of population in Barisal Sadar Upazila increases from 340841 to 527017 (BBS 1981, 2011) person. So in 40 years, almost 65% population increased. With this increasing population demand for cattle and goats is also increased. But during 1983-1984 to 2008, the

Figure 11. Temporal change of total annual fish production in inland water areas of Barisal District

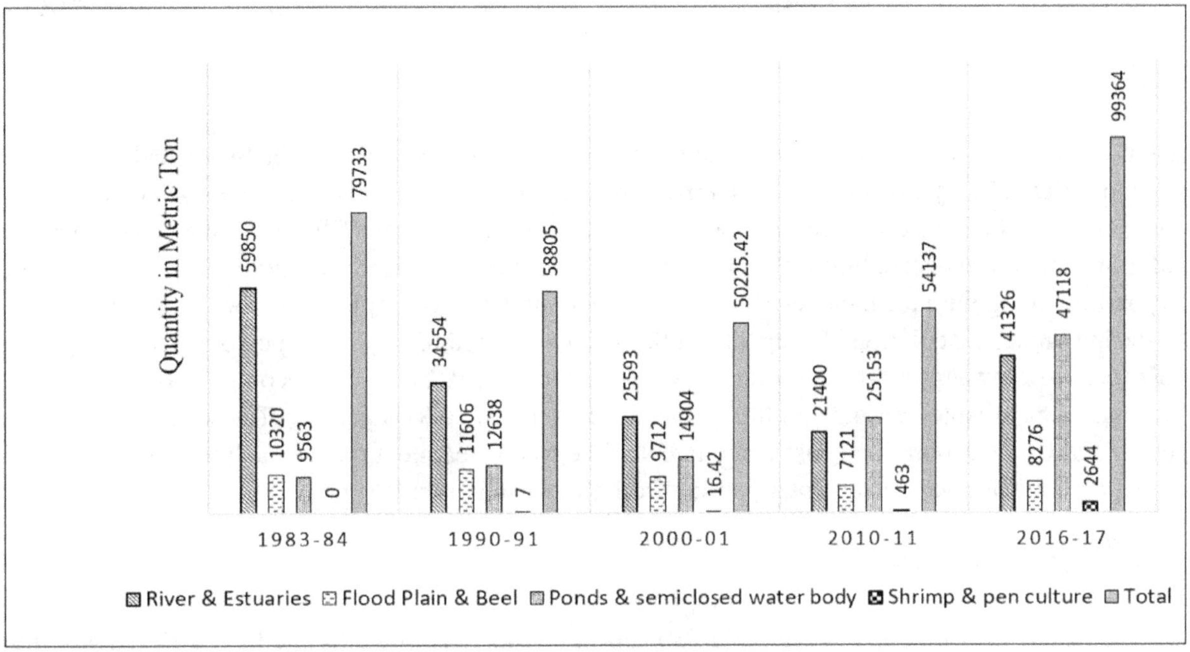

number of cows and buffalo decreases from 49220 to 44446; goats and sheep decrease from 21910 to 15598. Continuous reduction of pasture was found as the main reason for the reduction of livestock in the study area during the questionnaire survey. Once people raised livestock due to the easy availability of natural cattle food; over time new methods and techniques were introduced to raising livestock. These new procedures not only increased productivity but also increased cost. They had to buy hybrid food which

Figure 12. Temporal change of total annual catch of Hilsa in Barisal District

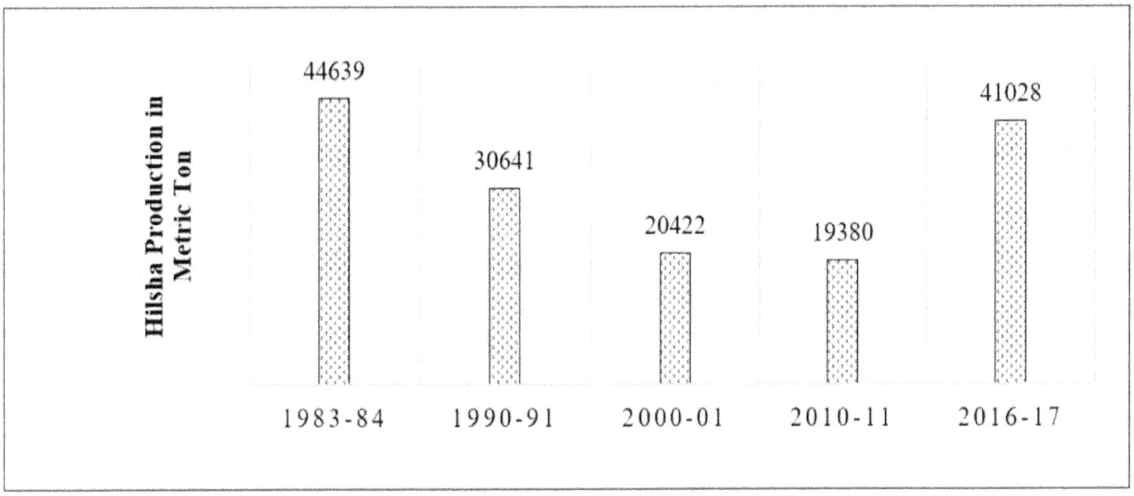

is expensive. Moreover, there is no facility for processing milk and meat, so they could not make up the cost and lost interest in raising livestock. Contrarily fowls and ducks increase from 210005 to 391264.

Freshwater

Land use change has a significant impact on fresh water in Barisal Sadar Upazila. Specifically, the urban area under Barisal City Corporation was remarkably affected by these land use changes. The canals and ponds, most of which were surface water sources in Barisal City, had been filled up and silted by invaders, land grabbers and real state builders. So, people were heavily dependent on groundwater. Barisal City Corporation along with local citizens used to use deep tubewell to supply water for domestic purposes. But the unplanned installation of deep tubewell, shortage of rainfalls and filling up of surrounding water bodies caused a lower groundwater level. As a result, Barisal City Corporation could supply only 4.0-4.5 million gallons of water per day but the demand is 12 million gallons per day. To overcome this shortage, Barisal City Corporation constructed two surface water treatment plants, each having the capacity to purify 4.0 million gallons of water per day from the Kirtankhola River.

Livelihoods of Societies

The livelihoods of people in Barisal Sadar Upazila were vastly affected by the land use change. Many people had left the agricultural profession and fishing work to find out other work. Agricultural work is a very hard-working job, so when the farmers became weak with age, they tried to open a tea stall or do something else rather than engage in agricultural work. Agricultural work also does not consider a very decent profession in society, even a farmer himself believes it and did not want their child to involve in this profession. So the young generation, especially the educated ones ages 18 to 35 years are not interested in this profession. They also encourage their parents to leave it. Moreover, the dimension of agriculture has changed such as farmers now want to grow vegetables, fruits, etc. rather than rice cultivation. Now sugarcane and jute cultivation is almost nil. Various non-agricultural livelihood groups such as auto-rickshaw drivers, shopkeepers, saw millers, small businessmen private and public services, entrepreneurship, etc. showed a significant increase in the town.

Transportation Facilities

The transportation system especially the road network was vastly improved during the last two decades in Barisal Sadar Upazila. Many muddy and brick-built road was paved, and bridges and culvert were constructed during this period. These improved road networks give them the advantage of selling and buying their needed equipment, seeds, and other materials easily and quickly. All of the interviewees mentioned their past suffering due to a lack of a proper road network and how satisfied now they are. After completing the survey with 60 distinctive individuals it was established undoubtedly that major portions of the families have benefitted from this development.

Educational System

Land-use changes have a huge positive impact on the educational system in Barisal Sadar Upazila. With the increasing population, improved road network, and urban and rural settlement many new schools, colleges, and universities were established and rebuilt.

Women's Contribution

Women's contribution to society is remarkable. They are now contributing economically to their family along with their husbands. They are producing vegetables for commercial purposes. During harvest time, women dry up the harvest and do rice scoops without expecting any return. Some women took training from NGOs on family planning, gardening, sewing, raising livestock, fisheries, etc. They go to every home to make the rural people aware of family planning. They are the pillar of the success of birth control in this locality.

Finally, the overall lifestyle of the community has vastly improved during the past two decades. Transportation facilities, educational facilities, and treatment facilities are within the reach of the community. People are very satisfied with them. Along with this improvement, some social problems arise also. Many landless farmers are now without agricultural land. Many joint families are broken into single ones leaving the aging parents in fragile conditions, the playing field reduced and many young people became drug addicted, almost in every house satellite channels are available and people prefer different foreign channels to Bangladeshi channels. Land-use changes also affect the environment of the study area adversely. Many trees were cut down to make space for infrastructure development and to fulfill the need for fuel.

CONCLUSION

There is an interplay between land use change and ecosystem services. The relationship between land use change and ecosystem services was explored in Barisal Sadar Upazila, a small coastal town in Bangladesh. Land use changes in this municipality were analyzed using Landsat satellite images from 1980 to 2010 at 10 years intervals. It was found that the town experienced a rapid change in its ecosystem services in proportion to land use change especially the conversion of agricultural land to urban settlement. Sometimes these changes brought economic solvency, better education and transportation facilities with the improved lifestyle of the local community. But it affected the various ecosystem services adversely and some of the damages are not recoverable. Even though the land use change had brought economic development to the people of the town, there had been a negative impact on ecosystem services which calls for a balancing trade-off between economic development and conservation of ecosystem services for the greater benefit of the society as well as sustenance of the ecosystem.

REFERENCES

Bangladesh Bereau of Statistics. (2011). *Agricultural Statistics Yearbook of Bangladesh*. BBS.

Bangladesh Bureau of Statistics. (2011). *Population Census 1981, 1991, 2001 and 2011*. BBS.

Bangladesh Bureau of Statistics. (2007). *Census of Agriculture 1983-84, 1996 and 2006*. BBS.

BBS. (2018). *Fisheries Statistical Yearbook of Bangladesh 1983-84, 1990-91, 2000-01, 2010-11, 2016-17*. BBS.

Chen, J., Sun, B. M., Chen, D., Wu, X., Guo, L. Z., & Wang, G. (2014). Land use changes and their effects on the value of ecosystem services in the small Sanjiang plain in China. *TheScientificWorldJournal*, *2014*, 2014. doi:10.1155/2014/752846 PMID:24741356

De Groot, R. S., Wilson, M. A., & Boumans, R. M. (2002). A typology for the classification, description and valuation of ecosystem functions, goods and services. *Ecological Economics*, *41*(3), 393–408. doi:10.1016/S0921-8009(02)00089-7

Dewan, A. M., & Yamaguchi, Y. (2009). Land use and land cover change in Greater Dhaka, Bangladesh: Using remote sensing to promote sustainable urbanization. *Applied Geography (Sevenoaks, England)*, *29*(3), 390–401. doi:10.1016/j.apgeog.2008.12.005

Dubois, O. (2011). *The state of the world's land and water resources for food and agriculture: managing systems at risk*. Earthscan.

Eva, H.D., Brink, A. and Simonetti, D. (2006). *Monitoring land cover dynamics in sub-Saharan Africa*. Institute for Environmental and Sustainability, Tech. Rep. EUR, 22498.

Fisher, B., Turner, R. K., & Morling, P. (2009). Defining and classifying ecosystem services for decision making. *Ecological Economics*, *68*(3), 643–653. doi:10.1016/j.ecolecon.2008.09.014

Hasan, M. N., Hossain, M. S., Islam, M. R., Bari, M. A., Karim, D., & Rahman, M. Z. (2013). Trends in the availability of agricultural land in Bangladesh. Soil Resource Development Institute (SRDI), Ministry of Agriculture. http://www. nfpcsp.org/agridrupal/sites/default/files/Trends -in-the-availability-of agricultural-land-in-Bangladesh-SRDI -Supported-by-NFPCSP-FAO. pdf

Hossain, M. S., Dearing, J. A., Rahman, M. M., & Salehin, M. (2016). Recent changes in ecosystem services and human well-being in the Bangladesh coastal zone. *Regional Environmental Change*, *16*(2), 429–443. doi:10.100710113-014-0748-z

Islam, G. T., Islam, A. S., Shopan, A. A., Rahman, M. M., Lázár, A. N., & Mukhopadhyay, A. (2015). Implications of agricultural land use change to ecosystem services in the Ganges delta. *Journal of Environmental Management*, *161*, 443–452. doi:10.1016/j.jenvman.2014.11.018 PMID:25516384

Kabba, V. T. S., & Li, J. (2011). Analysis of land use and land cover changes, and their ecological implications in Wuhan, China. *Journal of Geography and Geology*, *3*(1), 104. doi:10.5539/jgg.v3n1p104

Kumar, P., & Yashiro, M. (2014). The Marginal Poor and Their Dependence on Ecosystem Services: Evidence from South Asia and Sub-Saharan Africa. In J. von Braun & F. Gatzweiler (Eds.), *Marginality*. Springer. doi:10.1007/978-94-007-7061-4_11

Liu, J., Liu, M., Tian, H., Zhuang, D., Zhang, Z., Zhang, W., Tang, X., & Deng, X. (2005). Spatial and temporal patterns of China's cropland during 1990–2000: An analysis based on Landsat TM data. *Remote Sensing of Environment*, *98*(4), 442–456. doi:10.1016/j.rse.2005.08.012

Mendoza-González, G., Martínez, M. L., Lithgow, D., Pérez-Maqueo, O., & Simonin, P. (2012). Land use change and its effects on the value of ecosystem services along the coast of the Gulf of Mexico. *Ecological Economics*, *82*, 23–32. doi:10.1016/j.ecolecon.2012.07.018

Millennium Ecosystem Assessment (MA). (2005). *Ecosystems and Human Well-Being: Synthesis "Archived copy" (PDF). Archived (PDF) from the original on 3 December 2013*. Island Press.

Musa, K.B. (2008). *Identifying Land Use Changes and Its Socio-economic Impacts: A Case Study of ChacoriaSundarban in Bangladesh*. BBS.

Pereira, E., Queiroz, C., Pereira, H. M., & Vicente, L. (2005). Ecosystem services and human well-being: A participatory study in a mountain community in Portugal. *Ecology and Society*, *10*(2), art14. doi:10.5751/ES-01353-100214

Prakasam, C. (2010). Land use and land cover change detection through remote sensing approach: A case study of Kodaikanaltaluk, Tamil Nadu. *International Journal of Geomatics and Geosciences*, *1*(2), 150.

Rahman, S. (2010). Six decades of agricultural land use change in Bangladesh: Effects on crop diversity, productivity, food availability and the environment, 1948–2006. *Singapore Journal of Tropical Geography*, *31*(2), 254–269. doi:10.1111/j.1467-9493.2010.00394.x

Roebeling, P. C., Costa, L., Magalhães-Filho, L., & Tekken, V. (2013). Ecosystem service value losses from coastal erosion in Europe: Historical trends and future projections. *Journal of Coastal Conservation*, *17*(3), 389–395. doi:10.100711852-013-0235-6

Sayed, M. B., & Haruyama, S. (2015). Dynamics of Land Use/Cover Change in Manikganj District, Bangladesh: A Case Study of ManikganjSadarUpazila. *Journal of Geographic Information System*, *7*(06), 579–587. doi:10.4236/jgis.2015.76046

Vihervaara, P., Marjokorpi, A., Kumpula, T., Walls, M., & Kamppinen, M. (2012). Ecosystem services of fast-growing tree plantations: A case study on integrating social valuations with land-use changes in Uruguay. *Forest Policy and Economics*, *14*(1), 58–68. doi:10.1016/j.forpol.2011.08.008

Wang, Y., Yu, X., He, K., Song, S., Jia, G., & Huang, Z. (2012). Ecosystem services value based on land use change in Huairou reservoir watershed. *Nongye Gongcheng Xuebao (Beijing)*, *28*(5), 246–251.

Chapter 2
Tidal River Management in the Southwest Region of Bangladesh Using Mathematical Modelling

A.K.M Saiful Islam
https://orcid.org/0000-0002-2435-8280
Bangladesh University of Engineering and Technology, Bangladesh

Abu Saleh Khan
https://orcid.org/0000-0002-3415-6455
Institute of Water Modelling, Bangladesh

G M Tarekul Islam
Bangladesh University of Engineering and Technology, Bangladesh

Biswa Bhattacharya
UNESCO-IHE, The Netherlands

Dr. Md. Hazrat Ali
Chittagong University of Engineering Technology, Bangladesh

Md. Saiful Hossain
Bangladesh Water Development Board, Bangladesh

ABSTRACT

In the low-lying coastal areas of Bangladesh, a series of embankments, also known as polders, were built in the 1960s to protect these areas from tidal floods. However, these polders have adversely impacted social and coastal ecosystems by causing excessive siltation in rivers. To solve this problem, an initiative known as tidal river management (TRM) has been implemented since the 1990s. With this initiative, polder embankments are periodically cut in selected locations to allow the sediment-laden water to reach the polders, allowing sedimentation to occur. Subsequently, these cuts are closed. The authors assessed the suitability of the TRM initiative for the polder in East Beel Kuksia. The research was carried out using coupled 1D and 2D hydrodynamic models, built with the software tool MIKE11 and coupled with MIKE21. Erosion was calculated based on information on bed shear stress from the hydrodynamic model and erosional critical shear stress. With the implementation of TRM, the drainage capacity of the rivers downstream of the studied polder consequently increased.

DOI: 10.4018/978-1-6684-5619-4.ch002

INTRODUCTION

The coastal region of Bangladesh is enriched with vast natural resources and is home to 36.8 million people who depend on these resources for their livelihoods (Islam and Gnauck, 2017). The estuarine environment of the coastal region, which is composed of mangroves, salt marshes, seagrass, and seaweeds, serves as the feeding, breeding and nursery grounds for a variety of fish, birds, and animals. The ecology of the area depends on the upland freshwater and downstream tidal flows (Khan et al., 2011). Due to the presence of flat topography, the region is highly vulnerable to coastal flooding and sea level rise (Rose and Bhaskaran, 2016). Sixty percent of the Sundarbans, the world's largest mangrove forest, which has an area of 1 million hectares, is in the Bangladesh delta, and the remainder lies in India (Rahman and Salehin, 2013). The Meghna Estuary is the habitat of 53 fish species and the largest estuarine ecosystem in coastal Bangladesh (Hossain et al., 2012b).

The southwest coastal region of Bangladesh is characterized by numerous morphologically active tidal rivers, which comprise the main drainage network of the low-altitude Land and low-lying water bodies (known as beels) in the region (**Figure 1**). The main climatic disasters that occur are due to cyclones, storm surges, riverine floods, tidal floods, and salinity intrusions (Mondal et al., 2006). Moreover, the increasing population, in addition to environmental hazards, puts pressure on Land and water resources (Brammer, 2014). To increase the amount of arable land free from waterlogging, the Bangladesh Water Development Board took up the Coastal Embankment Project (CEP) to build a series of embankments during the 1960s and early 1970s that enclosed low-lying coastal areas which led to the formation of the so-called polders. After the development of the polders, area inhabitants enjoyed the benefits for many years. Sedimentation in rivers and waterlogging on adjoining lands are major concerns in the entire southwestern region. The entire river system of the southwest region is vulnerable to excessive sedimentation due to silts that come in with the high coastal tides, especially during the dry season. Despite increasing food production, polders in the southwest coastal region of Bangladesh have failed to deliver the intended results due to a lack of understanding of the regional hydro-morphology as well as improper maintenance and a lack of consideration of the social and cultural roles of the area inhabitants (Nowreen et al., 2016). The river morphology of the area is characterized by the active deposition of sediment and silt, which is carried into the area by high tides. The presence of the polders has aggravated the sedimentation process in rivers. This occurred as the polders restricted tidal flows to low-lying areas and prevented the deposition of sediment in these low-lying areas. As a result, the silt carried upstream by tides was gradually deposited on riverbeds and near the outlets of drainage sluices. On the other hand, the freshwater flow from upstream during the dry season has reduced over the past few decades, which has expedited the siltation process. Because of this continuing sedimentation process over the years, many of the rivers/channels/canals in the area lost their conveyance, which led to severe drainage congestion in the polders.

In the late 1980s, the problem of drainage congestion became very acute in the coastal districts of Khulna and Jessore. Dredging of rivers to solve the sedimentation and drainage congestion problems was not effective since the backfilling rate was quite significant, which led to 100% siltation within 3 to 4 months. Siltation in rivers due to *polderization* caused the rise of riverbeds, and polders became waterlogged (Talchabhadel et al., 2016). The lack of silt removal from canals, reduction of freshwater supply to canals and encroachment of canals are the main causes of waterlogging (Alam et al., 2017). Waterlogging problems in the coastal region resulted in damage to crops, which eventually affected the livelihoods of the coastal people (Alam et al., 2017). The implementation of existing management practices, such as the CEP, adversely impacted the social and coastal ecosystems (Roy et al., 2017b). A comprehensive understanding of societal and ecological systems in the coastal region of Bangladesh can help to provide an effective solution. Roy et al. (2017b) advocated for the implementation of integrated water resources management (IWRM) and integrated coastal zone management (ICZM). ICZM, with the adoption of Bangladesh's Coastal Zone Policy of 2005, was developed to support the coastal ecosystem and improve the livelihoods of coastal inhabitants (Iftekhar, 2006). ICZM helped develop capabilities and take priority actions for coastal development in Bangladesh. For the sustainable environmental and ecosystem management of the coastal regions, a major reform in policy was deemed essential (Roy et al., 2017a). Dewan et al. (2017) revealed that there is an increasing demand for the involvement of politically elected local governmental institutions in water management. Equitable and sustainable water management is extremely important, as a large proportion of the rural poor is dependent on natural water bodies, and the health of these water bodies depends on healthy aquatic ecosystems (Rasul et al., 2010). Basin-scale management through cooperation with transboundary countries and addressing social justice and equity will be required to solve complex water issues in Bangladesh (Gain et al., 2017b). To minimize water-sharing conflicts between Bangladesh and India, a multi-nation watershed management approach was suggested by Sood and Mathukumali (2011). Bangladesh should shift its policy from disaster management towards disaster preparedness for natural disasters such as floods (Mallick et al., 2005). Flood management in coastal Bangladesh requires plans, policies and projects that take into consideration the complex interactions among social processes, environmental dynamics, and technological interventions (Staveren et al., 2017). Thomas (2017) emphasized building cooperation regarding water issues among countries sharing transboundary rivers. Minar et al. (2013) emphasized the need to develop and implement effective adaptation policies and mitigation measures to minimize climate-related impacts in Bangladesh. Rouillard (2014) suggested increasing the adaptation of water resources to the impacts of climate change. He urged that present policy principles and implementation favoring integrated water resource management could be successful. Providing training in capacity building can benefit the integrated flood risk management policy (Samuels et al., 2006).

Before the implementation of coastal polders in the 1960s, tidal flood plains in the southwestern region were inundated by high tides twice a day, and normal agricultural practices were hampered by salinity intrusion during the dry season. To solve this problem, a large construction and maintenance program for permanent polders was implemented during the 1960s by the government at the time, which was known as the Coastal Embankment, and the aim was to increase agricultural production and improve the regional ecosystem. The CEP and subsequent flood control and irrigation projects in the region converted wetlands to dry land to facilitate the introduction of a variety of high-yield rice. Following the disastrous floods in 1987 and 1988, the Government of Bangladesh took the Flood Action Plan (FAP) initiative to study the causes and nature of floods in Bangladesh and to prepare guidelines for managing them. The objective of the FAP was to identify, plan, design, and test those high-priority flood-control projects feasible in technical, economic, environmental, and social terms. Though the FAP was focused on structural solutions for flood management, this plan was not able to meet expectations due to an inability to construct large engineering works and a lack of public participation (Thompson et al., 2006). Structural measures of flooding (i.e., construction of embankments, drainage channels, sluices, and regulators) yielded many positive impacts, such as an increase in agricultural production and the economy by reducing the impacts of natural disasters. However, such flood control, drainage, and irrigation (FCDI) projects were, in general, not able to achieve the desired goals due to a lack of consideration of the interdependence of land, water, ecosystems and socioeconomic development (Rah-

Figure 1. Project area in the southwestern region of Bangladesh

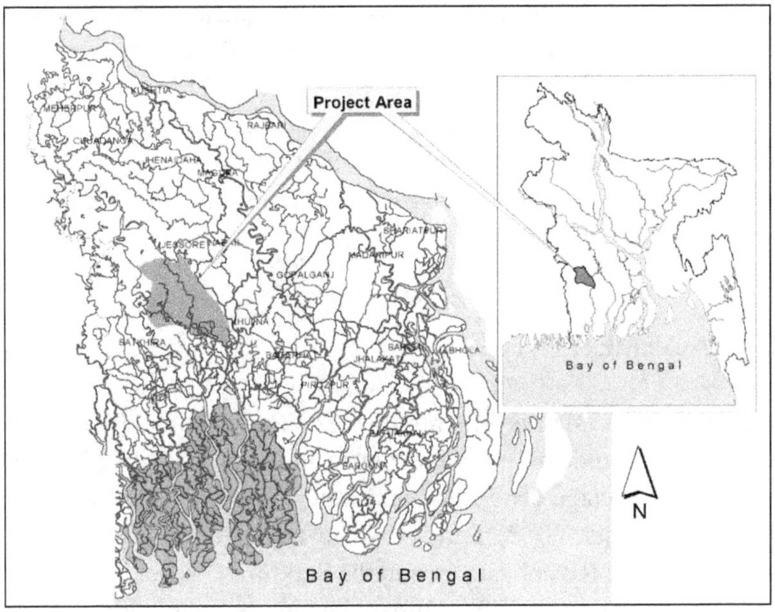

man and Salehin, 2013). The Bangladesh FAP was not able to achieve its goal of protecting the country from monsoon floods for many political, social, economic, and institutional reasons (Brammer, 2010). However, a recent study shows that due to increased flood risks, the non-structural measures of flood risk management are becoming less feasible (Wesselink et al., 2015).

In the southwestern region, a significant reduction in dry season flow caused by water withdrawal operations of the Farakka barrage in the Indian Ganges has contributed to increased salinity and, subsequently, more sedimentation in tidal rivers, which form the drainage systems of the polders. Shortages of freshwater flow in the Ganges and Gorai Basins have severe impacts on the socioeconomic and mangrove ecosystems in the Sundarbans in Bangladesh (Islam and Gnauck, 2017). The construction of polders also reduced the tidal prism because it prevented tidal flooding onto the tidal plain and accelerated river siltation. These tidal rivers cannot effectively drain water from nearby lands; thus, this causes waterlogging. In 1984, Beel Dakatia, which is part of polder 25, became waterlogged for the first time due to the rapid siltation of the Solmari, Hamkura and Hari Rivers. Later, this problem extended further to other polders. Moreover, land in the greater Jessore district, which is outside the polders, was also flooded. This problem is gradually creeping into the northern and southern parts of the embankment area.

To resolve the longstanding issues of drainage congestion and sedimentation, a series of studies were conducted. The tidal basin approach was first suggested in a feasibility study conducted by Haskoning (1993) and IWM (2005). During their study late in 1994, SMEC (1997; 2002) recommended a tidal basin approach. This approach is based on the concept that the river section at any location is a direct relation to the tidal reservoir capacity above that point (Williams, 1919).

Twenty years after the implementation of coastal embankments, waterlogging began to emerge in the polders upstream in the southwest region. The construction of polders reduced the tidal flood plains, and the reduction of freshwater flow from the upstream peripheral rivers during the dry season after the construction of the Farraka Barrage caused sedimentation in the peripheral river of the polder.

Consequently, waterlogging appeared in the southwestern region, which mainly constitutes the districts of Khulna, Jessore, Satkhira and Bagerhat. Williams (1919) indicated that if tidal flows can be freed, the navigability of the rivers will be restored, enclosed lands will be free from waterlogging, silt will accumulate inside the polders, and as a result, the level of land will rise. This approach was fulfilled in September 1990 when the polder of Beel Dakatia was breached in four places. Although there was no name for this process initially, the concept was later called the tidal river management system. In this study, an innovative investigation of removing drainage problems and the effects of future sea level rise in aggravating/improving the situation is presented.

TIDAL RIVER MANAGEMENT (TRM)

The TRM concept has been practiced in several regions of coastal Bangladesh (**Figure 1**) since the 1990s by periodic cutting and closing polders for land reclamation (Auerbach al., 2015). TRM can effectively solve the problems of severe drainage congestion and waterlogging in southwestern Bangladesh (Amir et al., 2013). Shampa and Pramanik (2012) argued that the drainage congestion problem could be solved by applying TRM to the Kobadak River basin area in the Satkhira District. TRM can be an effective tool for successfully combating issues induced by climate change by managing risks of waterlogging and sea level rise (Jakarya et al., 2016). Waterlogging effects can be removed by raising the plinth levels of physical infrastructures and homesteads, which are considered short-term measures. However, TRM can be considered a long-term or permanent solution for filling the lowlands (Awal, 2014). Khadim et al. (2013) showed that the introduction of TRM in combination with IWRM could provide a safeguard and lead to considerable improvement in regional livelihoods. Gain et al. (2017a) argued that TRM could be effective with the application of a transdisciplinary framework supported by the involvement of key institutions and local stakeholders. Khadim et al. (2013) suggested restoring open approaches, such as the tidal river management concept, which follows natural flood dynamics. Haque et al. (2015) emphasized that involving civil society organizations (CSOs) could be useful by illustrating the 'Paani (Water) Committee's (PC)' movement for the mainstreaming of TRM. Karim and Modal (2017) found the local community had mixed judgments regarding the implementation of the TRM project in Beel Kapalia in the Monirampur Subdistrict of Jessore. The TRM project helped to mitigate waterlogging problems by settling down sediment, which adversely affected the shrimp culture and wetland ecosystems.

STUDY AREA

The coastal region of Bangladesh covers approximately 20% of the total land area and over 30% of the cultivable lands in the country. Using hydrodynamic modeling, 54% of the area was found to be in the medium flood hazard zone, 26% in the high flood hazard zone and 20% in the low flood hazard zone (Tingsanchali and Karim, 2005). The study area is dominated by numerous low-lying water bodies (known as beels), creeks and canals, flat terrain, agricultural land and homesteads. Topographically, the study area has no definite slope. The land elevations of the study area vary from -0.50 m PWD (Public Works Datum) to 7.50 m PWD. The lowest levels in different beel areas range from -0.50 m PWD to 1.50 m PWD. Like most parts of the coastal area in Bangladesh, the study area is also governed by a semi-diurnal tide. Tides generated from the Bay of Bengal propagate through the Pussur-Sibsa system to the Gengrail-Telegati-Hari-Mukteswari system, which is shown in **Figure 2.**

Due to the absence of freshwater flow during the dry season, the tidal influence is the predominant factor of river sustainability in the study area. Surface water sources in the study area have seasonal variations due to upstream flow characteristics and climatic conditions. In general, surface water sources during the dry season are dependent on saline tidal water from the sea, as the rivers of the project area suffer from a shortage of upstream freshwater. In contrast, during monsoons, rainfall becomes the predominant source that feeds the river with catchment runoff water, although the tide is still an influencing factor. The runoff that occurs during the months of January to April is negligible; thus, the salinity level in the rivers of the study area increases during this period. The usual range of fluctuation in the water level is 3.0 m during neap tides and 4.0 m during spring tides. The moving average of the half-hourly water level shows a 1.2 m seasonal variation in the water level of the Gengrail River, which is one of the rivers in the southern coastal zone. The land in Bangladesh affected by salinity was 83.3 million hectares in 1973, which had increased to 105.6 million hectares in 2009 (Mahmuduzzaman et al., 2014). Soil salinity in the southwest region began increasing during the post-monsoon season and reached its peak value during the pre-monsoon season (Hossain et al., 2012a). The main reasons that soil salinity has been identified as critical in certain geographical locations in the country during low-flow conditions of the river are due to the construction of a barrage in the upstream neighboring country, improper management of coastal polders, sea level rise, cyclone and storm surge, backwater effects, precipitation and shrimp culture (Mahmuduzzaman et al., 2014). Dasgupta et al. (2016) found a high significance in salinity exposure during the last month of pregnancy as one of the determinants of infant mortality. Salinity impacts on households reveal that the critical zone for inundation risk is within approximately 4 km of the coastline (Dasgupta et al., 2014).

The sediment concentration in the river of the study area is quite high, especially during the dry season. It is evident from the field measurements that the depth-averaged sediment concentration during a spring tidal cycle varies from 800 mg/l to 16000 mg/l. The sediment concentration is higher during flood tide compared to the ebb tide. Field measurements of suspended sediment concentration in the Hari River show that the concentration during the spring tide is approximately 5 to 15 times higher than that of the neap tide. Grain-size analysis of the historical sediment data shows that the sediment type is predominantly fine silt with clay. The average particle size of sediment varies from 0.002 mm to 0.07 mm. The value of D_{50} (median size) varies from 15 microns to 25 microns.

Salinity levels in the coastal region have increased over the years due to tides, sea level rise, cyclonic storm surges and withdrawal of upstream freshwater, and as a result, there is a shortage of drinking water in the region (Khan et al., 2011). Changes in land use in coastal Bangladesh due to polderization and shrimp farming, as well as natural disasters, have significant impacts on agriculture, food security, freshwater supply, and coastal livelihoods (Parvin, 2017). In a recent study of the coastal district of Khulna, Mondal et al. (2013) found that the duration of sunshine hours decreased with increasing humidity. They reported that the number of rainy days and the magnitude of rain have increased in the Khulna District. They also found that the temperature is increasing at an alarming rate, particularly in recent years. Coastal shrimp aquaculture in Bangladesh contributes significantly to rural employment, even though it has an adverse social and environmental impact (Azad et al., 2009). The mangrove area of the Sundarbans has been adversely impacted by overfishing, increased shrimp farming and excess exploitation of plant and wildlife species (Hoq, 2006).

Figure 2. Gengtail-Telegati-Hari-Mukteswari River system

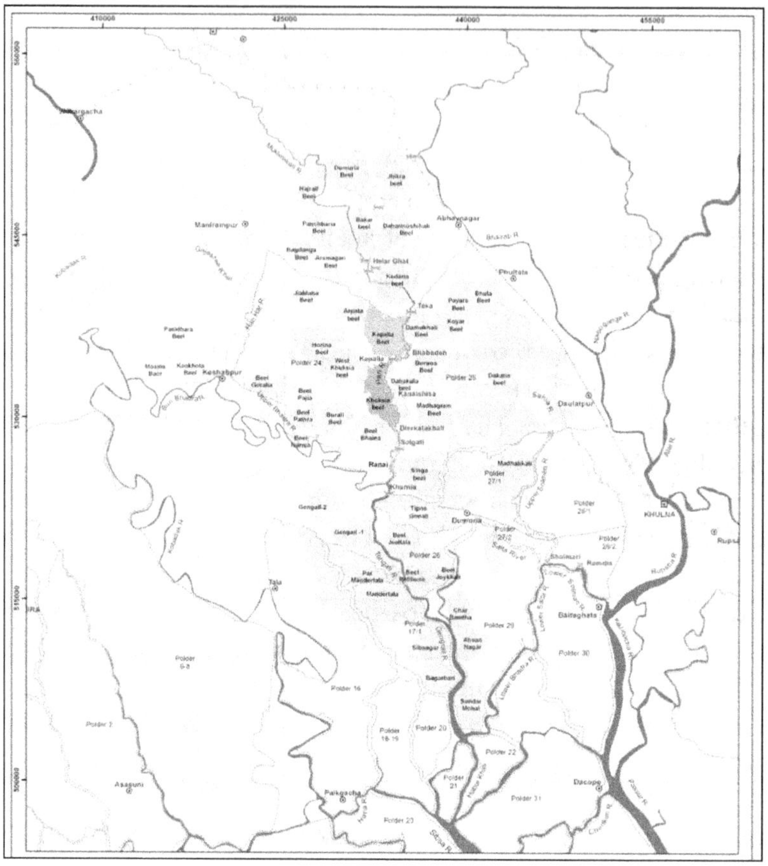

Data and Methodology

The present investigation is based on the collection of field data, data analysis, problem identification involving the participation of local stakeholders, the establishment of relationships between the tidal prism and mean cross-sectional areas of the river and the application of numerical modeling techniques to assess the effectiveness of the tidal basin. The performance of the tidal basin is assessed based on the monitoring of water level and discharge in both the river and basin.

At the upstream end of a river branch under the tidal influence, water velocities decrease in the upstream direction as the water volume that can be stored upstream is reduced. When the velocity falls below the critical deposition velocity, sedimentation occurs, which further decreases the storage capacity of the river and results in a further decrease in water velocities and ultimately leads to the siltation of the entire branch. Sedimentation rates are high in the upstream end, and these high rates eventually propagate downstream. Dead-end obstructions in a river create similar conditions and often become the starting point of severe sedimentation, which leads to the death of the river. The tidal basin can be regarded as a facility external to the river drainage system that guarantees a minimum tidal flow to keep the downstream portion of the river open and alive, which occurs if the capacity of the basin to store the tidal prism is not reduced by sedimentation.

According to the Dhaka University Earth Observatory (DUEO), subsidence in the southern delta is approximately 9.55 mm/year. Coastal polders suffer from a sediment deficit because the embankment prevents tidal flooding and results in the siltation of the river. To overcome the subsidence, it is essential to sequentially build a tidal basin that allows for free tidal movement for sedimentation.

The hydrodynamic and morphological performance of a tidal basin can be evaluated using mathematical models. Through the application of the models covering the entire area, it is easier to compare alternative design and operating strategies and evaluate their effectiveness in reducing drainage congestion. A hydraulic modeling investigation was carried out to assess the performance of a tidal basin prior to implementation in TRM. The location and required area of the tidal basin for TRM was selected based on the modeling results. The MIKE 11 modeling system (DHI, 2017) has been applied in this study for simulating tidal flow and water level. The hydrodynamic model is based on the southwest regional model, and project features, i.e., regulators, polders and tidal basin, have been incorporated. The model has been simulated both for dry and monsoon periods. The model has also been calibrated against flow and water level.

The impact of sea level rise on the sedimentation process inside the polder has been assessed by a 2-dimensional morphological model. In this study, the MIKE 21 FM modeling system (DHI, 2019) is used, which is based on an unstructured flexible mesh that consists of linear triangular elements. The hydrodynamic model is coupled with a sediment transport and bed change module. The morphological model includes the tidal basin of East Beel Kuksia and the adjacent canal, which connects the tidal basin with the Hari River. The boundary condition of the morphological model has been generated using the MIKE 11 model. As the sediments are cohesive in nature, a cohesive sediment transport model is developed. Bed erosion or deposition in the tidal basin is computed using the MIKE 21 FM modeling system. Erosion is calculated based on information about the bed shear stress from the hydrodynamic module and critical shear stress for erosion. The deposition of weak or strong flocks is calculated based on bed shear stress, critical shear stress for deposition and the settling velocity of suspended sediment.

Projection of Sea Level Rise and Calculation of Relative Mean Sea Level Rise (RMSLR)

The level of the sea observed along the coast changes in response to a wide variety of astronomical, meteorological, climatological, geophysical, and oceanographic forcing mechanisms. *Global* (eustatic) sea level change is often caused by a global change in the volume of water in the world's oceans in response to three climatological processes:

1. Ocean mass change associated with the long-term forcing of the ice ages, which are ultimately caused by small variations in the orbit of the earth around the sun;
2. Density changes from total salinity; and
3. The heat content of the world's oceans, which recent literature suggests may be potentially accelerating due to global warming.

One of the most important impacts of climate change for low-lying countries like Bangladesh is sea level rise. A recent study in the Beel Dakatia region at polder no. 25 in coastal Bangladesh showed that a 30.5 cm sea level rise would reduce the resilience of human, physical, natural, financial and socio-economic capitals by 55%, 53%, 87%, 60% and 64%, respectively (Ali et al., 2017). The occurrence of

Table 1. Projected change in global mean sea level rise (m) for the mid and late 21st century relative to the reference period of 1986–2005

Scenario	2046-2065		2081-2100	
	Mean	Likely range	Mean	Likely range
RCP 2.6	0.24	0.17 to 0.32	0.40	0.26 to 0.55
RCP 4.5	0.26	0.19 to 0.33	0.47	0.32 to 0.63
RCP 6.0	0.25	0.18 to 0.32	0.48	0.33 to 0.63
RCP 8.5	0.30	0.22 to 0.38	0.63	0.45 to 0.82

a storm surge after a 2 °C rise in sea surface temperature and a 30 cm rise in sea level will potentially flood 15.3% more areas within 20 km of the coastline (Karim et al., 2008). A sea level rise of 59 cm can induce a 0.9 ppt salinity in rivers up to 80 km upstream of the coastline (Bhuiyan et al., 2012). Sea level rise is caused by global warming, which affects two main processes. First, higher ocean water temperature causes thermal expansion, and second, atmospheric warming results in the melting of land ice. In this study, global sea level rise was reviewed for different climate change scenarios. The global mean sea level rise in accordance with the IPCC AR5 is presented in **Table 1.**

Brown et al. (2014) studied regional future sea level rise in Bangladesh. Regional sea level scenarios for the Bay of Bengal were developed based on global scenarios by Hinkel et al. (2014). The results showed that the sea level rise in Bangladesh would be slightly higher than the global average mean. However, the differences between the global and regional scenarios in most cases ranged from 5%-10% (Brown et al., 2014). This means that sea level rise could be up to 4 cm higher than the global mean by 2050 and up to 10 cm higher by the end of the century.

Relative sea level change is the local change in sea level relative to the elevation of the land at a specific point on the coast. Relative sea level change is a combination of both global and local sea level change due to changes in estuarine and shelf hydrodynamics, regional oceanographic circulation patterns, hydrologic cycles (river flow) and local and/or regional vertical land motion (subsidence or uplift). Relative sea level change is variable along the coast of Bangladesh, depending upon local and regional factors. For most of the impacts of sea level rise, such as salinity and increased inundation risks of polders, a rise in relative sea level is important. Relative sea level rise is also affected by subsidence and sedimentation. Using tidal gauge analyses, relative means sea level rise ranges from 2.8 mm to 8.8 mm per year (Pethick and Orford, 2013). However, due to deltaic subsidence and an increase in the tidal range inside a polder zone, the water levels will increase at a rate of up to 17.2 mm per year (Pethick and Orford, 2013).

In this study, relative sea level rise has been established based on a review of the recent literature, IPCC AR5, and a study report on relative sea level rise along the coast, which was carried out by Bangladesh University of Engineering and Technology, Center for Environmental and Geographic Information Services and Institute of Water Modelling through the Department of Environment. The relative mean sea level rise is established based on global mean sea level rise and local effects. This computation was completed for the central coastal area. The IPCC projected the global mean sea level rise (GMSL) with respect to the year 2005. **Figure 3** shows the GMSL at a different time. **Figure 4** shows subsidence in this delta, which was published by Higgins et al. (2014).

Figure 3. GMSL projection for the RCP8.5 scenarios

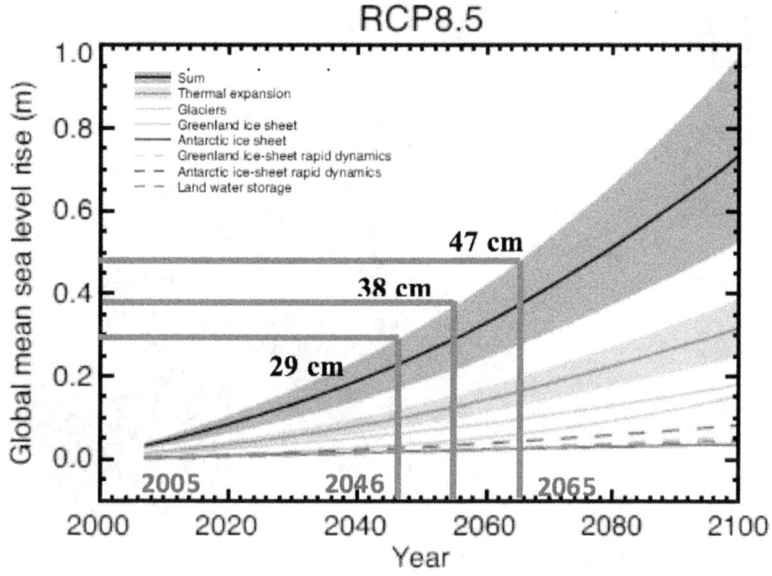

Syvitski et al. (2009) proposed a formula to calculate the relative mean sea level, which considers the contributions of global mean sea level, subsidence, sedimentation and tectonic movement as follows:

$\Delta RSL = A - \Delta E - CN - CA \pm M$ (1)

Where A = delta aggradation rate; ΔE = global sea level rise/(eustatic sea level); CN, CA = natural compaction (CN) and accelerated compaction (CA) of deltaic deposit, and M= vertical land surface movement (plate tectonics).

If only the vertical land elevation change is considered, the equation can be simplified as follows:

$\Delta RSL = \Delta E + C - A$ (2)

where C = subsidence.

Global sea level rise in 2030 with respect to 2015 is 9 cm when considering the upper line. The likely range of land subsidence rate **= is** 1-4 mm/year (Higgins et al., 2014) and is considered to be 2 mm/year. Sedimentation rate (inside polder area is prohibited) = 0 mm/year. Relative mean sea level rise by 2030 = global sea level rise + subsidence for 15 years during 2030-2015= 9 cm + (0.2 cm×15 years) = 9 cm+ 3 cm= 12 cm. Therefore, the relative mean sea level rise by 2030 is projected to be 12 cm. Similarly, other relative mean sea level rise values have been calculated to be 38 cm and 104 cm for 2050 and 2100, respectively.

RESULTS AND DISCUSSION

Impacts of Sea Level Rise on the Tidal Prism

Accordingly, the water level at the downstream boundary has been changed, and the tidal prism has been calculated for different scenarios. **Table 2** shows the changes in the tidal prism under TRM operation for a changing relative sea level. It is evident from the analysis that with sea level rise, the tidal prism

Figure 4. Locations of subsidence assessments

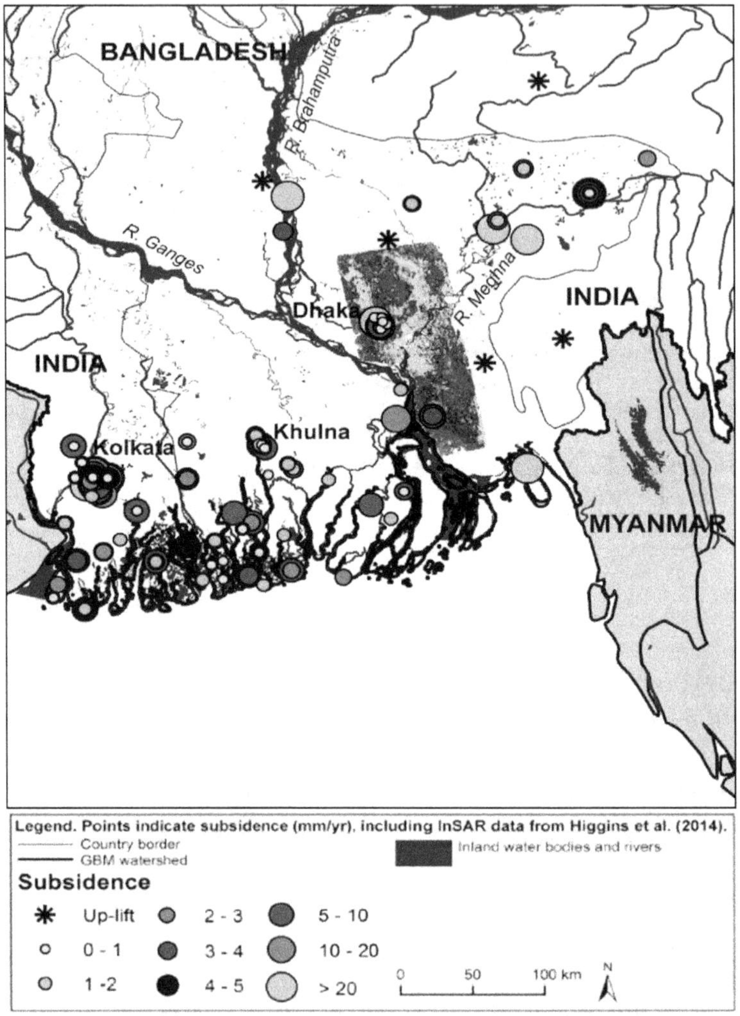

under TRM operation of the corresponding river will be increased in 2030, which will be almost double and 4 times that of the base conditions in 2050 and 2100, respectively.

Impacts of Sea Level Rise on the Sedimentation Process in the Tidal Basin

Sedimentation in the tidal basin is assessed considering relative mean sea level rise projections of 12 cm, 38 cm and 104 cm. Bed level change is assessed for different sea level rise scenarios. In **Figure 5**, B2 is the bed level change under sea level rise conditions; B1 is the bed level change without sea level rise, and (B2-B1) is the net bed level change under the effects of sea level rise.

Model results show that sedimentation will occur more in the tidal basin due to sea level rise. Sea level rise will increase the tidal prism, and an increased tidal prism will bring a higher volume of sediment into the tidal basin. The analysis shows that more sediment will be deposited at the entrance of the basin rather than at the other locations of the basin.

Table 2. Comparison of tidal prism in different scenarios of Sea Level Rise (SLR)

SL No	Condition	Tidal prism (Million m^3)	Remarks
1	Without SLR	1.29	
2	With SLR for 2030	1.57	12 cm RMSL
3	With SLR for 2050	2.29	38 cm RMSL
4	With SLR for 2100	4.24	104 cm RMSL

Tidal Prism and Volume of Sediment

Field data collection shows that rivers become silted up to 2 to 3 m in upstream portions when there is no upland flow, and sediment-laden water is not allowed to flood the low-lying area. Monitoring results also show that the backfilling rate is significant after dredging. Many scientists opine that there are some relationships between the general dimensions of a tidal estuary or river entrance on a sandy coast and the volume of the tidal prism, but no previous attempt has been made to achieve a definite correlation. Le Conte (1905) proposed an equilibrium area concept for tidal inlets. O'Brien (1931, 1969) examined field data from the tidal inlets of sandy barriers on the west coast of the United States and determined there was a relationship between the minimum cross-sectional flow area of the entrance channel and the observed tidal prism; thus, an equation was established in the form: $Ac = CP^n$, where Ac is the minimum inlet cross-sectional area during the equilibrium condition, C is an empirically determined coefficient, P is the tidal prism (typically during the spring tide/mean tide) and n is an exponent that is usually slightly less than unity. With an aim to establish a consistent relation, the measured discharge and cross-sectional area data of different tidal rivers were used to develop this relationship, which is a highly tide-dominated river in the southwest region. In doing so, an excellent relationship was found between cross-sectional area and tidal prism, which is expressed by the following equation:

$$Ac = 43.42\ P^{0.9}$$

where P = mean tidal prism (ebb + flood) in million m^3 and Ac = cross-sectional area below the mid-tide level in m^2.

Figure 5. Difference in bed level within 6 months, both with and without sea level rise conditions

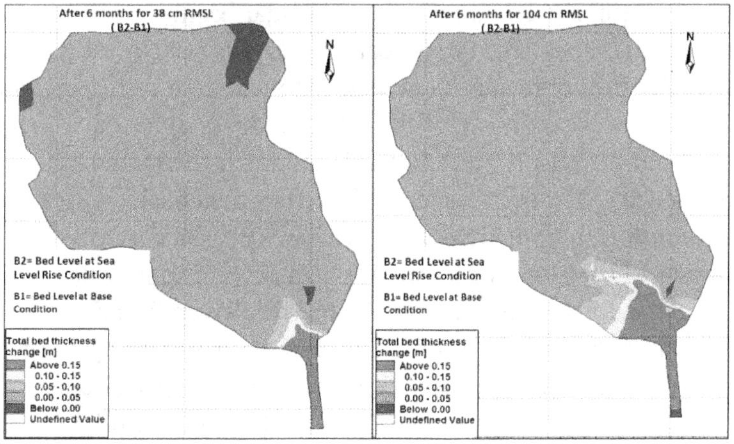

Figure 6. Comparison of tidal volume and cross-sectional area of the river

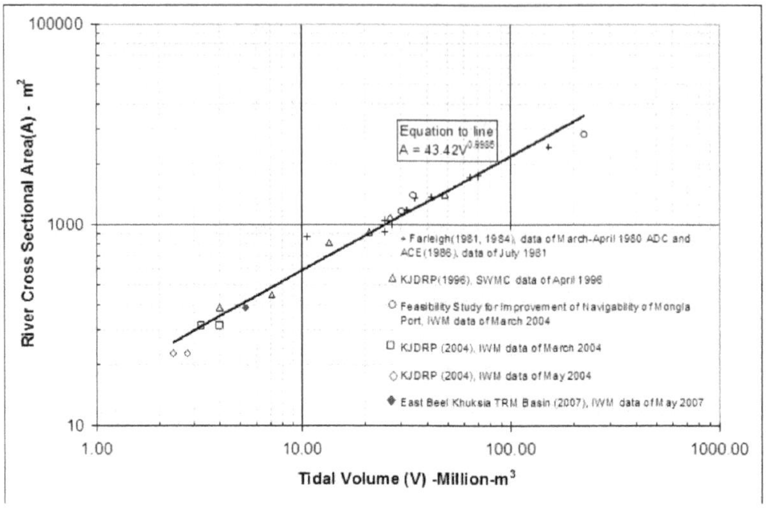

Model results show that one tidal basin generates a more than 4.8 million m^3 tidal prism, and other basins generate less voluminous tidal prisms (**Figure 6**). Two-dimensional sediment modeling has also been carried out to examine the distribution of sediment inside the tidal basin as well as the rate of sedimentation. Based on model results and field observations, the best-suited tidal basin was selected and implemented.

Performance of Tidal Basin

The effectiveness of the tidal basin in terms of maintaining navigability, proper drainage capacity of the river and siltation inside the tidal basin was assessed by monitoring the tide level, tidal volume, river cross-sections and siltation in the basin. Monitoring results show that the tidal range varies from 30 to 44 cm during spring tide and 18 to 30 cm during neap tide in the tidal basin. The tidal prism (volume) is considered to be an important indicator of the performance of a tidal basin. Tidal volume was calculated over a tidal cycle in the Hari River, which used measured discharge data for both spring and neap tides during different time periods. The tidal volume was increased by 6 times (0.9 to 5.3 million m^3) in 5 months after the beginning operation of the 3rd tidal basin, which is known as the East Beel Khuksia for tidal river management. The cross-sectional area was also increased by approximately 3 times after five months. Monitoring results show that sedimentation inside the tidal basin increases as the velocity reduces because of the entrance of the tide into the tidal basin, which increases the area of impact. During ebb tides, comparatively clear water flows downstream and erodes the riverbed to meet the sediment transport capacity. The performance of the first tidal basin, which is named Beel Bhaina, and the second tidal basin, which is named Beel Kedaria, are described as follows. Operation of the Beel Bhaina tidal basin generated 11.0 Mm^3 average tidal prisms over a tidal cycle, which caused significant morphological changes in the downstream reaches of the Hari River. Riverbed scour in the downstream reaches ranged from 5-7 m, and the cross-sectional area was increased by 2-3 times compared to the design section. This implies that the tidal basin can maintain a proper drainage capacity for the river in the downstream reaches. The total sedimentation in the Beel Bhaina tidal basin was estimated to be approximately 6.46 Mm^3 during the four-year operation of the basin (Nov 1997 to Dec 2001). Approximately 48% of the sediment came from riverbed scour, and the remainder came from downstream rivers. After the closure of the tidal basin, sedimentation downstream of the beel in the Hari River was 2.0 m over only two months, from January to March 2002. The high rate of siltation in the river indicates the immediate change in the morphology of the river as soon as the tidal basin was closed for operation. The closure of the Beel Bhaina tidal basin resulted in a remarkable decrease in tidal flow in the Hari River. While the average tidal volume was 10.50 Mm^3 in January 2001, the tidal volume in January 2002 after the closure decreased to 2.48 Mm^3. The reduction in discharge also caused a reduction in flow velocity, which ranged from 1.20 m/s to only 0.50 m/s downstream of the bezel.

The entire exercise was repeated by model runs using MIKE 11, which considered an SLR of 12 cm in 2030, 38 cm in 2050 and 104 cm in 2100. The tidal prism increases with SLR, which indicates that the situation will not be aggravated in the future due to SLR. The functioning of the existing hydraulic structures will be more problematic as the *Sill* levels are fixed, and higher water levels in rivers will lead to more drainage congestion inside the polders. The TRM exercise will be beneficial in the future, but the polders without TRM will have operational problems with the existing structures.

The social dimension of the TRM needs to be the focus of future research. During the implementation of the TRM, agricultural production was hampered due to the flooding of polders that were selected for TRM, and the plot owners did not agree to allow flooding of their land without proper crop compensation. Thus, TRM can only be implemented if the local populace agrees to allow the flooding of their land through social mobilization and proper crop compensation.

CONCLUSION

Long-term siltation in the rivers in the southwest coastal region and the subsequent drainage congestion during the early 1980s to late 1990s resulted in the suffering of the local people. Many studies and measures were undertaken to improve drainage congestion, but they failed to be effective. The introduction of TRM appeared to be a technically feasible, environment-friendly, socially acceptable, and overall effective solution. Implementation of the TRM increased the tidal prism, which resulted in an increased drainage capacity of the rivers downstream and allowed sediments to settle inside the basin. The tidal volume was increased by 6 times (0.9 to 5.3 million-m^3) in 5 months after beginning operation in East Beel Khuksia and increased by 2-3 times (11.0 Mm^3) in 4 months for Beel Bhaina. Even considering an SLR of 12 cm in 2030, 38 cm in 2050 and 104 cm in 2100, the tidal prism increased with SLR. TRM is an effective approach to overcoming subsidence and making measurements under the influence of climate change.

In this study, TRM is found to be an innovative and effective measure for drainage and sediment management in the southwest area of Bangladesh. However, TRM is an intermediate solution to the problem, and a long-term solution requires further study. Finally, numerical modeling was found to be an effective tool for analyzing characteristics of the tidal prism and drainage conditions, which were examined per different scenarios.

REFERENCES

Alam, M. S., Sasaki, N., & Datta, A. (2017). Waterlogging, crop damage and adaptation interventions in the coastal region of Bangladesh: A perception analysis of local people. *Environmental Development, 23*, 22–32. doi:10.1016/j.envdev.2017.02.009

Ali, M. S., & Syfullah, K. (2017). Effect of sea level rise induced permanent inundation on the livelihood of polder enclosed beel communities in Bangladesh: People's perception. *Journal of Water and Climate Change, 8*(2), 219–234. doi:10.2166/wcc.2016.236

Amir, M. I., Khan, M. S. A., Khan, M. K., Rasul, M. G., & Akram, F. (2013). Tidal river sediment management–a case study in southwestern Bangladesh. *International Journal of Civil Science and Engineering, 7*(3), 861–871.

Auerbach, L. W., Goodbred, S. L. Jr, Mondal, D. R., Wilson, C. A., Ahmed, K. R., Roy, K., Steckler, M. S., Small, C., Gilligan, J. M., & Ackerly, B. A. (2015). Reply to 'Tidal river management in Bangladesh'. *Nature Climate Change, 5*(6), 492–493. doi:10.1038/nclimate2620

Awal, M. A. (2014). Water logging in south-western coastal region of Bangladesh: Local adaptation and policy options. *Science Postprint, 1*(1), e00038. doi:10.14340pp.2014.12A0001

Azad, A. K., Jensen, K. R., & Lin, C. K. (2009). Coastal aquaculture development in Bangladesh: Unsustainable and sustainable experiences. *Environmental Management, 44*(4), 800–809. doi:10.100700267-009-9356-y PMID:19701597

Bhuiyan, M. J. A. N., & Dutta, D. (2012). Assessing impacts of sea level rise on river salinity in the Gorai river network, Bangladesh. *Estuarine, Coastal and Shelf Science, 96*, 219–227. doi:10.1016/j.ecss.2011.11.005

Brammer, H. (2010). After the Bangladesh flood action plan: Looking to the future. *Environmental Hazards, 9*(1), 118–130. doi:10.3763/ehaz.2010.SI01

Brammer, H. (2014). Bangladesh's dynamic coastal regions and sea-level rise. *Climate Risk Management, 1*, 51–62. doi:10.1016/j.crm.2013.10.001

Brown, S., Nicholls, R. J., Hanson, S., Brundrit, G., Dearing, J. A., Dickson, M. E., & Jimenez, J. A. (2014). Shifting perspectives on coastal impacts and adaptation. *Nature Climate Change, 4*(9), 752–755. doi:10.1038/nclimate2344

Dasgupta, S., Hossain, M., Huq, M., & Wheeler, D. (2014). *Facing the hungry tide: climate change, livelihood threats, and household responses in coastal Bangladesh*. World Bank.

Dasgupta, S., Huq, M., & Wheeler, D. (2016). Drinking water salinity and infant mortality in coastal Bangladesh. *Water Economics and Policy, 2*(01), 1650003. doi:10.1142/S2382624X1650003X

Dewan, C., Mukherji, A., & Buisson, M. C. (2015). Evolution of water management in coastal Bangladesh: From temporary earthen embankments to depoliticized community-managed polders. *Water International, 40*(3), 401–416. doi:10.1080/02508060.2015.1025196

DHI. (2017). *MIKE 11: A Modelling System for Rivers and Channels*. Danish Hydraulic Institute.

DHI. (2019). *MIKE 21 Flow Model FM*. Danish Hydraulic Institute.

Gain, A. K., Benson, D., Rahman, R., Datta, D. K., & Rouillard, J. J. (2017a). Tidal river management in the south west Ganges-Brahmaputra delta in Bangladesh: Moving towards a transdisciplinary approach? *Environmental Science & Policy*, *75*, 111–120. doi:10.1016/j.envsci.2017.05.020

Gain, A. K., Mondal, M. S., & Rahman, R. (2017b). From Flood Control to Water Management: A Journey of Bangladesh towards Integrated Water Resources Management. *Water (Basel)*, *9*(1), 55. doi:10.3390/w9010055

Haque, K. N. H., Chowdhury, F. A., & Khatun, K. R. (2015). Participatory environmental governance and climate change adaptation: mainstreaming of tidal river management in south-west Bangladesh. In *Land and disaster management strategies in Asia* (pp. 189–208). Springer India. doi:10.1007/978-81-322-1976-7_13

Haskoning (1993) *Second Coastal Embankment Rehabilitation Project, Main Report.* Bangladesh Water Development Board.

Higgins, S. A., Overeem, I., Steckler, M. S., Syvitski, J. P., Seeber, L., & Akhter, S. H. (2014). InSAR measurements of compaction and subsidence in the Ganges-Brahmaputra Delta, Bangladesh. *Journal of Geophysical Research. Earth Surface*, *119*(8), 1768–1781. doi:10.1002/2014JF003117

Hinkel, J., Lincke, D., Vafeidis, A. T., Perrette, M., Nicholls, R. J., Tol, R. S., Marzeion, B., Fettweis, X., Ionescu, C., & Levermann, A. (2014). Coastal flood damage and adaptation costs under 21st century sea-level rise. *Proceedings of the National Academy of Sciences of the United States of America*, *111*(9), 3292–3297. doi:10.1073/pnas.1222469111 PMID:24596428

Hoq, M. E. (2007). An analysis of fisheries exploitation and management practices in Sundarbans mangrove ecosystem, Bangladesh. *Ocean and Coastal Management*, *50*(5), 411–427. doi:10.1016/j.ocecoaman.2006.11.001

Hossain, M. L., Hossain, M. K., Salam, M. A., & Rubaiyat, A. (2012a). Seasonal variation of soil salinity in coastal areas of Bangladesh. International Journal of Environmental Science. *Management and Engineering Research*, *1*(4), 172–178.

Hossain, M. S., Das, N. G., Sarker, S., & Rahaman, M. Z. (2012b). Fish diversity and habitat relationship with environmental variables at Meghna river estuary, Bangladesh. *Egyptian Journal of Aquatic Research*, *38*(3), 213–226. doi:10.1016/j.ejar.2012.12.006

Iftekhar, M. S. (2006, August). Conservation and management of the Bangladesh coastal ecosystem: Overview of an integrated approach. [). Blackwell Publishing Ltd.]. *Natural Resources Forum*, *30*(3), 230–237. doi:10.1111/j.1477-8947.2006.00111.x

Islam, S. N., Reinstädtler, S., & Gnauck, A. (2017). Degraded Coastal Wetland Ecosystems in the Ganges-Brahmaputra Rivers Delta Region of Bangladesh. In *Coastal Wetlands: Alteration and Remediation* (pp. 187–213). Springer. doi:10.1007/978-3-319-56179-0_6

IWM. (2005) *Monitoring of Hydrological and Hydraulic Parameters on Tidal River Management under KJDRP Areal, Final report.* Bangladesh Water Development Board.

Jakarya, M., Sikder, M. T., & Yousuf, A. I. (2016). Review of an Integrated Strategy of Climate Change Adaptation Using Tidal River Management. *Journal of Health and Environmental Research*, *2*(1), 1–4. doi:10.11648/j.jher.20160201.11

Karim, M. F., & Mimura, N. (2008). Impacts of climate change and sea-level rise on cyclonic storm surge floods in Bangladesh. *Global Environmental Change*, *18*(3), 490–500. doi:10.1016/j.gloenvcha.2008.05.002

Karim, R., & Mondal, R. (2017). Local Stakeholder Analysis of Tidal River Management (TRM) at BeelKapalia and the Implication of TRM as a Disaster Management Approach. *Hydrology*, *5*(1), 1. doi:10.11648/j.hyd.20170501.11

Khadim, F. K., Kar, K. K., Halder, P. K., Rahman, M. A., & Morshed, A. M. (2013). Integrated water resources management (IWRM) impacts in south west coastal zone of Bangladesh and fact-finding on tidal river management (TRM). *Journal of Water Resource and Protection*, *5*(10), 953–961. doi:10.4236/jwarp.2013.510098

Khan, A. E., Ireson, A., Kovats, S., Mojumder, S. K., Khusru, A., Rahman, A., & Vineis, P. (2011). Drinking water salinity and maternal health in coastal Bangladesh: Implications of climate change. *Environmental Health Perspectives*, *119*(9), 1328–1332. doi:10.1289/ehp.1002804 PMID:21486720

LeConte, L. J. (1905). Discussion on the paper, "Notes on the improvement of river and harbor outlets in the United States" by D. A. Watt, paper no. 1009. *Trans. ASCE*, *55*(December), 306–308.

Mahmuduzzaman, M., Ahmed, Z. U., Nuruzzaman, A. K. M., & Ahmed, F. R. S. (2014). Causes of salinity intrusion in coastal belt of Bangladesh. *International Journal of Plant Research*, *4*(4A), 8–13.

Mallick, D. L., Rahman, A., Alam, M., Juel, A. S. M., Ahmad, A. N., & Alam, S. S. (2005). Case study 3: Bangladesh floods in Bangladesh: A shift from disaster management towards disaster preparedness. *IDS Bulletin*, *36*(4), 53–70. doi:10.1111/j.1759-5436.2005.tb00234.x

Minar, M. H., Hossain, B., & Shamsuddin, M. D. (2013). Climate change and coastal zone of Bangladesh: Vulnerability, resilience and adaptability. *Middle East Journal of Scientific Research*, *13*(1), 114–120.

Mondal, M. K., Tuong, T. P., Ritu, S. P., Choudhury, M. H. K., Chasi, A. M., Majumder, P. K., & Adhikary, S. K. (2006). Coastal water resource use for higher productivity: participatory research for increasing cropping intensity in Bangladesh. *Environment and Livelihoods in Tropical Coastal Zones: Managing Agriculture-Fishery-Aquaculture Conflicts*, 72-84.

Mondal, M. S., Jalal, M. R., Khan, M., Kumar, U., Rahman, R., &Huq, H. (2013). Hydro-meteorological trends in southwest coastal Bangladesh: perspectives of climate change and human interventions.

Nowreen, S., Jalal, M. R., & Khan, M. S. A. (2014). Historical analysis of rationalizing South West coastal polders of Bangladesh. *Water Policy*, *16*(2), 264–279. doi:10.2166/wp.2013.172

O'Brien, M. P. (1931). Estuary and Tidal Prisms Related to Entrance Areas. *Civil Engineering (New York, N.Y.)*, *1*(8), 738–739.

O'Brien, M. P. (1969). Equilibrium flow areas of inlets on sandy coasts. *Journal of Waterway, Port, Coastal, and Ocean Engineering*, *95*(l), 43–52.

Parvin, G. A., Ali, M. H., Fujita, K., Abedin, M. A., Habiba, U., & Shaw, R. (2017). Land Use Change in Southwestern Coastal Bangladesh: Consequence to Food and Water Supply. In Land Use Management in Disaster Risk Reduction (pp. 381-401). Springer Japan.

Pethick, J., & Orford, J. D. (2013). Rapid rise in effective sea-level in southwest Bangladesh: Its causes and contemporary rates. *Global and Planetary Change*, *111*, 237–245. doi:10.1016/j.gloplacha.2013.09.019

Rahman, M. R., & Asaduzzaman, M. (2013). Ecology of Sundarban, Bangladesh. *Journal of Science Foundation*, *8*(1-2), 35–47. doi:10.3329/jsf.v8i1-2.14618

Rahman, R., & Salehin, M. (2013). Flood risks and reduction approaches in Bangladesh. In *Disaster risk reduction approaches in Bangladesh* (pp. 65–90). Springer Japan. doi:10.1007/978-4-431-54252-0_4

Rasul, G., & Chowdhury, A. M. J. U. (2010). *Equity and social justice in water resource management in Bangladesh*. Bangladesh Water Development Board.

Rose, L., & Bhaskaran, P. K. (2016). The role of environmental forcing on tidal dynamics along complex near-shore waters off Bangladesh. *Ocean Engineering*, *116*, 68–81. doi:10.1016/j.oceaneng.2016.02.031

Rouillard, J. J., Benson, D., & Gain, A. K. (2014). Evaluating IWRM implementation success: Are water policies in Bangladesh enhancing adaptive capacity to climate change impacts? *International Journal of Water Resources Development*, *30*(3), 515–527. doi:10.1080/07900627.2014.910756

Roy, A. K. D., & Roy, A. K. D. (2017a). An investigation into the major environment and climate change policy issues in southwest coastal Bangladesh. *International Journal of Climate Change Strategies and Management*, *9*(1), 123–136. doi:10.1108/IJCCSM-03-2016-0028

Roy, K., Gain, A. K., Mallick, B., & Vogt, J. (2017b). Social, hydro-ecological and climatic change in the southwest coastal region of Bangladesh. *Regional Environmental Change*, *17*(7), 1–12. doi:10.100710113-017-1158-9

Samuels, P., Klijn, F., & Dijkman, J. (2006). An analysis of the current practice of policies on river flood risk management in different countries. *Irrigation and Drainage*, *55*(S1), S141–S150. doi:10.1002/ird.257

Shampa, M. I. M. P., & Pramanik, I. M. (2012). Tidal river management (TRM) for selected coastal area of Bangladesh to mitigate drainage congestion. *International Journal of Scientific & Technology Research*, *1*(5), 1–6.

SMEC. (1997). *Khulna-Jessore Drainage Rehabilitation Project, Feasibility Study Report for Overall Drainage Plan*. Bangladesh Water Development Board.

SMEC. (2002). *Khulna Jessore Drainage Rehabilitation Project, Project Completion Report (Final)*. Bangladesh Water Development Board.

Sood, A., & Mathukumalli, B. K. P. (2011). Managing international river basins: Reviewing India–Bangladesh transboundary water issues. *Intl. J. River Basin Management*, *9*(1), 43–52. doi:10.1080/15715124.2011.553832

Stocker, T. F., Qin, D., Plattner, G. K., Tignor, M., Allen, S. K., Boschung, J., & Midgley, P. M. (2013). *Climate Change 2013: The Physical Science Basis. Working Group 1 (WG1) Contribution to the Intergovernmental Panel on Climate Change (IPCC) 5th Assessment Report (AR5).* Cambridge, United Kingdom and New York, NY. Cambridge, UK and New York, New York, USA.

Syvitski, J. P., Kettner, A. J., Overeem, I., Hutton, E. W., Hannon, M. T., Brakenridge, G. R., Day, J., Vörösmarty, C., Saito, Y., Giosan, L., & Nicholls, R. J. (2009). Sinking deltas due to human activities. *Nature Geoscience*, *2*(10), 681–686. doi:10.1038/ngeo629

Talchabhadel, R., Nakagawa, H., & Kawaike, K. (2016). Tidal River Management (TRM) and Tidal Basin Management (TBM): A case study on Bangladesh. In E3S Web of Conferences (Vol. 7, p. 12009). EDP Sciences.

Thomas, K. A. (2017). The Ganges water treaty: 20 years of csooperation, on India's terms. *Water Policy*, *19*(4), wp2017109. doi:10.2166/wp.2017.109

Thompson, P. M., & Sultana, P. (1996). Distributional and social impacts of flood control in Bangladesh. *The Geographical Journal*, *162*(1), 1–13. doi:10.2307/3060212

Tingsanchali, T., & Karim, M. F. (2005). Flood hazard and risk analysis in the southwest region of Bangladesh. *Hydrological Processes*, *19*(10), 2055–2069. doi:10.1002/hyp.5666

van Staveren, M. F., Warner, J. F., & Khan, M. S. A. (2017). Bringing in the tides. From closing down to opening up delta polders via Tidal River Management in the southwest delta of Bangladesh. *Water Policy*, *19*(1), 147–164. doi:10.2166/wp.2016.029

Wesselink, A., Warner, J., Syed, M. A., Chan, F., Tran, D. D., Huq, H., & Wester, P. (2015). Trends in flood risk management in deltas around the world: Are we going'soft'? *International Journal of Water Governance*, *3*(4), 25–46. doi:10.7564/15-IJWG90

Williams, C. A. (1919). *History of the Rivers in the Gangetic Delta (1750-1918). Netherlands Engineering Consultants (NEDECO).* Bengal Secretariat Press.

Chapter 3
Land Subsidence in Ho Chi Minh City and the Mekong Delta Region, Vietnam:
Causes, Challenges, and Solutions

Trung Van Le
Faculty of Environment and Natural Resources, Ho Chi Minh City University of Technology, Vietnam & Vietnam National University Ho Chi Minh City, Vietnam

Phu L. Vo
Faculty of Environment and Natural Resources, Ho Chi Minh City University of Technology, Vietnam & Vietnam National University Ho Chi Minh City, Vietnam

Quang Khai Ha
Faculty of Environment and Natural Resources, Ho Chi Minh City University of Technology, Vietnam & Vietnam National University Ho Chi Minh City, Vietnam

Van Tran Thi
Faculty of Environment and Natural Resources, Ho Chi Minh City University of Technology, Vietnam & Vietnam National University Ho Chi Minh City, Vietnam

Hiep Dinh Luu
Faculty of Environment and Natural Resources, Ho Chi Minh City University of Technology, Vietnam & Vietnam National University Ho Chi Minh City, Vietnam

ABSTRACT

Land subsidence and its consequences are at an alarming level in the Southern and low-lying areas of the Mekong Delta. This chapter will discuss the situation of groundwater uses and urban development activities that caused the negative impacts of land subsidence and sea level rise in Ho Chi Minh City and the Mekong Delta region of Vietnam. The PSInSAR technique was applied for land subsidence monitoring and showed that an average subsidence rate at 15 mm/year which is 5 times greater than the effects resulting from the rise of sea level. This implies that land subsidence coupled with sea level rise have been as a major factor contributing to increasing the depth of annual flood and causing inconveniences to the daily lives of residents that suffered waterlogging of between 0.4 – 0.5m. Apart from the restriction of groundwater abstraction, rainwater harvesting – an alternative water sources – is a humanitarian engineering solution to minimize groundwater extraction which in turn to prevent from flood risks caused by land subsidence and the rise of sea level.

DOI: 10.4018/978-1-6684-5619-4.ch003

INTRODUCTION

Land subsidence and its consequences have posed profound challenges in many Asian megacities over the past decades. Particularly, land subsidence coupled with sea level rise have been identified at an alarming level in low-lying areas and coastal cities such as Jakarta, Bangkok, Ho Chi Minh City, Manila, and Tokyo (Erkens et al., 2015; Anh Cao et al., 2021). This phenomenon has increasingly drawn much attention from academies and managers because of triggering serious flood risks and causing infrastructure damage and economic loss.

Ho Chi Minh City, a megacity with a total area of 2,095 km^2 and more than 9 million of inhabitants, is the hub of socio-economic development with a fast urban expansion and rapid development of industrial, commercial, and service activities. The Vietnamese Mekong Delta region is a rice bowl of Vietnam and a home to 18 million people. This region occupies a total area of 4 million ha, and has witnessed a fast population growth, urbanization and land use changes over the past several decades. However, both Ho Chi Minh City (HCMC) and Mekong Delta (MD) are subsiding due natural and human-induced factors. Naturally, these areas locate at the soft soil and the low elevation of the land, rivers and canals form a complex network that is affected by tide. In addition, due to the geographical location, this delta system is also one of the most vulnerable areas to global climate change and SLR. Several evidence from human impacts are demonstrated in the relation between land subsidence, land use change and sea level rise. The rapidly developing urbanization since 1990 has led to increase in built-up areas and constructions zones at a very low elevation. Especially, the major factors can be caused land subsidence such as extensive conversion of prime agricultural areas into industrial parks and residential areas with high building density on soft soil, as well as the groundwater use was in a large scale for socio-economic development activities resulting the decline of water table (>20m from 1990 until now) which caused the deformation of soil making the ground level goes down each year. As a result, the impacts of the land subsidence and SLR have been increased frequency of tidal flooding that residents in these areas suffered waterlogging of between 0.4 – 0.5m. Many projects have been implemented in flood prevention plan of provinces/cities authorities such as: mapping land deformation, determining control measures on groundwater extraction in areas severely affected by land subsidence and modifying the impacts of the land subsidence and SLR in the master plan of 2030 vision in order to achieve the sustainable urban development.

Results derived from mapping land deformation by using the PSInSAR technique contribute to demonstrate the directly impacts towards urban infrastructure and show that an average subsidence rate at 15 mm/year can be considered as greater 5 times than the effects resulting from rising sea level. This problem pointed out the land subsidence coupled with sea level rise have been as a major factor contributing to increasing the depth and duration of annual flooding for the MD and coastal metropolitan cities including HCMC, Can Tho and adjacent delta areas.

The purpose of this paper is to discuss the situation and causes of land subsidence as well as the impacts of land subsidence and SLR in the context of urbanization and climate change. PSInSAR technique is deployed for monitoring land subsidence and mapping land deformation in HCMC and the MD region of Vietnam. Furthermore, the relation between land subsidence, groundwater uses, and urban development activities will be also elucidated. Some measures are proposed as humanitarian engineering solutions to enhance the resilience and adaptive capacity for the ongoing subsiding and the rise of sea level in these areas.

Figure 1. DEM of Mekong Delta and Ho Chi Minh City (m)

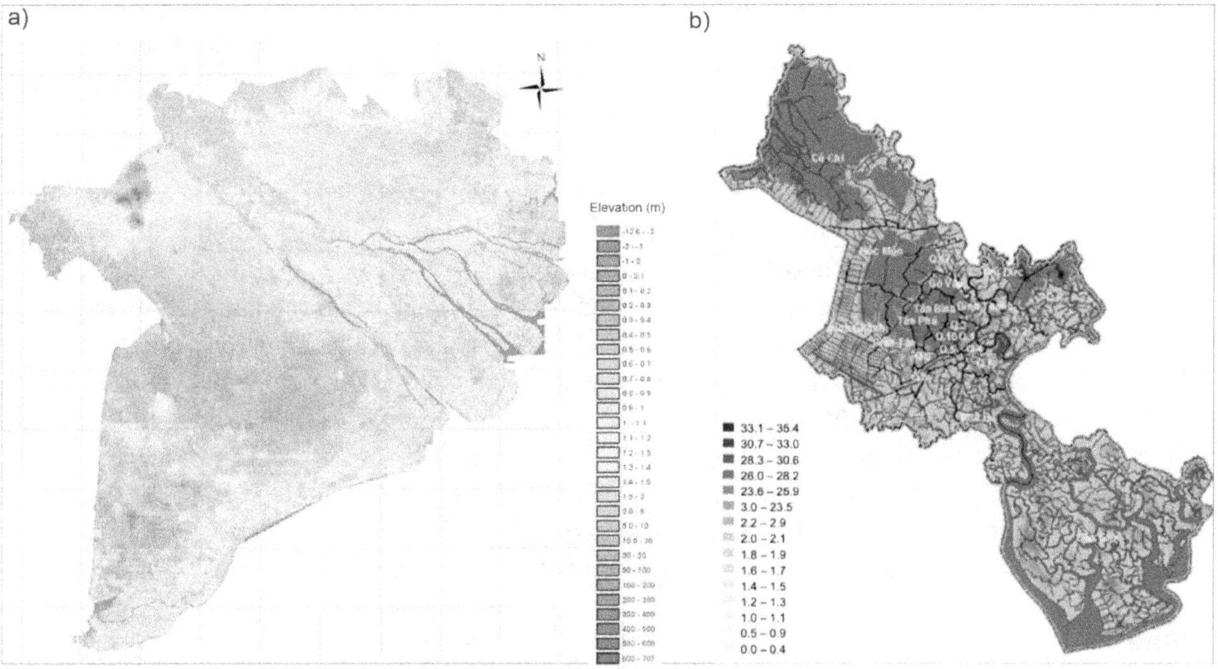

BACKGROUND AND ISSUES

Natural Characteristics

Floods in the MD and HCMC are intricately caused by heavy rain, high tide, rain, and tide combination. Natural characteristics of research area show major factors that contribute to cause the impacts of land subsidence and SLR. The situation of tidal flood for coastal cities will be increased frequency in which floodwater depth and duration of inundation for years depend on these natural factors.

Topography

Topography map of Mekong Delta (MD) and Ho Chi Minh City (HCMC) are created a Digital Elevation Models (DEM) as depicted in Figure 1 that use to analyze the combined effect of a large number of urbanized areas with the low elevation of the land, to illustrate the flood on low-lying land due to the combined effect of tide and land subsidence.

The MD has a flat topography, with an average elevation of 0.7 – 1.2 m, except for some hilly areas in the northern delta in An Giang province. Along the Cambodian border, the terrain is higher, with an elevation of 2.0 – 4.0 m. It gradually lowers to the centre of the delta at an elevation of 1.0 – 1.5 m and only 0.3 – 0.7 m in coastal intertidal areas. In the left region of the Tien River, deposited by alluvium from the Tien River, a low-lying trough topography sloped from the Northwest to the Southeast. In the area between the two Tien - Hau rivers, river alluvium is also accreted to form two high riverbanks and then lower into the field, forming a low-lying trough in the middle. The Hau River is generally sloping

East-West, gradually lowering from the Hau River towards the West Sea. The coastline is often caused by currents, wind and river alluvium, forming high coastal dunes with an arc convex to the sea, interspersed with low-lying tidal areas. The North-East region, the lower reach of Cai Lon - Cai Be River basin, and U Minh Thuong and U Minh Ha national parks are the lowest land areas, with elevations from 0.3 – 0.7 m, constantly flooded due to high tide, rainwater in the field and flood water upstream.

DEM of HCMC shows that the city is primarily flat, with few hilly areas in the North and Northeast sides. The ground level tends to slope from the North and Northwest to the South and Southeast. Based on the natural terrain and the current drainage infrastructure system, ground elevation planning area is divided into 4 large regions for choosing the height of construction works: (1) In the North - Northeast and part of the Northwest, ground elevation is divided 2 parts: from 4 -10 m accounts for about 19% of the total area, and the part > 10 m accounts for 11%; (2) Low flat land with an elevation of approximately 2 – 4 m, relatively favourable drainage conditions, distributed mainly in the inner city, accounting for 15% of the total area; (3) Low-lying, swampy form in the Southwest with a typical height of 1 to 2 m, accounting for about 34% of the total area; (4) Newly formed coastal marsh low-lying form, with a height of about 0 – 1 m. Generally, HCMC dominantly consists of plains of which the percentages are 60% is that ground elevation level is below +1.5 m while current max tide level: +1.72 m. Rivers and canals form a complex network that is affected by tide by daily tides, accounting for about 21% of the area.

Climate

The MD is distinguished into two distinct seasons. The rainy season coincides with the southwest monsoon season, from May to November, lasting 6-7 months, and the dry season coincides with the northeast monsoon season, from December to April next year, also lasts 5-6 months. The average monthly temperature in the whole region varies from 25.3-27.0°C, relatively evenly in space and time. The Mekong Delta has a very high number of sunny hours, from 2,200 - 2,400 hours per year. In the dry season, there are approximately 8 hours of sunshine every day. Annually, the average rainfall of the whole MD is about 1,400-1,800 mm. In the Western coastline of the MD, the rainfall is more significant (2,000 - 2,400 mm) the rainfall in the Eastern coastline tends to be lower (1,400 - 1,600 mm). The least amount of rain in the year from January to March and the highest rainfall from August to October. In the case of HCMC, the rainy season from May to November, the average annual rainfall is 1,979 mm. The dry season from December this year to April in the following year, the average temperature is 27.55°C.

Hydrology

The hydrological-hydraulic regime in the MD is very complicated. Strong effect of flood flow from the upstream Mekong countries during the rainy season from July-December and strong effect of tide from the East Sea during the dry season from January-June. Based on the influence of the source factors in space and time, in general, the Mekong Delta can be divided into three different hydrological regions: (a) the main flood-affected area (north delta, including part of the area of two provinces, An Giang and Dong Thap, an area of about 300,000 ha); (b) combined flood-tidal zone of influence (limited by Cai Lon river - Xeo Chit canal - Lai Hieu canal - Mang Thit river - Ben Tre river - Cho Gao canal to the limit of the region (a), with an area of about 1.6 million ha); and (c) the central tidal zone (including the entire coastal area, with an area of about 2.0 million ha). During the flood season, the alluvial content can reach from 500 -1,000 g/m^3 and the alluvium content is playing an important role in the agricultural system. During the dry season, however, under the influence of tides, salinity, and the lack of freshwater in coastal areas have a major impact on the flow regime in the Mekong Delta. The hydrological regime of HCMC is characterized rivers and canals that located downstream of the Dong Nai – Sai Gon River system. Recent studies show the canal system in the inner city of HCMC has a total length of about 76 km with 5 main sub-basins to help Ho Chi Minh City in irrigation. However, due to the influence of the semi-diurnal tides of the East Sea, the deep tidal penetration has caused negative impacts on agricultural production and limited water drainage.

Geology

Geology in the MD and HCMC are characterized by thick of unconsolidated sediments layer overlying the bedrock. In the Mekong Delta, the sediment thickness is about 200 m from the Cambodian border and become thicker to the coastal zone. As a summary of Ha et al. (2019a), in HCMC area, unconsolidated sediment layer thickness is about 200 m and become thinner to the mountainous areas. The unconsolidated sediment was deposited during Miocene to Holocene epoch. Miocene sediments are normally observed at depth 300 m bellow land surface. The Pliocene sediment and Pleistocene sediments distribute from 150 to 300 m and 20 to 150 m below land surface, respectively. The Holocene sediment thickness is up to 70 m with average of about 20 m, covered all area with elevation less than 5 m including Mekong Delta and southern part of HCMC. The sediment in the area is formed under estuarine, marine environment. Therefore, it characterized by fine particle such as fine sand, clay and silt and high proportion of organic matter. According to geotechnical survey in HCMC by Division for Water Resources Planning and Investigation for the south of Vietnam in 2010 (DWRPIS, 2010), Holocene formation is characterized by soft sediment such as very soft to soft clay and very loose to loose sand. The standard penetration test (SPT) in Holocene sediment is normally less than 2 in clay and less than 9 in the sandy layer (Thoang and Giao, 2015). Meanwhile, the older sediment layers of Pleistocene to Miocene formations are more compacted and consist of stiff to very stiff clay and medium to very dense sand.

Hydrogeology

The sediment layers in MD and HCMC are classified into eight aquifers including Holocene (qh), Upper Pleistocene (qp^3), Upper–Middle Pleistocene ($qp^{2,3}$), Lower Pleistocene (qp^1), Middle Pliocene (n_2^2), Lower Pliocene (n_2^1), Upper Miocene (n_1^3), and Upper–Middle Miocene ($n_1^{2,3}$). These aquifers are interlayered by the aquitards of the same formation (SVGMD, 2004). Based on some pumping tests, the horizontal hydraulic conductivity of aquifer ranges from 10 to 40 m/day (SVGMD, 2004; DWRPIS, 2018).

Groundwater resource is a crucial water supply source for the MD and HCMC, contributing to 40 and 60% of domestic uses in HCMC and MD, respectively. Particularly, groundwater abstraction is increasing overtime. The National Groundwater Monitoring Network System (NGMNS) wells located in Ca Mau province show declines of up to 11.1 m in the $qp^{2,3}$ aquifer from 1995 to 2014. The groundwater level decline has generally been even more serious in populated areas such as HCMC and Ca Mau city (Vo, 2007; Bui et al., 2015). The declining rates of groundwater level reflected in the NGMNS data in the period 1995-2014 are 0.03, 0.09, 0.26, 0.22, 0.36, 0.34, and 0.48 m/year on average for the qh, qp^3, $qp^{2,3}$, qp^1, n_2^2, n_2^1, and n_1^3 aquifers, respectively. This depicts that rates of groundwater withdrawal generally increase with depth. Residents now prefer to tap deep aquifers for domestic consumption with better groundwater quality in which qh and qp^3 are shallow aquifers with the limitation of freshwater resources in good quality.

Land Use Change

The MD – occupying more than 4 million hectares – is the main agricultural production region of Vietnam which supports large diversity of agricultural systems and aquatic products (Tran Thanh Be et al., 2007). However, land use and agricultural practice have been changed rapidly during the past decades, due to the population growth and urbanization of 13 provinces as well as climate change caused salt intrusion, exacerbated by drought conditions during the dry season. The area of the forest-cover in the MD region has decreased with the average cover rate from 10% of the natural area in 1997 to 4% in 2016 (Le et al., 2021). The area is mostly used for agriculture (62%) that is considered as the rice granary and produces 56% of total rice production of the country. According to General Statistics Office (GSO) of Vietnam, the proportion of land use in 2019 (https://www.gso.gov.vn/) shows that 15.6% of the total natural land area is the non-agricultural land area, which is divided into three major land types, including: residential land (6%), construction land and other purposes (9%), and the unused land area is about 0.6%.

Can Tho is the biggest city in the Mekong Delta region and locates on the south bank of the Hau River with the soft soil and the low elevation (0,8 - 1m above mean sea level). Results derived from analyzing the urbanization trend through the impervious surface change show that the urban area increased with average growth rate of 14% /year (Le et al., 2021).

In the case of HCMC, the rapid pace of urbanisation over the past 25 years has caused the loss of space for water and the creation of impermeable surfaces. Despite encompassing agricultural land is 54.2% of the total natural area, land use change analysis shows that residential land in the urban area covers 1,250 hectares and the land is used for infrastructure development in the city is estimated to total around 35,000 hectares, an increase of 15,233 hectares in comparison with 2015. The master plan of land use in the period 2020 – 2025 (http://www.donre.hochiminhcity.gov.vn/) indicates that half of the land in HCMC will be used for urbanization. The increase in population and urban development activities in HCMC leads to increase in built-up areas, decrease in green areas and increase impervious surfaces that water cannot infiltrate. These surfaces are primarily associated with transportation (streets, highways, parking lots, sidewalks) and buildings.

Groundwater Exploitation

The process of urbanization with some new districts, people and the industries used ground water as the main water resource. Rapid increase of ground water use started from the year 1990 for activities in the development of agriculture and industry and the level of groundwater has been lowered quickly in many areas of MD and HCMC (annual drawdown: 2m depth at the heavy ground water pumping stations). At present, industrial boreholes are still widespread and many farmers have attached water systems from water supply company, but do not use piped water that still primarily using groundwater due to free of charge. The municipal authorities are taking measures to restrict overexploitation which has led to exceed the permissible level induced subsidence and water pollution.

Ground water exploitation in HCMC is estimated at about 800,000 m^3/day from three of five aquifers play an important role in water supply: Pleistocene aquifer (20–50 m), upper Pliocene aquifer (50–100 m) and lower Pliocene aquifer (100–140 m). The City's authorities plan to reduce the exploitation of groundwater from 151 groundwater exploitation stations that belong to companies, industrial parks, and export processing zones to 100,000 m^3/day by 2025 and encouraged households to fill up about 100,000 bored wells in areas where clean tap water is available to prevent land subsidence and protect residents' health.

Figure 2. Distribution of pumping stations (red points) and groundwater is exploited in Holocene and Pliocene aquifers

Note: Dark blue locations are the exploitation with volume greater than 100,000 m^3/day. The left photos show the effects of over-pumping caused land subsidence.

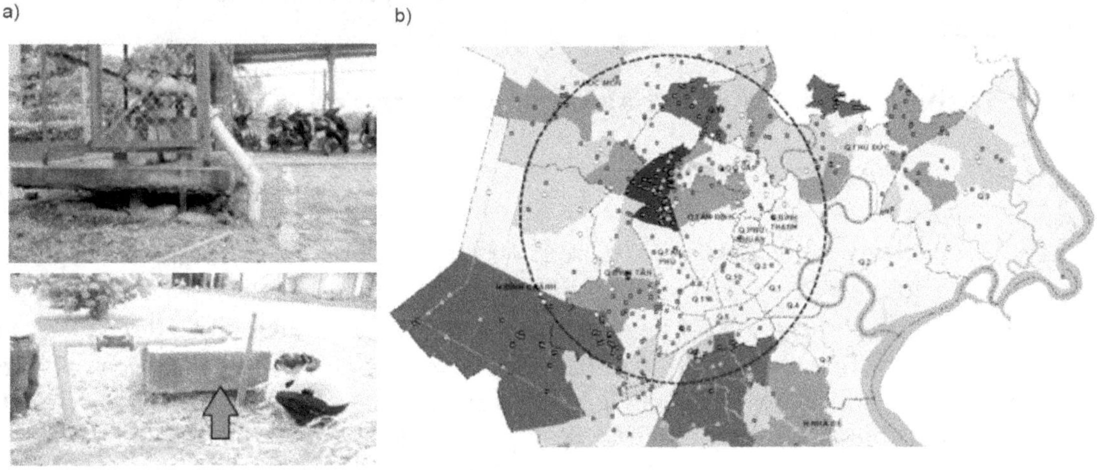

Many studies show that a strong relation between ground water overexploitation and causing ground deformation in MD. However, restriction of ground water use is challenged for this region because drought events are increasing in frequency and severity impacting livelihoods and agricultural production, as well as surface water is contaminated and salinization which have resulted in more groundwater extraction. The studies note that groundwater abstraction is increasing overtime and abstraction rate is about 1.9 million m^3/day and monitoring wells indicate the level of groundwater has been lowered with annual averages 26 cm/year. However, residents now prefer to tap deep aquifers for agriculture, aquaculture, and food processing industry in the dry season. Vietnam authorities also tries to reduce groundwater abstraction rate by importing more surface water and expand water supply system. Recent studies show that some cones of depression can be controlled by the permissible level of groundwater mining at a rate not exceeding 95,000 m^3/day. When groundwater is extracted at the heavy ground water pumping stations, pore pressures are reduced and sedimentary layers undergo compaction that results land subsidence (Minderhoud et al., 2017) and seawater intrusion (Ha et al., 2019a; Tran et al., 2021). Wells installation is also suggested as the cause for groundwater acidification and subsequently the mobilization of trace metals (Ha et al., 2019b).

Human Impacts on Land Subsidence

Anthropogenic activities are contributing to create extensive conversion of land use type that leads to increasing the density of high buildings in soft land areas, decreasing green areas as well as increasing impervious surfaces. Especially, industrial boreholes are widespread and groundwater overexploitation keeps on rising in the development of agricultural and industrial activities. Why subsidence is happening and what can be done to reduce impacts of anthropogenic activities concerning urban expansion and rapid economic development are much more challenging. It is difficult to directly answer the relationship among these factors and land subsidence rate in different locations. However, it is clear that anthropogenic factors are responsible for land subsidence-induced flood risks in coastal cities and low-lying areas. The following sections provide an overview of

PS InSAR Technique for Land Subsidence Assessment

Overview of PS InSAR

SAR images (Synthetic Aperture Radar) can be used for providing information on ground displacement based on the phase difference between two multi-temporal SAR images acquired over the same region. The two images which are merged point by point to form the image of their phase differences, called an interferogram and the first idea of the use of radar signal phase produces an InSAR (Interferometric Synthetic Aperture Radar) for measuring ground displacements that was suggested in the early 1970s. After several years, the differential SAR Interferometry (DInSAR) method is developed and commonly used for monitoring land subsidence in some areas (Ferreti et al., 2000). However, accuracy of the DInSAR method depend on the main errors in pairs of images taken from space at different moments: atmospheric delay, temporal, and spatial decorrelation, choosing the baseline of image pairs and the influence of the digital terrain model (DTM) that is used for eliminating the topographic phase. PSInSAR (Persistent Scatterer InSAR) is an extension of this InSAR technique which offers a practical way to reduce these main errors by using a series of images. The main characteristics of PSInSAR method are that it utilizes a single master in a stack of differential interferograms, and that only time-coherent pixels, i.e., "Permanent Scatterers" (PS) are considered (Ferretti et al., 2001). These pixels are also referred to as point targets or persistent scatterers points (Hooper et al., 2004). PS points typically corresponds to objects on man-made structures such as buildings, bridges, dams, water-pipelines, antenna, as well as to stable natural reflectors. In recent years, PSInSAR method is commonly used for measuring ground subsidence in urban areas at high accuracy and low costs in comparison with methods of precise leveling or DGPS (Differential Global Positioning System).

SAR Data

A large number of SAR images are needed to be collected to ensure the best results in using PSInSAR method. Multi-temporal SAR dataset often are commercial images, acquired from the sensors on the satellites such as: JERS (1992-1997) ERS-1 (1991-2000), ERS-2 (1995-2010), RADARSAT (1995-present), ENVISAT (2002-2012), ALOS (2006-2011), ALOS-2 (2014-present), COSMO SkyMed (2007-present). Especially, Sentinel-1 (2014 - present) is free data that has been applied for continuous updating and monitoring of Mekong Delta and HCMC. Evaluating results for case study in HCMC during the period

Figure 3. The phase difference of the same pixel in InSAR Geometry

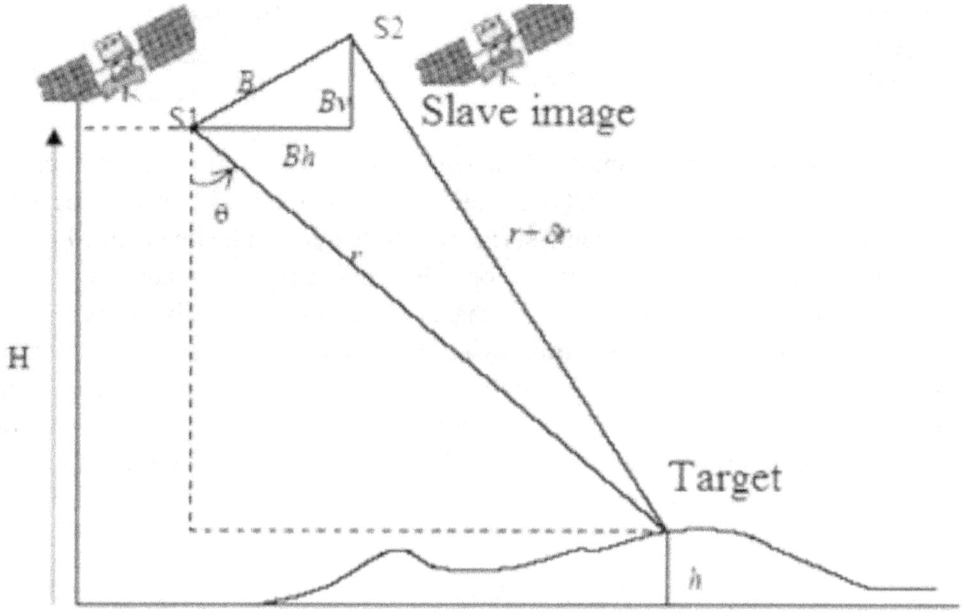

of 2014 – 2016 showed that Sentinel-1 image can be replaced COSMO SkyMed in measuring ground subsidence. This is to create new opportunities of using PSInSAR method with free data and open-source software to determine subsidence rates at the large areas. Sentinel-1 images were also applied to update and contribute to the subsidence information of HCMC with continuous temporal coverage from 2017 to 2019 (Duffy et al., 2020). Recently, an advanced Persistent Scatterers and Distributed Scatterers (PSDS) InSAR technique was also applied to Sentinel-1 images that results reinforced the remarkable suitability SAR Sentinel-1 for monitoring subsidence over the HCMC area from 2014 to 2020 (Le et al., 2021). Therefore, the capacity of using free images can provide an efficient improvement on cost for continuous monitoring land subsidence on a large scale of big cities or delta-wide such as Mekong Delta.

METHODOLOGY

Phase differences between two SAR images are chiefly caused by orbital effects. The selection of the images is made on the basis of baseline length and the time period between two image acquisitions. S1 (for Master image, used as a geometrical reference) and S2 (for slave) represent the two SAR images in between two orbits during acquisition.

The phase difference ϕ between two radar signals acquired from the same surface target at the two satellite positions (Li and Goldstein, 1990) can be calculated as follows:

$$\phi = \frac{4\pi(\delta r)}{\lambda} = \frac{4\pi(B_h \sin\theta - B_v \cos\theta)}{\lambda}$$

Whereas λ: is the wavelength; δr: is the range difference; *Bh*: baseline horizontal component; *Bv*: baseline vertical component and θ: is the look angle.

Based on the range difference and the look angle changes through between two SAR images, the phase difference (interferometric phase) between two sensor positions and the target terrain point can be generated as follows:

$$\Delta h = \frac{\lambda r \sin\theta}{4\pi B}\Delta\phi$$

Whereas *r*: the range in the master image; *B:* normal base line; $\Delta\phi$: the change in phase

This equation shows the relationship between the heights and the phase differences that also makes an interferogram image (reflects topographic surface). The height can be determined for each pixel of SAR image to form the necessary DEM in raster form. For measuring ground displacements, PSInSAR method is an extension of DInSAR technique which can reduce main errors by using a series of images such as the contribution from the interferogram by modeling the orbits, atmospheric and relief effects. Value of PS (permanent/persistent scatterer) points can be identified with high accuracy to show land subsidence rate of every pixel that includes man-made object in urban areas. Temporal variations of value of PS points are used to determine the averaged vertical velocity, to identify the displacement history and to apply a joint spatial and temporal analysis for land mapping.

Figure 4. PSInSAR method for measuring ground displacements: (a) – the principle for extraction of PS; (b) – differential interferograms images

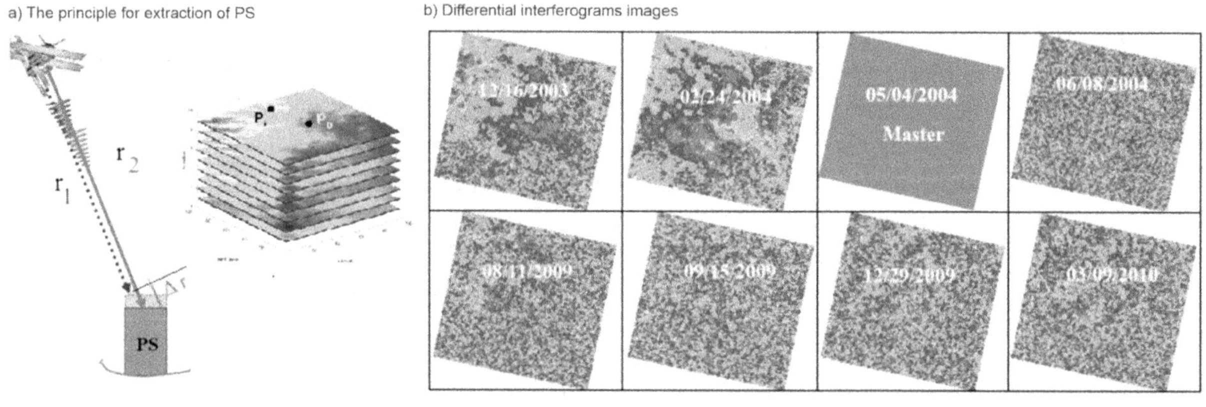

APPLICATION OF PSInSAR METHOD FOR LAND SUBSIDENCE ASSESSMENT

PSInSAR technique has been applied widely in many cities around the world to measure the ground deformation. However, the first time in Vietnam, this technique has been developed for HCMC that the average subsidence velocity map has been retrieved by the PS processing, and the validation indicates good agreement with leveling data. The land subsidence was discovered in 2003, there were several areas appeared land subsidence especially around water pumping stations of district 6 (10 – 22 cm), district 12 (20cm), district Binh Tan (3 – 30 cm) and Binh Chanh (4 – 19 cm). SAR commercial data of satellite JERS-1, ERS-1 and ERS-2, ENVISAT, COSMO SkyMed, ALOS PALSAR and free data of satellite Sentinel-1 have been used for monitoring the land subsidence over the urban area from 1996 to 2019.

Figure 5. The study site and SAR image using for mapping land subsidence of center region

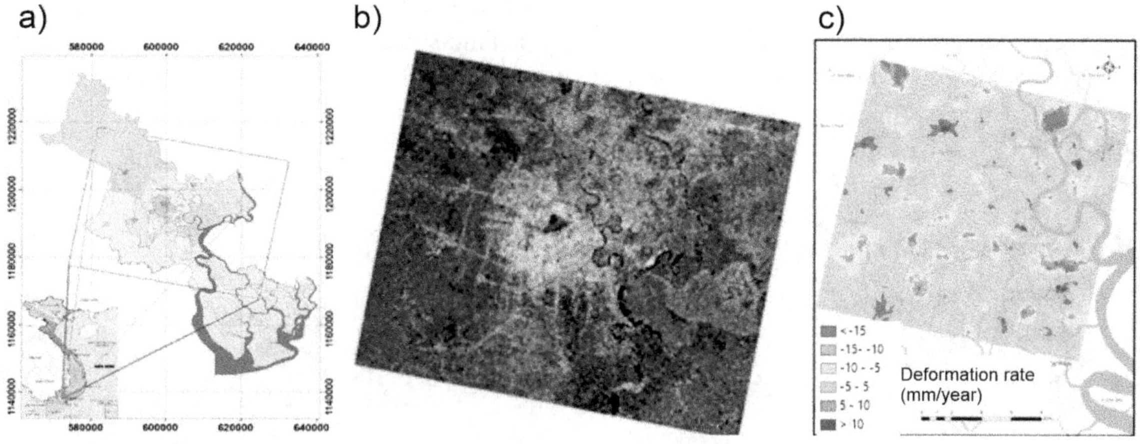

Land subsidence in HCMC can be caused by the excessive groundwater extraction, load of buildings and constructions, and the combined effect of a large number of urbanized zones are situated at a very low elevation and along the canals with alluvium soil. Results analysis in the period of 1996 to 2010 showed the rate of subsidence is 10 – 15 mm/year and covers 140 km^2 (0.67% area of HCMC). From 1997, several subsidence areas have formed and widen with 5 districts inducing the high subsidence trend (greater than 15mm/year). The maximum subsidence value during this period is about 150 mm and the below figure shows the affected area (blue color) and percentage of the area that each district is suffered by land subsidence (red color).

Figure 6. Subsidence rate at districts and affected areas in Ho Chi Minh City

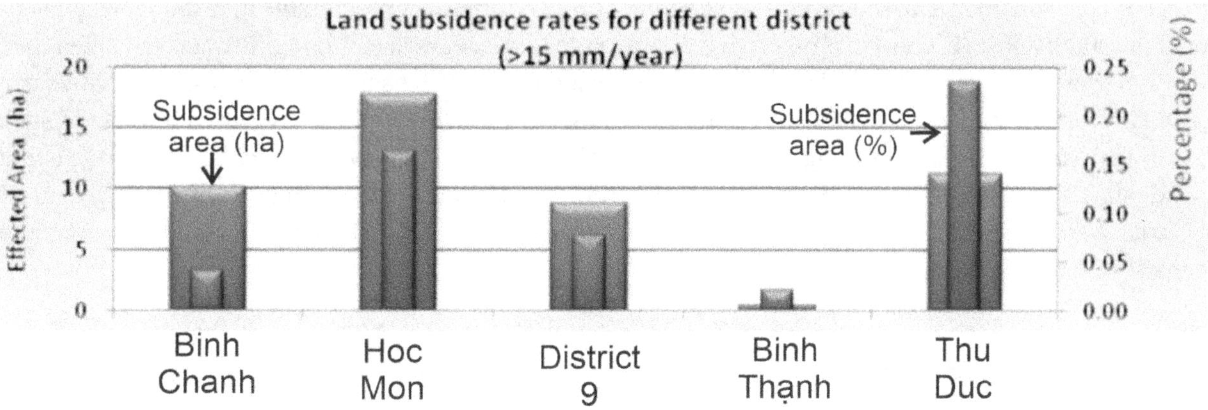

In the period of 2010 – 2019, Department of Natural Resources and Environment (DONRE) has been applied control measures on groundwater extraction in areas severely affected by land subsidence and re-measured accurate altitudes of the elevation benchmark system that have been affected by land subsidence. In addition, every year land deformation map is also established by using PSInSAR methodology with satellite data for monitoring, updating database, warning, and informing local communities about the high rates of land subsidence resulting in more frequent flooding. Evaluating the results in this period showed that the central area is not suffered by the land subsidence from 2017 – 2019. However, some districts on soft soil still get high subsidence rate (greater than 15mm/year) such as district 12 in the North, district 08, Binh Chanh and Nha Be in the South of the city. The total amount of subsidence value had increased to 400mm at the heavy ground water pumping stations by the year 2018.

Figure 7. Subsidence rate in District 8 (HCMC) from 1996 to 2018

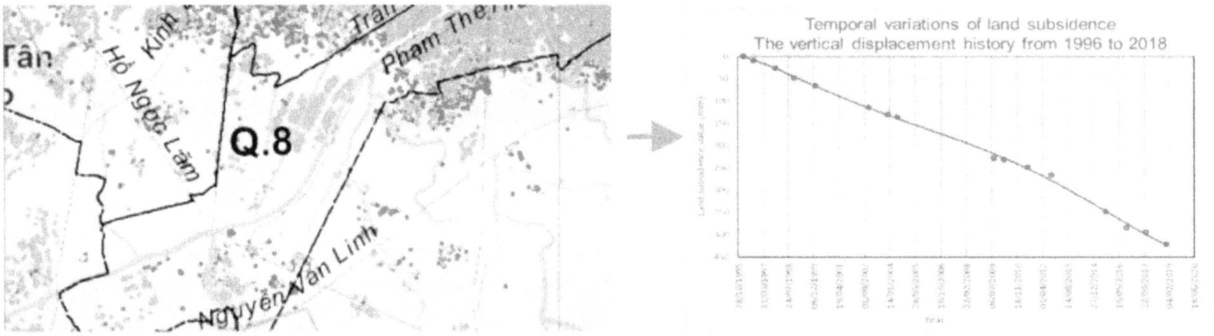

In mapping land deformation of HCMC on the different time periods, legend of map shows colors that represent the averaged vertical velocity of land subsidence area (mm/year) based on 3 levels: (i) Very high (subsidence rate > 15 mm/year) (ii) High (subsidence rate 5 – 10 mm/year) and (iii) Medium (subsidence rate < 5 mm/year). These maps are established every year that allow to retrieve precise and create the way for continuously monitoring for each district. In addition, maps also show the affected area, the high-vulnerability region and stabilization of subsidence is observed in this same region. The trend of land subsidence based on spatial and temporal variations can predictable and should be considered in establishing the construction ground elevation for drainage zones. Fig.8 illustrates the developed inner-city area severely affected by land subsidence in the year 2010 against (a), and shows that severe land subsidence in 2019 is observed in rural residential area and new urban area at suburban (b) such as: in the Northeast, the low-lying land of District 12 along Sai Gon river; in the Southwest, existing residential along Kenh Doi canal of District 08 and Binh Chanh; in the South the low-lying land of District 7 and Nha Be. These are urbanised zones that are situated at a very low elevation and along the canals with alluvium soil.

IMPACT OF LAND SUBSIDENCE IN HO CHI MINH CITY AND MEKONG DELTA

Land Subsidence in Ho Chi Minh City

Being located in the soft soil and the low elevation of the land, rivers and canals in HCMC form a complex network that is affected by tidal regime. The rapidly developing urbanization since 1990 has led to increase in groundwater extraction due to development of industrial activities and caused several

Figure 8. Comparison of the land deformation map (a) October 2010 against (b) December 2019

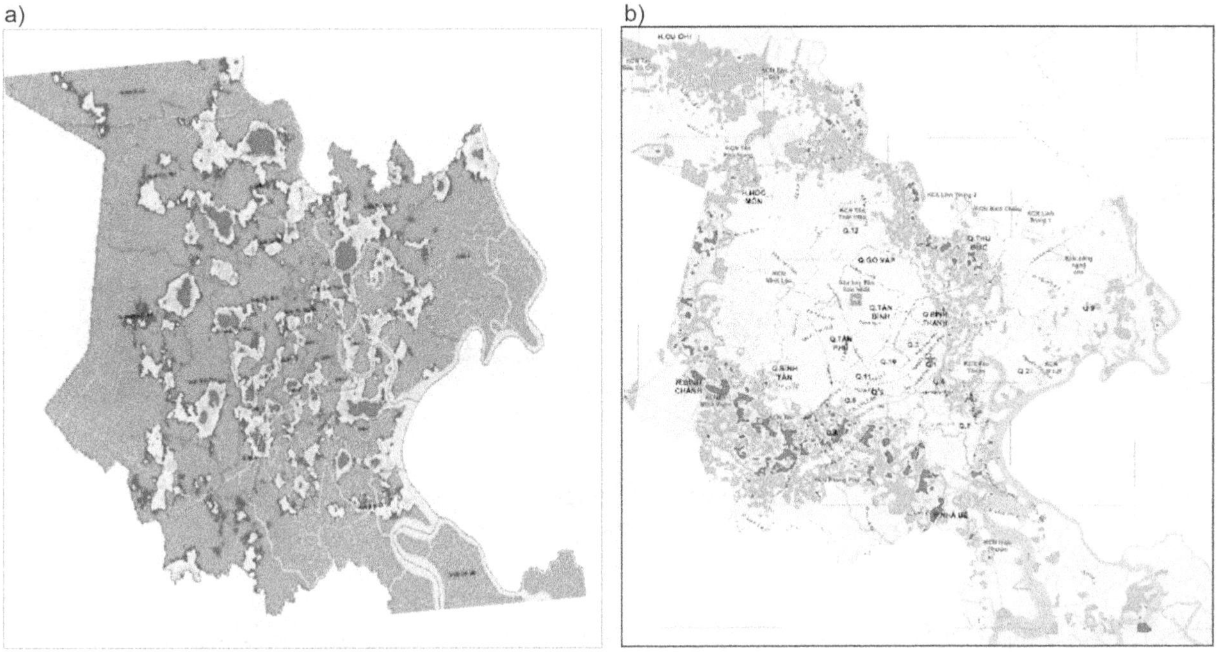

environmental problems such as: extensive conversion of prime agricultural areas into residential and industrial areas; the rapidly increasing the population, buildings and constructions zones are situated at a very low elevation and along the canals with alluvium soil, etc. Especially, HCMC ranks among the top 20 cities in the world to be severely affected by climate change. Land subsidence combined with a mean sea level rise has been increased frequency of flooding that residents in many wards suffered waterlogging of between 0.4-0.5m. Many projects have been implemented such as: construction of new sewers, repair of existing drainage systems, excavation of canals and reinforcement of the banks, etc. but in fact most of them are not effective, because urban development of HCMC and its surrounding areas has grown rapidly, and flood prevention has been one of the challenging tasks of Ho Chi Minh City authorities.

Ground elevation level is below +1.5m that covers 60% area of HCMC while current max tide level: +1.72 m. Each year climate change could cause sea levels to rise by 3 mm and impact of land subsidence (10 mm/year) makes this situation even worse. Technical planning on ground elevation based on the trend and the average subsidence rate plays an important role towards flood prevention plan in order to achieve the sustainable urban development. The land subsidence makes the ground level goes down and has resulted in tides moving into low-lying areas that were previously above high-tide levels. This situation causes residents in many wards suffered waterlogging of between 0.4-0.5m and many streets becomes on-ground river.

Figure10 illustrates the flood on low-lying land due to the combined effect of tide and land subsidence. The left scale is value of tide that highest level is predicted based on scenarios of Intergovernmental Panel on Climate Change (IPCC). The right scale is land subsidence (10 mm/year) that ground elevation is going down about 0.5m in 2030. Baseline (red color) shows the impact of flooding in some areas in

Figure 9. The impact of the tidal flood on (a) public transportation system and (b) activities of city residents

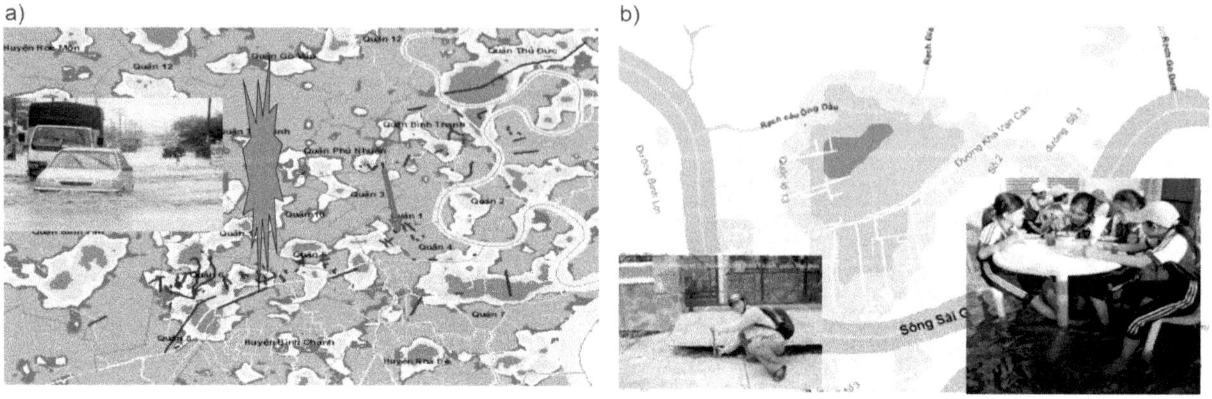

2010. Therefore, authority of HCM City needs to conduct a risk assessment and to modify the master plan for the urban drainage system in view of the ongoing effects of land subsidence on drainage.

Figure 10. Flood hazard caused by the land subsidence combined with high tide levels

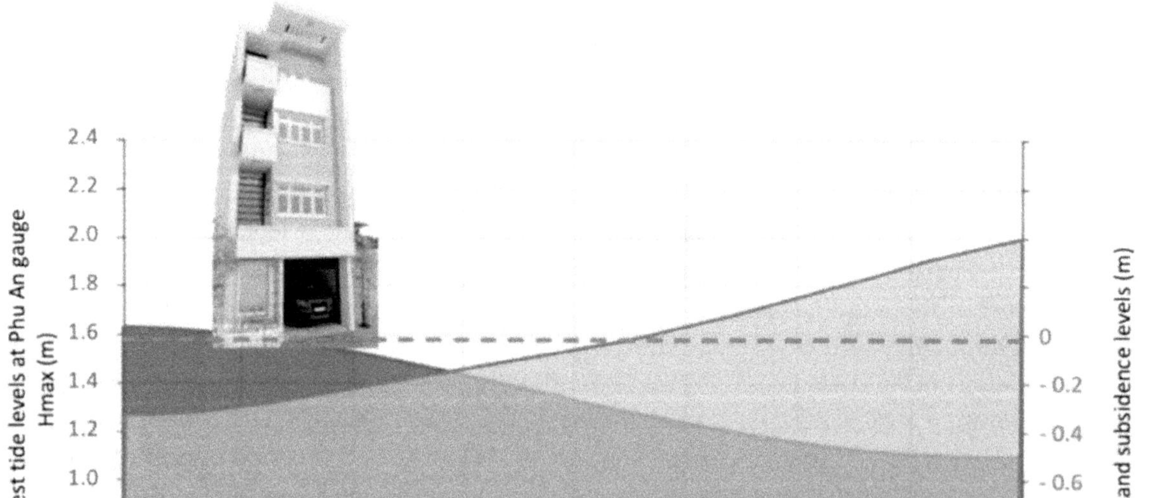

Implication of ground subsidence for flooding in HCMC allows providing an unprecedented spatial extent and continuous temporal coverage of the impact of the land subsidence combined with high tide. Severe land subsidence is observed and analyzed for modifying the 2030 vision of the master plan under

Figure 11. Digital Elevation Model (DEM) of Can Tho city

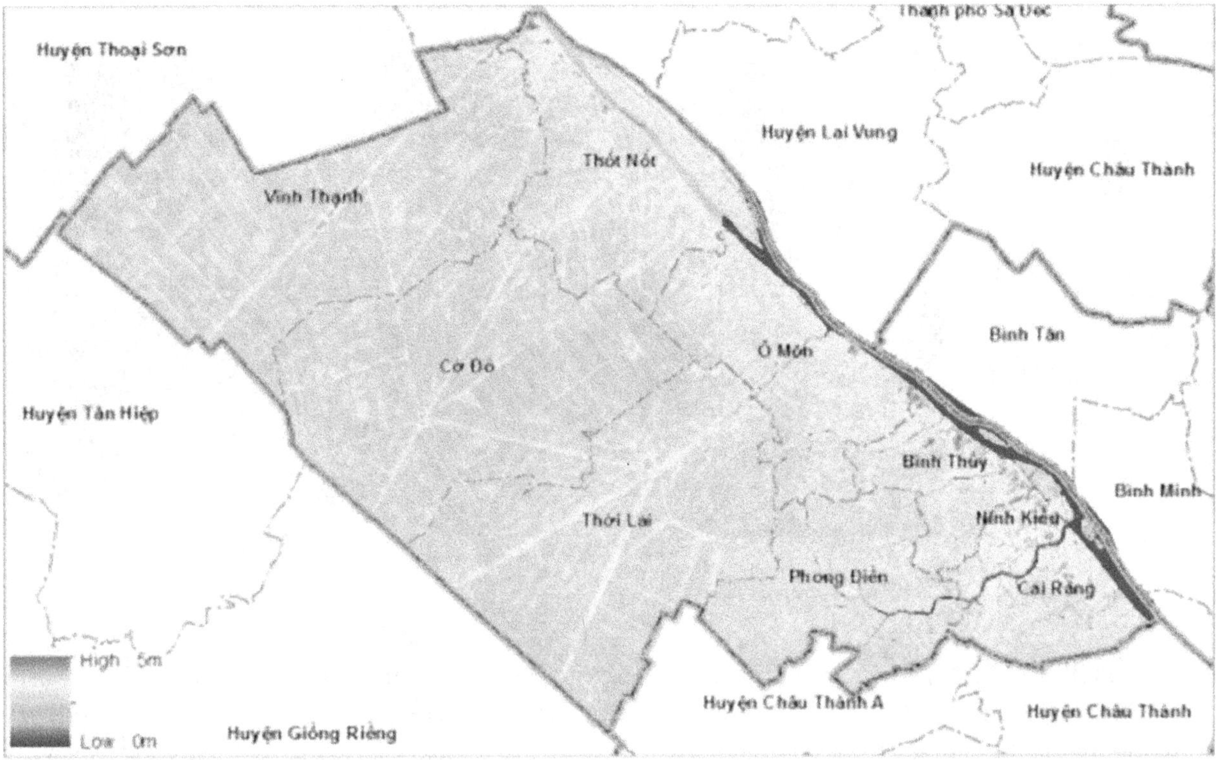

the climate change as well as determining control measures on groundwater extraction in areas severely affected by land subsidence.

Land Subsidence in the Mekong Delta Region

MD area locates at the soft soil and the low elevation as well as rivers and canals form a complex network that are often threatened by sea level rise (SLR). Impacts of land subsidence and SLR in this region are well known and analyzed by multiple scientific publications. InSAR study is the first mapped estimates of delta-wide land subsidence due to groundwater pumping (Erban et al., 2014) 78 ALOS PALSAR images are used to estimate average rates of land subsidence (1.6 cm yr^{-1} in the period of 2006–2010) and these results confirmed the increasing threat of land subsidence and potential impacts to infrastructure, as well as increases in the depth and duration of annual flooding. The results of the recent of Rise and Fall research project (2014 – 2019) have pointed out the impacts of land Subsidence and SLR as a major factor contributing to increasing vulnerability of many provinces of the Delta area. The land-use-based approach also predicted the spatially varying subsidence rates (Minderhoud et al., 2018) with the spatial patterns visible in the observed subsidence can largely be explained by land use. Therefore, negative impacts of subsidence and SLR can increase flood risk, aggravate other hazards such as coastline erosion and saltwater intrusion. In addition, Sentinel-1 images also are used for monitoring land subsidence that can be found in the final report of EMSN-057: Ground subsidence analyses, Mekong Delta, Vietnam. This service is provided under COPERNICUS by the European Union and ESA that "Mekong Urban

Figure 12. The land deformation map of Can Tho City

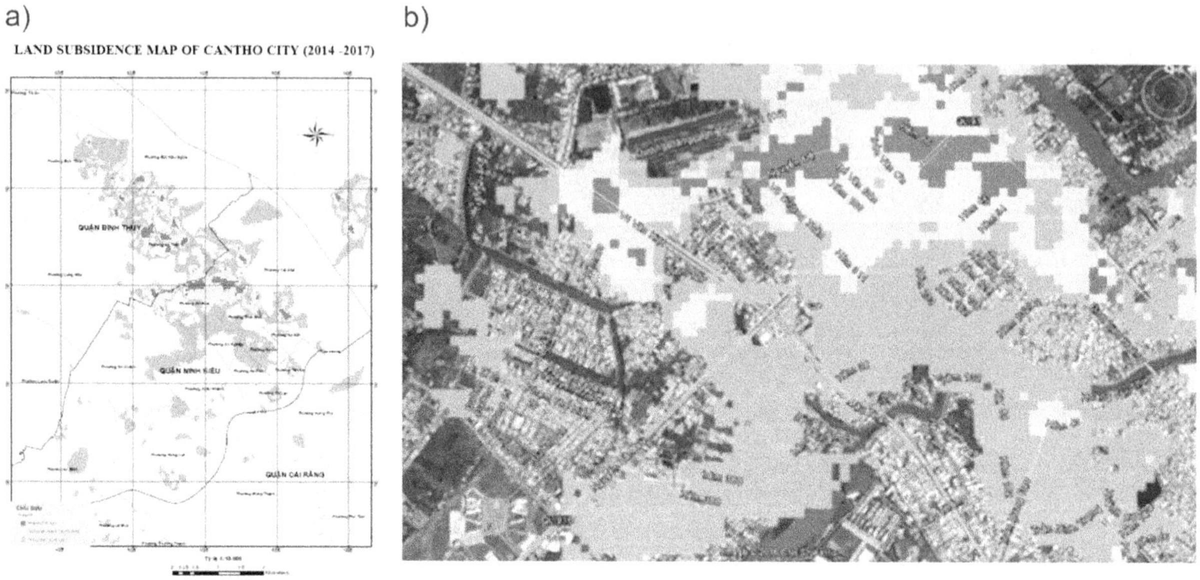

Flood Resilience and Drainage Program" aims to provide geospatial information facilitating assessment of drivers of ground subsidence and supporting analysis of relation between detected ground subsidence and land use changes in Ca Mau, Long Xuyen and Rach Gia areas. At present, Germany's GIZ is very interested in the further application of InSAR analysis in its flood risk mitigation and strengthening capabilities in MD area to this effect. The Rise and Fall project (2014-2019) is the first important study to understand land subsidence in Can Tho city, which shows the rate of average sea level rise is high and rising, but the rate of land subsidence is currently about ten times higher. Can Tho and adjacent delta lands which are derived from previous and ongoing studies on significantly exceeds the rate of sea level rise (SLR).

Similar to the case of HCMC, impacts of subsidence can be recognized: (i) increased flood risk (due to increased flood frequency, floodwater depth, and duration of inundation) and more frequent rainfall-induced floods due to ineffective drainage systems, and (ii) damage to buildings, foundations, infrastructure (Erkens et al., 2015). PS InSAR technique is also applied for measuring subsidence rates and the relationships between land use history (Minderhoud et al., 2018). Impacts of increased flooding due to the effects from ongoing sea level rise combined with land subsidence, largely caused by excess groundwater withdrawal (Takagi et al., 2016).

Generally, land subsidence map must be done for providing the rate of sinking land, the root causes and finding ways to mitigate subsidence or options for adaptation. Department of Science and Technology (DOST) also offered the project "GIS solutions for Urban management of Can Tho City, 2017) and land deformation map was also established by processing 21 images of Sentinel-1 (from Nov.2014 to March 2017) with PSInSAR technique as shown in Figure 12. This map is a useful tool for displaying subsidence to be distributed in some places, monitoring, warning, and informing local communities about the rates of land subsidence resulting in more frequent flooding.

In general, impacts of subsidence and SLR depend on the location where groundwater extraction rate continues to increase; urbanization problem triggers intensive conversion of land use type; the buildings and constructions zones are situated at a very low elevation.

FUTURE RESEARCH DIRECTION AND RECOMMENDATIONS

Improvement Technique for Land Subsidence Monitoring

The PSInSAR analysis results of case studies illustrated the significant effects of land subsidence as a major factor in the increasing vulnerability of the MD region and HCMC. Almost provinces and cities locate in the soft soil and the low elevation of the land, rivers and canals form a complex network that is affected by tides. For low-lying areas where have higher tidal flood risks and negative impacts of subsidence and SLR affect, it is hard to identify the minimum platform level for each area towards flood prevention plan and the sustainable urban development strategy.

In the context of urbanization and climate change, land subsidence can be caused by factors such as the excessive groundwater extraction, high building density, soft structure soil and land use change of urbanized zones, etc. Many scientific documents contribute to demonstrate the feasibility of using InSAR technique in monitoring land subsidence and determining directly impacts towards urban structures and infrastructure. However, there is no information yet about the contribution of each factor on the subsidence at each location and their spatial (contribution) variation that aims to support suitable measures in areas severely affected by land subsidence.

The main challenges of using PSInSAR technique for land subsidence monitoring are how to get the right software and free SAR image in Viet Nam. PSInSAR processing is a new technique and required complex procedure. Therefore, applications and manipulating of this approach need professional trainings which can create an efficient improvement on cost and accuracy in comparison with the existing terrestrial methods. In addition, value of PS points is usual measured and analyzed by various algorithms, if the lack of reliable and consistent data on land subsidence, hence the different expertise gives solutions, mitigating measures and the different trend of provinces that have a considerable impact on subsidence and SLR in future against flooding. However, research orientation and development of Web solution (an automated INSAR Platform as a Service) for monitoring impacts on subsidence and SLR at regional level is not developed yet. In order to support the implementation of sustainable development strategy of HCMC and MD region, the web based PSInSAR solution can be used not only in monitoring subsidence with higher frequency but also sharing all thematic maps and updating datasets (GIS and Remote Sensing image). The other challenges, all SAR images are processed and analyzed that need to convert onto the standard Vietnam map projection VN-2000 and a local DEM must be referenced. However, previous international researchers used the global DEM that did not correct for this difference, so almost results of Vietnamese researchers get higher accurate of average subsidence rates and the spatial and temporal variations. Because they have better data sources and land deformation rate depend on the corresponding urban development, hence a more accurate estimate of average rates of land subsidence (mm per year in the same period). Therefore, the web based PSInSAR solution is a necessary approach for sharing spatial data among administrative agencies and implementation of the international projects or projects. Web-database can be used in managing natural resources and environment at different levels, in order to provide fully integrated information of updated on the status of land cover changes and analyzing the trend of negative impacts of subsidence and SLR in this region. Because the land is sinking faster than the rise of sea level. Additionally, an early warning system for flood risk must be established for mitigating negative impacts on local communities.

Furthermore, the improvement of PSInSAR technique should be paid attention on what are the spatial accuracy achieved by the algorithms. How to evaluate the results (both in terms of product quality, speed of delivery and ease of use) of the different PSInSAR approaches? How to support local organisations in the training and technology transfer? How to combine both PSInSAR and geomatics expertise as well as water security and safety expertise in order to cope with impacts of subsidence and SLR? How to determine control measures on groundwater extraction in areas severely affected by land subsidence? and how to make policies for mitigation these impacts?

Groundwater Management

Over-exploitation of HCMC and MD's unmonitored groundwater resources, results in falling groundwater levels, which induced land subsidence, saline intrusion to aquifers further caused significant damage to houses, infrastructure and reduces agricultural productivity. Therefore, some measures need to maintain and recovery to minimize impacts of human induced land subsidence must be proposed as humanitarian engineering solutions to enhance the resilience and adaptive capacity. If no effective action is taken for the ongoing subsiding and the rise of sea level in these areas, then maximum tidal flood depth will drastically increase in the next 10–20 years. The following section are the proposed strategies for groundwater management in response to land subsidence.

- *Short-term (2021-2025):* The development of Web PSInSAR solution is an imperative for identifying areas which are severely affected by land subsidence, and exploitation of groundwater must be prohibited at existing urban areas where water supply system is available. Urban and rural areas must expand surface water supply and encourage households to fill up bored wells. In the case of groundwater exploitation for irrigation, industrial parks and export processing zones, etc. local government must have a plan with timeframe to reduce exploitable groundwater based on the permissible level of groundwater mining (100,000 m^3/day for each province). Planning ground level 2m for tidal flooding areas and anti-flooding projects should focus on various measures such as elevating ground level of roads, expanding the drainage systems, reinforcing current dikes, and constructing tidal sluice gates, etc.
- *Mid-term (2025-2030):* ground water can be sustainably exploited is assessed at a regional rather than province or city level. Therefore, hydrogeology map and land use planning map must be used to estimate of the groundwater allocation and the total annual water availability across regions and seasons. Local organizations fostering greater collaboration between the public, private and civil society on ground water resources management. In addition, design measures to minimize urban flooding under the negative impacts of subsidence and SLR by non-structural measures and drainage work that tidal sluice gate system are established, and artificial recharge (AR) should be a humanitarian engineering solution to minimize land subsidence rate. In the MD and HCMC, the aquifers are confined, and therefore, AR by using well injection method may be the most suitable option.
- *Long-term (with vision for the year 2050):* tidal flood depths will also become more significant with the increase in SLR and subsidence. The main scenario is that both SLR and land subsidence

will start to estimate flood situations in order to create preventative measures in administration, protection, and utilization of water resources in response to climate change, whereby minimizing effects caused by natural disasters. In addition, another scenario is the case that land subsidence stops at 2030, while SLR will further continue until 2050.

In general, impacts of subsidence and SLR depend on the location where groundwater extraction rate continues to increase. Therefore, maintain and recovery of groundwater level such as reduction of groundwater abstraction rate, saving groundwater use, relocation of abstraction wells, importing surface water to aquifers, finding alternative water supply sources, and using legal actions for minimizing land subsidence rate. At present, groundwater abstraction in HCMC is decrease from 800,000 m^3/day in 2012 to 500,000 m^3/day in 2015 and 100,000 m^3/day by the year 2025. In contrast, groundwater level in MD and Can Tho City is still drawing down.

Rainwater Harvesting

Ironically, the restriction of groundwater abstraction is an impossible case for the MD region because surface water is contaminated and salinization. Given this context, rainwater harvesting should be revisited as an engineering option bringing a wide range of benefits for the local water management practices, including: (i) reduced environmental impacts associated with reduced demands on centralized water systems and water sources; (ii) improved urban stormwater quality; (iii) reduced erosion and flooding associated with high rainfall; and (iv) deferred or reduced requirements for centralized infrastructure and operations for water, wastewater and stormwater..

The practice of rainwater harvesting dates to the earliest days of civilization. Rainwater was collected and utilized as an adaptation option to climate change by people living in various geographical and climatic regions. Archaeological investigations found many ancient man-made earthworks of rainwater harvesting developed by Mayan people in South America, large-scale earthworks in the Bolivian Amazon, and check-dams (water-impounding structures) in Wadi Sana in United Arab Emirates dating back 15,000 years ago (Pandey et al., 2003). Many cities around the world have made the adoption of rainwater harvesting as an alternative option for skyscrapers (Zhang et al., 2009). This application has commonly acknowledged as one of the best practices for designing cities of the future because of gaining various benefits in the age of global climate change.

In numerous cities, rainwater collected from roofs or other impermeable surfaces is a viable source of water for outdoor irrigation, and indoor use for toilet flushing. Such indoor and outdoor purposes do not require high quality drinking water and these utilizations instigate a ready opportunity for substitution of potable water by reusing rainwater. In fact, this practice has been adopted in many countries, including North America, Europe and Asia Pacific region. The collection of rainwater from a 20-ha area is enough to supply for toilet flushing for the whole campus of Nanyang Technological University, meeting the demand for water consumption up to 12.4% (Appan, 1999). Meanwhile, Malaysia has introduced guidelines on rainwater harvesting since 1999 to cope with accelerating water demand (Nor, 2011). Many investigations showed that roof-water collection was used for several household's purposes such as garden irrigation, toilet flushing and bus-washing in Australia (Coombes, 2003; White, 2010). Multi-story buildings in Muhlheim and Freiburg cities of Germany were provided with rainwater for domestic uses (McCann, 2008). In Hong Kong, high-rise buildings collect and store rainwater for residential water demand; and the island of Bermuda has relied on rainwater cistern systems as the primary source of household's water supply for some 300 years ago (Waller et al., 1998). In Canada, about 40% to 50% of homeowners' water consumption can be saved if rainwater is utilized with average precipitation rates ranging from 260 to 1500 mm/year (Brandes et al., 2006).

Most importantly, collecting rainwater for domestic uses is a crucial practice to close the urban water cycle because of considering rainwater and stormwater as a resource. Municipal water uses typically follow a loop from the source to the drinking water utility, off to the user, onward to the sewage treatment plant and back to the environment to become someone else's supplies. Most municipalities spend much cost on bringing source water to a standard of quality suitable for drinking. Therefore, rainwater harvesting systems for residential use are gaining acceptance and are already well-established in Australia to deal with water shortages and climate-related events (Chanan et al., 2010).

In Ho Chi Minh City, rainwater was long been collected for household's use at different scales depending on homeowners' income. For poor households, rainwater was collected from coconut or cottage roof and stored in clay or ceramic jars for drinking purpose only with average capacity of 300-600 liters/household. For households with medium income, rooftop rainwater was collected from corrugated iron or tile roofs and contained in porcelain or cement tanks around $6m^3$/household. Meanwhile, high-income residents and industrial enterprises, rainwater was captured at large scale and stored in concrete tanks with more than $10m^3$ and self-sufficient for 2-3 months. The rooftop-collected rainwater was used for drinking, washing, gardening and farming activities. Some households install pre-treatment steps (such as filter or first flush diverter) before using.

In the Mekong Delta, the collection of rooftop rainwater has been adopted for many purposes. A guidebook for harvesting rainwater is available for households and famers in the Mekong Delta region (Can Tho University, 2014). This practice has emerged as a potential alternative water source for minimise the vulnerability of the region which will be affected global climate change (sea level rise) causing serious saltwater intrusion.

REMARKS AND CONCLUSIONS

Remarks

Mekong Delta and Ho Chi Minh City locate in the low-lying area with the soft soil characteristics, where rivers and canals form a complex network that is affected by tidal regime. In the context of urbanization and climate change, this region is considered as a hotspot of vulnerability to the negative impacts of sea level rise and land subsidence – induced flooding as well as other hazards such as coastal erosion and saltwater intrusion.

Over the past 25 years, the rapidly developing urbanization has led land use changes such as extensive conversion of prime agricultural areas into residential and industrial areas; urban development activities lead to an increase in built-up areas, decrease in green areas, and increase impervious surfaces. These major factors are also contributing to induce the annual flood and the impacts have caused great inconvenience to community activities, leading to negative impacts on people and infrastructure damage as well as economic loss. At present, both the government and the residents are attempting to adapt using formal and informal countermeasures to minimize urban flooding under the negative impacts of subsidence and SLR. However, there is a lack of communication between them, which leads to decoupled actions and social conflicts during implementation.

Many studies show that land subsidence can be caused by the excessive groundwater extraction, load of buildings and constructions, and the combined effect of a large number of urbanized zones are situated at a very low elevation and along the canals with alluvium soil. Subsidence rate directly impacts urban structures and infrastructure as well as flood prevention plan that aims to achieve the sustainable urban development. However, a large portion of the population is unaware of the future flood risk, therefore industrial boreholes are still widespread, and many farmers do not use piped water that still primarily using groundwater due to free of charge. Beside regulations on groundwater extraction and building load control, local government needs more effort during implementation in order to increase public awareness, especially focusing on encouraged households to fill up bored wells in areas where clean tap water is available.

PS InSAR technique shown ability to map ground deformation on a large spatial scale for monitoring land subsidence in the MD and HCMC. Evaluating the preliminary results showed that the effects of land subsidence exhibit spatial and temporal variations (with the rates of about 15 mm/year) and can be considered as greater 5 times than those resulting from rising sea level (about 3 mm/year). This pointed out the impacts of land subsidence and SLR as a major factor contributing to increasing the depth and duration of annual flooding of HCMC and many provinces of the Delta region.

The possible and effective approach to the reduction of subsidence is to prohibit groundwater pumping in the subsidence region as well as undertaking aquifer restoration and artificial recharge. However, the restriction of groundwater abstraction is a challenging issue for the MD region because surface water is contaminated and intruded by seawater. This resulted in intensive water overexploitation activities of underground reservoirs in the development of agricultural and industrial activities.

Rainwater harvesting is an engineering applicable practice in both the urban and rural areas. Rainwater harvesting should be revisited as an alternative option for HCMC – the large urban area and the Mekong Delta region – mostly rural provinces. Rainwater collected from rooftops of buildings is usually considered as the cleanest water source available, requiring little treatment before being suitable for a wide variety of purposes. Further, the collection of rooftop rainwater will augment water demands and minimize groundwater abstraction. In turn, this practice will support to redress the vulnerability of land subsidence.

CONCLUSION

Intensive urban development activities and excessive exploitation of groundwater exacerbated the rate of land subsidence and flooding consequence over the past 25 years in HCMC and the MD region. The advanced multi-temporal InSAR technique was introduced to monitor land subsidence. Several evidence from human impacts are demonstrated in the relation between land subsidence, land use change and sea level rise.

Restriction of groundwater abstraction is challenged for this region because surface water is contaminated and salinization which have resulted in more groundwater extraction. In order to continuously monitor the ground subsidence phenomena at these large areas, the low-cost solution is also proposed that the utilization of PSInSAR method with free Sentinel-1 data and development of Web PSInSAR solution for identifying areas are severely affected by land subsidence.

Rainwater harvesting is proposed as a humanitarian engineering solution to augment the water source, and to minimize groundwater abstraction. In turn, this practice will indirectly help to prevent from land subsidence. The guidelines for collecting rooftop rainwater are available for households to apply in HCMC and the MD region.

ACKNOWLEDGEMENTS

The authors would like to thank Ho Chi Minh City University of Technology (HCMUT), VNU-HCM for the support of time and facilities for this study.

REFERENCES

Appan, A. (1999). A dual-mode system for harnessing roofwater for non-potable uses. *Urban Water*, *1*(4), 317–321. doi:10.1016/S1462-0758(00)00025-X

Be, T. T., Sinh, B. T., & Miller, F. (Eds.). (2007). *Challenges to Sustainable Development in the Mekong Delta: Regional and National Policy Issues and Research Needs*. The Sustainable Mekong Research Network.

Brandes, O. M., Maas, T., & Reynolds, E. (2006). *Thinking Beyond Pipes and Pumps: Top 10 Ways Communities can Save Water and Money, POLIS Project*. University of Victoria.

Bui, T. V., Dang, T. L., & Le, T. M. V. (2015). Groundwater issues and hydrogeological survey of the Mekong River basin in Vietnam. In K. Ha, T. M. N. Nguyen, E. Lee, & R. Jayakumar (Eds.), *Current Status and Issues of Groundwater in the Mekong River Basin* (pp. 93–121). KIGAM-CCOP-UNESCO Bangkok Office.

Can Tho University. (2014). *Rainwater Harvesting Guidebook for the Mekong Delta. College of Environment and Natural Resources*. DRAGON Institute, Can Tho University.

Cao, A., Esteban, M., Valenzuela, V. P. B., Onuki, M., Takagi, H., Thao, N. D., & Tsuchiya, N. (2021). Future of Asian Deltaic Megacities under sea level rise and land subsidence: Current adaptation pathways for Tokyo, Jakarta, Manila, and Ho Chi Minh City. *Current Opinion in Environmental Sustainability*, *50*, 87–97. doi:10.1016/j.cosust.2021.02.010

Chanan, A., Vigneswaran, S., & Kandasamy, J. (2010). Valuing Stormwater, Rainwater, and Wastewater in the Soft Path for Water Management: Australian Case Studies. *Water Science and Technology*, *62*(12), 2854–2861. doi:10.2166/wst.2010.544 PMID:21123915

Coombes, P. (2003). Rainwater Tanks Revisited: New Opportunities for Urban Water Cycle Management. The University of Newcastle, Australia.

Dinh, H. T. M., Vuong, Q. V., Le, V. T., & Ngo, Y.-N. (2016). Comparison Stripmap Cosmos SkyMed X-Band and TOPS Sentinel-1C Band in Estimating Ground Subsidence Using Irstea TomoSAR Platform: Ho Chi Minh City Case Study. In *Living Planet Symposium, ser. ESA Special Publication,* (*vol. 740*, p. 73).

Duffy, C. E., Braun, A., & Hochschild, V. (2020). Surface Subsidence in Urbanized Coastal Areas: PSI Methods Based on Sentinel-1 for Ho Chi Minh City. *Remote Sensing*, *12*(24), 4130. doi:10.3390/rs12244130

DWRPIS (2004). Relation of groundwater and mineral resources with geological formation and structure. In H. D. Nguyen (Ed.), *Research of Geological Structure and Classification of Neogene and Quaternary Sediments in the Southern Vietnam* (p. 145). Southern Geological Mapping Division. (in Vietnamese)

DWRPIS (2010). Internal report on the compilation of geological, hydrogeological and engineering geological maps of HCMC on the scale of 1/50,000. Division for Water Resources Planning and Investigation for the south of Vietnam, Ho Chi Minh City, Vietnam.

DWRPIS (2015). *Annual Report on the National Groundwater Monitoring Network for the Southern Vietnam*. Division of Water Resources Planning and Investigation for the South of Vietnam. (in Vietnamese)

Erban, L. E., Gorelick, S. M., & Zebker, H. A. (2014). Groundwater extraction, land subsidence, and sea-level rise in the Mekong Delta, Vietnam. *Environmental Research Letters*, *9*, 084010. doi:10.1088/1748-9326/9/8/084010

Erkens, G. T., Bucx, T., Dam, R., de Lange, G., & Lambert, J. (2015). Sinking coastal cities. *Proc. IAHS*, 372, 189–198.

Ferretti, A., Fumagalli, A., Novali, F., Prati, C., Rocca, F., & Rucci, A. (2011). A new algorithm for processing interferometric data-stacks: SqueeSAR. *IEEE Transactions on Geoscience and Remote Sensing*, *49*(9), 3460–3470. doi:10.1109/TGRS.2011.2124465

Ferretti, A., Prati, C., & Rocca, F. (2000). Nonlinear subsidence rate estimation using permanent scatterers in differential SAR interferometry. *IEEE Transactions on Geoscience and Remote Sensing*, *38*(5), 2202–2212. doi:10.1109/36.868878

Ferretti, A., Prati, C., & Rocca, F. (2001). Permanent scatterers in SAR interferometry. *IEEE Transactions on Geoscience and Remote Sensing*, *39*(1), 8–20. doi:10.1109/36.898661

Ha, Q. K., Choi, S., Phan, N. L., Kim, K., Phan, C. N., Nguyen, V. K., & Ko, K. S. (2019b). Occurrence of metal-rich acidic groundwaters around the Mekong Delta (Vietnam): A phenomenon linked to well installation. *The Science of the Total Environment*, *654*, 1100–1109. doi:10.1016/j.scitotenv.2018.11.200 PMID:30841385

Ha, Q. K., Kim, K., Phan, N. L., Phung, T. H., Lee, J., Nguyen, V. K., & Phan, C. N. (2019a). A hydrogeological and geochemical review of groundwater issues in southern Vietnam. *Geosciences Journal*, *23*(6), 1005–1023. doi:10.100712303-019-0021-z

Ha, Q.K., Phan, N.L., and Kim, K. (2018). *Mobilization of Fe, Mn, and as in saline confined aquifer in response to rainwater injection*. 45th IAH 2018 CONGRESS Groundwater and life: Science and Technology into Action (abstract), Daejeon.

Hooper, A. (2008). A multi-temporal InSAR method incorporating both persistent scatterer and small baseline approaches. *Geophysical Research Letters*, *35*(16), 1–5. doi:10.1029/2008GL034654

Hooper, A., Zebker, H., Segall, P., & Kampes, B. (2004). A new method for measuring deformation on volcanoes and other non-urban areas using InSAR persistent scatterers. *Geophysical Research Letters*, *31*(23). doi:10.1029/2004GL021737

Le, V. T. & Ho Tong Minh, D. (2008). Measuring Ground Subsidence in Ho Chi Minh City using Differential InSAR Techniques. *J. Sci. Technol. Vietnam Natl. Univ. Ho Chi Minh City*, *11*, 121–130.

Le, V. Trung Le Ngoc Lam and Dao Minh Tam, D. (2021). Web GIS and Remote sensing solutions for monitoring and thematic mapping in the Mekong delta region. *IOP Conference Series. Earth and Environmental Science*, *652*(1), 012023. doi:10.1088/1755-1315/652/1/012023

McCann, B. (2008). Global prospect for rainwater harvesting. *Water (Basel)*, *21*, 12–14.

Minderhoud, P. S. J., Coumou, L., Erban, L. E., Middelkoop, H., Stouthamer, E., & Addink, E. A. (2018). The relation between land use and subsidence in the Vietnamese Mekong delta. *The Science of the Total Environment*, *634*, 715–726. doi:10.1016/j.scitotenv.2018.03.372 PMID:29649716

Minderhoud, P. S. J., Erkens, G., Pham, V. H., Bui, V. T., Erban, L., Kooi, H., & Stouthamer, E. (2017). Impacts of 25 years of groundwater extraction on subsidence in the Mekong delta, Vietnam. *Environmental Research Letters*, *12*(6), 064006. doi:10.1088/1748-9326/aa7146 PMID:30344619

Minderhoud, P. S. J., Erkens, G., Pham, V. H., Vuong, B. T., & Stouthamer, E. (2015). Assessing the potential of the multi-aquifer subsurface of the Mekong Delta (Vietnam) for land subsidence due to groundwater extraction. *Proc. IAHS*, (*vol. 92*, pp. 1–4). 10.5194/piahs-372-73-2015

Minderhoud, P. S. J., Middelkoop, H., Erkens, G., & Stouthamer, E. (2020). Groundwater extraction may drown mega-delta: Projections of extraction-induced subsidence and elevation of the Mekong delta for the 21st century. *Environ. Res. Commun.*, *2*(1), 011005. doi:10.1088/2515-7620/ab5e21

Minh, D. H. T., Van Trung, L., & Toan, T. L. (2015). Mapping ground subsidence phenomena in Ho Chi Minh City through the radar interferometry technique using ALOS PALSAR data. *Remote Sensing*, *7*(7), 8543–8562. doi:10.3390/rs70708543

Ngo, M. T., Lee, J. M., Lee, H. A., & Woo, N. C. (2015). The sustainability risk of Ho Chi Minh City, Vietnam, due to saltwater intrusion. *Geosciences Journal*, *19*(3), 547–560. doi:10.100712303-014-0052-4

Nguyen, V. K., & Nguyen, D. T. (2011). The pilot test for assessment supplement capacity for groundwater by rainfall resource in dormitory area of Vietnam National University, Ho Chi Minh City (VNU-HCM) [in Vietnamese]. *Science & Technology Development Journal*, *14*, 16–24. doi:10.32508tdj.v14i4.2003

Nor, N. M. (2011). Rainwater Harvesting System for Link Houses. *American Journal of Scientific Research*, *19*, 51–57.

Pandey, D. N., Gupta, A. K., & Anderson, D. M. (2003). Rainwater Harvesting as an Adaptation to Climate Change. *Current Science*, *85*, 46–59.

Rise and Fall. (2020). *Copernicus project: Emergency Management Service – Risk and Recovery Mapping (2014-2019 Copernicus –EU)*. Rise & Fall project (2015-2019, the Netherlands)

SVGMD. (2004). Classification and characteristics of Neogene and Quaternary sediments in the southern Vietnam. In H. D. Nguyen (Ed.), *Research of Geological Structure and Classification of Neogene and Quaternary Sediments in the Southern Vietnam. Report, South Vietnam Geological Mapping Division* (p. 220). SVGMD. (in Vietnamese)

Takagi, H., Thao, N. D., & Anh, L. T. (2016). Sea-level rise and land subsidence impacts on flood projections for the Mekong. Delta's largest city. *Sustainability*, *8*(9), 1–15. doi:10.3390u8090959

Dao Minh Tam, Vuong Quoc Viet, Le Van Trung (2010). ERS-ENVISAT dual sensor CTM processing for monitoring subsidence in Ho Chi Minh City. *Proceedings of Map Asia*.

Thoang, T. T., & Giao, P. H. (2015). Subsurface characterization and prediction of land subsidence for HCM City, Vietnam. *Engineering Geology*, *199*, 107–124. doi:10.1016/j.enggeo.2015.10.009

Tran, D. A., Tsujimura, M., Pham, H. V., Nguyen, T. V., Ho, L. H., Le Vo, P., Ha, K. Q., Dang, T. D., Van Binh, D., & Doan, Q. V. (2021). Intensified salinity intrusion in coastal aquifers due to groundwater overextraction: A case study in the Mekong Delta, Vietnam. *Environmental Science and Pollution Research International*, 1–15. doi:10.100711356-021-16282-3 PMID:34498189

van Asselen, S., Erkens, G., Stouthamer, E., Woolderink, H. A., Geeraert, R. E., & Hefting, M. M. (2018). The relative contribution of peat compaction and oxidation to subsidence in built-up areas in the Rhine-Meuse delta, The Netherlands. *The Science of the Total Environment*, *636*, 177–191. doi:10.1016/j.scitotenv.2018.04.141 PMID:29704713

Vo, L. P. (2007). Urbanization and water management in Ho Chi Minh City, Vietnam-issues, challenges, and perspectives. *GeoJournal*, *70*(1), 75–89. doi:10.100710708-008-9115-2

Waller, D. H., Mooers, J. D., Samostie, A., & Sahely, B. (1998). *Innovative Residential Water and Wastewater Management*. Canada Mortgage and Housing Corporation.

Werner, C., Wegmuller, U., Strozzi, T., & Wiesmann, A. (2003). Interferometric point target analysis for deformation mapping. In *International Geoscience and Remote Sensing Symposium*, Toulouse, France. 10.1109/IGARSS.2003.1295516

White, I. (2010). Rainwater harvesting theorising and modelling issues that influence household adoption. *Water Science and Technology*, *62*(2), 370–377. doi:10.2166/wst.2010.891 PMID:20651442

Zhang, Y., Chen, D., Chen, L., & Ashbolt, S. (2009). Potential for rainwater use in high-rise buildings in Australian cities. *Journal of Environmental Management*, *91*(1), 222–226. doi:10.1016/j.jenvman.2009.08.008 PMID:19744767

KEY TERMS AND DEFINITIONS

Groundwater overexploitation: A situation in which average withdrawal (pumping) rate is greater that recharge rate. This causes a drawdown of water table and leads to the depletion of groundwater resources.

Humanitarian Engineering: The application of engineering to improve the living conditions and well-being of marginalized people and disadvantaged communities, usually in the developing world. Humanitarian engineering typically focuses on programs that are affordable, sustainable, and based on local resources. Projects are typically community-driven and cross-disciplinary, and they focus on finding simple solutions to basic needs (such as close access to clean water; adequate heat, shelter, and sanitation; and reliable pathways to markets).

Land Subsidence: A gradual settling or sudden sinking of the Earth's surface due to the removal of subsurface materials. Land subsidence occurs when the large amounts of groundwater have been withdrawn from certain types of rocks, such as fine-grained sediments. The rock compacts because the water is partly responsible for holding the ground up. When the water is withdrawn, the rocks fall in on itself. The main causes of land subsidence include aquifer-system compaction associated with groundwater withdrawals; drainage of organic soils; underground mining; natural compaction or collapse, such as with sinkholes or thawing permafrost.

PSInSAR: (Persistent Scatterer Interferometric Synthetic Aperture Radar)**:** A remote sensing technique that uses radar signals from a satellite to accurately measure ground displacement. PSInSAR is an improved and more accurate analysis algorithm compared to the DInSAR (Differential Interferometric Synthetic Aperture Radar).

Rainwater Harvesting: The harvesting of rainwater refers to the collection of water from surfaces on which rain falls, and subsequently storing this water for later use. Rainwater harvesting can occur in multiple contexts and at a variety of scales. The collection of rainwater from the rooftops of buildings is suitable for urban areas and requires little investment or technology, as only roof gutters and storage tanks are needed to capture rainwater.

Sea Level Rise: An increase in the mean level of the ocean due to melting ice and thermal expansion related to global warming and may have consequences for life on Earth. Generally, sea-level rise is a climate change phenomenon through which the ocean water volume increases, mostly as a consequence of icesheets and glaciers melting and water thermal expansion. As a consequence of sea-level rising, coastal life around the world is threatened and adaptation policies need to be developed to avoid further and faster catastrophic impacts.

Urbanization: The process by which a large number of people become permanently concentrated in relatively small areas, forming cities or built-up areas. This phenomenon is an emerging trend in developing countries that poses a wide range of urban environmental issues such as the contamination of water sources (bodies), groundwater depletion, air pollution, and municipal solid wastes.

Waterlogging: A form of natural flooding when underground water rises to surface level as the result of over-irrigation simply the saturation of soil with water, either temporarily or permanently. When there is too much water in an area, the soil is unable to absorb the water as it should ordinarily.

Chapter 4

Lifting the Veil of Time:
Natural Hazards and Structural Problems Affecting Cultural Heritage Sites and Scientific Methods for Their Preservation

Elena Kapogianni
National Technical University of Athens, Greece

ABSTRACT

Cultural heritage sites are fundamental elements of national and international heritage and are regarded as sensitive and critical structures of high and of national importance, exposed to a range of natural hazards, leading often structural failures. Cultural heritage sites include structures and monuments with effectively limitless lifetime, while admitting only limited intervention options. Therefore, the efficient assessment of their structural risk forms a very demanding task in comparison to similar analyses applied to modern structures. The current study is involved with natural hazards and natural disasters that have affected cultural heritage sites in the past and their consequences regarding their structural integrity. After a description of the aforementioned, specific scientific tools and advanced methods for the investigation of the geophysical/hydrological and climatological/meteorological natural hazards and their consequences at cultural heritage sites, regarding their structural integrity are presented and discussed.

INTRODUCTION

Natural hazards and natural disasters have always accompanied human history. Natural hazards are naturally occurring physical phenomena caused either by rapid or slow onset events which can be geophysical/hydrological such as earthquakes, landslides, tsunamis, rockfalls, volcanic activity, avalanches, and floods, and/or climatological/meteorological events, such as extreme temperatures, droughts, wildfires, cyclones and storms. Natural hazards have negative effects on humans and the environment (including the man-made environment), and in some cases they may escalate into natural disasters.

In around 1600 B.C., the volcano of Santorini (Thera) erupted with such force that this single event caused the collapse of the Minoan civilization (Figure 1, source: Hellenic Ministry of Culture, Greece),

DOI: 10.4018/978-1-6684-5619-4.ch004

Figure 1. Akrotiri: The Ancient Town Buried by a Volcano. The cataclysmic eruption of Thera split the island of Santorini into three smaller ones

110km away, at the island of Crete (Athanassas et al., 2018). The Thera eruption is considered to be the largest one on Earth in the last 10.000 years, with subsequent earthquakes of great magnitude and tsunamis of approximately 50m height. It is also estimated that around 40 km^3 of magma and rock had erupted from the volcano. In 730 B.C., according to the Roman geographer Strabo, the city of Chalcis, located at the island of Euboea in Greece, suffered a severe drought and 10% of the city's young men were sent as an offering to Delphi (Radt, 2011). In 464 B.C., an earthquake struck Sparta, at a time when the relations of Sparta and Athens were in a delicate state. It took Sparta by surprise, killing more than 20,000 Spartans, leading also to internal and external uprisings (Holladay, 1977). In 373 B.C., the ancient Greek city of Helike, the flourishing capital of the Achaean League, was destroyed during a winter's night. The city sank into the earth and the inhabitants perished without a trace, probably due to a rapid and catastrophic landslide, and/or due to a massive tsunami (Psarropoulos & Gkantona, 2020). The dramatic destruction of Helike was widely discussed by many Greek and Roman authors and may have inspired Plato to contrive the myth of Atlantis.

In Europe, a very well-known case of an extreme natural disaster was the Great Lisbon Earthquake of 1755 A.D., in Portugal, where approximately 60,000 people died due to an 8.5 magnitude earthquake and the subsequent seismic activity (Figure 2 is for a depiction of this natural disaster). A series of earthquakes rushed-up the Tagus River, causing tsunamis of around 20m high, while in parallel destructive fires broke out in the city, leading to a significant devastation of Lisbon. The Great Lisbon Earthquake occurred on All Saints' Day when a large part of the population was in the city's Churches at the moment the earthquake struck. The Churches were unable to withstand the extreme seismic activity and collapsed, killing thousands of worshippers. The disaster was widely discussed by European Enlightenment philosophers, leading to the birth of modern seismology and earthquake engineering (Reinhardt & Oldroyd, 1983). That was the time when earthquake phenomena were studied scientifically for the first time.

Figure 2. A depiction of the 1755 Lisbon earthquake as seen from across the Tagus River, showing Lisbon in flames and a tsunami overwhelming the ships in the harbor

In East Asia, the 1556 A.D. Shaanxi earthquake in China is considered to be the deadliest earthquake in (recorded) history and according to imperial records approximately 830,000 people lost their lives. Around 100 cities in the provinces of Shaanxi were affected, and an 800 km^2 wide area had destruction. Around 38 centuries before the Shaanxi earthquake, at around 2300-2200 B.C., a major flood occurred in ancient China, known as the Great Flood of Gun–Yu, resulting in many casualties and great population displacement. Even though the Great Flood of Gun-Yu is believed to be a myth, recent archaeological evidence suggests that it happened (Montgomery, 2016). More specifically, sediments along the Yellow River (i.e. the second-longest river in China) indicate that an earthquake-induced landslide may have blocked an upper section of the river, creating a natural dam. It is considered that during the Great Flood, following the natural dam collapse, water reached an elevation of approximately 250m above current river level.

Apart from disasters, natural hazards in combination with man-made hazards and the passing of time have affected important cultural heritage sites, leading to smaller or larger structural failures.

Cultural Heritage Sites and Structural Problems

The integrity of any cultural heritage site may be threatened by one or more natural hazards during its lifetime, and this can be expressed in terms of structural risk. Structural risk is directly related to the hazards and to the characteristics of the exposed systems, in the form of their vulnerability, which depends on the design, construction and operational conditions. The overall risk is related to the anticipated corresponding loss, in terms of human lives and of the corresponding socio-economic cost.

Figure 3. The Acropolis Hill and Circuit Wall (left), and the central area of the Acropolis Wall with significant part detachments (right)

Cultural heritage sites are fundamental elements of national and international Heritage and are regarded as sensitive and critical structures of high and of national importance, because they can be exposed to a range of natural hazards, leading often to structural failures. They feature an elevated overall risk level, since serious damage, and, especially, a total failure may be irreversible, and thus unacceptable. In addition, they attract large numbers of visitors and spectators, and a potential failure or insufficient risk management may have a direct impact on their health and safety. These aged structures and monuments have effectively limitless lifetime, while admitting only limited intervention options. The cost-efficient assessment and management of their overall structural risk forms a very demanding task in comparison to similar analyses applied to modern systems, and in parallel an issue of paramount importance to the communities-at-large.

A characteristic case of a very well-known cultural heritage site that has been damaged through the passing of time is the Acropolis of Athens. During the last twenty-five centuries, numerous structural damages over the Site have been attributed to various natural hazards, and to the human interference, such as war invasions and air pollution, increasing in parallel the risk of local or extensive structural failures (Figure 3). Furthermore, various studies at the Acropolis Hill and its monuments have resulted in the documentation of seismic events, such as those occurring in 1705 and in 1805 in Athens, and in 1837 in Troezen (Ambraseys, 2010). Another example of an important cultural heritage site, vulnerable to natural hazards, is the Pan-Hellenic Sanctuary of Delphi, lying at a very complicated geomorphology at the mount Parnassos in Greece (Figure 4). The wider region of Delphi includes a seismic fault, while occurrence of earthquakes is also related to the seismic zones of the Corinthian and Saronic Gulf. The Site suffers also from rockfall phenomena, related to the complicated geomorphology of the region.

Another characteristic type of cultural heritage sites, threatened by natural hazards and structural failures are the Ancient Theatres. Over one hundred Ancient Theatres are related to the wider Hellenic region (including areas at Asia Minor and Ionia) and were constructed at different periods of time, such as the Ancient Greek and the Roman eras. The Cavea (Koilon) of many Ancient Theatres is usually shaped into natural cavities, at the foot of mountains or hills, and often with extensions over an earth fill, supported by retaining walls (Figure 5). The critical locations at Ancient Theatres (i.e., the locations that are prominent to structural problems and failures) are often those related to the retaining walls, facing the common problems of such masonry structures. Furthermore, the possibility of slope instability at a plane

Figure 4. The Tholos of the Athena Pronaia at the Sanctuary of Delphi (left) and a failure of a masonry retaining wall at Delphi Archaeological Site (right)

from the top of the Cavea (which is usually identified as the crown of a natural cavity) to the Orchestra (which is usually identified as the toe of a cavity), is another site-specific engineering problem, which should be investigated thoroughly. Slope instability is also prominent at the locations of the retaining walls, and for the same direction of movement as for the potential Cavea-Orchestra instability. The aforementioned locations are considered to be very vulnerable and in combination with natural hazard occurrences, significant part detachment failures and/or other damages may take place. In Figure 5, the Ancient Theatre of Dodona, an integral part of the Dodona Sanctuary in Epirus can be seen, which is considered as one of the largest and best preserved ancient Greek theatres, able to accommodate about 18,000 spectators. The Ancient Theatre of Dodona is a very dominant monument due to its curvilinear surfaces and imposing retaining walls and its construction, using two kinds of local Epirus limestone, dating to the 3rd century B.C., when King Pyrrhus of Epirus implemented an ambitious building project, aiming to reshape this Pan-Hellenic Sanctuary. Over 2,300 years, numerous structural failures at this Site have been attributed to various natural hazards (either geological or meteorological), and to the human interference, increasing in parallel the risk of local or extensive structural failures.

ADVANCED METHODS AND SCIENTIFIC TOOLS

Various scientific tools and advanced methods can be utilized for the investigation of the geophysical/hydrological and climatological/meteorological natural hazards and their consequences at cultural heritage sites, regarding their structural integrity. In this framework, the assessment and documentation of common natural hazards that have occurred during the lifetime of the sites (i.e., since their construction to the present day), and of common failures that have occurred in the past (in regard also to their structural vulnerability) are indispensable. Furthermore, the sufficient identification of the main structural problems the sites may encounter now is also very important, as well as the suggestion of specific mitigation measures regarding their preservation from a structural safety point of view, aiming in parallel to assist future restoration works and cost-efficient interventions. It should be noted that high structural

Figure 5. View of the Ancient Theater of Dodona at Tomaros Mountain in Epirus (left) and the Theatre hosting a theatrical play (right)

vulnerability in combination with natural hazards (either moderate or extreme) may lead to catastrophic results regarding the integrity at cultural heritage sites.

More specifically, it is observed that advanced computational methods can be utilized in order to provide useful information about the structural response and distress of any structure of a cultural heritage site. Numerical/computational simulations require the geometrical and mechanical properties of the super-structures and their sub-structures under examination, and the anticipated static and dynamic loadings. To achieve a realistic simulation of the structural response, the local site conditions (i.e., soil, geomorphology, topography) and the corresponding potential soil-structure interaction need to be taken into consideration. Furthermore, uncertainties of the loading conditions can be limited by a reassessment based on the monitoring of the loading conditions and measuring the mechanical and geometrical properties of any structure. A typical method of numerical modelling is the finite-element stress analysis method (Kapogianni et al, 2016). The finite-element method is a widely used approach for numerically solving differential equations, arising in engineering. The method subdivides the model under investigation into smaller parts (these are the finite elements), creating in parallel a mesh, taking into consideration the required geometrical and mechanical characteristics, the interface properties between the model materials and the loading conditions.

In Figures 6 and 7, an example of the numerical simulation of the Ziggurat of Ur in Ancient Mesopotamia (nowadays Iraq) can be seen. The Ziggurats were built by the Sumerians, Babylonians, and Assyrians in ancient Mesopotamia, from the 22nd to the 5th century B.C. (Ratchet, 1983, Kapogianni, 2021). They consisted of soil interiors with protective stone walls, covered with clay bricks and layers of mats made of baskets, which were placed in the mud between the bricks to provide horizontal reinforcement. Additional reinforcement was included in some of the Ziggurats in the form of ropes, about 50 mm in diameter each, placed perpendicular to the wall at even spacings, in both horizontal and vertical directions. It is noteworthy that this technique resembles the reinforced–soil applications that are used nowadays in geotechnical engineering (Koerner, 2012).

The numerical analysis, utilizing the finite element method shows that soil reinforcement plays an important role in the structural safety of the Ziggurats, increasing significant their Safety Factor. Further-

Figure 6. Representation of Ziggurat (left) and after extensive reconstruction works (right)

more, the increased length of the reinforcement layers used was another factor that benefited such high, inclined soil-mass structures (i.e. structures with soil materials as their main volume). It is evident that ancient civilizations had awareness of advanced construction techniques for structural integrity and applied reliable methods to minimize risk of failures when their structures were subjected to the structural loadings and various natural hazards.

Another example of the numerical analysis of an important cultural heritage site is the simulation of the masonry Circuit Wall of the Acropolis in Athens. The Acropolis is one of the most important archaeological sites in the world and attracts annually 1 million visitors. The Circuit Wall serves a geotechnical engineering purpose and retains the backfill materials forming the plateau around the Parthenon, and the other great monuments (such as the Erechtheion, the Propylaia and the Arrephorion). It is founded on inclined limestone, and retains soft soil materials (i.e. for the backfill) and the geometrical properties (i.e. wall height and width) vary substantially around the Acropolis Hill. In the past, the southern

Figure 7. Strains and deformations predicted from a computational simulation of a Ziggurat, with (on left-side of the model) and without (on right-side of the model) showing the influence of having the reinforcement

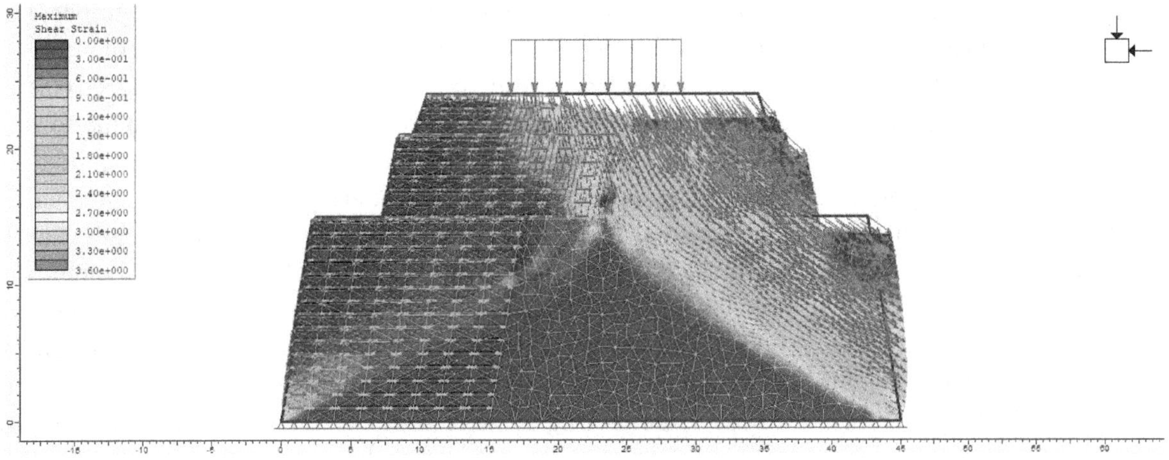

Figure 8. Typical cross section of the Acropolis Circuit Wall (left). Two-dimensional finite element model of the southern Wall (right): 1. limestone, 2, artificial earth fill, 3, retaining structure

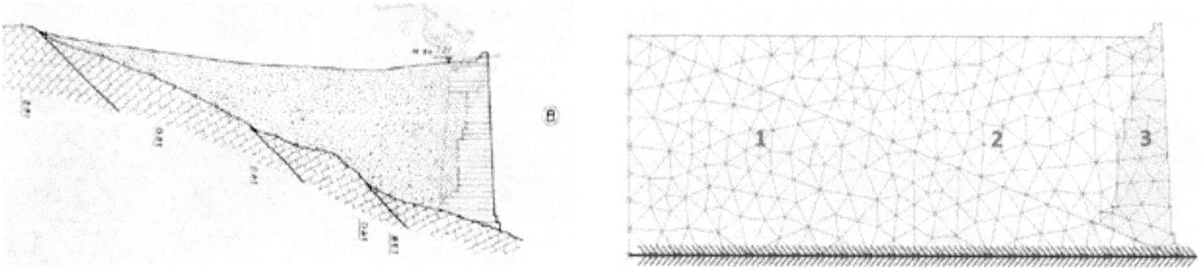

part of the Wall has exhibited some serious structural damages (i.e. significant detachments), while few notable cracks are currently being observed at the South-East corner of the Wall. The numerical simulations shown in Figure 8 have been performed by simulating the dynamic response of two-dimensional finite element models subjected to seismic excitations (Trikkalinos, 1977, Psarropoulos et al., 2018). A key finding from this numerical simulation is that the seismic response of a retaining structure, such as the Circuit Wall of Acropolis, depends, not only on the geometrical and material properties of the wall and the retained soil materials, but also on the acceleration levels and the spectral characteristics of the seismic excitations.

Smart structural monitoring is another useful scientific method that can be utilized for remote hazard assessment (i.e. assessment of the loading conditions) and for early detection of damages and/or failures at cultural heritage sites, contributing to optimal risk management and to their overall safety. More specifically, structures and monuments of cultural heritage sites can be instrumented with sensors and devices, providing important real-time information, and early warning systems can be developed, activated in the case of exceeding the predefined limits, leading thus to self-control of a system under examination. Over the last decades, optical fiber sensors that measure strains, temperatures, and accelerations have been used for smart monitoring. Optical fiber sensors combine precision with resistance to mechanical stress and multiplexing of signals with a relatively low cost (Kapogianni et al., 2017). An important advantage is that they are portable systems and can be attached at various positions, on (almost) any structure, monitoring hazards (i.e., the loading conditions) and structural response/distress at critical locations. Furthermore, modern instruments of strong ground motion (i.e., accelerographs) have significant advantages compared to the ones employed, even 15 years ago. This is due to their higher-resolution digitizers, their longer bandwidth (recording more precisely), their internet connectivity and their higher storage capacity. They are, therefore, able to record not only the strong ground motions but the weak ground motions as well. A case of complex smart structural monitoring can be found at the Acropolis of Athens, where in 2015 and 2016, an optical-fibre-sensor array was installed at the Circuit Wall. It consists of eight active strain sensors, two temperature sensors and one acceleration sensor, and has continuously transmitted real-time data since June 2016 (Figure 9a). This structural monitoring application is combined with an already established monitoring system consisting of 10 high-quality broadband accelerographs installed at specific locations over the Acropolis Hill. Recordings are continuous and 24-bit digitizers transmit the data in real-time for evaluation (Kapogianni et al., 2019, Kalogeras et al., 2012). The locations for the structural monitoring of the Acropolis Hill were selected taking (also) into consideration various known geotechnical conditions. In Figure 9(b), typical real-time monitoring recordings from the Acropolis Hill, utilizing optical fibre sensors can be seen, after processing of the recorded measurements.

Figure 9. Optical fibre terminals (left), and Smart Rod Optical Fibres installed on the South Wall of the Acropolis Hill

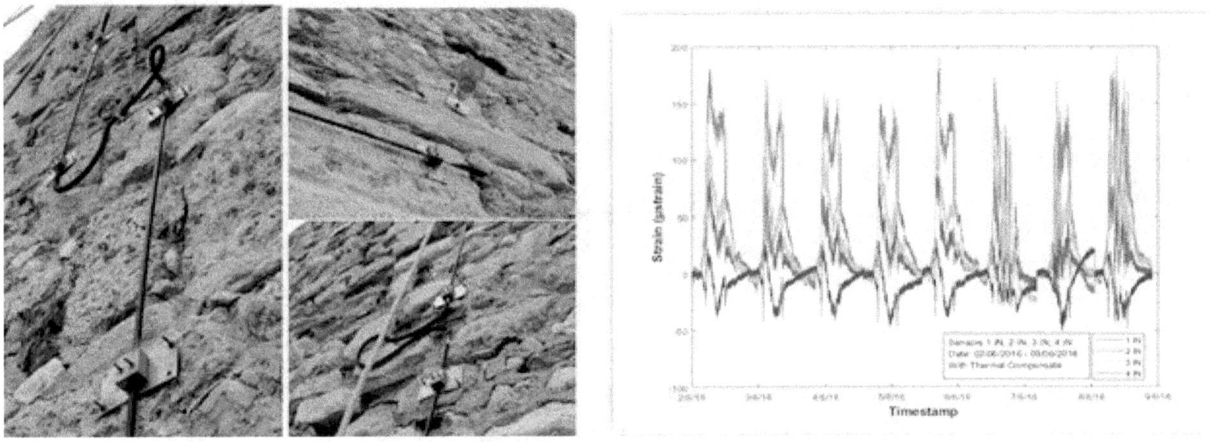

Figure 10. Real-time strain recordings, after processing of the initial measurements, from the Acropolis Hill

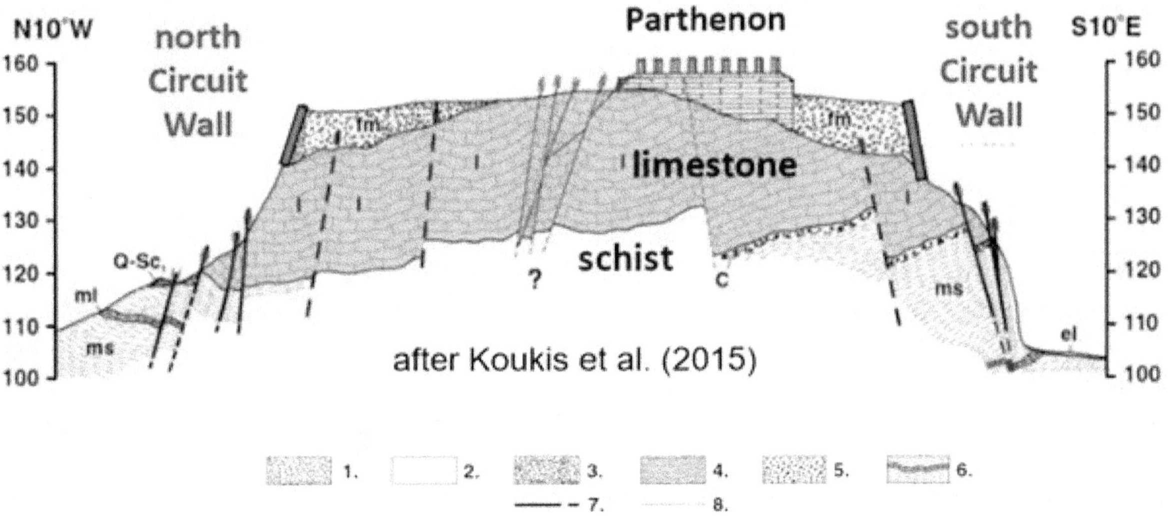

Numerical modelling and smart structural monitoring can be combined to provide further useful information regarding the structural response that may affect cultural heritage sites (and other important structures/systems). An example in this regard is the ground response analyses that have been performed at the Acropolis Hill, utilizing real-time monitoring data, in order to assess the dynamic excitation of the hill when subjected to an earthquake (see Figures 10 and 11) (Kapogianni et al., 2020a). It is evident

Figure 11. Geological section of the Acropolis Hill. 1: artificial earth-fill(fm), 2: eluvial mantle (el), 3: talus (Q-Sc), 4: limestone (l), 5: conglomerates (C), 6: schist-sandstones-marl series (ms-ml), 7: fault, 8: main fractures on the Acropolis plateau (after Koukis et al., 2015). Cross-section showing the locations (in red) of four accelerographs on the hill

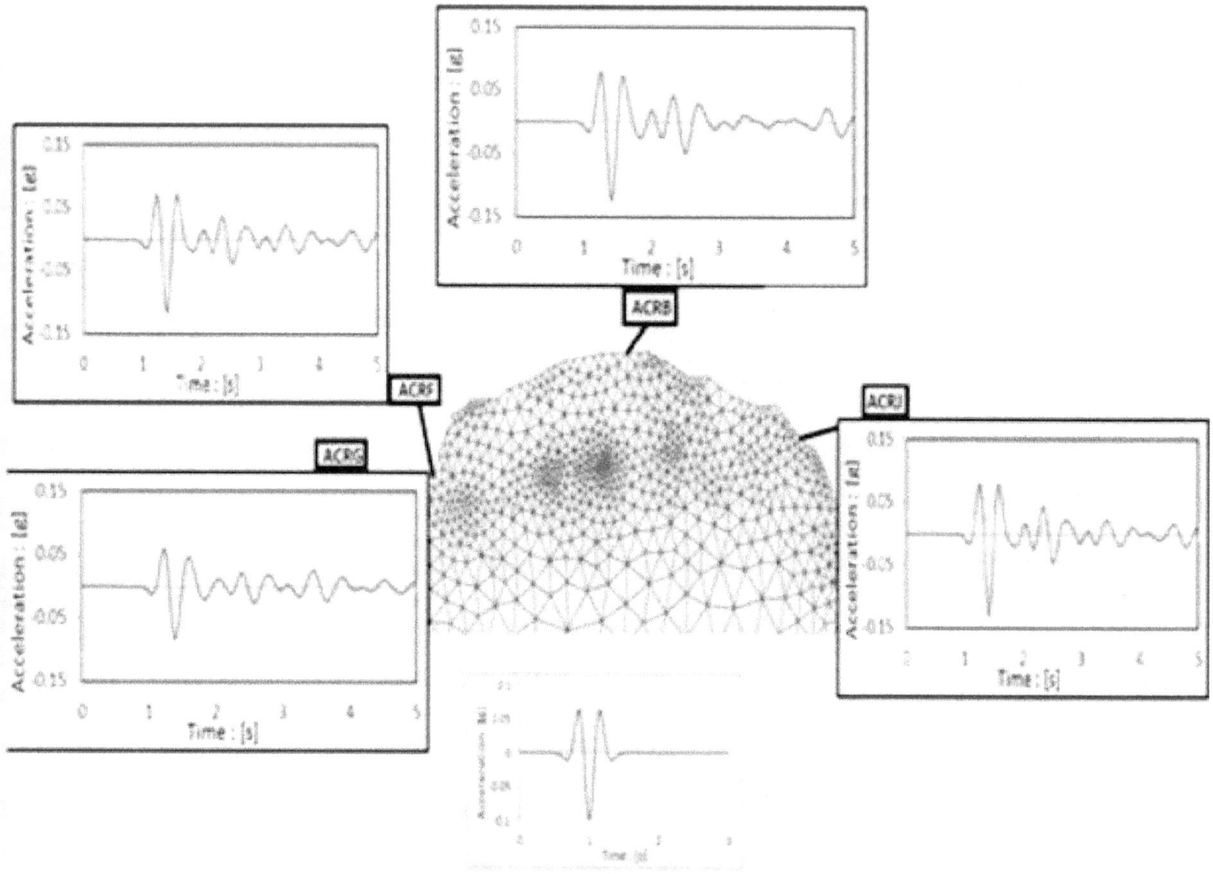

that the aforementioned scientific approaches can be applied to monitor any structure or monument, provided that the geometrical and physical properties of a structure's/monument's materials have realistically been determined.

In an environment where various networks of sensors have been deployed, the need for intelligent management and processing of the recorded and transmitted data has emerged (Chatzi, 2017). An integral part to the data processing is the combination of data fusion between heterogeneous networks (Sun, 2016) and the support of additional meta-data in order to efficiently handle the available measurements. In addition, data obtained from measuring devices and sensors can be used as a digital reliable information database that could be incorporated into a Building Information Model (BIM), for decision-making. The preservation of cultural heritage sites could be supported effectively via the digital representation of their physical and functional characteristics, especially at critical locations. The combination of remote reviewing and BIM ought to enable the study of a facility in the context of the environment (at differ-

Figure 12. Two-dimensional finite-element model developed for the ground response analyses at the Acropolis Hill

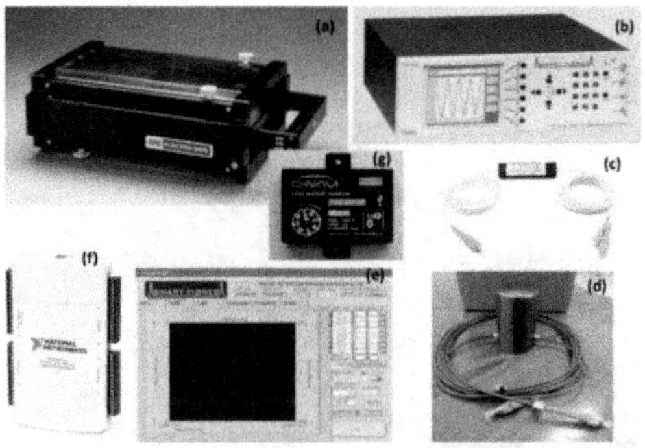

ent periods of time) and could be powerful in providing us with knowledge on materials, construction techniques, building pathologies and for important repair methods/strategies.

PHYSICAL MODELLING AND LABORATORY TEST SET UP

Physical Modelling is another scientific activity that can provide useful information regarding the investigation of the structural safety of cultural heritage sites. More specifically, structures/monuments on sites can be constructed as models at small or large-scale in the laboratory, and be subjected to specific loading conditions (i.e., the corresponding natural hazards and structural loadings). In addition, real-time monitoring techniques can be incorporated during the Physical Modeling (Springman et al., 2014) and useful test results can be obtained regarding the structural vulnerability and behavior of structures/ monuments. Furthermore, the monitoring techniques, and especially those applied for the first time, can be examined for proof of concept in the laboratory, prior to field applications (i.e., at the cultural heritage sites), ensuring, thereby, their successful operation and the minimum disturbance at the sensitive structures/monuments of the sites.

In this section an indicative laboratory test set-up and equipment that can be used for Physical Modelling of scaled models, related to cultural heritage site structures or monuments is presented. More specifically, a force generator can be utilized to impose predefined static and seismic loadings, and several sensors and monitoring devices can be incorporated into the model to record loading levels and structural responses. Furthermore, Image Analysis Techniques can be applied for the determination of strains and deformations during transient loading schemes, including recording data on potential failure mechanisms. It should be noted that scaling laws of the materials used in the models and for the loadings imposed needs to be taken into consideration. In Figure 12, indicative laboratory equipment is shown and more specifically: a) is for an earthquake simulator for the seismic loading impose; b) is showing an optical fiber interrogator for the real time data gathering; (c) is an optical fiber sensors recording strain; (d) is showing a single axis optical fiber acceleration sensor; (e) represents the requirement of software for optical fiber recordings acquisition; (f) is a data acquisition card for the system operation; (g) is showing a sensor for real-time acceleration recordings. In Figures 13 and 14, examples of the use of the aforementioned equipment are shown, at the laboratory and in the field, and more specifically a

Figure 13. Indicative laboratory equipment for the test setup

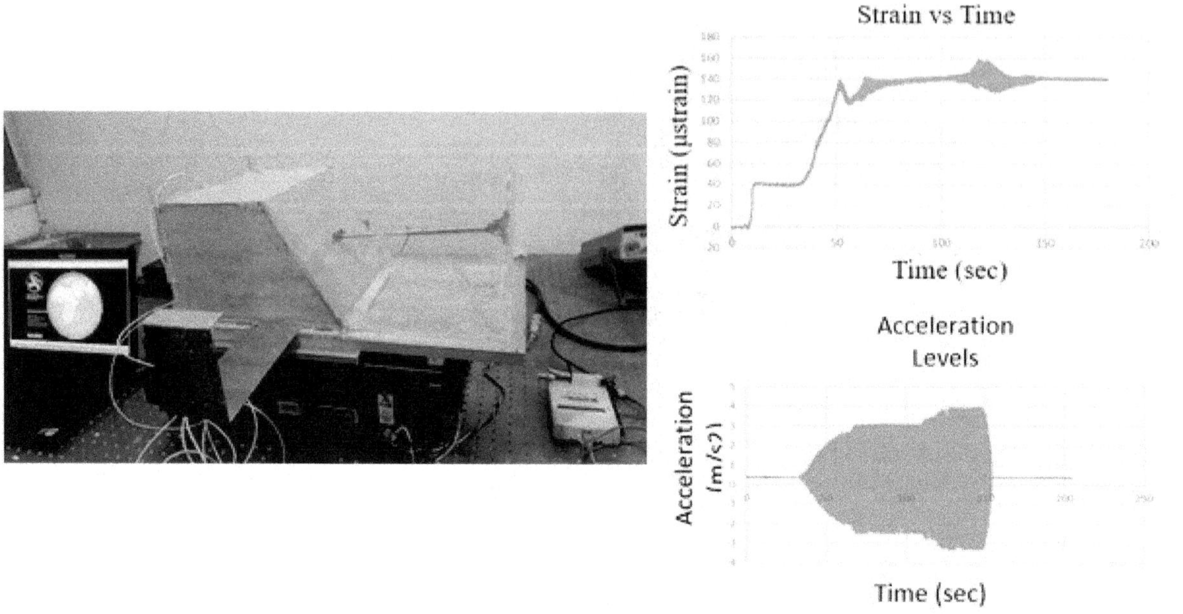

scaled slope model tested at an earthquake simulator, incorporating real-time structural monitoring with the use of optical fiber sensors and an accelerograph, including indicative results (Figure 13), and the use of similar equipment at the Acropolis of Athens for real-time remote structural monitoring (Figure 14).

Image Analysis Techniques (IATs) are used for the determination of the strains/deformations and the failure mechanisms which developed in scaled physical models during the loading procedure. IATs (Adrian, 1991) are optical methods that use correlation analysis in order to examine digital images of models taken during various loading scenarios, such as static or seismic, and can be internal or external. Consecutive images are captured with a digital camera during the deformation period and the changes of surface characteristics are recorded and evaluated. The small or large-scale models must be properly prepared to be able to apply the method with reliability. A 'zero-deformed' reference image of the model before loading is impose is necessary to be taken, followed by a series of digital images during the deformation process. By using the appropriate computer software, the deformed images are compared to the reference image by correlating all the pixels and strain distribution maps can be created. Two very well-known ITAs are the Digital Image Correlation (DIC) Technique and the Particle Image Velocimetry (PIV) Analysis. In Figure 15 results from DIC analysis during the physical modelling of scaled models imposed to seismic loading are shown. An important advantage of these techniques is that they can be applied to different types of models for a sensitivity assessment, by constructing the models with materials having various mechanical and geometrical properties (Kapogianni et al., 2020b).

Figure 14. Scaled slope model tested at the laboratory, with the use of an earthquake simulator, incorporating optical fiber sensors and one accelerograph and indicative results

Figure 15. Real-time remote structural monitoring at the Acropolis of Athens

Figure 16. Stabilized scaled model on an earthquake simulator and strains developed (using the DIC method)

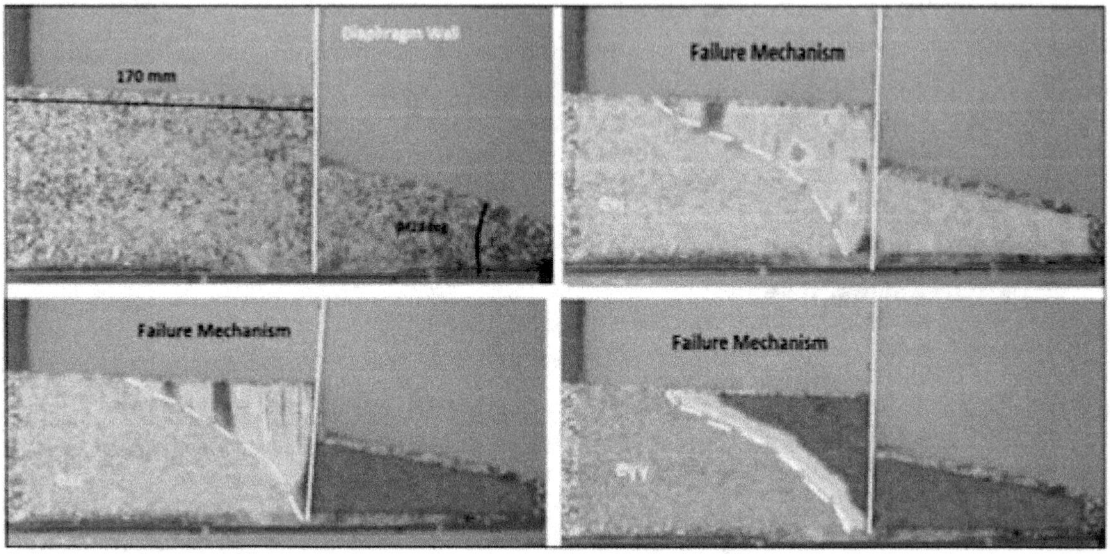

CONCLUSION

The interdisciplinary combination of advanced methods and technologies can be rather useful to the development of integrated solutions that will contribute to structural risk management and to optimal (rather limited) interventions options at cultural heritage site worldwide. Since cultural heritage sites are characterized by a variety of individual conditions (e.g., topographic, geological, structural), a major challenge is to achieve adaptability of the corresponding methods and tools, in order to successfully identify the critical locations in structures/monuments at these sites (i.e., the locations that are prominent to damage and structural failures).

Regarding the application of computational analyses, the finite-element method is adopted to generate useful information on the structural responses and distresses of structures/monuments. Successful numerical simulations do however require reliable knowledge of the geometrical and material properties of the super-structure and sub-structure under examination, and of the anticipated static and dynamic loadings. In order to achieve a realistic simulation for the structural response, the local site conditions (i.e., soil, geomorphology, topography) and the corresponding potential soil-structure interaction need to be taken into consideration. Furthermore, uncertainties in the loading conditions can be limited by a reassessment based on real-time monitoring with sensors of loadings and mechanical properties.

Smart structural monitoring can provide important real-time information (records) on both the actual hazard(s) and the corresponding vulnerability of structures/monuments, in terms of structural responses and their distress. Monitoring can contribute to the dynamic (i.e., time-dependent) structural risk assessment and can be fused with numerical modelling to investigate the states of structures/monuments. The hybrid output of such research (informed by real data and the physics underlying the modelling methodologies) can deliver an estimation of the expected response and distress of structures/monuments under examination, when subjected to the anticipated natural hazard(s).

In an environment where various networks of sensors have been deployed, the need for intelligent management and processing of the recorded and transmitted data has emerged. An integral part to the data processing is the combination of data fusion between heterogeneous networks and the support of additional meta-data, in order to efficiently handle the available measurements. In addition, data obtained from measuring devices and sensors can be used as a digital reliable information database and can be incorporated in a Building Information Model (BIM), for important decision-making. Smart structures could be supported effectively via the digital representation of their physical and functional characteristics, especially at critical locations where failures can be initiated. The combination of remote reviewing and BIM in the assessment process could enable the study of a facility in the context of the environment (at different periods of time), and could provide knowledge on materials, construction techniques, building pathologies and for future repair methods.

Furthermore, early warning systems can be developed, which will activate in the case of exceeding predefined limits for the likelihood of structural failure. Early warning systems can lead to self-control of structures under examination, enhancing the requirement for early intervention on sites, and contributing to the safety of visitors and employees, and to overall public safety.

Physical modeling for the investigation of structures and monuments can also be applied to provide useful information concerning their behavior during various loading scenarios. It can be performed either by means of full-scale testing or by using small scale physical models. Static or seismic loadings can be applied in order to generate define stress-strain levels and kinematically admissible collapse mechanisms. During the physical modelling procedure, apart from scaling laws and model preparation, attention needs to be paid to measurement systems that are utilized. Image analysis methods, such as the Particle Image Velocimetry (PIV) and Digital Image Correlation (DIC) are known to provide full-field strain measurements with experimental models and contour maps of strains can be created over entire areas of interest.

It is evident that ancient civilizations had awareness of advanced engineering techniques and applied reliable methods to deal with structural loadings and various hazards. The remains of such structures, including geotechnical structures, can be found today at numerous cultural heritage sites worldwide and are threatened by static and dynamic loadings. Advanced scientific tools and emerging technologies can be utilized in order to quantify the structural loadings and hazards and to estimate the vulnerability of these structures, aiming to the reduction of the structural risk of the sites.

REFERENCES

Adrian, R. J. (1991). Particle imaging techniques for experimental fluid mechanics. *Annual Review of Fluid Mechanics*, *23*, 261–304.

Ambraseys, N. (2010). On the long-term seismicity of the city of Athens. *Proceedings of the Academy of Athens*, (Vol A, pp. 81-136). Academy of Athens.

Athanassas, C. D., Modis, K., Alcicek, M. C., & Theodorakopoulou, K. (2018). 'Contouring the Cataclysm: A Geographical Analysis of the Effects of the Minoan Eruption of the Santorini Volcano. *Environmental Archaeology*, *23*(2), 160–176.

Chatzi, E. (2017). A Monitoring Approach to Smart Infrastructure Management. *Proceedings*, *1*(8), 755.

Holladay, A. (1977). Sparta's role in the First Peloponnesian War. *The Journal of Hellenic Studies, 97*, 54–63.

Kalogeras, I., Evangelidis, C., Melis, N., & Boukouras, K. (2012). *'The Athens' Acropolis Strong Motion Array', EGU General Assembly, Geophysical Research Abstracts* (Vol. 12). EGU.

Kapogianni, E. (2021). Geotechnical Structures in the Ancient World: The case of the Ziggurat of UR. *12th International Conference on Structural Analysis of Historical Constructions SAHC 2020*. P. Roca, L. Pelà, & C. Molins (Eds.)

Kapogianni, E., Kalogeras, I., Psarropoulos, P., Michalopoulou, D., Eleftheriou, V., & Sakellariou, M. (2019). 'Suitability of Optical Fibre Sensors and Accelerographs for the Multi-disciplinary Monitoring of a Historically Complex Site: The Case of the Acropolis Circuit Wall and Hill'. Geotechnical and Geological Engineering. *Springer International Publishing, 37*(5), 4405.

Kapogianni, E., Psarropoulos, P. N., Kokoris, D., Kalogeras, I., Michalopoulou, D., Eleftheriou, V., & Sakellariou, M. G. (2020a). *Impact of Local Site Conditions on the Seismic Response of the Athenian Acropolis Hill. Geotechnical and Geological Engineering*. Springer International Publishing.

Kapogianni, E., Sakellariou, M., Laue, J., & Springman, S. (2016). Investigation of the mechanical behaviour of the interface between soil and reinforcement, via experimental and numerical modelling. *Procedia Engineering, Elsevier, 143*, 419–426.

Kapogianni, E., Sakellariou, M.G., & Laue, J. (2017). *Experimental investigation of reinforced soil slopes in a geotechnical centrifuge, with the use of optical fibre sensors, 35*(2), 585-605. Geotechnical and Geological Engineering, Springer International Publishing.

Kapogianni, E., Tsafou, S., & Sakellariou, M. G. (2020b). Image Analysis and Structural Monitoring of Physical Geotechnical Models Subjected to Seismic Loading, pp. 674-686. Springer Series in Geomechanics and Geoengineering.

Koerner, R. M. (2012). *Designing with Geosynthetics* (6th ed.). Xlibris Publishing Co.

Montgomery, D. (2016). Emperor Yu's Great Flood. *Science, 353*(6299), 538–539. PMID:27493166

Psarropoulos, P., & Gkantona, A. (2020). *Information Technology and a New Theory for the Verification of the Location of the Ancient City of Helike in Achaia, Greece*. Springer Series in Geomechanics and Geoengineering.

Psarropoulos, P., Kapogianni, E., Kalogeras, I., Michalopoulou, D., Eleftheriou, V., Dimopoulos, G., & Sakellariou, M. (2018). *Seismic response of the Circuit Wall of the Acropolis of Athens: Recordings versus numerical simulations*. Journal Soil Dynamics and Earthquake Engineering.

Radt, S. (2011). *Strabons Geographika* (C. Apparatus, Trans.). Vandenhoeck & Ruprecht.

Ratchet, G. (1983). *Dictionnaire L' Archéologie* (S. A. Robert Laffont, Ed.).

Reinhardt, O., & Oldroyd, D. R. (1983). 'Kant's theory of earthquakes and volcanic action'. *Annals of Science, 40*(3), 247–272.

Springman, S. M., Laue, J., Herzog, R., & El-Hamalawi, A. (2014). The evolution of a physical modelling course over two decades. *Proceedings of the 8th International Conference on Physical Modelling in Geotechnics*, Perth, Australia.

Sun, D., Lee, V., & Lu, Y. (2016). An intelligent data fusion framework for structural health monitoring. *IEEE 11th Conference on Industrial Electronics and Applications*. IEEE. 10.1109/ICIEA.2016.7603550

Trikkalinos, I.K. (1977). *Remarks on the published study on the geology of the Acropolis*. Proc Acad Athens, Athens.

Chapter 5
Antibiotic Resistance: A Global Threat to Humanity

Mohammad Shoeb
University of Dhaka, Bangladesh

Rafiza Islam
University of Dhaka, Bangladesh

Nargis Parvin
University of Dhaka, Bangladesh

ABSTRACT

Antibiotics are certainly miracle drugs and have comprehensive usages in humans and livestock for the treatment of various bacterial diseases. However, inappropriate or misuse of antibiotics can lead to the development of antimicrobial-resistant bacteria. Antibiotic resistance happens when bacteria change in responding to the usage of these medicines. These bacteria can also be disposed of from animals to humans via contact between humans and animals directly, or through the food chain and the environment. Antibiotic resistance is a great threat to global humanity as it can prolong treatment in hospitals, increase medical costs, increase mortality, and burden the family. Residual antibiotics above the maximum residue limit (MRL) are harmful to human health and animals and need to be monitored. By increasing national policies, international declarations, agreements for enough chemical and analytical instrumental facilities, and above all, creating awareness about the harmful effect of resistance, the situation of antibiotics resistance can be improved.

INTRODUCTION

Antibiotics are synthetic, semi-synthetic, or naturally occurring compounds with antimicrobial activity to treat and prevent diseases. Antimicrobial agents are chemotherapeutic, prophylactic, synthetic, and therapeutic agents used to prevent and treat infections in humans, animals, and plants (Patel, et al., 2020). They include antibiotics, antiviral, antifungal, and anti-parasites. In general, antibiotics are used to treat

DOI: 10.4018/978-1-6684-5619-4.ch005

bacterial infections. Antifungals are used to kill or prevent the growth of fungi. Viral infections are treated by antiviral drugs. Anti-parasites are used for the treatment of infectious diseases caused by parasites.

Usually, antibiotics possess sensitive strains which selectively inhibit the growth rate of target bacteria and stop multiplying the cell wall developing, destroying, or attacking the surrounding bacteria. Antibiotics are effective against the suspected infections caused by bacteria, fungi, or protozoa but not against viral diseases like the common cold, influenza, flu, etc. according to the World Health Organization (WHO, 2020).

HISTORY OF ANTIBIOTICS

Antonie van Leeuwenhoek, a Dutch microbiologist identified the first single-cell microorganism bacteria in the 1670s. Later, Louis Pasteur and Robert Koch worked on *tuberculosis mycobacteria* in 1882. Rudolph Emmerich and Oscar Low discovered the antimicrobial drug for cholera and typhus in 1890 and Paul Ehrlich introduced the bacteria responsible for syphilis in 1909 (Lambrin et al., 2018). However, the first natural antibiotic known as Penicillin which was derived from *Penicillium notatum* in 1928 by Alexander Fleming was effective against gram-positive and gram-negative bacterial infections (Patrick, 2013) (Table1). Within 14 years of discovery, in 1942 penicillin saved the life of Anne Sheafe Miller who was suffering from a streptococcal infection. The discovery of penicillin saved many more lives and inspired scientists to work on the discovery and development of new antibiotics. Later, many new and useful antibiotics of different classes were discovered from natural sources and synthesized in laboratories to save human lives. It has been an ongoing process of discovery until now.

CLASSIFICATION OF ANTIBIOTICS

Antibiotics can also be classified based on their chemical structure, the basis of origin, mechanism of action, the spectrum of action effects on their activity and route of administration (Singh, 2020; Kebede et al., 2014) as described in Table 2-7:

SELECTED ANTIBIOTICS

Antibiotics are regularly used in humans and livestock to treat various bacterial diseases. Their improper usage results in residues in animal tissues that cause antibiotic resistance via the food chain and environment. So, it is essential to investigate residual amounts of those familiar antibiotics used in humans and food-producing animals to make awareness of using antibiotics and stopping antibiotic resistance.

Amoxicillin

Amoxicillin ($C_{16}H_{19}N_3O_5S$; α-amino-p-hydroxybenzyl penicillin) is an antibiotic of the beta-lactam class. It is used for the treatment of bacterial infections (nose, ears, throat, lung, skin, and urinary tract), stomach ulcers by *Escherichia coli, Haemophilus influenzae, H. pylori, Neisseria gonorrhea,* pneumococci, streptococci, *Salmonella spp*, *Proteus mirabilis*, and certain strains of staphylococci (El-Sayed et

Table 1. Name of some antibiotics, class, and year of discovery with the inventor

Antibiotics	Class of Antibiotics	Inventor	Year of Discovery
Penicillin	Penicillins and penicillinase	Alexander Fleming	1928
Sulfonamides	Sulfonamides	Bayer's research team	1932
Sulfamethazine	Sulfonamides	W. T Caldwell & co-workers	1941
Patulin	Polyketide	Birknshaw et al.,	1943
Streptomycin	Aminoglycosides	A.I. Schatz	1943
Cephalosporin	β-lactam	Giuseppe Brotzu	1948
Chlortetracycline	Tetracycline	Benjamin Duggar	1948
Chloramphenicol	Antimicrobials	Mildred C. Rebstock	1948
Oxytetracycline	Tetracycline	Francis A. Hochstein	1950
Tetracycline	Tetracycline	Lloyd H. Conover	1952
Amoxicillin	β-lactam	Alexander Fleming	1958
Ciprofloxacin	Fluoroquinolones	Patented by WHO	1987

al., 2014). This antibiotic may cause headache, fever, vomiting, diarrhea, indigestion, nausea, stomach pain, and sore throat. Amoxicillin acts against susceptible bacteria when they multiply and inhibits the cell wall's biosynthesis, leading to the death of bacteria (Dowling et al., 2017).

Chloramphenicol

Chloramphenicol ($C_{11}H_{12}Cl_2N_2O_5$) is the antibiotic that was first isolated as the metabolism product of bacteria of the "*Streptomyces venezuelae*" type but now can be synthesized artificially in the lab. It is feasible against many pathogenic bacteria, rickettsia, and mycoplasma showing the effect by entangling

Table 2. Classification of antibiotics based on their chemical structure

Type of Antibiotics	Examples
Alicyclic	Cycloheximide
Aliphatic amine	Spermidine
Aminoglycosides	Streptomycin
Aromatic (Nitrobenzene)	Chloramphenicol
Carbohydrate containing	Aminoglycosides, streptomycin, kanamycin, amikacin
Macrocyclic lactone	Erythromycin
N-containing heterocyclic	β-lactam
N/O glycosides	Chromomycin
O-containing heterocyclic	Cycloserine
Peptide antibiotics	Polymyxin, bacitracin, gramicidin
Pure saccharides	Streptozotocin
Quinolones	Fluoroquinolones

Table 3. Classification of Antibiotics based on origin

Type of Antibiotics	Examples
1. Microbial origin	i. Actinomycetes origin: *S. venezuelae*: Chloramphenicol : *S. ervthreus*: Erythromycin : *S. mediterranae*: Rifampicin : *Streptomyces griseus:* Streptomycin ii. Bacterial origin : *Bacillus polymvxa*: Polymyxin : *Chromobacter violaceum:* Bacitracin : *Micromonospora spp*: Gentamycin iii. Fungal origin : *Penicillium notatum:* Penicillin : *Cephalosporin spp:* Cephalosporin
2. Semi-synthetic antibiotics	Ampicillin, amoxycillin, doxycycline, sulfonamide, tigecycline
3. Synthetic antibiotics	Chloramphenicol, sulfonamide, 4-quinolones

Table 4. Classification of antibiotics based on mechanism of action

Type of Antibiotics	Examples
Amphenicols	Chloramphenicol, florfenicol, thiamphenicol
Aminoglycosides	Neomycin, spectinomycin, streptomycin
β-lactam	Amoxicillin, flucloxacillin, penicillin G
Carbapenems	Biapenem, doripenem
Fluoroquinolones/ Quinolones	Ciprofloxacin, enrofloxacin, levofloxacin, norfloxacin
Lincosamides/ Forest Amine	Clindamycin, lincomycin, pirlimycin
Macrolides	Azithromycin, clarithromycin, erythromycin, spiramycin, tylosin
Nitrofuran	Furazolidone, furaltadone, nitrofurazone, nitrofurantoin
Tetracyclines	Chlorotetracycline, doxycycline, oxytetracycline, tetracycline
Polyketide	Amphotericin
Sulfonamides/ Sulfanilamide	Sulfadiazine, sulfadimethoxine, sulfamethazine, Sulfamerazine, sulfamethoxazole, sulfamethoxypyridazine

Table 5. Classification of antibiotics based on their spectrum of action

Type of Antibiotics	Examples
Narrow spectrum (Effective towards relatively fewer microorganisms)	Fidaxomicin, vancomycin, sulfonamides, macrolides, polymyxin
Moderate spectrum (Effective towards Gram-Positive bacteria as well as some systemic and UTI-causing Gram-Negative bacteria)	Aminogycosides, sulfonamides
Narrow-Board spectrum (Effective against gram-positive and gram-negative bacteria)	β-lactam

Table 6. Classification of Antibiotics based on effects on their activity

Type of Antibiotics	Examples
Bacteriostatic (Interdicts the growth of bacteria)	Chloramphenicol, tetracyclines, sulfonamides, spectinomycin, trimethoprim, macrolides
Bactericidal (Kills bacteria)	Aminoglycosides, cephalosporin, penicillin

the protein synthesis in microorganisms. Chloramphenicol is not a highly promoted antibiotic for treatment because of its many adverse effects like bone marrow, depression, anemia, leukemia, allergic reactions, cholera, and typhoid (Vuran, 2021). It is just applied for treating severe infections. It is currently a banned antibiotic due to producing many infectious diseases and is identified as having an adverse effect on humans and animals. Now it is globally banned for regular usage for all types of food-producing animals improving food safety. The FAO and WHO combinedly reported it as a banned antibiotic and did not estimate any acceptable limit of consumption for livestock (Mou et al., 2021).

Patulin

Patulin (4-hydroxy-4H-furo[3,2-c] pyran-2(6H)-one) is an organic compound, a member of the polyketide (Wikipedia.org), and is used as an antibiotic. It is used against gram-positive and gram-negative bacteria (Mahunu et al., 2018). Patulin is a mycotoxin produced by various molds, particularly *Aspergillus, Byssochlamys,* and *Penicillium* (Mahato, 2021). Patulin is neurotoxic (Anene et al., 2016), immunotoxic (Liu et al., 2019), genotoxic (Bagheri et al., 2018), toxic to reproduction (Li et al., 2018), ototoxic, and hepatulinotoxic (Yang et al., 2018). Patulin acts as an inhibitor of potassium uptake with antibacterial and introducer of ion flux across cell membranes that involves Na+, K+ dependent adenosine triphosphate (ATP). It induces intramolecular and intermolecular protein crosslinking (Dowling et al., 2017).

Sulfa Drugs

Sulfa drugs are classified according to chemical groups, based on bacterial spectrum, chemical functional group, route of administration and activity, mode of action, or effectiveness. They can treat life-threatening bacterial diseases. Sulfa drugs are also available based on their functional group like sulfamethoxazole (hepatically metabolized and renally excreted), sulfadiazine (half-life is 12 hours), sulfadoxine (half-life of 7–9 days), sulfamethazine, sulfaquinoxaline, sulfamethazine, sulfamerazine, etc (Stanley Jr, 2010).

These drugs are regularly used in animals for growth-promoting purposes (Cheong et al., 2010) and in veterinary medicine and humans due to their synthetic, wide spectrum against gram-positive and negative microorganism properties. Sulfa drugs kill bacteria and fungi by interfering with their metabolism

Table 7. Classification of antibiotics based on the route of administration

Type of Antibiotics	Examples
Oral antibiotics (Acid stable antibiotics)	Penicillin V
Parental route (Intravenous administration)	Penicillin G

and inhibiting the rapid multiplication of resistant strains of bacteria (Tirmazi et al., 2021). They also have broadly effective antibacterial activity (Zimdahl, 2015), but extensive usages have been paid to the potential human health risk of being carcinogenic (Cheong et al., 2010). Sulfa drugs inhibit the growth and multiplication of bacteria because of their bacteriostatic properties. Sulfa drugs act by interfering with a member of the vitamin B complex named folic acid synthesis from more straightforward starting materials which is common in all living cells and inhibits the growth of bacteria without harming humans or animals. When trimethoprim and sulfamethoxazole work together as an inhibitor of sequential blockage of folic acid synthesis. Sulfa drugs accumulate in inflamed tissues and skin, persist for a long time and restrict the growth of attacking microorganisms without damaging the host, so these drugs are very responsive to treating urinary tract infections and leprosy (Britannica, 2021).

The overall incidence of drug reactions like allergy is common; on the other hand, severe reactions may be triggered, including urinary tract disorders, hypersensitivity, bone marrow depression, aplastic anemia, early stages of intestinal and acute infections, digestive and respiratory tract diseases (Zimdahl, 2015).

Tetracyclines

Tetracycline ($C_{22}H_{24}N_2O_8$) is a compound with a four-ring arrangement and is abundantly used in the neurology and oncology field (Fuoco, 2012) due to its antibacterial properties, and cheapness. It is commercially available for the treatment of bacterial infections. It works equally against eukaryotes and prokaryotes bacteria and prevention of growth of gram-positive or gram-negative bacteria. It also enacts against intracellular chlamydiae, mycoplasmas, and rickettsiae (Nelson & Ismail, 2007).

Broad-spectrum antibiotic, oxytetracycline ($C_{22}H_{24}N_2O_9$) belongs to the tetracycline group and can interfere with essential proteins of resistant bacteria and save the immune system of animals and humans. It is used to treat pneumonia, chlamydiae, gastrointestinal and respiratory system diseases, etc.

Chlortetracycline ($C_{22}H_{23}ClN_2O_8$) is a broad-spectrum antibiotic belonging to tetracyclines and a bacteriostatic agent inhibiting bacterial growth and reproduction. It contends with the bacterial ribosome (tRNA) and finally inhibits the reproduction of bacterial cell synthesis.

Tetracycline antibiotics inhibit bacterial growth by inhibiting translation and are considered short-acting antibiotics. They prevent the amino-acyl tRNA from binding to the A site of the ribosome by binding the 30S ribosomal subunit. They also bind to the 50S ribosomal subunit due to different methods of action to aminoglycosides. Tetracycline may change the cytoplasmic membrane of bacteria and cause leakage of nucleotides from the cell. They prevent proof-reading of proteins from being produced by bacteria and lead to the death of bacteria by effectively preventing the binding of tRNA to the bacterial ribosome (Dowing et al., 2017).

STRUCTURE OF ANTIBIOTICS

Antibiotics are a group of chemical compounds belonging to a specific structure (Table 8) and different types of antibiotics have a base skeletal form with a variety of functional groups. Each antibiotic has a specific class, physical and chemical properties, and the ability to fight against distinctive bacterial infections with a particular mode of action.

Table 8. Some antibiotics and their chemical structures

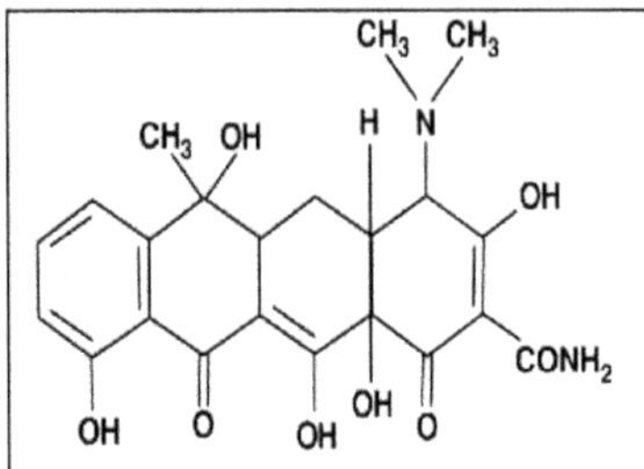	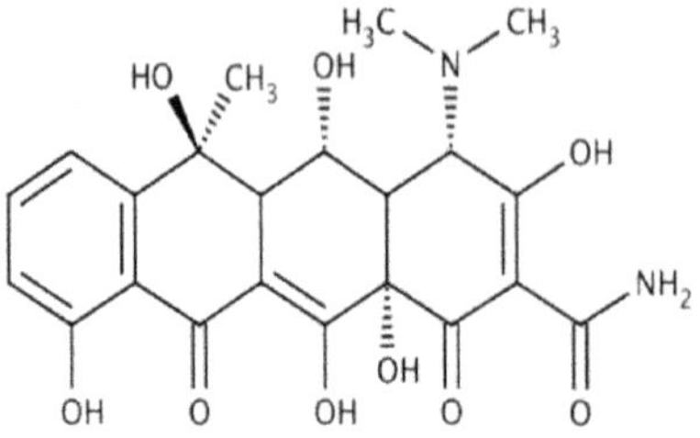	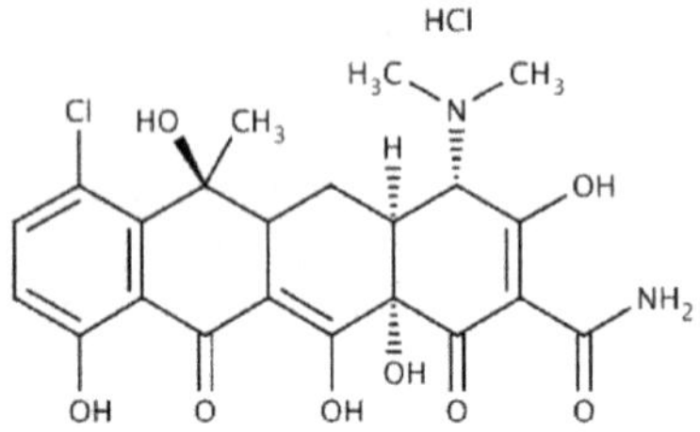
Tetracycline, $C_{22}H_{24}N_2O_8$	Oxytetracycline, $C_{22}H_{24}N_2O_9$	Chlortetracycline, $C_{22}H_{24}Cl_2N_2O$
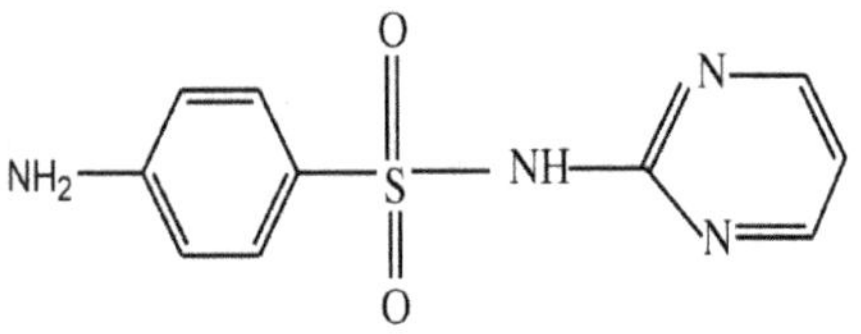	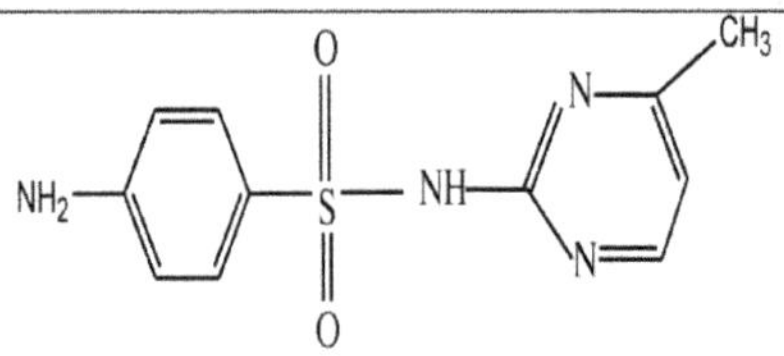	
Sulfadiazine, $C_{10}H_{10}N_4O_2S$	Sulfamerazine, $C_{11}H_{12}N_4O_2S$	
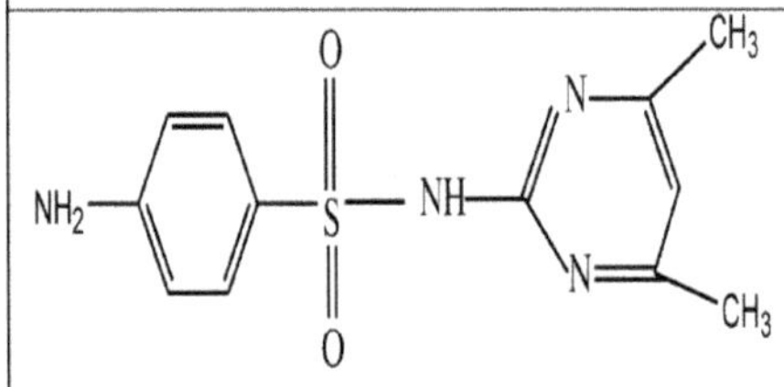	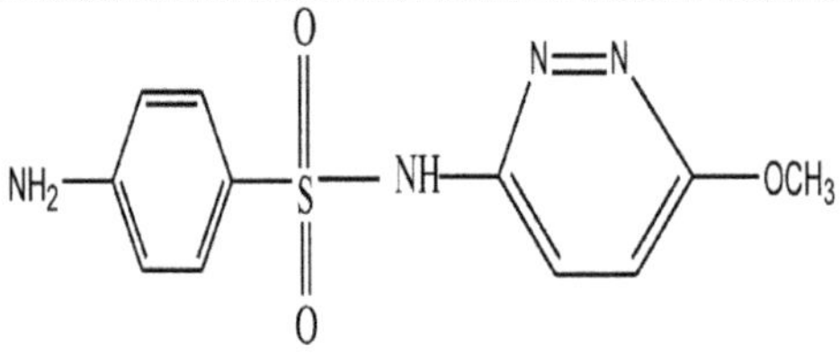	
Sulfamethazine, $C_{12}H_{14}N_4O_2S$	Sulfamethoxypyridazine, $C_{11}H_{12}N_4O_3S$	
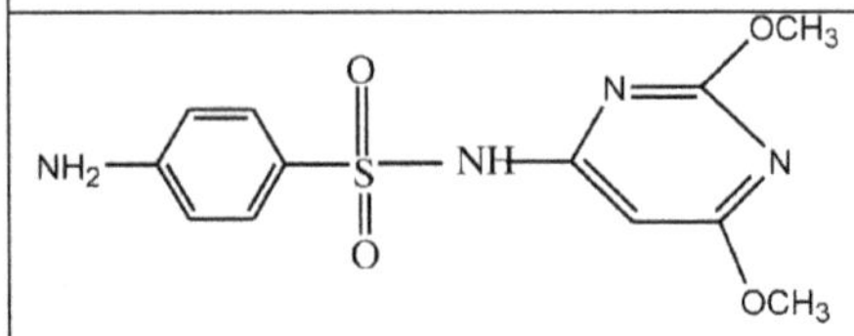	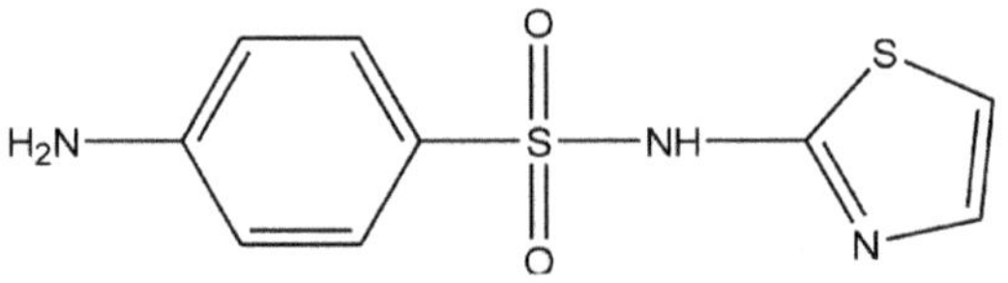	
Sulfadimethoxine, $C_{12}H_{14}N_4O_4S$	Sulfathiazole, $C_9H_9N_3O_2S_2$	
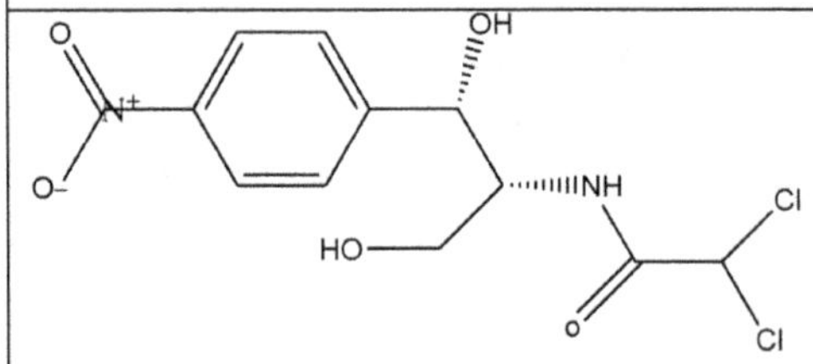		
Chloramphenicol, $C_{11}H_{12}Cl_2N_2O_5$		

ANTIBIOTICS USAGES IN HUMAN

Antibiotics are generally used for the treatment of various types of diseases caused by bacteria when average recovery is impossible. Usually, it takes a long time to heal any infection. If antibiotics are not used and the disease spreads all over the body, and serious complications start. Only antibiotics are helpful in combating different types of infectious diseases (Britannica, 2021). The common usages of some antibiotics in humans are outlined in (Table 9).

ANTIBIOTICS USAGES AS VETERINARY DRUGS

Most of the antibiotics are used in livestock as a veterinary drug for the treatment of clinical diseases such as anthrax, chronic respiratory difficulties, fowl cholera, fowl typhoid, gumboro, bird flu, pneumonia etc. caused by bacteria, and to control aggressive bacterial infections of a single animal or large group of animals administrated by farmer or veterinarian at different times in the life cycle of food-producing animals (Landers et al., 2012). Antibiotics are administered orally or by injection frequently in cattle and poultry farms as a growth promoter (Chattopadhyay, 2014) to increase meat and milk production. Still, it may negatively affect animals' welfare and food quality.

ADVERSE EFFECTS OF ANTIBIOTICS ON HUMAN

Antibiotics are very important for the rapid and specific treatment of different types of bacterial infections for both humans and livestock. But antibiotics not only fight against bacterial diseases but also show some adverse effects on humans and livestock. For example, aminoglycosides show ototoxicity confusion, gastrointestinal irritation palpitations while cephalosporins show gastrointestinal disturbances, headache, dizziness, lethargy, paresthesia, nephrotoxicity etc., amoxicillin shows diarrhea, nausea (Bartlett, 2002). Antimycobacterials, fluoroquinolones, lincosamides, penicillins and penicillinase, macrolides, monobactams, sulfonamides, and tetracyclines have different adverse effects of their own (Table 10).

IMPACTS OF ANTIBIOTICS ON THE ENVIRONMENT

Antibiotics are used in humans and livestock for the treatment of bacterial diseases and enhancement for the production of eggs, milk and meat to meet the demand worldwide. Antibiotics in animal feed increase efficiency and growth rate, prevent or reduce the incidence of infectious diseases, and treat clinically sick animals. For example, penicillin and tetracycline display actual development in egg production, hatchability and feed proficiency, but do not show a significant effect on death Chlortetracycline and oxytetracycline also display an improvement in the growth rate but a small effect on death. After feeding antibiotics for a long period, they preserve strains of bacteria that are unsusceptible to antibiotics. Those bacteria multiply in the animals. The resistant bacteria are dispatched to the other animals through interaction and bacteria prolong the intestinal flora of the animals as well as in muscle by forming colonization of antibiotic resistant bacteria. As a result, animal feces often contain resistant bacteria and can enter the environment by excretion. Currently, toxicity to humans via water or soil pathways becomes visible implausibly. For example, feces used as manure that contain antibiotics is used in the soil for growing crops and plants. Plants uptake antibiotics from manure which is applied on fields that become a concern to humans and animals feeding on these plants and plant products. It is one of the

Table 9. Common usages of some antibiotics

Antibiotic class	Name	Common usages
Aminoglycosides (fend protein synthesis)	Gentamicin	abdominal cavity, pelvic inflammatory disease
	Tobramycin	pelvic inflammatory disease, respiratory and urinary tracts, blood, abdominal cavity
Cephalosporins (interdiction of cell wall synthesis)	Cefazolin	pestilence of the genitourinary tracts, skin, bone and joints, and blood, endocarditis disease
	Cefaclor	skin disease, otitis media
	Cefamandole	peritonitis, bone and joints, respiratory and urinary tracts, skin disease
	Ceftriaxone	meningitis, gonorrhea, abdominal cavity, bone and joints, pelvic inflammatory disease
	Cefuroxime	pestilence of the respiratory and urinary tracts, skin, blood, bone and joints
Chloramphenicol (fend protein synthesis)	Chloramphenicol	eyes, ears, and skin disease, cystic fibrosis, prevention of infection in minor wounds
Fluoroquinolones (interfere with DNA synthesis)	Ciprofloxacin	diarrhea, gonorrhea, sinusitis, pneumonia, prostatitis, anthrax
	Norfloxacin	STDs caused by Neisseria gonorrhoeae, urinary tract infections, prostatitis
Macrolides (interdiction of protein synthesis)	Azithromycin	chronic obstructive pulmonary disease, pneumonia
	Clarithromycin	otitis media
	Erythromycin	acne, diphtheria, intestinal amebiasis, STDs, pertussis, Legionnaire disease, prevention of infection in minor wounds
Nitrofurans (indolent essential cell components)	Nitrofurantoin	urinary tract infections
Penicillins (fend cell wall synthesis)	Amoxicillin	various streptococcal and staphylococcal infections
	Ampicillin	endocarditis, gonococcal infections, meningitis
	Penicillin G	staphylococcal and streptococcal infections
	Ticarcillin	gastrointestinal tract infections, gonorrhea, impetigo, lyme disease, otitis media, meningitis, streptococcal and pseudomonas infections, tonsillitis
	Piperacillin	respiratory and genitourinary tract infections, abdominal cavity, skin, blood, bone and joints infections
Tetracyclines (fend protein synthesis)	Achromycin, Sumycin	acne, rickettsia, chlamydia, pneumonia, intestinal amebiasis
Lincosamides (interdiction of protein synthesis)	Clindamycin	pelvic inflammatory disease, pestilence of the respiratory tract, acne and skin

indirect routes of occurring antibiotic resistance through food plants (Figure 1). Another indirect route is the presence of antibiotics in water sources. Antibiotics can be released into the water sources by the feces of food-producing animals through rainwater which also persists in the environment and can cause resistance through interaction. Antibiotic resistance also can occur by the direct consumption of antibiotics in humans and other animals. Antibiotic toxicity to humans is implausible to be a concern from antibiotics that may appear in animal products or food crops receiving manure (Dibner & Richards, 2005).

Table 10. Antibiotics and their adverse effects on human

Antibiotics	Adverse Effects on Human
Aminoglycosides	■ Blood pressure changes Hypersensitivity reactions ■ Disorientation renal toxicity ■ Gastrointestinal irritation Palpitations ■ Ototoxicity confusion
Amoxicillin	■ Nausea ■ Diarrhea
Antimycobacterials, leprostatics	■ Gastrointestinal effects ■ Neuritis, malaise, drowsiness, hallucinations
Cephalosporins	■ Gastrointestinal disturbances ■ Lethargy, paresthesias ■ Nephrotoxicity ■ Pseudomembranous colitis ■ Superinfections
Fluoroquinolones	■ Bone marrow depression ■ Insomnia, depression ■ Gastrointestinal effects ■ Fever, photosensitivity, rash
Lincosamides	■ Bone marrow depression ■ Gastrointestinal effects ■ Pseudomembranous colitis
Macrolides	■ Abnormal thinking, Confusion ■ Gastrointestinal effects ■ Hypersensitivity reactions ■ Pseudomembranous colitis ■ Superinfections
Monobactams	■ Gastrointestinal tract ■ Allergic reactions ■ Hepatotoxicity
Penicillins and penicillinase	■ Gastrointestinal effects, including sore mouth and furry ■ Hypersensitivity reactions, including anaphylaxis ■ Tongue Superinfections
Sulfonamides	■ Bone marrow depression ■ Dermatological effects, including hypersensitivity and photosensitivity ■ Ataxia, vertigo, depression, seizures, gastrointestinal effects ■ Hepatotoxicity ■ Nephrotoxicity
Tetracyclines	■ Dermatological reactions, including rash and photosensitivity ■ Hepatotoxicity ■ Hypersensitivity reactions ■ Teeth (staining) and bone damage ■ Superinfections

ANTIMICROBIAL RESISTANCE (AMR)

According to the WHO, antimicrobial agents develop resistance when bacteria, viruses, fungi, and parasites change over time (WHO, 2020). For this reason, it is no longer efficient to respond to medicines, developing infections difficult to treat and increasing the risk of expanding diseases, severe illness, and death of humans, animals and plants. That impedes the feasible prevention and treatment of an ever-

Figure 1. Fate of antibiotics in the environment

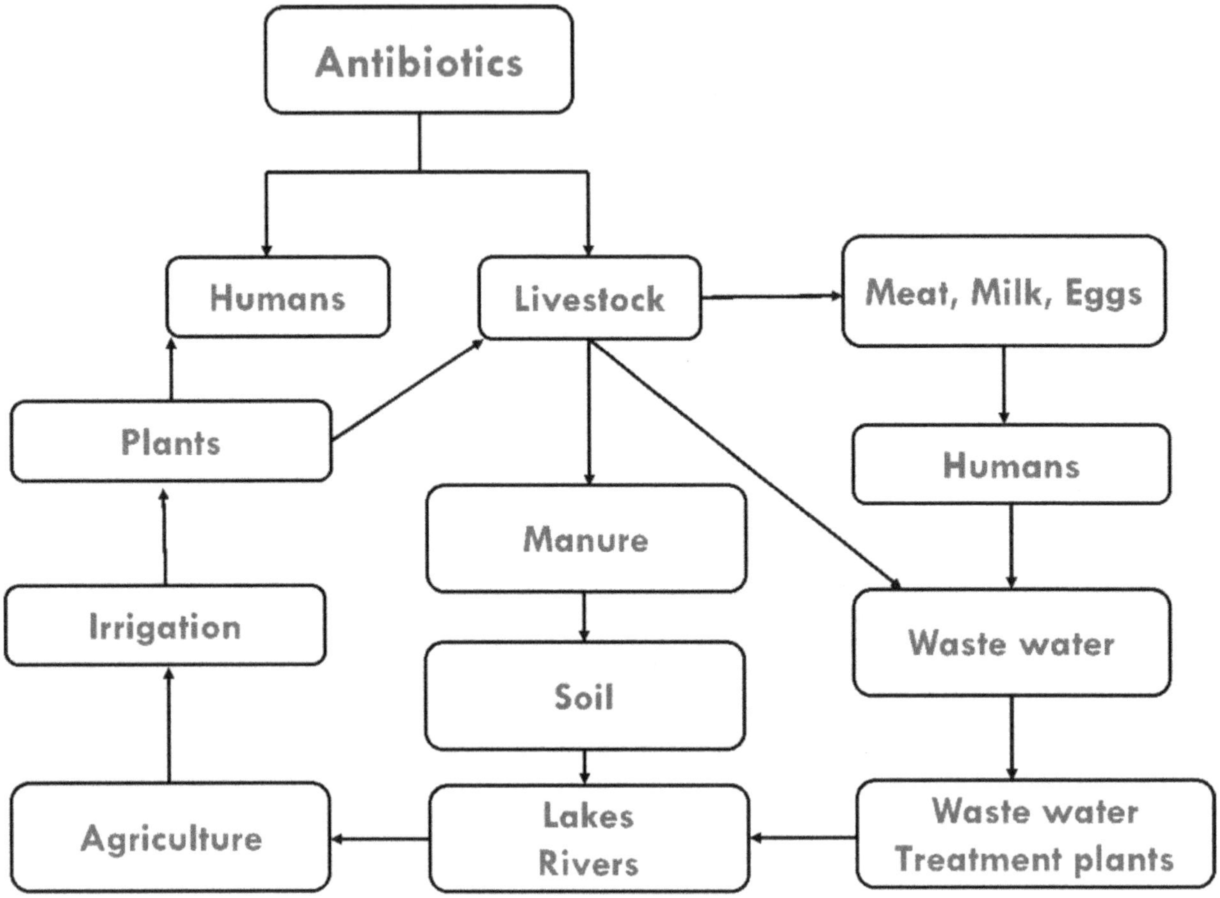

increasing range of infections in the body caused by bacteria, viruses, fungi, and parasites increasing the risk of spread to others.

Causes of Antibiotic Resistance

Antibiotics are powerful medicines for the prevention and treatment of bacterial infections. Antibiotic resistance happens when bacteria switch to protect themselves from the reaction to these antibiotics. These bacteria may infect humans and animals. The World Health Organization (WHO) has identified antibiotic resistance as one of today's biggest threats to global health, food security, and development. The infections caused by resistant bacteria are harder to treat than those caused by non-resistant bacteria. There are so many identical causes of occurring antibiotic resistance which are reported by WHO (2020) and (Ventola, 2015). Antibiotic resistance leads to higher medical costs, enhanced hospital stays, and tumid mortality. Several factors include misuse or overuse of antibiotics, sale without prescription, incomplete treatment, etc., responsible for the cause of antibiotic resistance (Figure 2).

Figure 2. Causes of antibiotic resistance

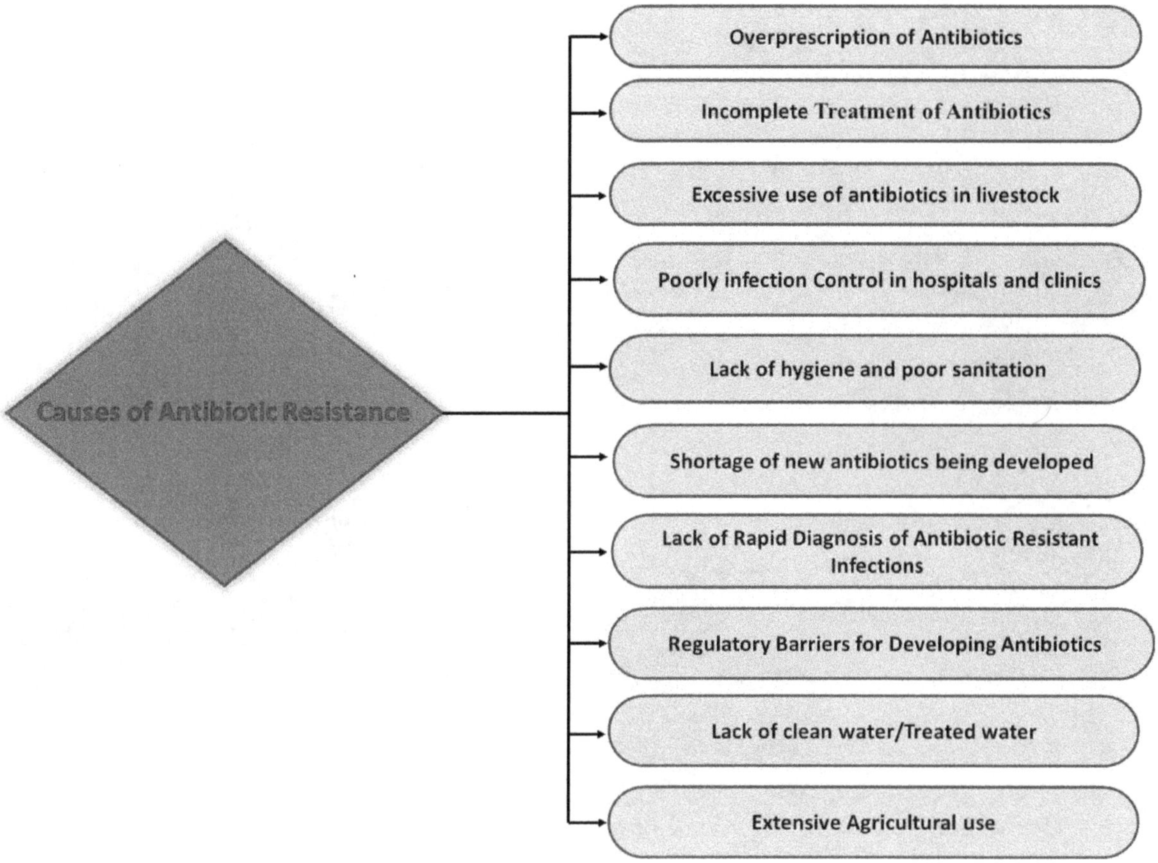

Spread of Antibiotic Resistance

Antibiotic resistance can happen when illuminated to antibiotics and susceptible bacteria are killed, while inappropriate and additional use of antibiotics causes bacteria to become resistant and continue to grow and accumulate. These resistant bacteria may expand and cause infections in other people (Figure 3). The major sources of antibiotic resistance are wastewater from large-scale animal farms, aquaculture, hospitals and municipalities and the excretion of antibiotics given to humans and animals unchanged in feces and urine can also cause antibiotic pollution in the environment (Sridhara et al., 2021; Hoque et al., 2020).

Antibiotic resistance is increasing day by day to high levels which is a great concern for humans, animals and the environment globally (Figure 4). Mechanisms of new resistance are arising and expanding, threatening the ability of humans to treat infectious diseases that are usually affected. An increasing list of commonly affected infections such as pneumonia, gonorrhea, blood poisoning and foodborne diseases are harder and impossible to treat as antibiotics become less efficient. Therefore, World Health Organization (WHO), 2021 and Centers for Disease Control and Prevention (CDC), 2019 have individually published a report on how antibiotic resistance is enhancing gradually all over the world and identified some causes which can spread antimicrobial resistance globally.

Figure 3. Antibiotics spreading to humans, animals and the environment

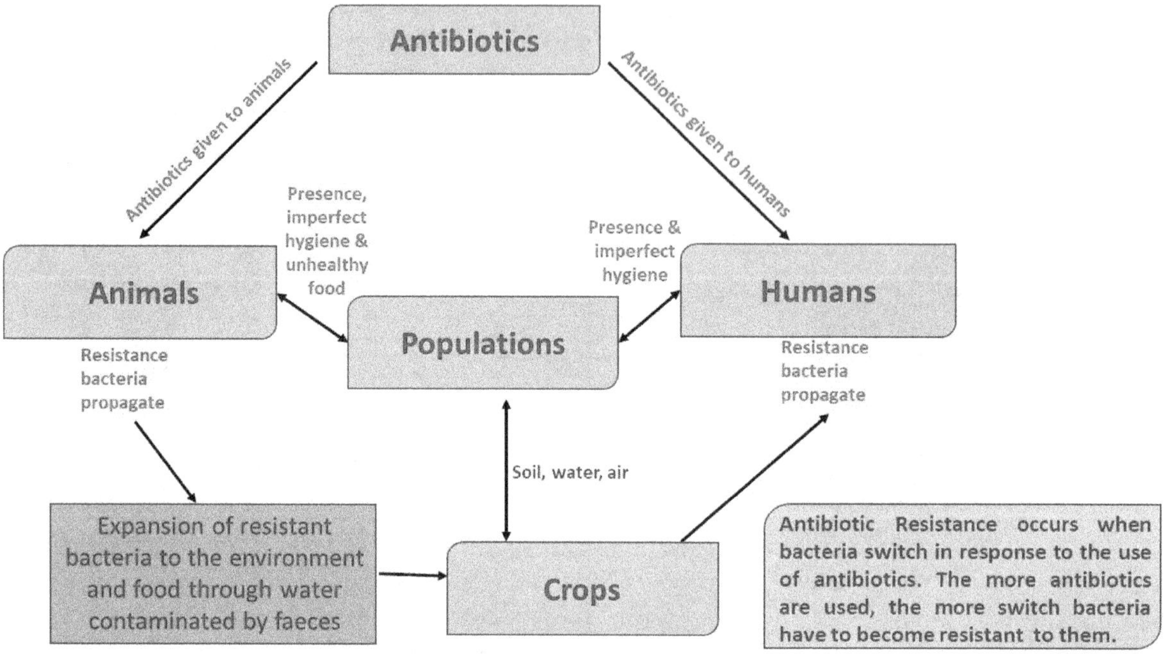

Figure 4. Global Statistics of Antimicrobial Resistance

Global Antimicrobial Resistance (AMR)

10 Million Deaths Each Year by 2050 due to drug-resistant diseases worldwide.	35,000 Patients Die Each Year **as a result of antibiotic-resistant infections in the** USA.	More than 33,000 people **in Europe die each year from antibiotic-resistant infections.**
700,000 People Die **each year due to drug-resistant diseases worldwide**.	More than 2.8 million **antibiotic-resistant infections occur in the USA each year.**	AMR **will cause a global loss of** US$100 trillion **and a decrease of global gross domestic product by** 3.5%.
By 2030 **antimicrobial resistance could force up to** 24 million people **into extreme poverty.**	11,000 Estimated Deaths caused by methicillin-resistant *Staphyococcus aures* MRSA each year in the USA.	10 million by 2050, with most of these deaths in low income countries in Asia and Africa due to antibiotic resistant.
230,000 People Die **from multidrug-resistant TB in** 2019 **worldwide.**	In USA 223,900 cases of *Clostridiodes. difficile* occurred in 2017 and at least 12,800 people died.	More than 58,000 children die every year in India alone, from antibiotic-resistant infections.

ANALYTICAL METHOD FOR THE ANALYSIS OF RESIDUAL ANTIBIOTICS

Many analytical methods including thin-layer chromatography (TLC), capillary electrophoresis, ultraviolet-visible spectrophotometry (UV-Visible) (Perez et al., 2002), gas chromatography (GC), and gas chromatography-mass spectrometry (GC-MS) (Wang et al., 2006), high-performance liquid chromatography (HPLC), liquid chromatography-tandem mass spectrometry (LC-MS/MS) are available for the analysis of different types of antibiotics. However, LC-MS/MS is considered one of the best methods nowadays for its high sensitivity & selectivity, linear range, accuracy & precision and recoveries. LC-MS/MS methods are widely used for qualitative and quantitative analysis (mainly determination and confirmation) of the residual level of antibiotics. (Mou et al., 2021).

LC-MS/MS with Electrospray Ionization (ESI) and Triple Quadrupole (QQQ) Mass Analyzer coupled with Ultra-Fast Liquid Chromatography (UFLC) used for antibiotic residue analysis. The triple quadrupole mass analyzer is combined with four parallel rods, and high voltage allows for a scan with mass-to-charge ratio (*m/z*), resulting in a mass spectrum (Figure 5). Macromolecules are produced using high voltage and high-temperature application to create droplets from liquid certified reference standards in ESI technique. It is also called the soft ionization technique.

Nitrogen (N_2) gas is used as drying and nebulizing gas. Air and argon (Ar) gas are used as heating and collision gas, respectively. Separations are usually performed on a C18 reversed-phase column and using the flow injection method in scan mode of the first quadrupole (Q1) and third quadrupole (Q3) mass analyzers for precursor ion identification. A protonated $[M+H]^+$ or deprotonated $[M-H]^-$ molecule is observed as the base peak ion in the mass spectra of the analyzed antibiotics and these ions are selected as precursor ions. Using collision gas in the second quadrupole (Q2) three product ions (m1, m2 and m3) are detected from precursor ion fragmentation by Multiple Reactions Monitoring (MRM) mode for quantification based on their *m/z* with high sensitivity (Table 11). Finally, the expected chromatogram is found in a fixed retention time based on the mobile phase composition and flow rate using gradient or isocratic elution.

LC-MS/MS methods for the analysis of residual sulfa drugs, tetracycline, oxytetracycline, chlortetracycline, chloramphenicol, patulin, amoxicillin were validated with linearity, limit of detection (LOD), limit of quantification (LOQ), standard and matrix match calibration curve, accuracy, precision (repeatability and reproducibility in intra- and interday), recovery percentage, high sensitivity, selectivity and specificity, matrix matched effects.

RETENTION TIME

Retention time is the time that depends on the interaction of the analyte with the stationary phase due to their affinity and usually the analyte comes out with a mobile phase and creates a chromatographic peak for identification. Retention time in high-performance liquid chromatography (HPLC) system changes depending on column temperatures, column length, flow rate, mobile phase composition, gradient elution programs and is expressed in minutes.

Figure 5. LC-MS/MS spectrum of Chloramphenicol (A) precursor ion (B) product ions

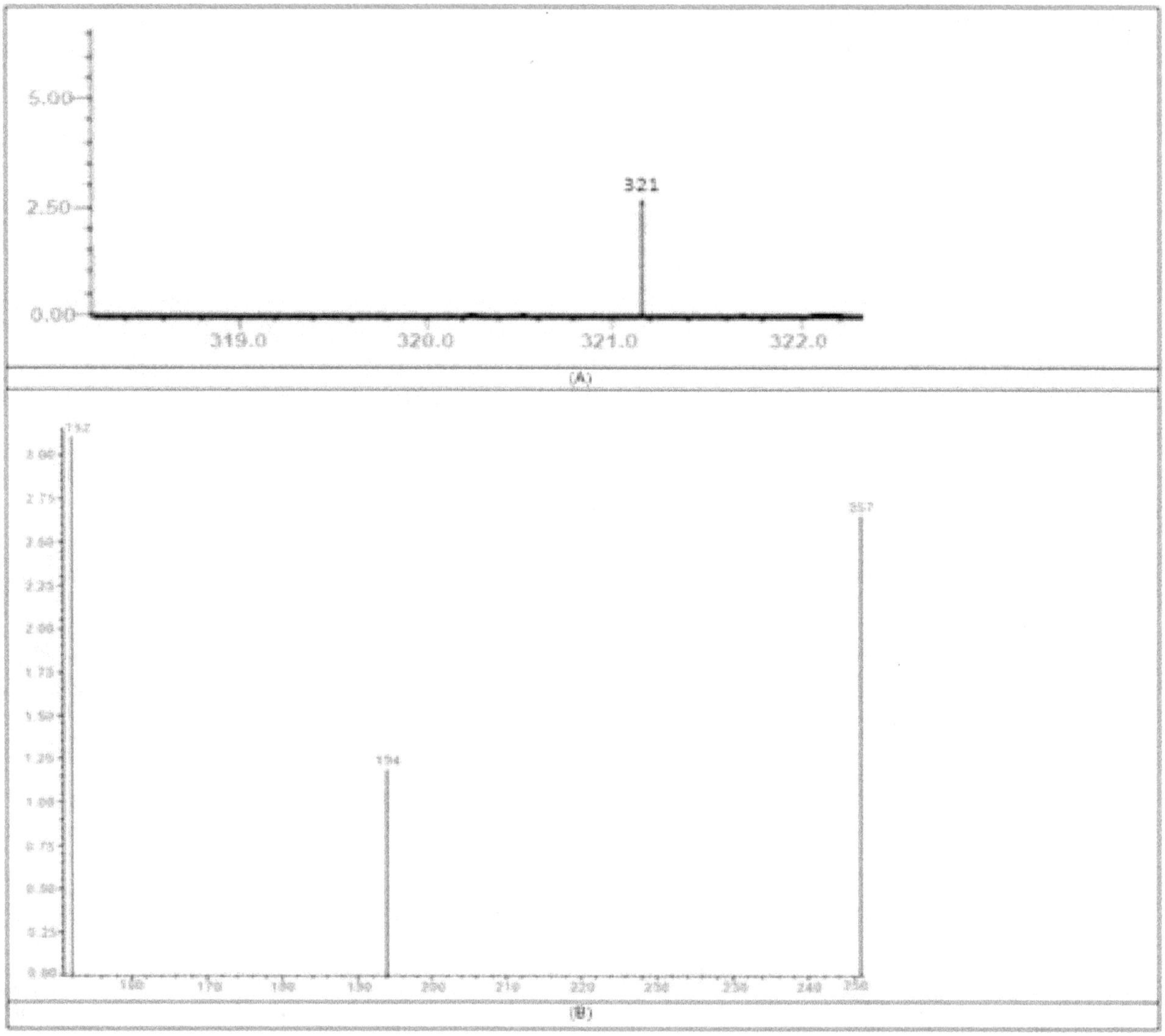

Limit of Detection (Lod) And Limit of Quantification (Loq)

The limit of detection is defined as the lowest amount of analyte that can be detected by an analytical instrument and the limit of quantitation is the lowest amount of analyte that can be quantified. LOD and LOQ can be measured based on a standard deviation of the blank, standard deviation of the response of the analytical instrument, and the slope of the straight line of the linearity of the areas for the known concentrations, signal to noise ratio according to ICH Q2 (2005). LOD and LOQ are important criteria for method validation. LOD and LOQ are usually expressed by µg/kg or ng/g.

Table 11. Antibiotics with molecular mass, precursor and product ions

Antibiotic Name	Molecular mass (g/mol)	Precursor ion, (m/z) (M-H) $^{+}$ or (M-H)$^{-}$	Product ion	References
Chloramphenicol	322.00	321.00	152.00 194.00 257.00	Mou, et al., 2021
Sulfathiazole	255.30	256.00	92.00 108.00 156.00	Clark, et al., 2005
Sulfadiazine	250.278	251.00	92.00 108.00 156.00	Clark, et al., 2005
Sulfamerazine	264.305	265.00	92.00 108.10 156.00	Clark, et al., 2005
Sulfamethazine	278.33	279.00	92.10 108.00 156.00	Clark, et al., 2005
Sulfamethoxypyridazine	280.31	281.00	92.10 108.00 156.05	Yu, et al., 2011
Sulfadimethoxine	310.33	310.90	92.10 107.90 156.00	Clark, et al., 2005
Tetracycline	444.435	445.20	154.00 410.00 427.00	Uekane et al., 2011
Oxytetracycline	460.434	461.20	337.00 426.00 443.00	Uekane et al., 2011
Chlortetracycline	478.88	479.00	154.00 444.00 462.00	Uekane et al., 2011

Maximum Residue Limit (Mrl)

The Maximum Residue Limit (MRL) is the maximum allowable concentration of a chemical compound (Menkem et al., 2018) that is legally tolerated in a feed or food (milk, meat, egg, vegetables) for safe consumption by humans (Table 12) according to Bangladesh food safety authority (BFSA, 2017), CODEX Alimentarius (2018), European Commission (EU, no 37/2010).

CAUTIONS AND PRECAUTIONS OF ANTIBIOTICS APPLICATION

Antibiotics are widely used to safeguard animal and human health (Savic, 2018). However, they have several adverse effects on human health and animals, and can lead to antibiotic resistance. Special precautions are needed to consider before applying:

Table 12. Maximum residues limit (MRL) (mg/kg)

Antibiotics	Species	Tissue	MRL (µg/kg)		
			CODEX	EU	BFSA
Amoxicillin	Cattle	Muscle	50	50	50
	Cattle	Liver	50	50	50
	Cattle	Kidney	50	50	50
	Cattle	Milk	4	4	4
Chloramphenicol	-	-	NA	Not Established	-
Enrofloxacin	Poultry, Cattle	Muscle	NA	100	100
	Poultry, Cattle	Liver	NA	300	300
	Poultry, Cattle	Kidney	NA	200	200
	Bovine	Milk	NA	100	-
Ciprofloxacin	Poultry	Muscle	NA	100	100
	Poultry	Liver	NA	300	300
	Poultry	Kidney	NA	200	200
	Bovine	Milk	NA	100	-
Patulin	-	-	NA	-	-
Sulfonamides	Not Specified	Muscle	100	100	100
	Not Specified	Liver	100	100	100
	Not Specified	Kidney	100	100	100
Tetracycline Oxytetracycline Chlortetracycline	Cattle	Muscle	200	100	200
	Cattle	Liver	600	300	600
	Cattle	Kidney	1200	600	1200
	Not Specified	Egg	-	200	400
	Not Specified	Milk	100	100	0.1
Tylosin	Cattle	Muscle	100	100	-
	Cattle	Liver	100	100	-
	Cattle	Kidney	100	100	-
	Not Specified	Milk	-	50	-
	All types of food-producing species	Egg	-	200	-

1. Antibiotics should be applied in humans based on clinical diagnosis by registered doctors.
2. Good animal husbandry practices are required for all kinds of livestock.
3. Antibiotic residue levels can be comprehensively monitored by a routine analysis and kept in a record file to register.
4. Narrow-spectrum antibiotics can be used for the first trial for all antibacterial diseases except serious issues or ineffective functions.
5. Developed a monitoring program with state of art equipment.
6. The withdrawal period of antibiotics should be maintained properly.

7. More Research laboratories with available facilities are required with high sensitivity and rapid performance for analyzing residual antibiotics in food.

Withdrawal Period

In general, the withdrawal period counts as the minimum time from administering the last dose of medication or the drug concentration falling below the tolerance in humans and animals. The time between taking the last dose of antibiotic and food animals slaughtered for food safely. For example, if anybody suffers anxiety, insomnia, fatigue, headaches, etc., after consuming food, it is a common sign of withdrawal period of prescribed medication or food. The withdrawal period of each antibiotic is varied (Khatun et al., 2018; Menkem et al., 2018) one from another (Table 13). So, maintenance of appropriate withdrawal periods is a part of food safety.

Table 13. Withdrawal periods of antibiotics

Antibiotics	Withdrawal periods (days)
Ampicillin	6–15
Erythromycin	2–14
Oxytetracycline	15–35
Tetracycline	5
Enrofloxacin	7
Ciprofloxacin	7
Sulfonamide	4
Furazolidone	28
Chloramphenicol	14
Gentamicin	14

EFFECT OF ANTIBIOTIC RESISTANCE IN HUMAN AND LIVESTOCK

Antibacterial resistance is becoming a global health burden and challenge for a serious health issue all over the world including in Bangladesh. The inappropriate application of antibiotics in humans and the livestock industry has expanded different types of diseases as well as adverse effects. Moreover, humans can be affected by antibiotic resistance for the abundant uses of antibiotics in humans and animals frequently which makes it difficult to investigate their direct or indirect entrance through medicinal consumption for sickness, via food animals, consuming cross-contaminated food including fruits and vegetables or affecting spreading resistance from others. It has a worsened impact on human health and livestock. Antibacterial resistance is spreading from animal to human day by day, fixing the extreme illness and rate of death. In hospitals, proper doses of antibiotics do not respond in critical care points because of converting ineffective drugs into superbugs. Most of the time, antibacterial resistance occurs due to inappropriate and incomplete doses of antibiotics, misuse and overuse of antibiotics, application of poor-quality drugs, unrestricted access to antibiotics, use of non-human and unregistered antibiotics.

Antibiotics are not directly treated in fish, but they are controlling the population growth of bacteria in fish which promotes their immune system. It is the route of contaminations of antibiotics through unusual usages (Savic, 2018). The direct adverse impact of antibacterial resistance has been observed in poultry and cattle firms like production quality losses, resulting in diminished food security, increasing the risk of spreading diseases, and persistence of infections in the animal's body.

Antibacterial resistance leads to

1. Increase higher medical costs and mortality
2. Prolonged hospital stays
3. Ineffective treatments
4. Higher number of infections
5. Biggest threats to global health and food insecurity
6. Long-term hazards affect human health
7. Suffer allergic or hypersensitivity reactions
8. Indirect microbiological effects
9. Reproductive disorders in humans and animals
10. Allergic reactions, hypersensitivity, fatigue
11. Bone marrow toxicity
12. Damage to the targeted organ (kidney and liver)
13. Become harder to treat
14. Causes of occurring dermatitis, gastrointestinal indications
15. Occurrence of typical lymphocytes, lung congestion and toxic granulation of granulocytes, damage to calcium-rich organs (teeth and bones), increased Inflammation, and asthmatic attack
16. Causes of depression, impaired hearing, vision, taste, smell, etc.

CONCLUSION

A humanitarian threat is an event or series of events regarding the crisis of safety and healthy life, security, or natural or man-made disasters within a national and international range. As a result, it may create short- or long-term negative effects on the earth. So, every humanitarian threat requires a particular targeted response to achieve a better solution. Antibiotic resistance is a major threat worldwide and food safety is a great responsibility for all. They are the milestone invention for both basic and highly specialized medicines of all health sectors. However, antibiotic resistance is an alarming humanitarian threat to achieving antibiotic benefits, which is reflected in the medical sector and a burden on the sustainable health-related economy. Therefore, we must manage sustainable access to effective antibiotics and fundamental changes by creating awareness for all, especially the scientific community, prescribers, regulators, national and international policymakers, political circles, stakeholders, and consumers. More investment in the health sector, international declarations, agreements, and initiatives will increase instrumental facility and monitoring for residual antibiotic analysis. Moreover, the threat of antibiotic resistance can be mitigated through awareness of antibiotic usage, improving farm management, paying attention to the health status of farms to supplied animals, not engaging in excessive usage of antibiotics by consumers, and other new generation remedies approach for antibacterial treatments. Therefore, careful and responsible usage of antibiotics is a vital part of the excellent approach to enhancing animal health and food safety by reducing harmful effects and improving positive responses to antibiotics.

ACKNOWLEDGMENT

The authors thank the International Science Programme (ISP), Uppsala University, Sweden and University Grant Commissions of Bangladesh for financial support. Rafiza Islam is thankful to Bangabandhu Science and Technology Fellowship Trust, Ministry of Science and Technology, Government of the People's Republic of Bangladesh for Ph.D. Fellowship.

REFERENCES

Anene, A., Hosni, K., Chevalier, Y., Kalfat, R., & Hbaieb, S. (2016). Molecularly imprinted polymer for extraction of patulin in apple juice samples. *Food Control*, *70*, 90–95. doi:10.1016/j.foodcont.2016.05.042

Bagheri, N., Khataee, A., Habibi, B., & Hassanzadeh, J. (2018). Mimetic Ag nanoparticle/Zn-based MOF nanocomposite capped with molecularly imprinted polymer for the selective detection of patulin. *Talanta*, *179*, 710–718. doi:10.1016/j.talanta.2017.12.009 PMID:29310298

Bartlett, J. G. (2002). Antibiotic-Associated Diarrhea. *The New England Journal of Medicine*, *346*(5), 334–339. doi:10.1056/NEJMcp011603 PMID:11821511

Britannica, The Editors of Encyclopaedia. (2021). Sulfa drug. *Encyclopedia Britannica*. https://www.britannica.com/science/sulfa-drug

Centers For Disease Control and Prevention (CDC) Report 2019. https://www.cdc.gov/about/24-7/2019-Progress-Report.html

Chattopadhyay, M. K. (2014). Use of antibiotics as feed additives: A burning question. *Frontiers in Microbiology*, *5*. Advance online publication. doi:10.3389/fmicb.2014.00334 PMID:25071747

Cheong, C. K., Hajeb, P., Jinap, S., & Ismail-Fitry, M. R. (2010). Sulfonamides determination in chicken meat products from Malaysia. *International Food Research Journal*, *17*, 885–892.

Clark, S. B., Turnipseed, S. B., Madson, M. R., Hurlbut, J. A., Kuck, L. R., & Sofos, J. N. (2005). Confirmation of Sulfamethazine, Sulfathiazole, and Sulfadimethoxine Residues in Condensed Milk and Soft-Cheese Products by Liquid Chromatography/Tandem Mass Spectrometry. *Journal of AOAC International*, *88*(3), 736–743. doi:10.1093/jaoac/88.3.736 PMID:16001847

Codex Alimentarius Commission no CX/MRL 2-2018, updated 2020. https://www.fao.org/fao-who-codexalimentarius/codex-texts/dbs/vetdrugs/en/

Commission Regulation (European Union) no 37/2010, updated 2018. https://ec.europa.eu/health/sites/default/files/files/eudralex/vol-5/reg_2010_37/reg_2010_37_en.pdf

Dibner, J. J., & Richards, J. D. (2005). Antibiotic growth promoters in agriculture: History and mode of action. *Poultry Science*, *84*(4), 634–643. doi:10.1093/ps/84.4.634 PMID:15844822

Dowling, A., Dwyer, J. O., & Adley, C. C. (2017). *Antibiotics: Mode of action and mechanisms of resistance.* In A. Mendez-Vilas (Ed.), *Antimicrobial research: Novel bio knowledge and educational programs* (pp. 536–545). https://www.researchgate.net/publication/317381477_Antibiotics_Mode_of_action_and_mechanisms_of_resistance

El-Sayed, M. G. A., El-Komy, A. A. A., El-barawy, A. M., & Mustafa, G. E. A. (2014). Pharmacokinetical Interactions of Amoxicillin and Amprolium in Broiler Chickens. *Journal of Physiology and Pharmacology Advances*, *4*(12), 515–524. doi:10.5455/jppa.20141208013007

Food safety regulation, Bangladesh food safety Authority, 2017. https://www.fas.usda.gov/data/bangladesh-bangladesh-issues-regulation-food-safety-2017

Fuoco, D. (2012). Classification Framework and Chemical Biology of Tetracycline-Structure-Based Drugs. *Antibiotics (Basel, Switzerland)*, *1*(1), 1–13. doi:10.3390/antibiotics1010001 PMID:27029415

Hoque, R., Ahmed, S. M., Naher, N., Islam, M. A., Rousham, E. K., Islam, M. A., & Hassan, S. (2020). Tackling antimicrobial resistance in Bangladesh: A scoping review of policy and practice in human, animal and environment sectors. *PLoS One*, *15*(1), e0227947. doi:10.1371/journal.pone.0227947 PMID:31986167

ICH Q2. (2005). *Validation of Analytical Procedures: Text and Methodology*. https://www.ema.europa.eu/en/ich-q2-r1-validation-analytical-procedures-text-methodology

Kebede, G., Zenebe, T., Disassa, H., & Tolos, T. (2014). Review on Detection of Antimicrobial Residues in Raw Bulk Milk in Dairy Farms. *African Journal of Basic & Applied Sciences*, *6*(4), 87–97. doi:10.5829/idosi.ajbas.2014.6.4.8642

Khatun, R., Howlader, A. J., Ahmed, S., Islam, N., Alam, K., Haider, S., Mahmud, M. S., & Hasan, M. A. (2018). Validation of the Declared Withdrawal Periods of Antibiotics. *Universal Journal of Public Health*, *6*(1), 14–22. doi:10.13189/ujph.2018.060103

Lambrin, K., Areti, K., Kourkourikos, K., Iliadis, C., Plati, P., & Dimitriadou, A. (2018). History of Antibiotics. *Sumerianz Journal of Medical and Healthcare*, *1*(2), 51–54.

Landers, T. F., Cohen, B., Wittum, T. E., & Larson, E. L. (2012). A Review of Antibiotic Use in Food Animals: Perspective, Policy, and Potential. *Public Health Reports*, *127*(1), 4–22. doi:10.1177/003335491212700103 PMID:22298919

Li, M., Chen, W., Zhang, Z., Zhang, Z., & Peng, B. (2018). Fermentative degradation of Patulin by Saccharomyces cerevisiae in aqueous solution. *Lebensmittel-Wissenschaft + Technologie*, *97*, 427–433. doi:10.1016/j.lwt.2018.07.040

Liu, M., Wang, J., Yang, Q., Hu, N., Zhang, W., Zhu, W., Wang, R., Suo, Y., & Wang, J. (2019). Patulin removal from apple juice using a novel cysteine-functionalized metal-organic framework adsorbent. *Food Chemistry*, *270*, 1–9. doi:10.1016/j.foodchem.2018.07.072 PMID:30174021

Mahato, D. K., Kamle, M., Sharma, B., Pandhi, S., Devi, S., Dhawan, K., Selvakumarf, R., Mishra, D., Kumar, A., Arora, S., Singh, N. A., & Kumar, P. (2021). Patulin in food: A mycotoxin concern for human health and its management strategies. *Toxicon*, *198*, 12–33. doi:10.1016/j.toxicon.2021.04.027 PMID:33933519

Mahunu, G. K., Zhang, H., Apaliya, M. T., Yang, Q., Zhang, X., & Zhao, L. (2018). Bamboo leaf flavonoid enhances the control effect of Pichia caribbica against Penicillium expansum growth and patulin accumulation in apples. *Postharvest Biology and Technology*, *141*, 1–7. doi:10.1016/j.postharvbio.2018.03.005

Menkem, Z. E., Ngagom, B. L., Shinwin, S., Tamunioh, A., & Boyom, F. (2018). Antibiotic residues in food animals: Public health concern. *Acta Ecologica Sinica*, *39*(5), 411–415. doi:10.1016/j.chnaes.2018.10.004

Mou, S. A., Islam, R., Shoeb, M., & Nahar, N. (2021). Determination of chloramphenicol in meat samples using liquid chromatography–tandem mass spectrometry. *Food Science & Nutrition*, *9*(10), 1–6. doi:10.1002/fsn3.2530 PMID:34646535

Nelson, M.L. & Ismail, M Y. (2007). Comprehensive Medicinal Chemistry. *The Antibiotic and Nonantibiotic Tetracyclines,* 597–628. doi:10.1016/B0-08-045044-X/00221-2

Patel, S. J., Wellington, M., Shah, R. M., & Ferreira, M. J. (2020). Antibiotic Stewardship in Food-Producing Animals: Challenges, Progress, and Opportunities. *Clinical Therapeutics.* doi:10.1016/j.clinthera.2020.07.004

Patrick, G. L. (2013). *An Introduction to Medicinal Chemistry* (5th ed.). Oxford University Press.

Perez, N., Gutierrez, R., Nao, M., Diaz, G., Luna, H., Escobar, I., & Muniev, Z. (2002). Liquid chromatographic determination of multiple sulfonamides, nitrofurans, and chloramphenicol residues in pasteurized milk. *Journal of AOAC International*, *85*(1), 20–24. doi:10.1093/jaoac/85.1.20 PMID:11878601

Savic, S. (2018). Introductory Chapter: Antibiotic Use in Animals Today. In Antibiotic Use in Animals. doi:10.5772/intechopen.73076

Singh, G. S. (2020). Carbohydrate-based antibiotics: Opportunities and Challenges. In *Carbohydrates in Drug Discovery and Development* (pp. 523–559). Synthesis and Application. doi:10.1016/B978-0-12-816675-8.00013-0

Sridhara, S., Turbett, S. E., Harrisd, J. B., & Larocque, R. C. (2021). Antimicrobial-resistant bacteria in international travelers. *Infectious Diseases*, *34*(5), 423–431. doi:10.1097/QCO.0000000000000751 PMID:34267046

Stanley, S. L., Jr. (2010). Antiparasitic agents. Infectious Diseases, 2, 1490–1507. doi:10.1016/B978-0-323-04579-7.00150-7

Tirmazi, S. A. A. S., Qadir, M. A., Ahmed, M., Imran, M., Hussain, R., Sharif, M., & Muddassar, M. (2021). Levofloxacin and sulfa drugs linked via Schiff bases: Exploring their urease inhibition, enzyme kinetics and in silico studies. *Journal of Molecular Structure*, *1235*, 130226. doi:10.1016/j.molstruc.2021.130226

Uekane, T. M., Aquino Neto, F. R., & Gomes, L. N. F. (2011). Development and validation of a method for the analysis of tetracyclines in chicken-muscle by liquid chromatography-electrospray-mass spectrometry in tandem (LC-ESI-MS/MS). *Quimica Nova*, *34*(1), 43–48. doi:10.1590/S0100-40422011000100009

Ventola, C. L. (2015). *The Antibiotic Resistance Crisis. Part 1: Causes and Threats*. https://pubmed.ncbi.nlm.nih.gov/25859123/

Vuran, B., Ulusoy, H. I., Sarp, G., Yilmaz, E., Morgül, U., Kabir, A., Tartagla, A., Locatelli, M., & Soylak, M. (2021). Determination of chloramphenicol and tetracycline residues in milk samples by means of nanofiber coated magnetic particles prior to high-performance liquid chromatography-diode array detection. *Talanta*, *230*, 122307. doi:10.1016/j.talanta.2021.122307 PMID:33934773

Wang, S., Zhang, H. Y., Wang, L., Duan, Z. J., & Kennedy, I. (2006). Analysis of sulphonamide residues in edible animal products: A review. *Food Additives and Contaminants*, *23*(4), 362–384. doi:10.1080/02652030500499359 PMID:16546883

WHO. (2020). *Antibiotics Resistance*. https://www.who.int/news-room/fact-sheets/detail/antibiotic-resistance

WHO. (2021). *Antimicrobial Resistance*. https://www.who.int/news-room/fact-sheets/detail/antimicrobial-resistance

Wikipedia.org. (n.d.). *Patulin*. https://en.wikipedia.org/wiki/Patulin

Yang, G., Bai, Y., Wu, X., Sun, M., Liu, X., Yao, X., Zhang, C., Chu, Q., Jiang, L., & Wang, S. (2018). Patulin induced ROS-dependent autophagic cell death in Human Hepatoma G2 cells. *Chemico-Biological Interactions*, 24–31. Advance online publication. doi:10.1016/j.cbi.2018.03.018 PMID:29604266

Yu, H., Tao, Y., Chen, D., Wang, Y., Huang, L., Peng, D., Dai, M., Liu, Z., Wang, X., & Yuan, Z. (2011). Development of a high-performance liquid chromatography method and a liquid chromatography–tandem mass spectrometry method with the pressurized liquid extraction for the quantification and confirmation of sulfonamides in the foods of animal origin. *Journal of Chromatography. B, Analytical Technologies in the Biomedical and Life Sciences*, *879*(25), 2653–2662. doi:10.1016/j.jchromb.2011.07.032 PMID:21840270

Zimdahl, R. L. (2015). Antibiotics. *Six Chemicals That Changed Agriculture*, 165–182. doi:10.1016/b978-0-12-800561-3

ADDITIONAL READING

Barnes, J., Anderson, L. A., & Phillipson, J. D. (2007). *Herbal Medicine* (3rd ed.). Chicago Pharmaceutical Press.

Cosgrove, S. E., & Carmeli, Y. (2003). The Impact of Antimicrobial Resistance on Health and Economic Outcomes. *Clinical Infectious Diseases*, *36*(11), 1433–1437. doi:10.1086/375081 PMID:12766839

Friedman, N. D., Temkin, E., & Carmeli, Y. (2016). The negative impact of antibiotic resistance. *Clinical Microbiology and Infection*, *22*(5), 416–422. doi:10.1016/j.cmi.2015.12.002 PMID:26706614

Hanson, J. R. (2006). *Chemistry and Medicine: An Introductory Text.* The Royal Society of Chemistry. doi:10.1039/9781847555403

Kar, A. (2007). Medicinal Chemistry (4th ed.). New Age International (P) Ltd., Publishers.

Lemke, T. L., Williams, D. A., Roche, V. F., & Zitto, S. W. (2008). Foye's Principal of Medicinal Chemistry (6th ed.). Lippincott Williams & Wilkins, a Wulters Kluwer Business.

Neu, H. C. (1992). The Crisis in Antibiotic Resistance. *Science*, *257*(5073), 1064–1072. doi:10.1126cience.257.5073.1064 PMID:1509257

Sandle, T. (2016). *Pharmaceutical Microbiology, Essential for Quality Assurance and Quality Control.* Woodhead Publishing.

Thomas, G. (2003). *Fundamentals of Medicinal Chemistry*. John Wiley & Sons Ltd.

Ventola, C. L. (2015). The Antibiotic Resistance Crisis, Part 1: Causes and Threats. *P&T*, *40*(4), 277–283. PMID:25859123

KEY TERMS AND DEFINITIONS

Adverse Effects: Of antibiotics mean a negative effect that can be observed after treating with antibiotics if the proper doses are not maintained.

Antibiotic Resistance: Occurs when effectiveness of antibiotics is invalid for bacterial infectious diseases and resistant bacteria are also multiplying cell division.

Humanitarian Threat: Is also called humanitarian emergence which represents a critical threat to the health, safety, security, or wellbeing of a community usually over a large area.

Humanity: Means human well-being from natural or man-made disasters.

Maximum Residue Limit (MRL): Means the maximum acceptable amount of residual antibiotics in edible tissues of animals set by national and international authorities BFSA, CODEX, EU, FAO and WHO that must maintain for global food safety.

Veterinary Drugs: Are antibiotics usually used in livestock for treating various types of bacterial diseases and enhancing animal growth.

Withdrawal Periods: Of antibiotics mean a fixed time for each antibiotic that does not remain in the edible tissues of animals.

Chapter 6
A Review for Neuroimaging Techniques in Multimedia Learning:
An Experimental Framework

Pınar Ozel
Nevsehir Hacı Bektas Veli University, Turkey

Duygu Mutlu Bayraktar
https://orcid.org/0000-0002-2276-3768
Istanbul University-Cerrahpaşa, Turkey

Tugba Altan
Faculty of Education, Kahramanmaras Sutcu Imam University, Turkey

Veysel Coskun
https://orcid.org/0000-0002-7189-2363
Faculty of Education, Hatay Mustafa Kemal University, Turkey

Ali Olamat
Faculty of Engineering, Yıldız Teknik University, Turkey

ABSTRACT

The purpose of this study was to examine neuroimaging technologies for quantifying cognitive load in intelligent interactive multimedia systems for experimental applications by conducting a systematic review of all relevant papers published up to April 2020. The study's most striking finding is that electroencephalography, functional magnetic resonance imaging, functional near-infrared spectroscopy, and transcranial doppler ultrasonography are the most frequently used neuroimaging equipment in cognitive load research in multimedia learning. Forty papers were selected depending on the equipment that should be understood in the field of neuroimaging in examining cognitive load in multimedia learning, the benefits and drawbacks of neuroimaging devices, and the experimental protocol for cognitive load in multimedia research. The study's findings were analyzed, and numerous discrepancies in the research on cognitive load and multimedia learning were discovered.

DOI: 10.4018/978-1-6684-5619-4.ch006

INTRODUCTION

Cognitive Load (CL) Theory is a significant framework that gives instruction recommendations based on current knowledge about human cognition (Sweller, 2020). The theory intends to identify the ways information processing load inducted by learning tasks affects students' ability to process new information and construct it in long-term memory (Sweller, J., van Merriënboer, J. J. G., & Paas, F., 2019). According to the theory, any new information is primarily processed by working memory and stored in unlimited long-term memory (Anmarkrud, Ø., Andresen, A., & Bråten, I., 2019) (Sweller, J., van Merriënboer, J. J. G., & Paas, F., 2019). Working memory capacity is assumed to be limited (Baddeley A., 2012), and only limited items can be processed at a time (Cowan, N., 2001) (Miller, G. A., 1956). Cognitive overload occurs when the learner exceeds the working memory's capacity, restricted to processing the information. The CL theory thus aims to prevent this problem by efficiently designing the learning environment to optimize the limited working capacity and enhance the acquisition of knowledge (Sweller, J., van Merrienboer, J. J. G., & Paas, F., 1998).

CL Theory assumes three types of CL; extraneous CL, intrinsic CL, and germane CL (Paas, F., Renkl, A., & Sweller, J. C, 2003). The intrinsic CL increases with the large number of elements coming to the working memory for schema construction (element interactivity) to be processed simultaneously. The extraneous CL is explained with multimedia objects and design, and it results from inadequately designed learning materials. On the other hand, the degree of mental effort with schemas construction is related to germane load. And it is relevant to motivation and interest factors (Sweller, J., 2010) (Sweller, 2020). The intrinsic and extraneous CL reflect design factors, and those differ from the germane load, which would be explained in terms of subjective experiences. The revised CL theory model (Sweller, 2020) includes just two types: extraneous and intrinsic load. The deactivation of the germane load was because of the close interaction between the intrinsic and germane CLs, which resulted in the incapability to differentiate the distinctive impact of different factors on the overall CL. The germane CL is still known to be germane resources representing the amount of working memory capacity allocated to learning. Besides, CL triggered by the pertinent information processing and learning strategies is integrated into CL factor (Korbach, A., Brünken, R., & Park, B., 2017).

CL theory is primarily concerned with enhancing complex cognitive tasks by transforming current scientific knowledge to make the cognitive structures and process the guidelines for instructional design (Sweller, J., van Merriënboer, J. J. G., & Paas, F., 2019). Therefore, it has contributed to the field of multimedia learning as well. Multimedia learning happens when mental representations are constructed through pictures and words (Mayer, R., 2014b). While words could be in printed (e.g., on-screen text) or verbal (e.g., narration) forms, the pictures could be in static (graphs, illustrations, photos, charts, or maps) or dynamic (e.g., video, animation, or interactive illustrations) forms (Mayer, R. E., Moreno, R., 2003). The instructional design utilized in multimedia learning should be appropriate for the individual's cognitive processing, and it should avoid overloading memory during learning. The Cognitive Theory of Multimedia learning is built on CL theory and formed based on previous studies. The ways individuals process information and learn through multimedia approaches are addressed in theory (Mayer, R., 2014a) which includes three basic assumptions:

1) processing visual and audio information is performed through separate channels,
2) there is a limited amount of information per unit of time for each channel, and

3) active learning is realized by accessing related information, organizing the selected information via mental structures, and integrating these structures with previous ones.

The theory suggests the following for the realization of multimedia learning (Mayer, R., 2014a): sensory memory with unlimited capacity selects the words and pictures and then transfers it to working memory; working memory, which is limited, organizes the knowledge and integrates it in long-term memory.

The aim of this study is to examine cognitive load measurement methods in multimedia learning studies from a different perspective. Such that, differences in measurement methods due to technological developments can serve different purposes by showing different parameters in neuroimaging experiments. Knowing these differences facilitates the experiment process before starting the experiment design and ensures a conscious experiment. Therefore, designing an neuroimaging experiment by knowing the technological differences or purposeful differences in studies using neuroimaging methods will give the reader an idea for their own experimental design.

For this purpose, before the reader starts a neuroimaging experiment to measure the CL in the multimedia learning area, research questions are presented and answered in section 3 in order to find answers to the question marks in a reader's mind and to make comparisons between neuroimaging technologies. Accordingly, the research questions presented are as follows.

RQ1. What kind of operations and parameters are followed during the experimental process?
RQ2. What are the advantages and disadvantages of the neuroimaging devices used in CL in multimedia learning research?

METHODOLOGY

A literature review evaluates, identifies, and harmonizes the research studies that will be used to summarize the study topic based on existing research (Oakley, A., 2012). This outstanding experience uses a comprehensive review of the literature to address the research questions. The systematic assessment procedure was carried out in three phases: (1) preparation, (2) examination, and (3) presentation of the findings. During the preparation phase, the indexes for browsing for research studies, the inclusion and exclusion criteria, and the study categories were established. During the study's second phase, research studies were selected and investigated. After then, the data was analyzed, synthesized, and encoded. In the last phase, the study findings, debate of observations, trends, and evaluation opinions were all recorded.

Search Strategy

The research parameters, such as general, specific, and exclusive, are determined once the study's research questions are specified.

Inclusive Criteria

General Criteria

- The research should be documented in the form of articles or conference papers.

- The research should have included in Web of Science (WOS), Education Resources Information Center (ERIC), Scopus, and EBSCO databases.
- The full text ones among research should be considered.
- Articles and meetings papers should be written in English.

Selection Criteria

- The research should contain a multimedia learning environment.
- In the research, CL should be measured using neuroimaging techniques.

Search for papers should be done using keywords like "Multimedia," "Cognitive load," "Multimedia learning," "fMRI," "EEG," "Magnetoencephalography," "MEG," "fNIRS," "TDCS," "Transcranial doppler," "Transcranial electrical stimulation," "Transcranial magnetic stimulation," and "TMS."

Exclusive Criteria

The research should not involve a review, commentary article or meta-analysis. There should be no books, theses, or book chapters chosen.

The Process of Choosing a Study

The research was conducted in April 2020 and focused on keywords in selected databases, consisting of a total of 892 studies. The study's findings were recorded in an excel spreadsheet and stored. This list has been updated to remove 12 duplicate research from multiple databases. The 880 records on the list were examined using both inclusive and exclusive criteria. Because they did not follow the study's parameters, 773 studies were deemed unacceptable for the study. The research questions were used as a roadmap for the rest 107 studies. During this phase, 65 papers were deleted from the report because they did not meet the review criteria, and one study did not provide a full text accessible. As a consequence, for this extensive examination, 40 papers were picked. Among the papers were 23 journal articles and 17 conference papers.

Data Coding and Analysis Categories

There were multiple categories for each of the research questions. The researchers then used these categories to encode the study data before evaluating it. Moreover, the results of the examination were considered as the following research questions as stated in Section 1.

RESULTS

What Kind Of Operations And Parameters Are Followed During The Experimental Process?

Considering the experimental process, many parameters such as the number of people participating in the experiment, the average age of the participant, the recording time, which channels are used, which brain imaging method is used, the brand of the brain imaging device, and the sampling rate, and more have an influence on the outcomes. We summarized the operations and parameters that we identified in the experiment processes in Table 1 to make them easier to be interpreted.

What Are The Advantages And Disadvantages Of The Neuroimaging Devices Used In CI In Multimedia Learning Research?

Electroencephalography (EEG)

Due to the existing technological limitations, it may be difficult to analyze appropriately the gathered data based on activated brain regions in large samples. The data collection (e.g., utilizing electrodes gel) may not be pleasant for volunteers and may restrict data collection (Uysal, M. P., 2016). However, EEG has adequate time resolution to monitor changes in mental health as complex activities unfold. Additionally, EEG can be tracked outside specific laboratory environments due to the related equipment's compactness, although some functional neuroimaging tools involve great machines and equipment, large teams of specialists, and full immobilization of the participant. EEG is a non-invasive neuro-imaging tool and is widely utilized to measure and monitor the brain's numerous cognitive activities, workload, and behavioral state. EEG evaluates the electrical activity due to variation in voltage across the scalp in the brain region and is convenient to use it and records brain oscillations across the scalp, and its commercial presence is rapid and far less expensive. Not only does EEG show features of brain function, but it also represents simple neural processes (utilizing appropriate quantifiers). The temporal resolution of EEG is very high in comparison to many other neural technologies. The fMRI is not suitable for measuring time scales ranging from 1 ms to 100 ms due to its inadequacy in calculating many cognitive tasks (Mazher, M., Abd Aziz, A., Malik, A. S., Ullah Amin, H., 2017). Among neuroscience techniques, EEG is the most common and preferred approach since it can offer noninvasive measurements for brain activity. Not only fMRI demands subjects to lie in restricted positions, but also its evaluation process is long, and PET requires subjects to ingest possible dangerous materials (Mazher, Moona, Aziz, A. Abd., Malik, A. S., Qayyum, A., 2015a).

Functional Near-Infrared Spectroscopy(fNIRS)

There are many other optical mechanisms designed to measure changes in the human brain. MEG, fMRI, and PET are well-established techniques. However, these mechanisms can have major limitations. Participants can, for instance, be exposed to potentially hazardous machinery. Some of the technologies can be quite costly. Besides, participants are limited to restricted positions because the equipment is extremely sensitive to movement artifacts, and thus these mechanisms may not be appropriate in

Table 1. The most striking operations and parameters for the experimental process for neuroimaging data on CL in multimedia learning research

Authors	Experimental Process					
	Age range or mean	Number of Samples	Neuroimaging Tool And Brand of it.	Sampling Rate EEG/fNIRS	Recording Time	EEG/fNIRS Channels
Dan & Reiner (2017)	29.76	17	EEG (Mistar-EEG 201 wired monitoring device)		5 min	Fz, Pz
Zahabi et al. (2020)	63.1		fNIRS (Techen CW6 system, Techen Inc. MA, USA)			AF7 (C3,4,5) and FP1 (C6,7,8) in the left hemisphere and AF8 (C1,2,9) and FP2 (C10,11,12) in the right hemisphere
Fernandez Rojas et al. (2020)	22-50	16(4 females)	EEG (Emotiv EPOC)	128 samples per second.	1 min-Baseline, 2 min-Break 4 min-Each Session 50 min-Complete Session	AF3, F7, F3, FC5, T7, P7, O1, O2, P8, T8, FC6, F4, F8, AF4, M1, and M2.
Castro-Meneses et al. (2020)	25.11	35	EEG (Emotiv EPOC+, SensoMotoric Instruments (SMI) Experiment Centre software package.)	256Hz	4 min	AF3, AF4, F3, F4, F7, F8, T7, T8, P3, P4, P7, P8, O1, and O2.1
Liu et al. (2020)	1.session 18–24 2. session 18–24	1. session 30 2. session 48	fMRI (3T Siemens Trio scanner)	-	the first and second session, respectively: resting-state fMRI scan (6 min- 6 min), studying materials inside scanner (14 min) and fMRI scan with test questions (6 min), and a structural MRI scan (5 min- 5 min).	-
Wang et al. (2020)	18.36	60 (21 male, 39 female)	EEG (DSI-24 EEG system (Wearable Sensing, San Diego, CA).)	300 Hz	240 seconds	Fp1, Fp2, F3, Fz, F4, F7, F8, P3, Pz, P4, T3, T4, T5, T6, C3, Cz, C4, O1, and O2 with respect to mastoid electrodes (A1 and A2) in common reference.
Balamurugan et al. (2020)	22-33	38	EEG (NeuroSky MindWave EEG device)			
Eldenfria & Al-Samarraie (2019)	20-25	41(36 male and 5 female)	EEG (Emotive)			
Makransky et al. (2019b)	19-45 (23.49)	78	EEG (ABM EEG system)	256 Hz		(F3, Fz, F4, C3, Cz, C4, P3, POz, P4),
Bernhardt et al. (2019)	20.9	47 (two women and 45 men)	EEG (Advance Brain Monitoring (ABM) B-Alert X-24 wireless Bluetooth system)	256 Hz		Fp1, Fp2, F7, F3, Fz, F4, F8, T3, C3, Cz, C4, T4, T5, P3, Pz, P4, T6, O1, POz, and O2. Reference electrodes were placed at the mastoids
Makransky et al. (2019a)	19 to 42 (M=23.8)	52 (22 males and 30 females)	EEG	256 Hz	9 min	F3, F4, Fz C3, C4, Cz P3, P4, POz
Örün & Akbulut (2019)	22.27	129 (66 females 63 males)	EEG (A wireless, 14-channel Emotiv EPOC + headset)	256	9 min	F3, F4, F7, F8
Loftus et al. (2018)	29	29	TCD (Neurovision system, Multigon Industries, Elmsford, CA, USA)	1000 Hz	30 seconds	
Dan & Reiner (2018)	29.71	14	EEG (MITSAR equipment EEG device WinEEG software)			Fz, Pz
Reinerman-Jones et al. (2017)	19.62	38	EEG (Advanced Brain Monitoring (ABM) B-Alert X10 system)	256 Hz		Fz, F3, F4, Cz, C3, C4, Pz, P3, and P4 with reference electrodes placed at each mastoid
Mazher et al. (2017)	20-30	34	EEG (HydroCel Geodesic Sensor Net (Electrical Geodesic Inc., Eugene, OR, USA)	250 Hz	5 min	the frontal region channels (9, 11, 22, 24, 33, 122, and 124)
Loftus et al. (2017)			TCD ((Neurovision system, Multigon Industries, Elmsford, CA, USA))		30 seconds	
.Lee(2014		43	EEG (QEEG KNF-101 System)			F7 and T3
Kruger et al. (2014)		68	EEG (Emotiv™ Neuro-headset EEG)		14 minutes	
Ortiz De Guinea et al. (2013)		Study 1 17, Study 2 24	EEG	256 Hz	Study 1: the participants completed a standard 15-min baseline test Study 2: Each participant had 10 min to resolve this task.	(F3, F4, FZ, C3, C4, Cz, P3, P4 and POz)

continued on following page

Table 1. Continued

Authors	Experimental Process					
	Age range or mean	Number of Samples	Neuroimaging Tool And Brand of it.	Sampling Rate EEG/fNIRS	Recording Time	EEG/fNIRS Channels
Gerlič & Jaušovec (2001)	20.3	40	EEG	500 Hz		(Fp1, Fp2, F3, F4, F7, F8, T3, T4, T5, T6, C3, C4, P3, P4, O1, O2, Fz, Cz, and Pz).
Uysal (2016)	25-28	19	fNIRS. (The COBI software 16-channel sensor headband)			
Alex & Reiner (2017)	24,9	10				Fz, Pz
Loftus et al. (2017)			TCD (Neurovision system, Multigon Industries, Elmsford, CA, USA)			
Chang et al. (2016)	19-25	15	EEG (The NeuroSky brainwave headset)		20 minutes	
Qayyum, Faye, et al. (2018)		34	EEG device (HydroCel Geodesic Sensor Net) (Electrical Geodesic Inc., Eugene, OR, USA)	250 Hz	5 min	(9, 22, 33, 24, 11, 122, 124, 52, 62 and 92) numbered channels
Mazher et al. (2016)		34	The EEG device used for this data was EGI EEG 128 Channel; EGI Inc, USA.	250 Hz	five minutes -baseline five minutes -L1 five minutes -L2 five minutes -L3	The 128 channels EEG device was used for recordings.
Al Madi & Khan (2016)	20-40	16 (8 female, 8 male)				F3, F4, T7,T8,O1,O2
Mazher et al. (2015b)	18-30	10 males	EEG	250 Hz	5 minutes du- non-learning state.. 10 minutes du- EEG recording	(9, 11, 22, 24, 33, 122 and 124) numbered channels
Mazher et al. (2015a)	18-30	5 males	EGI EEG 128 Channel; EGI Inc, USA.	250 Hz	5 min- relaxing. • After 5 minutes, Multimedia Learning (2D animations) was shown to each participant in a controlled environment	F3, F4, F7,F8,Fz, T3,T4, T5,T6, P3,P4, O1,O2
Walter et al. (2013)	Median= 25,20-68	21	EEG ((actiCap, BrainProducts GmbH),)	512 Hz		F3, Fz, F4, FC1, FC2, CP1, CP2, P3, Pz, P4
Cao et al. (2019)		62				
Qayyum, Khan, et al. (2018)	24-30	34	The EEG device (HydroCel Geodesic Sensor Net (Electrical Geodesic Inc., Eugene, OR, USA)	250 Hz	5 Min.	(9, 22, 33, 24, 11, 122, 124, 52, 62 and 92) numbered channels
Zhang & Liu (2017)	27.1	34	EEG, NeuroSky's EEG device. The MindWave headset		15 Min.	
Conrad & Bliemel (2016)	23-30	4	EEG, Muse device		15 Min.	FpZ, AF7, Af8, TP9 and TP10
Díaz et al. (2015)	25-33	12	EEG, EMOTIV EPOC EEG system,			AF3, AF4, F3 and F4.
Crk & Kluthe (2014)	21.1	34	EEG, Emotiv Epoc headset,	128 Hz		
Bertolo et al. (2014)		17				AF3, F7, F3, and FC5
Jan-Louis Kruger et al. (2013)		46	EEG, EMOTIV EPOC EEG system,			

the classroom. fNIRS is, therefore, recognized as more practical than other physiological technologies (Uysal, M. P., 2016).

Furthermore, EEG measurements are subject to the various head or body motion artifacts, and PET and fMRI involve participants to lie supine and immobile during data acquisition. There is, thus, a need for sensitive, continuous, and reliable measures that can distinguish between different rates of mental effort. fNIRS is a reasonably exciting and innovative imaging technique that satisfies these measurement criteria and has the significant advantage of being compact and field-deployable. This mechanism estimates oxygenated (HbO2) and deoxygenated (HHb) hemoglobin in the brain's blood supply and has been shown to distinguish between different levels of mental effort. Compared to the more traditional neuroimaging technologies, fNIRS enables in vivo imaging under ecological conditions with natural, free movement and extreme environments like high-fidelity flight simulators (Castro-Meneses, L. J., Kruger, J.-L., Doherty, S., 2020).

Transcranial Doppler Ultrasonography (TCD)

This approach guided the choice to use transcranial Doppler ultrasonography (TCD) as it is more sensitive to temporal changes in cognitive processing compared to functional neuroimaging like fMRI (Uysal, M. P., 2016). TCD is an entirely noninvasive ultrasonographic technique seen on a beat-to-beat basis for real-time assessment of blood flow velocities in the major basal cerebral arteries. Portability, noninvasive nature, and considerably lower cost make TCD a favored tool for cerebral vascular pathology screening and diagnosis. The other benefits also include relative simplicity. Although the performance cannot be regarded as simple in terms of learning and mastery, comprehension is, in fact, rather uncomplicated. TCD is widely acknowledged as an extension of the stethoscope-like clinical examination. It offers useful structural hemodynamic knowledge with the use of CT or MR angiography. It provides comprehensive information for extended monitoring periods and helps the practitioner recognize these cerebrovascular pathologies' temporal evolution. TCD contains numerous serious drawbacks against some of these advantages.

The first one is that it does not provide clear anatomical details about cerebral vasculature. What is measured for each vessel is the depth, direction of flow, end-diastolic, peak systolic, mean flow velocities. These measurements are achieved on a spectral waveform display (sonogram) delivered through FFT or another kind of mathematical transformation of the signal. Secondly, TCD output is detected through the bone windows. Anterior TCD is conducted for anterior circulation via the temporal bone. In all patients, the bone is not generally absorbent to the sound. Another drawback is the fluctuation of TCD signals due to distal and extracranial proximal artery status and systemic and cardiac physiology and abnormalities (Topcuoglu,M.A., 2012).

Functional Magnetic Resonance Imaging (fMRI)

Thus, why is fMRI preferred in this way? Does it not have any advantage over other brain imaging methods? fMRI is a very helpful tool for mapping and estimating human brain activity intensity, offering knowledge regarding the brain's various functions, and recognizing cerebral pathology. This methodology uses two concepts, one of which is focused on altering the deoxyhemoglobin paramagnet to diamagnetic oxyhemoglobin. One of the fMRI research findings' primary priorities is to provide effectively a functional map of the human vision system, anatomical pathways, and what knowledge they hold. In fMRI studies, the change in the BOLD signal inferred the neural response to the visual stimulus. One disadvantage of this strategy is that fMRI's limited spatial resolution. Compared to multiunit activity (MUA), blood oxygen levels are very local; thus, the image signal is very low compared to statistical noise. Vascular processes such as cardiac activity and respiration that are not connected to the brain's neuronal function add unnecessary signals or physiological noise to the fMRI results. fMRI has focused on neural adaptation (fMRI-A), which determines the regions that are sensitive to various types of movements to resolve the resolution problem. Another approach is the multi-voxel pattern analysis (MVPA), which uses algorithms for pattern classification to discern a specific response to a given set of stimuli. It is smarter than other modalities despite the drawbacks of fMRI. It does not use ionizing radiation to quantify blood flow in the brain, as opposed to PET (Afrasiabi,M., Noroozian,N., 2015)

Compared to electrophysiology, it is noninvasive because it does not involve electrodes' usage inside or near the neurons. In contrast to BOLD fMRI, noninvasive techniques such as MEG and EEG have a higher temporal resolution; however, fMRI has a higher spatial resolution. Combining fMRI with MEG or

EEG could improve the Spatio-temporal resolution of the features captured (Afrasiabi,M., Noroozian,N., 2015) (Ashby, F. G., 2015).

The technologies for functional neuroimaging have strong relevance for cognitive processing studies. fMRI offers great potential in interpreting brain activity across different learning phases among neuroimaging techniques. The utilization of this technique for educational research is questionable and quite often consists of misunderstandings about the findings acquired (Uysal, M. P., 2016). However, it is not completely obvious whether fNIRS is as susceptible as fMRI, although fNIRS could seem more practicable than fMRI. EEG is used in extremely qualitative research and can be favored in CL experiments to fNIRS or fMRI.

CONCLUSION

When taking a closer look at the experimental process in multimedia learning environment studies for CL measurement, it encountered different parameters. Accordingly, 29 studies among 40 studies present their participants' age range or mean. In 37 studies, the number of participants is given; additional information is given about the gender of the participants in 4 studies. When the Neuroimaging tools and brand of them are examined, it is seen that most EEG devices and the Emotiv brand of them for recording systems are used the most. The Neurovision system device was used in TCD studies, which belong to the same author and are quite numerous. Apart from these, the 3T Siemens Trio scanner system is used in the fMRI study, which is also very few. Similarly, Techen CW6 system is used in the fNIRS study, which is also very few. The sampling rate is a parameter used usually for EEG signals. However, in this study, we also recorded the sampling rates to evaluate the differences and see that 250 Hz is used commonly. Regarding the recording time, additional information is given about the recording time in 22 studies, and 20 minutes is the longest recording time. In our EEG / fNIRS channel investigation, 29 studies provide the necessary information by sharing either channel names or channel numbers. It is observed that the frontal region channels are mostly preferred.

Experimental protocols are essential information frameworks that facilitate the definition of the process steps by which experimental research results are produced. So, the components of an experiment protocol should be identified clearly. Furthermore, it is critical to effectively share experimental methods in order to ensure that results can be replicated by others. In addition, it is thought that studies that show the experimental processes visually and clarify the experimental steps will contribute more to the understanding of the purpose of the study.

On the other hand, most of the neuroimaging techniques mentioned above have advantages or disadvantages. However, it is important to choose the work to be done according to the target and available opportunities. In addition, new neuroimaging methods that have emerged in recent years can be followed and new research papers can be created.

Our study is a very comprehensive and detailed study aimed at answering all the questions in the mind of the reader. Other research questions that the reader may have in mind can be found in our first two publications. We invite our researchers who want to have a comprehensive idea in the field of 'Neuroimaging tools in multimedia learning', including this study, to read our two previously published studies (Ozel, 2021, 2022).

REFERENCES

Afrasiabi, M., & Noroozian, N. (2015). Advantages and Limitations of Functional Magnetic Resonance Imaging (fMRI) of the Human Visual Brain. *Horizons in Neuroscience Research Series, 17.*

Al Madi, N. S., & Khan, J. I. (2016). Measuring learning performance and cognitive activity during multimodal comprehension. *7th International Conference on Information and Communication Systems (ICICS)*, 50-55. 10.1109/IACS.2016.7476085

Anmarkrud, Ø., Andresen, A., & Bråten, I. (2019). Cognitive Load and Working Memory in Multimedia Learning: Conceptual and Measurement Issues. *Educational Psychologist, 54*(2), 1–23. doi:10.1080/00461520.2018.1554484

Ashby, F. G. (2015). An Introduction to fMRI. In B. U. Forstmann & E.-J. Wagenmakers (Eds.), *An Introduction to Model-Based Cognitive Neuroscience* (pp. 91–112). doi:10.1007/978-1-4939-2236-9_5

Baddeley, A. (2012). Working memory: Theories, models, and controversies. *Annual Review of Psychology, 63*(1), 1–29. doi:10.1146/annurev-psych-120710-100422 PMID:21961947

Balamurugan, B., Mullai, M., Soundararajan, S., Selvakanmani, S., & Arun, D. (2020). Brain-computer interface for assessment of mental efforts in e-learning using the nonmarkovian queueing model. *Computer Applications in Engineering Education.*

Bernhardt, K. A., Poltavski, D., Petros, T., Ferraro, F. R., Jorgenson, T., Carlson, C., Drechsel, P., & Iseminger, C. (2019). The effects of dynamic workload and experience on commercially available EEG cognitive state metricsin a high-fidelity air traffic control environment. *Applied Ergonomics, 77*, 83–91. doi:10.1016/j.apergo.2019.01.008 PMID:30832781

Bertolo, D., Dinet, J., & Vivian, R. (2014). Reducing cognitive workload during 3D geometry problem solving with an app on iPad. *Science and Information Conference*, 896–900. 10.1109/SAI.2014.6918292

Cao, X., Cheng, M., Xue, X., & Zhu, S. (2019). Effects of Lecture Video Types on Student Learning: An Analysis of Eye-Tracking and Electroencephalography Data. In F. Xhafa, S. Patnaik, & M. Tavana (Eds.), *Advances in Intelligent* (pp. 498–505). Interactive Systems, and Applications. doi:10.1007/978-3-030-02804-6_66

Castro-Meneses, L. J., Kruger, J.-L., & Doherty, S. (2020). Validating theta power as an objective measure of cognitive load in educational video. *Educational Technology Research and Development, 68*(1), 181–202. doi:10.100711423-019-09681-4

Chang, H.-C., Hung, I.-C., Chew, S. W., & Chen, N.-S. (2016). Yet Another Objective Approach for Measuring Cognitive Load Using EEG-Based Workload. *IEEE 16th International Conference on Advanced Learning Technologies (ICALT)*, 501-502.

Conrad, C., Bliemel, M. . (2016). Psychophysiological Measures of Cognitive Absorption and Cognitive Load in E-Learning Applications. *ICIS 2016 Proceedings.*

Cowan, N. (2001). The magical number 4 in short-term memory: A reconsideration of mental storage capacity. *Behavioral and Brain Sciences, 24*(1), 87–114. doi:10.1017/S0140525X01003922 PMID:11515286

Crk, I., & Kluthe, T. (2014). Toward using alpha and theta brain waves to quantify programmer expertise. *36th Annual International Conference of the IEEE Engineering in Medicine and Biology Society*, 5373–5376. 10.1109/EMBC.2014.6944840

Dan, A., & Reiner, M. (2017). Real Time EEG Based Measurements of Cognitive Load Indicates Mental States During Learning. *Journal of Educational Data Mining, 9*(2), 31-44.

Dan, A., & Reiner, M. (2017a). EEG-based cognitive load of processing events in 3D virtual worlds is lower than processing events in 2D displays. *International Journal of Psychophysiology, 122*, 75–84. doi:10.1016/j.ijpsycho.2016.08.013 PMID:27592084

Dan, A., & Reiner, M. (2018). Reduced mental load in learning a motor visual task with virtual 3D method. *Journal of Computer Assisted Learning, 34*(1), 84–93. doi:10.1111/jcal.12216

Díaz, D., Ramírez, R., & Hernández-Leo, D. (2015). The Effect of Using a Talking Head in Academic Videos: An EEG Study. *IEEE 15th International Conference on Advanced Learning Technologies*, 367-369.

Eldenfria, A., & Al-Samarraie, H. T. (2019). Towards an Online Continuous Adaptation Mechanism (OCAM) for Enhanced Engagement: An EEG Study. *International Journal of Human-Computer Interaction, 35*(20), 1960–1974. doi:10.1080/10447318.2019.1595303

Fernandez Rojas, R., Debie, E., Fidock, J., Barlow, M., Kasmarik, K., Anavatti, S., Garratt, M., & Abbass, H. (2020). Electroencephalographic Workload Indicators During Teleoperation of an Unmanned Aerial Vehicle Shepherding a Swarm of Unmanned Ground Vehicle. *Frontiers in Neuroscience, 14*, 40. doi:10.3389/fnins.2020.00040 PMID:32116498

Gerlič, I., & Jaušovec, N. (2001). Differences in EEG power and coherence measures related to the type of presentation: Text versus multimedia. *Journal of Educational Computing Research, 25*(2), 177–195. doi:10.2190/YDWY-U3FJ-4LY4-LYND

Korbach, A., Brünken, R., & Park, B. (2017). Measurement of cognitive load in multimedia learning: A comparison of different objective measures. *Instructional Science, 45*(4), 515–536. doi:10.100711251-017-9413-5

Kruger, J.-L., Hefer, E., & Matthew, G. (2013). Measuring the impact of subtitles on cognitive load: Eye tracking and dynamic audiovisual texts. *Proceedings of the 2013 Conference on Eye Tracking South Africa*, 62-66. 10.1145/2509315.2509331

Kruger, J.-L., Hefer, E., & Matthew, G. (2014). Attention distribution and cognitive load in a subtitled academic lecture: L1 vs. L2. *Journal of Eye Movement Research, 7*(5). Advance online publication. doi:10.16910/jemr.7.5.4

Lee, H. (2014). Measuring cognitive load with electroencephalography and self-report: Focus on the effect of English-medium learning for Korean students. *Educational Psychology, 34*(7), 838–848. doi:10.1080/01443410.2013.860217

Liu, C., Wang, R., Li, L., Ding, G., Yang, J., & Li, P. (2020). Effects of encoding modes on memory of naturalistic events. *Journal of Neurolinguistics, 53*(100863), 100863. doi:10.1016/j.jneuroling.2019.100863

Loftus, J. J., Jacobsen, M., & Wilson, T. D. (2017). Learning and assessment with images: A view of cognitive load through the lens of cerebral blood flow. *British Journal of Educational Technology*, *48*(4), 1030–1046. doi:10.1111/bjet.12474

Loftus, J. J., Jacobsen, M., & Wilson, T. D. (2018). The relationship between spatial ability, cerebral blood flow and learning with dynamic images: A transcranial Doppler ultrasonography study. *Medical Teacher*, *40*(2), 174–180. doi:10.1080/0142159X.2017.1395401 PMID:29113527

Makransky, G., Terkildsen, T. S., & Mayer, R. E. (2019a). Adding immersive virtual reality to a science lab simulation causes more presence but less learning. *Learning and Instruction*, *60*, 225–236. doi:10.1016/j.learninstruc.2017.12.007

Makransky, G., Terkildsen, T. S., & Mayer, R. E. (2019b). Role of subjective and objective measures of cognitive processing during learning in explaining the spatial contiguity effect. *Learning and Instruction*, *61*, 23–34. doi:10.1016/j.learninstruc.2018.12.001

Mayer, R. (2014a). Cognitive Theory of Multimedia Learning. In R. Mayer (Ed.), *The Cambridge Handbook of Multimedia Learning* (2nd ed., pp. 43–72). doi:10.1017/CBO9781139547369.005

Mayer, R. (2014b). Introduction to Multimedia Learning. In R. Mayer (Ed.), *The Cambridge Handbook of Multimedia Learning* (2nd ed., pp. 1–24).

Mayer, R. E., & Moreno, R. (2003). Nine Ways to Reduce Cognitive Load in Multimedia Learning. *Educational Psychologist*, *38*(1), 43–52. doi:10.1207/S15326985EP3801_6

Mazher, M., Aziz, A. A., Malik, A. S., Qayyum, A. (2015a). A statistical analysis on learning and nonlearning mental states using EEG. *IEEE Student Symposium in Biomedical Engineering Sciences (ISSBES)*, 36-40. 10.1109/ISSBES.2015.7435889

Mazher, M., Aziz, A. A., Malik, A. S., Qayyum, A. (2015b). A comparison of brain regions based on EEG during multimedia learning cognitive activity. *IEEE Student Symposium in Biomedical Engineering Sciences (ISSBES)*, 31-35. 10.1109/ISSBES.2015.7435888

Mazher, M., Abd Aziz, A., Malik, A. S., & Ullah Amin, H. (2017). An EEG-Based Cognitive Load Assessment in Multimedia Learning Using Feature Extraction and Partial Directed Coherence. *IEEE Access: Practical Innovations, Open Solutions*, *5*, 14819–14829. doi:10.1109/ACCESS.2017.2731784

Mazher, M., Aziz, A. B. A., Malik, A. S., & Qayyum, A. (2015b). A comparison of brain regions based on EEG during multimedia learning cognitive activity. *IEEE Student Symposium in Biomedical Engineering Sciences (ISSBES)*, 31-35. 10.1109/ISSBES.2015.7435888

Miller, G. A. (1956). The magical number seven, plus or minus two: Some limits on our capacity for processing information. *Psychological Review*, *63*(2), 81–97. doi:10.1037/h0043158 PMID:13310704

Oakley, A. (2012). Foreword. In An Introduction to Systematic Reviews. SAGE Publications.

Ortiz De Guinea, A., Titah, R., & Léger, P.-M. (2013). Measure for Measure: A two study multi-trait multi-method investigation of construct validity in IS research. *Computers in Human Behavior*, *29*(3), 833–844. doi:10.1016/j.chb.2012.12.009

Örün, Ö., & Akbulut, Y. (2019). Effect of multitasking, physical environment and electroencephalography use on cognitive load and retention. *Computers in Human Behavior*, *92*, 216–229. doi:10.1016/j.chb.2018.11.027

Ozel, P. B.-M. (2022). A Review for Neuroimaging Techniques in Multimedia Learning: Methodological Framework. In *AI-Enabled Smart Healthcare Using Biomedical Signals* (p. 26). IGI Global.

Ozel, P. B.-M., Mutlu-Bayraktar, D., Altan, T., Coskun, V., & Olamat, A. (2021). Neuroimaging tools in multimedia learning: A systematic review. *Interactive Learning Environments*, 1–18. doi:10.1080/10494820.2021.1984255

Paas, F., Renkl, A., & Sweller, J. C. (2003). Cognitive Load Theory and Instructional Design: Recent Developments. *Educational Psychologist*, *38*(1), 1–4. doi:10.1207/S15326985EP3801_1

Qayyum, A., Faye, I., Malik, A. S., & Mazher, M. (2018). Assessment of Cognitive Load using Multimedia Learning and Resting States with Deep Learning Perspective. *IEEE-EMBS Conference on Biomedical Engineering and Sciences (IECBES)*, 600-605. 10.1109/IECBES.2018.8626702

Qayyum, A., Khan, M. K. A. A., Mazher, M., & Suresh, M. (2018). Classification of EEG Learning and Resting States using 1D-Convolutional Neural Network for Cognitive Load Assessment. *IEEE Student Conference on Research and Development (SCOReD)*, 1-5.

Reinerman-Jones, L., Barber, D. J., Szalma, J. L., & Hancock, P. A. (2017). Human interaction with robotic systems: Performance and workload evaluations. *Ergonomics*, *60*(10), 1351–1368. doi:10.1080/00140139.2016.1254282 PMID:28745552

Sweller, J. (2010). Element Interactivity and Intrinsic, Extraneous, and Germane Cognitive Load. *Educational Psychology Review*, *22*(2), 123–138. doi:10.100710648-010-9128-5

Sweller, J. (2020). Cognitive load theory and educational technology. *Educational Technology Research and Development*, *68*(1), 1–16. doi:10.100711423-019-09701-3

Sweller, J., van Merrienboer, J. J. G., & Paas, F. (1998). Cognitive architecture and instructional design. *Educational Psychology Review*, *10*(3), 251–296. doi:10.1023/A:1022193728205

Sweller, J., van Merriënboer, J. J. G., & Paas, F. (2019). Cognitive Architecture and Instructional Design: 20 Years Later. *Educational Psychology Review*, *31*(2), 261–292. doi:10.100710648-019-09465-5

Topcuoglu, M. A. (2012). Transcranial Doppler ultrasound in neurovascular diseases: Diagnostic and therapeutic aspects. *Journal ofNeurochemistry*, *123*(S2), 39–51. doi:10.1111/j.1471-4159.2012.07942.x PMID:23050641

Uysal, M. P. (2016). Evaluation of learning environments for object-oriented programming: Measuring cognitive load with a novel measurement technique. *Interactive Learning Environments*, *24*(7), 1590–1609. doi:10.1080/10494820.2015.1041400

Uysal, M. P. (2016). Evaluation of learning environments for object-oriented programming: Measuring cognitive load with a novel measurement technique. *Interactive Learning Environments*, *24*(7), 1590–1609. doi:10.1080/10494820.2015.1041400

Walter, C., Schmidt, S., Rosenstiel, W., Gerjets, P., & Bogdan, M. (2013). Using Cross-Task Classification for Classifying Workload Levels in Complex Learning Tasks. *Humaine Association Conference on Affective Computing and Intelligent Interaction*, 876–881. 10.1109/ACII.2013.164

Wang, J., Antonenko, P., Keil, A., & Dawson, K. (2020). Converging Subjective and Psychophysiological Measures of Cognitive Load to Study the Effects of Instructor-Present Video. *Mind, Brain and Education: the Official Journal of the International Mind, Brain, and Education Society*, *14*(3), 279–291. doi:10.1111/mbe.12239

Zahabi, M., Abdul Razak, A. M., Shortz, A. E., Mehta, R. K., & Manser, M. (2020). Evaluating advanced driver-assistance system trainings using driver performance, attention allocation, and neural efficiency measures. *Applied Ergonomics*, *84*, 84. doi:10.1016/j.apergo.2019.103036 PMID:31987518

Zhang, X., & Liu, S. (2017). Understanding Reading Comprehension in Multi-display Presenting System: Visual Distribution and Cognitive Effect. In C. Stephanidis (Ed.), *HCI International 2017 – Posters' Extended Abstracts* (pp. 207–214). doi:10.1007/978-3-319-58753-0_32

Chapter 7
Pesticides:
Impact on Environmental and Sustainable Management

Mohammad Shoeb
University of Dhaka, Bangladesh

Nilufar Nahar
University of Dhaka, Bangladesh

ABSTRACT

Pesticides are chemical compounds that are toxic and are used to control different types of pests and uncontrolled growth of weeds for plant protection in crop production and food storage. A large number of pesticides are being used all over the world. Pesticides play a great role in food security. Crop production has been increased by controlling insect attack, removing weeds by herbicides, controlling crop loss by rodenticides. To get a high yield of products, a higher amount of fertilizers is also needed. The growth of plants is associated with the growth of weeds by which nutrients for the plants are consumed by herbs. So higher amounts of herbicides are also needed to remove herbs as traditional weed removal by humans has become expensive. The residue of pesticides in the foodstuff is reducing food safety as well as a threat to the environment. Pesticides are ubiquitous in the environment and are found in air, soil, and water. Judicial and sustainable use of pesticides is required for food security and ensuring a healthy environment.

INTRODUCTION

Chemical compounds of different types such as pesticides, antibiotics, plant growth regulators, hormones, etc. are used in crop production, aquaculture, and human medicine. Pesticides are natural or synthetic chemical compounds mainly used in agriculture to protect plants from pests or weeds. They are designed to control pests and diseases of plants, eradicate weeds, to kill pests and microorganisms that also spoil agricultural products. Pesticides have a long history of utilization against vector disease and parasites as well. They are also used to control malaria, dengue fever, and schistosomiasis in many countries.

DOI: 10.4018/978-1-6684-5619-4.ch007

Insecticides, fungicides, herbicides, rodenticides, and growth regulators of plants are some examples of pesticides. The main intention of their application is plant protection also known as crop protection which protects plants from insects and pest attacks and control weeds. The Food and Agriculture Organization (FAO) has defined pesticides as:

Any substance or mixture of substances intended for preventing, destroying, or controlling any pest, including vectors of human or animal disease, unwanted species of plants or animals, causing harm during or otherwise interfering with the production, processing, storage, transport, or marketing of food, agricultural commodities, wood and wood products or animal feedstuffs, or substances that may be administered to animals for the control of insects, arachnids, or other pests in or on their bodies (FAO, 2013).

HISTORY OF PESTICIDES

People have been dependent on pesticides to protect crops from pest attacks since ancient times (Unsworth, 2010). Sulfur was the first recorded effective insecticide to control insects and mites and was used by Sumerians about 4500 years ago. Chinese people used mercury and arsenical compounds for controlling body lice about 3200 years ago. Bordeaux, a mixture of copper sulphate and hydrated lime was used as an effective fungicide and became widely used between 1860 and 1942. Pyrethrum, which was recognized as Pyrethrum daisies had been used as an insecticide for over 2000 years originated from the dried flowers of *Chrysanthemum cinerariaefolium*. Nitrophenols, chlorophenols, creosote, naphthalene, and petroleum oils were also popular for fungal and insect pest control. The use of synthetic pesticides such as aldrin, benzene hexachloride (BHC), captan, chlordane, dichlorodiphnyltrichloroethane (DDT), 2,4-dichlorophenoxy, dieldrin, endrin, and parathion started in the 1940s (Unsworth, 2010). These products were effective and inexpensive; among them, DDT had better broad-spectrum activity and was the most popular. Eventually, DDT was widely used and effective against vector diseases, like malaria, yellow fever, and typhus, and Dr. Paul Muller won the Nobel Prize in Medicine in 1949 for discovering its insecticidal properties (Delaplane, 1996). However, it was reported later that DDT harms non-target plants and animals, and leaves as residues. The publication of *Silent Spring* by Rachel Carson in 1962 describes how DDT can enter the food chain, accumulate in the fatty tissue of all animals, including humans, and cause cancer and genetic damage (Carson, 2002). She concluded that DDT and other chlorinated pesticides had irreversibly harmed birds, animals and negatively affected the world's food supply. Later, scientists paid attention to developing more environmentally friendly and effective pesticides. However, considering the adverse effects of halogenated pesticides scientists and policymakers around the world became concerned about these harmful compounds and decided to make a convention meeting in Sweden on May 23, 2001, known as the "Stockholm Convention" (UNEP, 2019). Twelve hazardous persistent polychlorinated compounds *i.e.* aldrin, chlordane, dieldrin, DDT, endrin, heptachlor, mirex, toxaphene, hexachlorobenzene, polychlorinated biphenyls (PCBs), polychlorinated dibenzodioxins (PCDDs), and polychlorinated dibenzofurans (PCDFs) were identified as persistent organic pollutants (POPs) and are commonly known as the 'Dirty Dozen' in the Stockholm Convention. The convention aims to protect human health and the environment by phasing out these hazardous pollutants from the environment. These chemicals are of particular concern due to their intrinsic characteristics, namely- wide spectrum persistency, bioaccumulation, biomagnification, transportability, and toxicity. Later, more pesticides were included in the list of POPs (UNEP, 2019). POPs are a poison that can spread without maintaining its boundary. There are several physical and chemical properties of POPs that are responsible for adverse effects like neurological disorders, immune suppression, reproductive disorders, and abnormal fetus development on humans, and persistence in the environment that are mentioned (Kutz, et al., 1991) below:

1) low water solubility
2) easily bioaccumulate in animal fat tissues because of their high lipid solubility
3) It has biomagnification properties that occur across various trophic levels in a food chain
4) lower biological and chemical degradation
5) long-term and long-range persistence
6) chemically stable
7) toxic to both humans and wildlife

CLASSIFICATION OF PESTICIDES

Pesticides can be classified in several ways to know and describe their chemistry and application. They can be the natural origin or synthetic compounds with inorganic and organic properties as below:

According to the Chemical Nature

1. **Inorganic pesticides:** Mercury, fluorine, copper, barium, sulfur, etc.
2. **Organic pesticides:** Organochlorine, organophosphate, pyrethroid, etc.

According to the Function of Different Types of Pests They Control

1. **Insecticide:** Substances that are used to control insects such as pyrethroids, organophosphorus, carbamates, organochlorine, manganese compounds, etc.
2. **Fungicide:** Substances that are used to control fungi such as thiocarbamates, mdithiocarbamates, Cupric salts, Tiabendazoles, Triazoles, dicarboximides, dinitrophenoles, Organotin compounds, etc.
3. **Herbicide:** Substances that are used to control weeds such as bipyridyls, chlorophenoxy, glyphosate, acetanilides, triazines.
4. **Bactericide:** Substances that are used to control bacteria such as Halogens, including chlorine, fluorine, and iodine, and chlorine compounds including sodium hypochlorite, chloramines, and chlorine dioxide.
5. **Algaecide:** Substances that are used to control algae such as oxidants and hydrogen peroxide compounds, copper salts and other copper compounds, Hydrogen peroxide compounds.
6. **Acaricide:** Substances that are used to control mites, ticks such as chlorinated hydrocarbons, organophosphorous compounds, carbamates, pyrethroids, formamidines, avermectins.

According to How They Enter the Target Organism

1. **Stomach Toxicant:** Enter an insect's body through the mouth and digestive tract, where they are absorbed into the insect's body.

2. **Contact Toxicant:** Enter the pest or plant's body either by exposure to water treated with the chemical or by exposure to some treated surface.
3. **Fumigant:** Volatile compounds that enter the bodies of insects in a gaseous phase.
4. **Systemic Toxicant:** Absorbed by plants, pets and disseminated throughout the organism via the vascular system.

FUNCTIONAL PESTICIDES

Fungicide

Fungicides are chemical substances that avert or impede fungal diseases such as scabs, blotches, rots, molds, rusts, and mildew when applied to plants or their seeds and bulbs.

Fungicides are categorized into two broad classes, systemic and non-systemic, according to their ability to move within the plant. Applying to the plant, it swiftly distributes itself in the plant tissue as a systemic fungicide. Therefore, the application of systemic fungicides on the stem of a plant shall assure that fungus would not deforest the branches, leave, or fruit. Thiabendazole and carboxin are systemic fungicides. But non-systemic fungicides protect only the stem from fungus growth by not proceeding from one part of the plant to the other after applying it to the stem of a plant. Therefore, non-systemic fungicides must be administered in a diligent manner and in large doses to provide an effective cover to plants against fungus. Captan and thiram are non-systemic fungicides (Sodhi, 2002).

Fungicides are extensively used to protect commercial crops from diseases, for conducting seeds of plants, to conserve forest trees, and to control slime during the processing of woods in the wood industry. They are also used in building products as mold inhibitors, in paints and varnishes as mildew prohibition (Belsky, 2020).

Herbicide

Herbicides are a class of compounds that destroy plants or interpose with their growth and are used to control unwanted weeds. Herbicides are categorized into two broad classes: Contact herbicides and systemic herbicides. Contact herbicides are rapid-acting substances that kill weeds by direct contact with their leaves. Diquat and bromacil are included in herbicides. Glyphosate, 2,4-D, dicamba (Banvel), imazaquin (Image), and sethoxydim (Vantage) are included in systemic herbicides that proceed from the treated foliage to other plant parts, including the roots.

The vegetation should be released from diseases and land preserved for commercial crops should be released from unwanted growth for the demand of successful large-scale agriculture, and the usage of fungicides and herbicides are given a share of credit for the green revolution. At present, global agricultural production has increased due to the use of fungicides and herbicides in farming. The benefits of these agrochemicals are much in terms of human welfare than their detrimental effect on the environment (Sodhi., 2002).

Insecticide

Insecticides are chemical compounds such as aldrin, dieldrin, parathion, diaxinon, etc. that can kill or destroy one or more species of insects as well as other pests and are broadly classified into four groups-organochlorine hydrocarbons, organophosphates, carbamates, and pyrethroids. The major usages of insecticides are controlling the insect pests during the harvesting of crops, stopping the spoilage of food crops at the time of storage, killing disease-carrying organisms, enhancing esthetic well-being as well as controlling insects like cockroaches, houseflies, and mosquitoes (Britannica, 2021).

Rodenticide

Rodenticides are chemicals or other agents used to destroy rats or rodent pests or impede them from damaging food, crops, etc.; rat poison is a familiar rodenticide.

These substances kill by impeding normal blood clotting and occurring internal hemorrhaging. Phosphorus paste, barium carbonate salt, and powders such as zinc phosphide, white arsenic, thallium sulfate, strychnine, strychnine sulfate, and calcium cyanide are mixed with enticement and settled where rodents will find and eat them. All these poisons are toxic to other animals, and most provoke death by a disorder of nervous system functions. Fumigants such as sulfur dioxide, carbon monoxide, hydrogen cyanide, and methyl bromide are also effective rodenticides (Britannica, 2021).

Fumigants

Fumigants are pesticides that stay as a gas or a vapor at room temperature and are treated as insecticides, fungicides, or rodenticides especially in closed storage places and kill every living organism. They are highly toxic compounds due to their physical properties, rapid environmental dissemination, and human or animal absorption. Aluminum and zinc phosphide, ethylene dibromide, cyanide, aluminum phosphate, and methyl bromide are used as fumigants (Britannica, 2022).

CHEMICAL PESTICIDES

Organochlorine Pesticides

Organochlorine pesticides (OCPs) are derived from chlorinated hydrocarbons and are known as familiarly used pesticides. They were extensively used in agriculture to protect plants from pest attacks. Aldrin, dieldrin, chlordane, DDT, endrin, heptachlor, lindane, benzenehexachloride, mixex, toxaphene are comprised of organochlorine pesticides. DDT was also extensively used to impede the expansion of malaria in developing countries including Bangladesh. Lindane was the second extremely used OCPs that influence the nervous system of pests and originated in the hormonal systems of the body. Therefore, they are known as endocrine smashing agents. OCPs are commonly persistent organic pollutants (POPs) and are deposited in the food chain (Wong et al., 2008). Nowadays, there is a rising concern that organochlorine compounds assert many toxic effects on the health of human and wildlife populations (Brody and Rudel, 2003; Safe, 2004). The toxic effects on humans include disorders in the female and male reproductive system and infertility, carcinogenicity, developmental toxicity, neurotoxicity, and immunotoxicity (Skakkebaek et al. 2001; Teilmann et al. 2002; Toft et al. 2005). Dioxins and especially tetrachlorodibenzodioxin (TCDD) is a known human carcinogens (EPA, 2001). The production and usage of OCPs have been banned by Stockholm Convention due to their toxicity (UNEP, 2019).

Organophosphorus Pesticides

Organophosphorus pesticides (OPPs) are chemical substances produced due to the reaction between phosphoric acid and alcohol. It contains a phosphate ester side chain. The central phosphorous atom is double-bonded to an oxygen or sulphur atom, and single bonded to two methoxy (–OCH_3) or ethoxy (–OCH_2CH_3) groups (Dyguda-Kazimierowicz et al., 2014). Currently, OPPs are heavily used in agricultural production. Certain nerves of insects function by releasing acetylcholine (ACh) into the intracellular space where the nerve cell sends a signal to a muscle cell. Acetylcholine stimulates the muscle cell for contraction. Acetylcholinesterase is an enzyme that stops this contraction that destroys the released ACh signal molecules. OPPs are highly toxic because they chemically bond to the acetylcholinesterase enzyme in such a way that it cannot destroy Ach. This affects the nervous system by inhibiting the action of the enzyme acetyl cholinesterase (AChE). This causes irreversible blockage leading to the deposit of the enzyme, resulting in the overstimulation of muscles. Then the insect dies with its muscles from prolonged contraction and its nervous system was in a state of sustained excitation (Bai, 2006). These pesticides are biodegradable, cause minimum environmental pollution and slow pest resistance. Malathion, parathion, diazinon, chlorpyrifos, etc are included in organophosphorus pesticides.

Carbamate Pesticide

Carbamate compounds consist of esters of carbamic acid with a common chemical formula RHN-COOR. They are generally used as insecticides. Aldicarb, carbaryl, carbofuran and captan are important carbamate (Bachman et. al., 1999). The mode of action is hindering acetyl cholinesterase to the organophosphates analogously. This bond formed for inhibition is less long-lasting and reversible. Carbamates are extremely soluble in water, relatively polar, and reactive. They can be degraded easily in a natural manner with minimum environmental pollution.

PYRETHROID PESTICIDES

Pyrethroids are the most recent group of insecticides. They are a synthetic version of pyrethrin, a natural insecticide. The structural analogues of pyrethrum are based on the naturally occurring insect repellent compound present in the chrysanthemum. These are derived from ketoalcoholic esters of chrysanthemic and pyrethroic acids. They are more substantial in sunlight than pyrethrins and are relatively more stable with more prolonged residual effects than natural pyrethrins. Synthetic-pyrethroid pesticides are highly toxic to insects but of only slight toxicity to mammals (Roberts and Hutson, 1999). They are non-systemic (in plants) contact and stomach poison too many insects and arachnids with an additional anti-feeding action. The compounds act on the central and peripheral nervous systems of target insects at very low doses (Roberts and Hutson, 1999). Synthetic pyrethroids are effective against a wide spectrum of pests. These are the most popular insecticides as they can easily pass through the exoskeleton of the insect. They are broadly used as pest control agents in the production of the agricultural field due to their particular insecticidal activity, biotransformation swiftly, discharge by the mammalian catabolic system, and their non-persistence in the environment. Pyrethroids have two important advantages: they are neither persistent nor toxic. Therefore, they rationalize about one-third of world insecticides usage regardless of their high cost. Pyrethrum, deltamethrin, cypermethrin, etc. are in the group of pyrethroids (IPCS, 2000-2002).

FORMULATION

Formulations mean a homogeneous and stable mixture of active and inert ingredients, making the final product simpler, safer, and more efficacious to apply to a target pest. The effective use of pesticides to control pests, plant diseases, and weeds not only depends on their toxicity but also to a considerable extent on the form of pesticide. Common formulations are:

1) **Emulsifiable Concentrate (EC):** Emulsifiable concentrates are a solution of a pesticide in oil comminated into fine drops coated with a protective layer of a surfactant, a liquid, that can be mixed with water to form a suspension in a spray tank.
2) **Granules (G):** The active ingredient is made into coarse particles with inert material like fired clay particles; these formulations are incorporated into the soil for controlling soil-dwelling pests, for intoxicating plants through their root system, and also for controlling surface pests by aerial dispersion.
3) **Wettable Powders (WP):** The active ingredient is combined with a fine powder that looks like dust that yields stable suspensions when diluted with water, the active ingredient is added to an edible or attractive substance and is often used to control slugs, snails, ground-dwelling insects, and rodents and an active ingredient in powder form is dissolved in water (soluble powders; SP).

Pesticides are also used for household pests; these are very low-concentrate solutions, usually applied as a fine spray or mist generally sold in aerosol cans for mosquitoes (aerosols; A), and made by adding the active ingredients to a fine, inert powder or talc (dust; D).

Pesticides are also formulated as liquid, gel, paste, chalk, powder, granules, pellets, baits and contain 2% to 80% of the active ingredient. Glass, plastic or metal flasks, bottles, drums, traps, plastic bags, or paper bags are employed as containers for carrying several types of pesticides.

ADVERSE EFFECTS OF PESTICIDE USE

Both acute and chronic effects can be occurred due to the disclosure of pesticides. Proximate effects of pesticide disclosure comprise headache, stinging of the eyes and skin, irritation of the nose and throat, skin itching, emergence of the rash and blisters on the skin, dizziness, diarrhea, abdominal pain, nausea and vomiting, blurred vision and blindness (Pesticides and Human Health, 2014). Chronic effects of pesticides can be lethal and may need several years to occur. These are long-term effects that originate with damage to multiple body organs. Pesticide disclosure for extended periods emerges in the following consequences:

1) Pesticide disclosure can provoke a range of neurological health effects such as loss of coordination and memory, reduced visual ability, and reduced motor signaling (Yadav et.al, 2020).
2) Long-term pesticide exposure damages the immune system (Culliney et al., 1992) and can cause hypersensitivity, asthma, and allergies.
3) Pesticide residues have been traced in the bloodstream of cancer patients compared to normal individuals. Pesticides have been associated with leukemia, brain cancer, lymphoma, cancer of the breast, prostate, ovaries, and testes (Pesticides and Human Health, 2014).
4) The presence of pesticides in the body for a longer time also affects reproductive capabilities by altering the levels of male and female reproductive hormones. Stillbirth, birth defects, spontaneous abortion, and infertility are consequences of these effects (Pesticides and Human Health, 2014).
5) Long-term pesticide disclosure also damages the liver, lungs and kidneys. Ingestion of organochlorines occurs hypersensitivity to light, sound, and touch, dizziness, tremors, seizures, vomiting, nausea, confusion, and nervousness (Yadav et al., 2020).

Disclosure of organophosphates and carbamates causes, symptoms similar to those of increased neurotransmitter-acetylcholine. These pesticides interfere with the normal nerve signal transduction and disclosure to them causes headaches, dizziness, confusion, nausea and vomiting, muscle and chest pain. Distress breathing, convulsions, coma, and death may occur in severe cases (Pesticides and Human Health, 2014). Pyrethroids can originate an allergic skin response, aggressiveness, hyper- excitation, reproductive or developmental effects in addition to generating tremors and seizures (Hua & Relyea, 2019). It is regarded that there is a relationship between pesticides and Parkinson's disease and Alzheimer's disease (Casida & Durkin, 2013).

EFFECTS ON HUMAN HEALTH

Pesticides have been linked to a wide range of human health hazards, ranging from acute impacts like headaches and nausea to chronic impacts like immune suppression, hormone disruption, diminished intelligence, reproductive abnormalities, and cancer (Best & Ruthven, 1995). Workers engaged in the production and management of various pesticides revealed a high occurrence of abdominal pain, dizziness, headache, nausea, vomiting, fatigue, and also irritation of skin and eyes. Pesticide disclosure is associated with respiratory problems, memory disorders, and depression. The risk of Parkinson's disease was reported to increase due to the disclosure of both insecticides and herbicides and the risk of increasing Parkinson's disease is 70% greater in those exposed to even low levels of pesticides. Scientists also warned that long-term disclosure of pesticides might increase the risk of dementia, birth defects, fetal death, and altered fetal growth. Pesticide disclosure resulted in reduced fertility, pregnancy loss, genetic alterations in sperm, a reduced number of sperm, and altered hormone function (Bretveld, 2006). Disclosure of pesticides is associated with cancer risk, including leukemia, lymphoma, brain, kidney, breast, prostate, pancreas, liver, lung, and skin cancers (Kachuri et al., 2017).

EFFECTS ON ENVIRONMENT

Pesticides are ubiquitous in the environment and their residues are found in soil, air, and in surface, and groundwater across the globe. Several environmental concerns appear due to the usage of pesticides. Over 98% of sprayed insecticides and 95% of herbicides get through a goal other than their target species, comprising non-target species, air, water, bottom sediments, and food. Stream of pesticides ensues when pesticides are suspended in the air as particles are sustained by the wind to other areas, dynamically contaminating them. Soil and water pollution is also occurred due to the presence of pesticide residues and some pesticides are persistent organic pollutants which is a major concern as they assert a momentous threat to biological communities comprising humans. There are various ways for entering pesticides into water such as accidental spillage, industrial effluent, surface runoff, and transport from pesticide-treated soils, washing of spray equipment after spray operation, drift into ponds, lakes, streams, and river water, aerial spray to control water-inhibiting pests (Carter and Heather, 1995; Singh et al., 2018; Mandal et al., 2012). Pesticides usually pass from fields to various water reservoirs by runoff or drainage founded by rain or irrigation (Larson et al., 2020). The availability of pesticides in the air can be occurred to several factors comprising spray drift, exhalation from the treated surfaces, and aerial application of pesticides. The range of drift relies on droplet size and wind speed. The rate of exhalation is relied on time after pesticide treatment, the surface on which the pesticide settles, the ambient temperature, humidity and wind speed, and the vapor pressure of the ingredients (Kips, 1985). The extension in the concentration of pesticides due to their persistent and non-biodegradable nature in the tissues of organisms at each continuous level of the food chain is known as biomagnification. Due to this occurrence, organisms at the higher levels of the food chain encounter greater impairment as compared to those at lower levels (Niethammer, 1984).

Analytical Methods of Pesticide Monitoring in Food and Environment

The use of pesticides in agriculture has increased after World War II to increase world food production. Since then there had been marked development of different types of pesticides belonging to various groups. The use of pesticides and additional environmental pollution due to industrial emissions during the production of pesticides has resulted in the occurrence of residues of these chemicals and their metabolites in every component of the environment, i.e. air, water and soil along with that in the crops, vegetables and fruits (Bai et al., 2006).

Many scientists have estimated the pesticide residues in various fruits, vegetables and environmental samples and reported the occurrence of pesticide residues to be even more than maximum residue limits (MRL) values recommended by European Union (EU) and Codex Alimentarius Commission (2013). The content of pesticides in various fruits and crops does not only depend on the sprayed amount over them but also the content present in soil or water used for irrigation.

Extraction methods of pesticides differ from the physical state of the matrix to matrix i.e., solid, semisolid, liquid, and samples to samples i.e., vegetables, soil, water, fruits, etc. as well as the chemical nature of the targeted pesticides i.e., polar and nonpolar. Recovery within an acceptable level is important irrespective of pesticides and matrices. Several extraction methods including solvent extraction and solid-phase dispersion for solid matrices, liquid-liquid extraction for liquid samples, and solid-phase extraction (SPE) for pre-concentration and clean-up (reversed-phase, bonded-phase, and normal phase) are traditionally used for residue analysis (Åkerblom 1995). Recently several economic and environmental friendly methods including solid-matrix solid-phase dispersion (MSPD), supercritical fluid extraction (SFE), solid-phase microextraction (SPME), and the quick, easy, cheap, effective, rugged and safe (QuEChERS) method have been developed (Anastassiades et al., 2003).

After the selection of an appropriate extraction technique, most of the resulting extracts need to be cleaned by solid-phase extraction (SPE), dispersive solid-phase extraction (DSPE), or gel permeation chromatography (GPC). Sulphuric acid treatment is also used for the clean-up of many halogenated compounds. Other clean-up methods such as florisil, silica gel, charcoal, etc. are also used. With the introduction of QuEChERS by Anastassiades, et al. (2003), the popularity of using DSPE clean-up techniques increased where the primary secondary amine (PSA) stationary phase is used.

Gas chromatography (GC) coupled with electron captured detector (ECD) and/or nitrogen phosphorus detector (NPD) is the most commonly used instrument for analysis of non-polar and semi-polar compounds. For polar and non-volatile compounds, high-performance liquid chromatography (HPLC) with ultraviolet (UV-Vis/PDA) and fluorescence detectors are used. One of the advantages of GC over HPLC is that a large number of compounds can be separated by GC with a capillary column. Recently, gas chromatography with mass spectrometry (GC-MS), gas chromatography with tandem mass spectrometry (GC-MS/MS), liquid chromatography-mass spectrometry (LC-MS), and liquid chromatography with tandem mass (LC-MS/MS) has been frequently used for the analysis and confirmation of pesticides in different matrices.

Maximum Residue Limit

The maximum residue limit (MRL) has been defined as "the maximum concentration of a residue that is legally permitted or recognized as acceptable in or on a food or agricultural commodity or animal feedstuff (FAO code of conduct, 2002). The MRL is deliberated mainly as a check that the use of pesticides is occurring according to authorized labels and Good Agricultural Practice (GAP). It implies that label directions and GAP have been properly followed when residues are found at or below the MRL. MRLs for pesticides are formed by national authorities or advisory bodies primarily on the basis of field residue for the observation that provides acuteness as to the levels of residues that may ensue under the worst-case scenario (i.e., maximum permitted application rate and several applications, minimum permitted pre-harvest interval). MRL is not set based on toxicology data but once proposed based on GAP they must be evaluated for safety. It is usually realized through a risk assessment process comparing dietary intakes which are estimated from expected residue concentrations in food consumed with the relevant health-related regulatory end-points, the acceptable daily intake (ADI) and the acute reference dose (ARFD). CODEX Alimentarius (FAO combined with WHO, European Union (EU), Bangladesh Food Safety Authority (BFSA), Canadian Authority have set up and monitored MRL for pesticides in food, vegetables, food animals to ensure safe food for humans (Ohkawa, 2007).

Every pesticide has a Pre-Harvest Interval (PHI) for residues to dissipate below the MRL established for that crop. PHI means, the interval between the last application of pesticide and the safe harvesting of edible crops for immediate consumption. Food products become only safe after the pre-harvest interval and it differs from pesticide to pesticide and crop to crop. The extent and rate of dissipations depend on the nature of the pesticides, crop, way of application, and various environmental conditions under which the crop is grown.

Management of Pesticides

Management of pesticides is very poor in many developing countries, and this is a great challenge to increase knowledge for sustainable and judicial management of pesticides. Management of pesticides is very poor in many developing countries and this is a great challenge to increase knowledge for sustainable and judicial management of pesticides. Farmers and family members are usually exposed to pesticides during application and storage. It is assumed that occupational risks to farmers can be reduced by taking personal protective measures during and after pesticide application (Yang et al., 2014). However, uses of personal protective equipment (PPE) including face mask, shoes, gloves and body covering with suits are not very common among the farmers in developing countries including Bangladesh. Usually, farmers, spray pesticides in field barefooted, without gloves and suits. Although they use long cotton cloth known as Gamsa in Bangladesh, they also dry their faces with it while sweats. Many farmers even do not wash their hands, take shower and change dresses after application of pesticides but take food and mix with community without taking safety measures. Farmers usually store pesticides in their houses living with their family members. Sometimes containers of pesticide are stored with other household stuffs which have the possibility to harm children and other family members of the family. Women are living in areas of recent high agricultural pesticide use experience higher rates of breast cancer (Reynolds et al., 2005). Bangladesh is a land of rivers and many water bodies close to agricultural fields. Pesticides used in the crops usually discharge and fall into the water systems. Again, farmers through pesticide containers, bags, and other toxic materials on field or in the water. Thus, the indiscriminate disposal of pesticides and their containers pollute the water bodies which are also the harmful aquatic environment. To increase knowledge about the management awareness program can be helpful.

HUMANITARIAN ENGINEERING IN PESTICIDE MANAGEMENT

Agricultural crops loss by pests, drought, irrigations and farmland inundated by water during heavy rain are the main hindrance in acquiring food security. Humanitarian engineers can be in touch with the cultivators to give guidelines on how to manage proper irrigation and pest management. It is also necessary for integrated pest management, cultivation of effects of pesticides in food production, poverty alleviation and toxic on health, environment and their rolls are inter-related. Humanitarian engineers can educate people involved with pesticide application mainly in the developing countries. Active work of humanitarian engineers in the field may make an understanding and safety in agricultural projects, to manage disaster and food storage, trading to reduce environmental contamination especially in water bodies. Contamination of water bodies due to discharge of chemicals from agricultural fields. High yielding varieties crops, vegetables, fruit and cultivation of crops which can be harvested before monsoon. Farmers should also be taught to take steps for production of safe food, not making any harmful effect on the non-targeted pest's as well as environmental safety, personal safety and safety as a whole community. Arrangement of an awareness program on safe food production and food storage are essential in this regard. Active work of humanitarian engineers in the field may make an understanding and safety in agricultural projects, to manage disaster and food storage, trading to reduce environmental contamination especially in water bodies.

Pesticide Use in Bangladesh

Bangladesh is an agricultural country, due to its numerous water bodies i.e., rivers, lakes, etc., and fertile lands. Currently, more than 70% of Bangladeshis are directly or indirectly dependent on agriculture for their livelihood. The warm and humid climatic conditions of the country increased modern high-yielding varieties of crops and more use of chemical fertilizers are highly favorable for the development and multiplication of pests and diseases (Aziz, 2005). The use of pesticides rises in Bangladesh during the middle of the 1950s (Rahman and Alam, 1997; Islam, 2000) and since then to protect the major crops such as rice (820 boro varieties), potato, bean, eggplant, cabbage, sugarcane and mango growers (Aziz, 2005; Rahman, 2003). Farmers are exposed to pesticides by different routes such as inhalation, ingestion, and dermal contact which eventually leads to acute and chronic health effects (cardiopulmonary disorders, neurological and hematological and skin diseases, etc.). The lack of training on proper management of pesticide use, MRL, pre-harvest residue limits, post-harvest residue limits, etc. not only pose health hazards to consumers (Ibitayo, 2006; The Environment Court Act, 2000) but also an inappropriate selection of insecticides and doses, improper spray scheduling and inadequate spray coverage may cause to the failure in controlling insect pests. There are plentiful profitable effects of pesticides including crop protection, preservation of food and material, and prevention of vector-borne disease. To meet the high food demand for the increasing high population, the production of crops needs to be increased substantially. Pesticides essentially improve the food quality, quantity, and variety by controlling the natural food poisons, increasing the production of fruits and vegetables, and thus lowering food prices. The National Environment Management Action Plan (NEMAP) of the Bangladesh government has identified an alarming rate of pesticide usage contributing to the pollution of potential surface and groundwater. Thus to know the extent of residual contamination and the magnitude of exposure to the human being, there is a need for analysis and estimation of pesticide residues for respective samples.

ANALYSIS OF VEGETABLE SAMPLES FOR RESIDUAL PESTICIDES

Methods And Materials

Chemicals and Reagents

Ethyl acetate ($C_4H_8O_2$), acetone (C_3H_6O), n-hexane, dichloromethane (DCM), methanol (MeOH), anhydrous magnesium sulphate ($MgSO_4$), sodium chloride (NaCl), and aluminum oxide (alumina) were purchased from E. Merck, Germany. Acetonitrile (ACN; HPLC grade) was purchased from Sigma Aldrich. Florisil and Charcoal were obtained from ACROS Organics, the USA, and Uni-Chem, China. Cypermethrin (91% purity), chlorpyrifos (99.5% purity), diazinon (99.5% purity), and fenvalerate (98.5% purity) standards were purchased from Dr. Ehrenstorfer, GmbH, Augsberg-Germany and stored at -20 ^{0}C in a freezer. Florisil and alumina were activated by heating at 105 ^{0}C for 3 h, charcoal was activated by washing with distilled water, n-hexane, MeOH in Buchner funnel & then dried at 105 ^{0}C and stored in desiccators.

Instruments

Gas chromatographic (GC; Shimadzu-2010) analyses were performed using an electron capture detector (ECD) with a split/splitless auto-injector. Injector and detector temperatures were set at 280 °C and 290 °C, respectively. Separations were performed on an HP-5 capillary column (30 m x 0.25 *i.d.* and film thickness, 0.25 µm) where nitrogen (N_2) was used as a carrier and make-up gas, and the flow rate was 1.78 mL/min. The oven temperature was programmed at an initial temperature of 120 °C (2 min hold) and rises at 10 °C/min up to 270 °C (hold 10 min).

Sample Collection

A total of 102 vegetable samples Bitter gourd (Korola), *Cucumis sativus* (Cucumber), Eggplant (Brinjal), Teasle gourd (Kakrol), Ladies finger (Okra), Pointed gourd (Potol), Ridge gourd (Jhinga), Snake gourd (Dhundol) and *Solanum lycopersicum* (Tomato) were collected in labeled Jip-lock polyethylene bags from cultivated fields and different markets of Comilla, Gazipur, Jhenaidah, Mymensingh, Manikgonj, Narsingdi, Savar, Sonargoan area and immediately kept in Chill box to transfer to the Department of Chemistry, the University of Dhaka during January-June 2013 and finally, stored at -20 ^{0}C until analysis was carried out (Figure 1).

ANALYSIS OF VEGETABLE SAMPLES

Extraction and Clean Up

QuEChERS method was used for the extraction of vegetable samples (Anastassiades, 2003). Briefly, samples (10 g) were blended with Kitchen Blender and ethyl acetate (20 mL) was added to 10 g homogenized sample in 50 mL Teflon tube and shaken for 1 minute in hand & vortex for 1 min. Anhydrous $MgSO_4$ (6 g) & NaCl (1.5 g) were added and vortexed for 1 min and then centrifuged for 5 minutes at 4000 rpm. 10 ml supernatant solution was taken in 100 mL round bottom flask, evaporated in a rotary vacuum evaporator, and then reconstituted in n-hexane (2 mL).

For clean up a glass column (40 cm long & 12 mm internal diameter) was packed with a 10.5 g mixture of aluminum oxide, florisil, and charcoal (10:10:1) in n-hexane. The column was equilibrated with 50 mL n-hexane and then the sample extract in n-hexane (2 mL) was transferred to the column. The column was washed with 20 mL of n-hexane and eluted with 100 mL of DCM at the rate of 1 mL-min^{-1}. The eluent was concentrated to dryness on a rotary evaporator & dissolved the residue in 2 mL of n-hexane and injected to GC-ECD for the residual amount of chlorpyrifos, cypermethrin, diazinon, and fenvalerate in vegetable samples.

Standard Calibration Curves

The calibration curves of chlorpyrifos, cypermethrin, diazinon, and fenvalerate were made by injecting solutions at concentrations 0.002, 0.005, 0.01, 0.025, 0.05, 0.125, 0.25, 0.5, and 1.0 µg/g of their standards into the GC-ECD, plotted integrated areas of the peaks against the standard concentration. Correlation coefficients (r^2) were 0.99, 0.99, 0.99 and 0.99 for them, respectively (Figure 2).

Figure 1. Sample collection

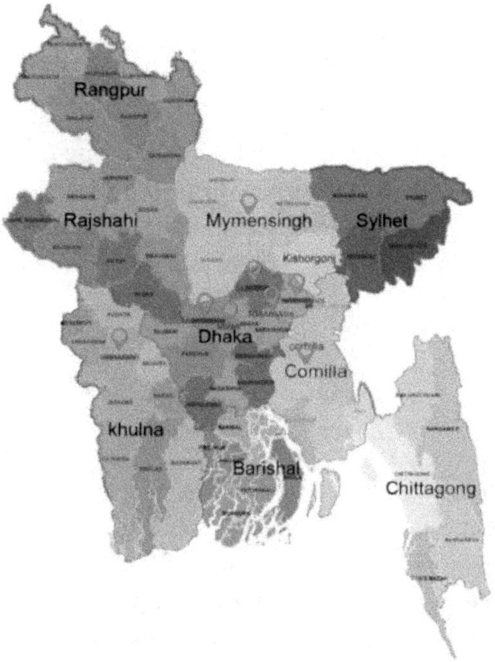

Figure 2. Standard calibration curve of (a) chlorpyrifos, (b) cypermethrin, (c) diazinon, (d) fenvalerate

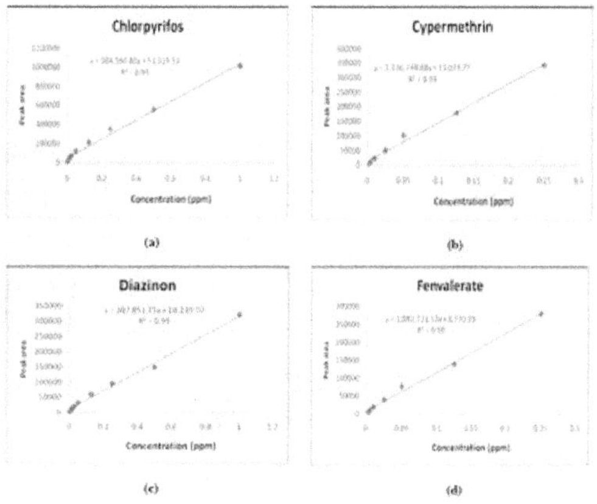

Limit of Detection (LOD) and Limit of Quantification (LOQ)

The limit of detection (LOD) and limit of quantification (LOQ) values were found to be 0.002, 0.002, 0.002 (Signal to noise ratio, S/N 3:1) and 0.0066, 0.0066, 0.0066 (S/N: 10:1) mg/kg for cypermethrin, diazinon, and fenvalerate, respectively. The LOD and LOQ were 0.0004 and 0.0013 mg/kg for chlorpyrifos, respectively.

Recovery

The recovery experiments were performed in three replicates at two fortified concentrations. The recovery experiment was carried out using the QuEChERS method. The sample extract was analyzed by GC-ECD following the same analytical conditions. Retention time and corresponding peak area were recorded. The recovery was calculated using a standard calibration curve. The percent recoveries from bitter gourd were varied from 72–98% for diazinon, 76–119% for chlorpyrifos, 87–105% for cypermethrin, and 84–107% for fenvalerate with relative standard deviations up to 12%, which is acceptable in an analytical purpose. The recovery percentage of pesticides from the bitter gourd sample revealed that the extraction method was acceptable.

RESULTS AND DISCUSSION

Out of twelve samples from the Comilla area, five samples were found to contain residual pesticides. Chlorpyrifos was found in bitter gourd (0.003 mg/kg), snake gourd (0.019 mg/kg). Diazinon was found in kakrol (0.046 mg/kg), snake gourd (0.009-0.02 mg/kg). Cypermethrin was found in snake gourd (0.006-0.041 mg/kg). Fenvalerate was found in bitter gourd (0.006 mg/kg), snake gourd (0.019 mg/kg). Out of twenty-five samples from Mymensingh area, fourteen samples were found to contain residual pesticides. Diazinon was found in kakrol in the range of 0.009-0.046 mg/kg, in bitter gourd (0.003 0.024 and mg/kg) and pointed gourd (0.006 and 0.019 mg/kg). However, chlorpyrifos was found in bitter gourd (0.002-0.007 mg/kg) and eggplant (0.002-0.006 mg/kg). Out of twenty-two samples from the Narshindi area, seven samples were found to contain residual pesticides. Chlorpyrifos was found in eggplant (0.002-0.02 mg/kg), tomato (0.005-0.020 mg/kg) but diazinon was detected in only korola (0.032-0.040 mg/kg). Out of eleven samples from the Gazipur area, six samples were found to contain residual pesticides. Chlorpyrifos was found in ladies' finger (0.005-0.012 mg/kg), tomato (0.002-0.007 mg/kg). Diazinon and fenvalerate were found in only ladies' finger samples. Twelve vegetable samples were analyzed from Jhenaidah area and eight samples were found to contain residual pesticides. Diazinon was found in Bitter gourd, Kakrol, snake gourd, Ladies finger and pointed gourd. Chlorpyrifos was found in Bitter gourd, kakrole, snake gourd, and ladies' fingers. Out of twelve samples from the Savar area, seven samples were found to contain residual pesticides. Chlorpyrifos was found in Kakrola (0.003-0.017 mg/kg), tomato (0.003-0.014 mg/kg). Diazinon was detected in only one sample of Korola and cypermethrin was found in only one sample of tomato. Out of six samples from the Sonargoan area, four samples were found to contain residual pesticides. Diazinon, chlorpyrifos, and fenvalerate were found in ladies fingers in the range of 0.012-0.014 mg/kg while chlorpyrifos was found in tomatoes in the range of 0.020-0.025 mg/kg. Residual pesticides were not found in cucumber.

Out of 102 analyzed vegetable samples, 54 samples (53%) were found to contain residual pesticides in varying amounts. Chlorpyrifos, diazinon, cypermethrin, and fenvalerate were detected in 50%, 22%, 6%, and 5% samples, respectively (Figure 3-4). Among four pesticides, chlorpyrifos was detected in the highest number of vegetable samples (50%) and it was found in all areas. It was detected in 7 samples of ladies fingers (n=11), 15 samples of tomato (n=21), 3 samples of kakrol (n=11), 9 samples of bitter gourd (n=13), 8 samples of eggplant (n=20), 5 samples of snake gourd (n=8), 1 sample of a pointed gourd (n=3) and 1 sample of ridge gourd (n=2).

Figure 3. Pesticides in vegetable samples (n=102)

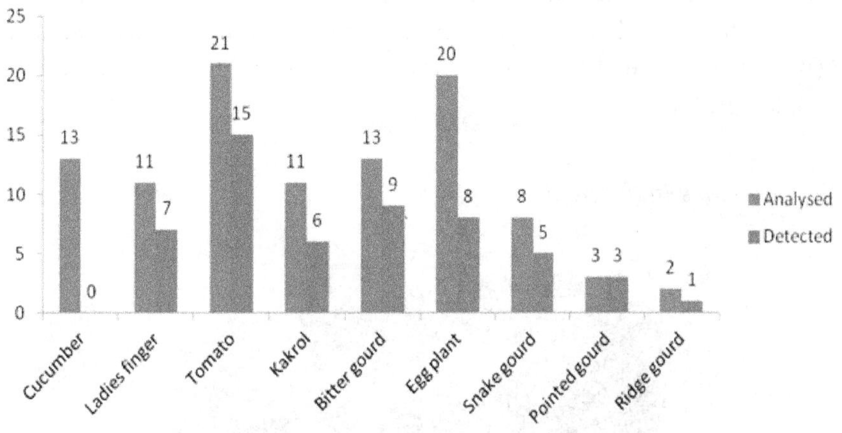

Figure 4. Comparison of detected pesticides in vegetables samples (n=102)

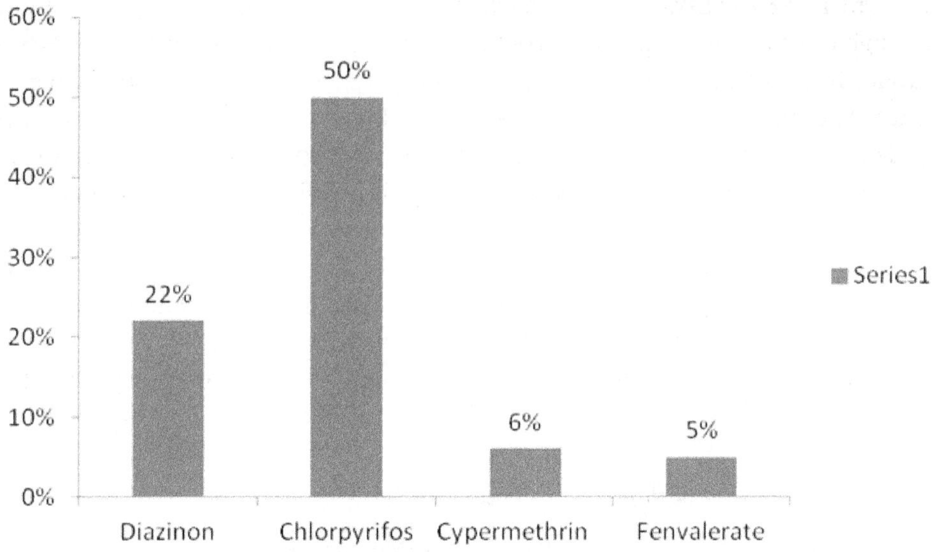

The level of residual pesticides is very low and it can be predicted that these vegetables are less contaminated with these four pesticides. The reasons for a low amount of pesticides can be explained that farmers may use other types of pesticides than chlorpyrifos, cypermethrin, diazinon, and fenvalerate. Or due to the sunshine pesticides may be dissipated over time or washed away from vegetables due to rain as all vegetables were collected in the summer season. The level of pesticides was below the respective MRL values set by Codex Alimentarius or BFSA.

A workshop was held on 23 June 2013 in Jhenaidah, one of the biggest vegetable cultivation areas in Bangladesh to create awareness about the toxic effect of chemicals on human health due to the overuse of pesticides among the farmer's community & other relevant authorities. About 40 farmers aged between 30-65 years were interviewed with the following questions: (i) How often do they use pesticides during vegetable cultivation? (ii) Do they get benefits after pesticide application? (iii) Where do they keep pesticides? (iv) When do they harvest crops after the application of pesticides? (v) Do they suffer from any health-related problems?

Figure 5. Spraying pesticides in the field

Most of the farmers (98%) answered that they get benefits after the application of pesticides and wish to use a wide range of pesticides to prevent crop loss from pest attacks (Figure 5). They use pesticides after getting advice from government agricultural officers, pesticide suppliers, shop owners, and friends. However, about 80% of farmers responded that they do not know the active ingredients in pesticides that they are using. Many of them keep pesticides in their house and some of them keep in the fields. About 60% of respondents told us that they use pesticides every day, 20% every 2 days interval, 30% 3-4 days interval, and 5% after 7 days interval. However, 50% of farmers harvest crops the next day after the application of pesticides. 30% harvest 3 days interval and 10% 7 days interval. 40% of farmers said that they suffered from respiratory problems, 10% allergy, 5% alsar, 10% have acidity and 5% did not have any problem. 60% of farmers know the side effects of pesticides but 90% of farmers do not know the integrated pest management. Intersecting, some local farmers informed that they do not bother about

Figure 6. Interactive discussion with farmers about safe use of pesticides

the toxicity of pesticide in vegetable samples as they sell and rich people in the city consume. Another two workshops were arranged one in Nuritola, Comilla on 09 January 2016 and another one in Nurundi, Jamalpur on 06 February 2016. Approximately, 100 farmers attended each workshop and were asked similar questions. Many of them also replied that they get to benefit from the use of pesticides but they do not know about the toxic and harmful effects of these chemicals. However, they feel that farmers' awareness program for toxicity of pesticide on health and Environment will be helpful and beneficial for them (Figure 6).

CONCLUSION

Pesticides are toxic compounds and old pesticides especially organochlorine pesticides are long persistent in the environment and have detrimental effects on human health. Bioaccumulation, biomagnification, and the food chain have a bad effect on human health. These chemicals possess teratogenic, mutagenic, carcinogenic, and other adverse on human health and are banned to further use. Organophosphorous, pyrethroid, and carbamates are less-lasting pesticides but residues above recommended value are harmful to human health. During application without proper protection are harmful to farmers' health. Storage of pesticides and disposal of the containers of pesticides and storage without cleaning sprayers again harmful to the living with surrounding areas.

ACKNOWLEDGMENT

Authors are grateful to International Science Programme (ISP), Uppsala University, Sweden; Higher Education Quality Enhancement Project (HEQEP), University Grant Commissions of Bangladesh and Ministry of Science and Technology, Government of the People's Republic of Bangladesh for Ph.D. for financial supports.

REFERENCES

Akerblom, M. (1995). *Environmental Monitoring of Pesticide Residues*. Guidance for the SADC Region.

Anastassiades, M., Lehotay, S. J., Stajnbaher, D., & Schenck, F. J. (2003). Fast and easy multiresidue method employing acetonitrile extraction/ partitioning and dispersive solid-phase extraction for the determination of pesticide residues in produce. *Journal of AOAC International*, *86*(2), 412–431. doi:10.1093/jaoac/86.2.412 PMID:12723926

Aziz, M. A. (2005). Country reports Bangladesh. In *Proceedings Asia Regional Workshop: Implementation Monitoring and Observance International Code of Conduct on the Distribution and Use of Pesticide*. FAO.

Bachman, J., Howard, H., & Patterson. (1999). Photodecomposition of the Carbamate Pesticide Carbofuran: Kinetics and the Influence of Dissolved Organic Matter. *ACS Environ. Sci. Technol.*, *33*, 874-881. doi:10.1021/es9802652

Bai, Y. (2006). Organophosphorus pesticide residues in market foods in Shaanxi area. *Food Chemistry*, *98*, 240–242. doi:10.1016/j.foodchem.2005.05.070

Belsky, J., & Joshi, N. K. (2020). Effects of Fungicide and Herbicide Chemical Exposure on Apis and Non-Apis Bees in Agricultural Landscape. *Frontiers in Environmental Science*, *8*(81), 1–10. doi:10.3389/fenvs.2020.00081

Best, G. A., & Ruthven, D. (1995). Analytical Division & Royal Society of Chemistry. In *Pesticides: Developments, impacts, and controls*. Royal Society of Chemistry.

Bretveld, R.W., Thomas, C.M., Scheepers, P.T., Zielhuis, G. A., & Roeleveld, N. (2006). Pesticide exposure: the hormonal function of the female reproductive system disrupted? *Reprod Biol Endocrinol.*, *31*(4), 30. doi:10.1186/1477-7827-4-30

Britannica, The Editors of Encyclopaedia. (2019). Insecticide. *Encyclopedia Britannica*. https://www.britannica.com/technology/insecticide

Britannica, The Editors of Encyclopaedia. (2021). Rodenticide. *Encyclopedia Britannica*. https://www.britannica.com/technology/rodenticide

Britannica, The Editors of Encyclopaedia. (2022). Fumigant. *Encyclopedia Britannica*. https://www.britannica.com/technology/fumigant

Brody, J., & Rudel, R. (2003). Environmental pollutants and breast cancer. *Environmental Health Perspectives*, *111*(8), 1007–1019. doi:10.1289/ehp.6310 PMID:12826474

Carson, R. (2002). Silent Spring. Houghton Mifflin Harcourt.

Carter, A. D., & Heather, A. I. J. (1995). Pesticides in groundwater. In G. A. Best & A. D. Ruthven (Eds.), *Pesticides- developments, impacts, and controls* (pp. 113–123). The Royals Society of Chemistry.

Casida, J. E., & Durkin, K. A. (2013). Neuroactive insecticides: Targets, selectivity, resistance, and secondary effects. *Annual Review of Entomology*, *58*(1), 99–117. doi:10.1146/annurev-ento-120811-153645 PMID:23317040

Codex Alimentarius Commission. (2013). *Codex Maximum Residue Limits for Pesticides*. https://www.who.int/entity/foodsafety/areas_work/chemicalrisks/IEDIcalculation0217clusterfinal.xlsm

Culliney, T. W., Pimentel, D., & Pimentel, M. H. (1992). Pesticides and natural toxicants in foods. *Agriculture, Ecosystems & Environment*, *41*(3-4), 297–320. doi:10.1016/0167-8809(92)90117-T

Delaplane, K. S. (1996). Pesticide Usage in the United States: History, Benefits, Risks, and Trends. Cooperative Extension Service, The University of Georgia College of Agricultural and Environmental Sciences.

Dyguda-Kazimierowicz, E., Szczepan, R., & Andrzej Sokalski, W. (2014). *Alkaline Hydrolysis of Organophosphorus Pesticides: The Dependence of the Reaction Mechanism on the Incoming Group Conformation*. ACS J. Phys. Chem. B. doi:10.1021/jp503382j

EPA. (2001). *Dioxin Reassessment*. https://cfpub.epa.gov/ncea/cfm/dioxin doi:10.5772/57399

FAO. (2013). *International Code of Conduct on the Distribution and Use of Pesticides*. https://www.fao.org/3/bt565e/bt565e.pdf

Food Safety Act, Bangladesh Food Safety Authority (BFSA), October 2013. https://puc.ac.bd/Content/PU-Journal/PU-Critical-Perspective-Article3.pdf

Hua, J., & Relyea, R. (2019). The effect of a common pyrethroid insecticide on wetland communities. *Environ. Res. Commun.*, *1*(1), 015003. doi:10.1088/2515-7620/aaebb3

Ibitayo, O. O. (2006). Egyptian farmers' attitudes and behaviours regarding agricultural pesticides: Implications for pesticide risk communication. *Risk Analysis*, *26*(4), 989–995. doi:10.1111/j.1539-6924.2006.00794.x PMID:16948691

IPCS. (2002). *The WHO recommended classification of pesticides by hazard and guidelines to classification 2000-02*. https://www.who.int/ipcs/publications/en/pesticides_hazard.pdf

Islam, M. A., Huq, S., Rahman, A., & Conway, G. R. (2000). *Consequences of increased pesticide use. In Environmental Aspects of Agricultural Development in Bangladesh*. The University Press.

Kachuri, L., Harris, M. A., MacLeod, J. S., Tjepkema, M., Peters, P. A., & Demers, P. A. (2017). Cancer risks in a population-based study of 70,570 agricultural workers: Results from the Canadian census health and Environment cohort (CanCHEC). *BMC Cancer*, *17*(1), 343. doi:10.118612885-017-3346-x PMID:28525996

Kips, R. H., & Haskell, P. T. (1985). *Environmental aspects. Pesticide application: principles and practice*. Clarendon Press. doi:10.5772/57399

Kutz, F. W., Wood, P. H., & Bottimore, D. P. (1991). Organochlorine pesticides and polychlorinated biphenyls in human adipose tissue. *Reviews of Environmental Contamination and Toxicology*, *120*, 1. PMID:1899728

Larson, S. J., Capel, P. D., & Majewski, M. (2020). *Pesticides in surface waters: Distribution, trends, and governing factors (No. 3)*. CRC Press.

Mandal, S., Rahman, A. M. H., Mamun, M. I. R., Shoeb, M., & Nahar, N. (2012). Method Development and validation for estimation of Commercially produced sex pheromones in Lure. *J. Bangladesh Chem. Soc.*, *25*(2), 180–185. doi:10.3329/jbcs.v25i2.15084

Niethammer, K. R., White, D. H., Baskett, T. S., & Sayre, M. W. (1984). Presence and biomagnification of organochlorine chemical residues in oxbow lakes of northeastern Louisiana. *Archives of Environmental Contamination and Toxicology*, *13*(1), 63–74. doi:10.1007/BF01055647 PMID:6703785

Ohkawa, H., Miyagawa, H., & Lee, P. W. (2007). Pesticide Residues in Food and International Trade: Regulation and Safety Considerations. Pesticide Chemistry: Crop Protection, Public Health, Environmental Safety, 29-41.

Pesticides and Human Health. (n.d.). *Californians for Pesticide Reforms*. http://www.Pesticidereform.org/section.php?id=38

Rahman, M. H., & Alam, M. J. B. (1997). Risk assessment of pesticides in Bangladesh. *J. Civil Engineering., 25*(1), 97–106.

Rahman, S. (2003). Farm-level pesticide use in Bangladesh: Determinants and awareness. *Agriculture, Ecosystems & Environment, 95*(1), 241–252. doi:10.1016/S0167-8809(02)00089-0

Reynolds, P., Hurley, S. E., Gunier, R. B., Yerbati, S., Quach, T., & Hertz, A. (2005). Residential proximity to agricultural pesticide use and incidence of breast cancer in california. *Environmental Health Perspectives, 113*(8), 993–1000. doi:10.1289/ehp.7765 PMID:16079069

Roberts, T., & Hutson, D. (1999). *Metabolic Pathways of Agrochemicals. Part 2. Insecticides and Fungicides*. The Royal Society of Chemistry.

Safe, S. (2004). Endocrine disruptors and human health: Is there a problem. *Toxicology, 205*(1-2), 3–10. doi:10.1016/j.tox.2004.06.032 PMID:15458784

Singh, S., Sharma, R., & Parween, T. (2018). Pesticide Contamination and Human Health Risk Factor. *Modern Age Environmental Problems and Their Remediation*, 49-68. doi:10.1007/978-3-319-64501-8_3

Skakkebaek, N., Rajpert-De, M., & Main, K. (2001). Testicular dysgenesis syndrome: An increasingly common developmental disorder with environmental aspects. *Human Reproduction (Oxford, England), 16*(5), 972–978. doi:10.1093/humrep/16.5.972 PMID:11331648

Sodhi, G. S. (2002). *Fungicides and Herbicides*. Fundamental Concepts of Environmental Chemistry.

Teilmann, G., Juul, A., Skakkebaek, N., & Toppari, J. (2002). Putative effects of endocrine disrupters on pubertal development in the human. *Best Practice & Research. Clinical Endocrinology & Metabolism, 16*(1), 105–121. doi:10.1053/beem.2002.0184 PMID:11987902

The Environment Court Act, 2000. extwprlegs1.fao.org/docs/pdf/bgd42277.pdf

The Environment Court Act, 2000. National Integrated Pest Management Policy, B. Distribution and Use of Pesticides in Bangladesh. http://www.doebd.org/ pop/pdf/Report

Toft, G., Axmon, A., Giwercman, A., Thulstrun, A. M., Rignell-Hydborn, A., Pedersem, H. S., Luwicki, J. K., Zyyezday, V., Zinchuk, A., Spano, M., Manicardi, G. C., Bonefeld-Jorgensen, E. C., Hagmar, L., Bonde, J. P., & INUENDO. (2005). Fertility in four regions spanning large contrasts in serum levels of widespread persistent organochlorines: a cross sectional study. *Environmental Health: A Global Access Science Source, 4*, 26-37. doi:10.1186/1476-069X-4-26

UNEP. (2019). *Stockholm Convention, Protecting human health and the environment from the persistent organic pollutants*. www.pops.int

Unsworth, J., (2010). *History of pesticide use, International union of pure and applied chemistry*. Academic Press.

Wong, F., Alegria, H. A., Jantunen, L. M., Bidleman, T. F., Salvador-Figueroa, M., Gold-Bouchot, G., Ceja-Moreno, V., Waliszewski, S. M., & Infanzon, R. (2008). Organochlorine pesticides in soils and air of southern Mexico: Chemical profiles and potential for soil emissions. *Atmospheric Environment, 42*(33), 7737–7745. doi:10.1016/j.atmosenv.2008.05.028

Yadav, K., Garg, L., & Kaur, L. (2020). Pesticides Usage: Effect on Human Health. *International Journal of Current Microbiology and Applied Sciences*, *9*(8), 2605–2615. doi:10.20546/ijcmas.2020.908.298

Yang, X., Wang, F., Meng, L., Zhang, W., Fan, L., Geissen, V., & Ritsema, C. J. (2014). Farmer and retailer knowledge and awareness of the risks from pesticide use: A case study in the Wei River catchment, China. *The Science of the Total Environment*, *497–498*, 172–179. doi:10.1016/j.scitotenv.2014.07.118 PMID:25128887

Chapter 8
Aligning Humanitarian Attributes and Needs Between Academia, Industry, and Quality Assurance in Graduate Engineering Programmes:
An EU-ASIA Survey

Georgia Kremmyda
University of Warwick, UK

Angelos Georgoulas
https://orcid.org/0000-0001-8845-3781
University of West Attica, Greece

Yiannis Koumpouros
https://orcid.org/0000-0001-6912-5475
University of West Attica, Greece

ABSTRACT

There is growing interest in enabling competences such as creativity, critical thinking, and problem solving to engineering students by expanding their engagement to complex, interdisciplinary problems linked to global challenges. Optimal curriculum design aims at meeting quality assurance requirements and delivering graduate attributes (knowledge, skills, behaviours) needed from industry and at the same time also from the society for tackling humanitarian challenges. Evidence on how the needs of stakeholders (academia, industry, quality assurance) align to the integration of humanitarian attributes to engineering curricula is missing. This chapter presents and discusses the findings of a comparative qualitative study undertaken among higher education institutions, industry, and quality assurance agencies in Europe and Asia. Findings reveal how graduate attributes are perceived by various stakeholders in an attempt to demystify the suitability and effectiveness of engineering education practices within an interdisciplinary, inter-professional humanitarian context.

DOI: 10.4018/978-1-6684-5619-4.ch008

INTRODUCTION

Community-based engineering education gives students the professional attributes to tackle humanitarian challenges with transformative impact in communities. Today, when various humanitarian crises are happening, we find that professionals with a set of integrated humanitarian attributes are essential to tackle societal challenges. This is even more critical in advancing skills and education related to engineering to previously disadvantaged or under-represented communities. Nevertheless, evidence on how the needs of stakeholders (academia, industry, quality assurance) align with the integration of humanitarian attributes into engineering curricula is missing.

This chapter presents and discusses the findings of a comparative qualitative study undertaken among higher education institutions, industry and quality assurance agencies in Europe and Asia. Findings reveal how various stakeholders perceive graduate attributes in an attempt to demystify the suitability and effectiveness of engineering education practices within an interdisciplinary, inter-professional humanitarian context.

Background and Rationale

21st-century humanitarian challenges are complex phenomena manifested at variable temporal and spatial scales, affecting society through disruption in economic, political, societal, cultural and institutional systems. This vulnerable situation is promoted in recent UN dialogues, notably the Hyogo Framework for Action 2005-2015 (UN, 2005), Sendai Framework for Disaster Risk Reduction 2015-2030 (UN, 2015) and the UN Sustainable Development Goals 2030 (UN, 2015); building resilience against humanitarian challenges is central to effective action. Engineering education programmes are considered to be uncreative, boring, male-dominated, having minimum relevance to real-life problems and humanitarian challenges. Inclusive, interdisciplinary education programmes which contribute to the students' global awareness via a programme of real-life-focused activities starting at first arrival and continuing beyond graduation are needed. In the last decade, education and research in the humanitarian field have often been limited to a disciplinary approach. However, as the Humanitarian contexts such as; resilient infrastructures, process safety, armed conflicts, epidemics, natural or man-made disasters, social exclusion etc., in which humanitarian interventions unfold are more complex than ever, an interdisciplinary perspective can bring new insights. A better understanding of the interplay of engineering, political, economic, social, and medical factors is therefore required.

Engineering programmes are long criticised for being male-dominated, infertile, having minimum relevance to real-life problems, industry and community needs (Gill *et al.*, 2008). Inclusive, interdisciplinary education programmes which enable the students' global responsibility via creative activities, nurtured even at pre-university level and continuing beyond graduation, are needed (Al Bawada, 2022). It is imperative that we build the capacity of our students and educators to meet the future demands for engineering skills to address them in the places where they are most needed (British Council, 2015), meeting the ever-evolving needs of students, employers and communities in challenging times.

Reliance on increasingly irrelevant curricula, divorced from real-life problems and global challenges, contributes to an exacerbated skills gap, especially in emerging economies, where saving lives, protecting livelihoods, and strengthening recovery from disasters and crises is mostly needed (Unesco, 2014). My practice overturns such outdated traditional educational models, which artificially separate knowledge, disciplines, and skills (Asian Development Bank, 2015).

To this respect, a contemporary educational programme targeting humanitarian challenges should aim at offering a co-evolutionary, innovative, conceptual framework (Kelley and Knowles, 2016) in graduate engineering education, ensuring: (a) explicit and implicit links to resilience and transformative change; (b) integration of highly diverse, yet complementary, expertise and skills; (c) promotion of innovative pedagogical developments across different disciplines and sectors to have a global humanitarian impact. Institutional reforms in the curriculum of graduate and engineering programmes, but also for society, may be necessary after re-evaluating the importance of humanitarian intervention.

The current Chapter presents the outcomes of a wide online survey in order to reveal the perception of Higher Educational Institutions in different countries and with different cultures for modern graduate engineering programmes targeting humanitarian challenges versus the actual needs and perceptions of the Industry sector and the Quality Assessment Authorities.

METHODOLOGY

In the framework of the ENHANCE project funded under the ERASMUS+, KA2 - Capacity Building in the higher education programme, an online survey was conducted focusing on interdisciplinary teaching and learning and skills/attributes needed for tackling complex interdisciplinary, inter-professional problems such as humanitarian challenges.

The survey was anonymous and followed ethics and deontology rules applied in the EU. An appropriate signed consent form was required before initiating the survey. Ethical approval for the research study was granted by the University of Warwick's Biomedical and Scientific Research Ethics Sub-committee (REGO-2019-2341).

The online survey was developed and run using Qualtrics software, Version November 2019 of Qualtrics. For the purposes of the study, we used the Statistical Package for Social Sciences (IBM SPSS version 19).

The aim of the online survey was to explore perceptions around the following questions:

1. To what extent engineering curricula offered by the Partner HEIs and around the world reflect industrial, community, national and global engineering needs;
2. What are the challenges faced in integrating industrial, community, national and global engineering needs in engineering education;
3. What are the current practices among the Partner HEIs and around the world in delivering graduate engineering education (undergraduate and postgraduate);
4. What are the emerging teaching and learning activities and assessment methods;
5. What is the contribution of industry in the delivery of engineering education;
6. What are the implications of Quality Assurance requirements in the development of engineering curricula;
7. What are the needed attributes (knowledge, skills, experience) of graduate engineers for tackling contemporary engineering problems such as humanitarian challenges.

The survey aimed at revealing and integrating through our taxonomy the perception of educators in Higher Education Institutions in different countries about all aspects of teaching and learning (content, assessment and delivery methods, feedback and quality assurance, student experience, contact time, intended attributes etc.) with a focus to tackling humanitarian challenges versus the actual needs and perception of the Industry sector and the Quality Assurance Agencies on the suitability and effectiveness of engineering education practices (Unger and Polt, 2017).

SURVEY RESULTS

The average time to complete the online questionnaire was around 11 minutes. The sample consisted of 734 respondents in total; 403 respondents from Higher Education Institutions (HEIs), 221 respondents from Industry, and 110 respondents from Quality Assurance Agencies (QAAs), mainly from UK, Greece, Bangladesh, Indonesia and Viet Nam; see Fig. 1 for a detailed view.

The results per organisation category (HEIs, Industry and QAAs) are presented next.

Industry Perception

The online survey, as aforementioned, was taken by 221 respondents from Industry, clustered in nine countries - majority of respondents from Greece, UK, Bangladesh, Indonesia, Viet Nam and a small number of respondents from USA, Malaysia, Germany and Canada. Even though the number of respondents outside of the project countries was small, the responses were maintained to provide an indication across the nine countries and across three continents (Europe, Asia, and North America). Nevertheless, the focus remains on the project countries.

The first question for the Industry target group aimed at exploring the confidence of the sector with regards to having enough in-country engineering graduates with the skills to fill high-skilled job vacancies, moving towards an economy driven by the Fourth Industrial Revolution (also known as Industry 4.0).

Regarding the industry's degree of confidence of having enough high-skilled graduate engineers (to fill the related high-skilled job vacancies), Figure 2 presents the overall degree of confidence (pie chart), with only 12.9% of the industry organisations respondents stated that they're not confident, as well as for each country (clustered bar chart). Moreover, Figure 2 shows the mean of the ratings of the responses (0: for the "Not confident" response, 1: for the "Somewhat Confident" and 2: for the "Very confident") across all countries (clustered bar chart with error bars for the means), and one for the continents.

Figure 3 illustrates the industry's mean response ratings (from 1: for "Poor" to 5: for "Excellent") concerning 10 capabilities that graduate engineers develop in their countries. The most developed overall capabilities are *Theoretical knowledge* followed by *Understanding of regulations/standards/practice codes,* while the least developed are *Experience in real-life applications* followed by *Project management*.

Analytical skills are (on average) equally developed by graduate engineers across all countries. This capability is equally developed across the three continents. Moreover, every country has analytical skills equally developed with the others in Asia and N. America, while UK graduate engineers have this skill more developed than the other European. *Creativity* is (on average) equally developed by graduate engineers across all countries. In general, the corresponding N. Americans have more developed their creativity. All countries in the three continents have the same level of importance regarding their industry's beliefs about culture preservation. German graduate engineers have more developed creativity among their European colleagues, while Asian and N. American have equally developed creativity with their colleagues in other countries of the same continent. *Experience in real-life applications of graduate engineers* (on average) seems to be the least developed skill across all countries and continents.

Figure 1. The three main categories of the participants per continent

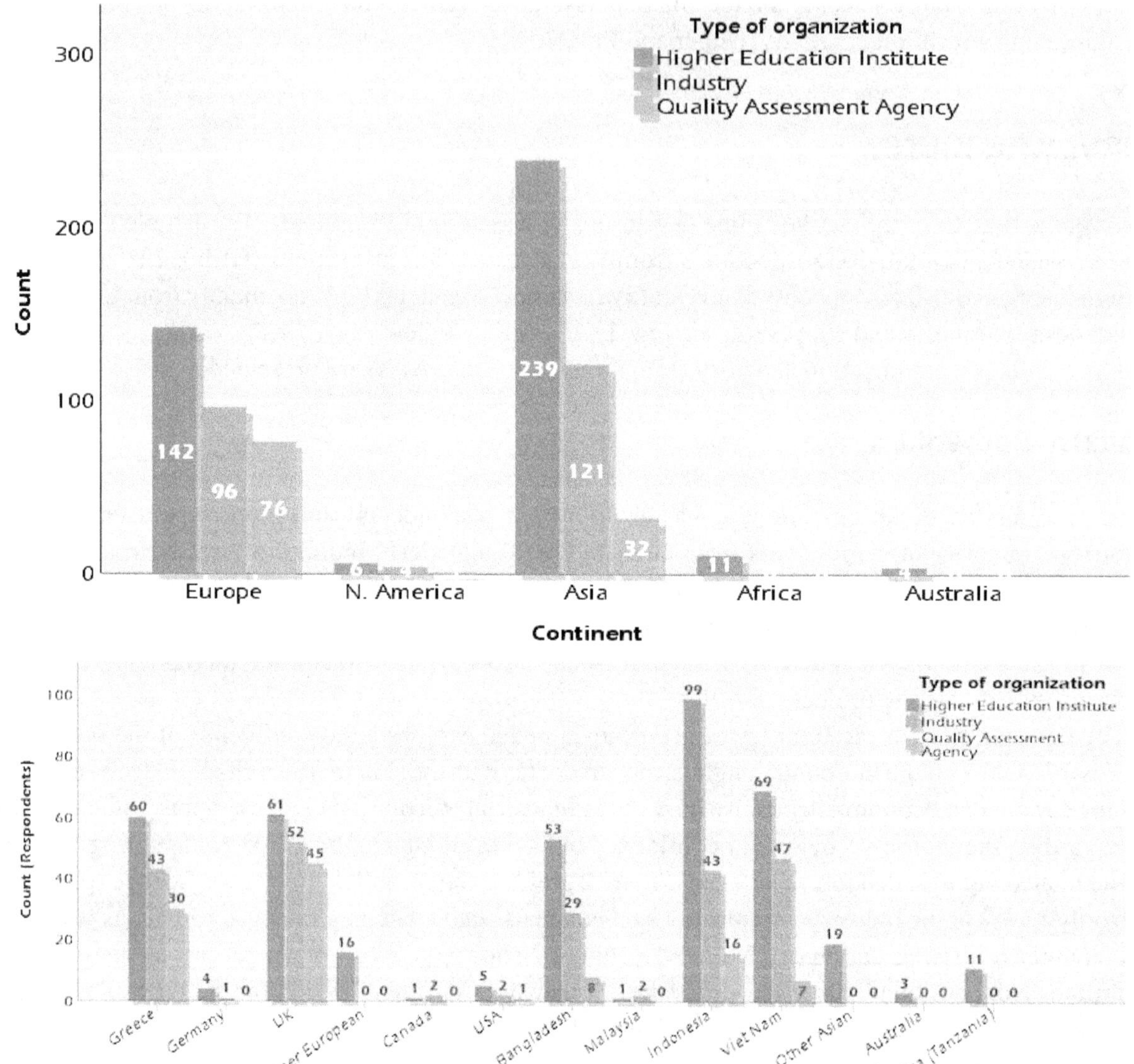

Regarding the characteristics which a "good" engineering programme should have, Figure 4 (first clustered bar chart) illustrates the "Yes" responses from industry for six given characteristics under study. For a more informative view, Figure 4 also presents the mean of the responses ratings (0: for "No" response, and 1: for "Yes") across countries (middle clustered bar chart). Therefore, the higher the bar (the clustered bar charts of Fig. 4) the most important the specific characteristic is.

Creativity in curricula and delivery method as well as *Opportunities to experience the workplace* are the most desired/important characteristics by the industry across all countries for a "good" engineering programme, while *Accreditation and quality assurance certification* and *Career opportunities* are the least desired.

Figure 2. The three main categories of the participants per main participating countries

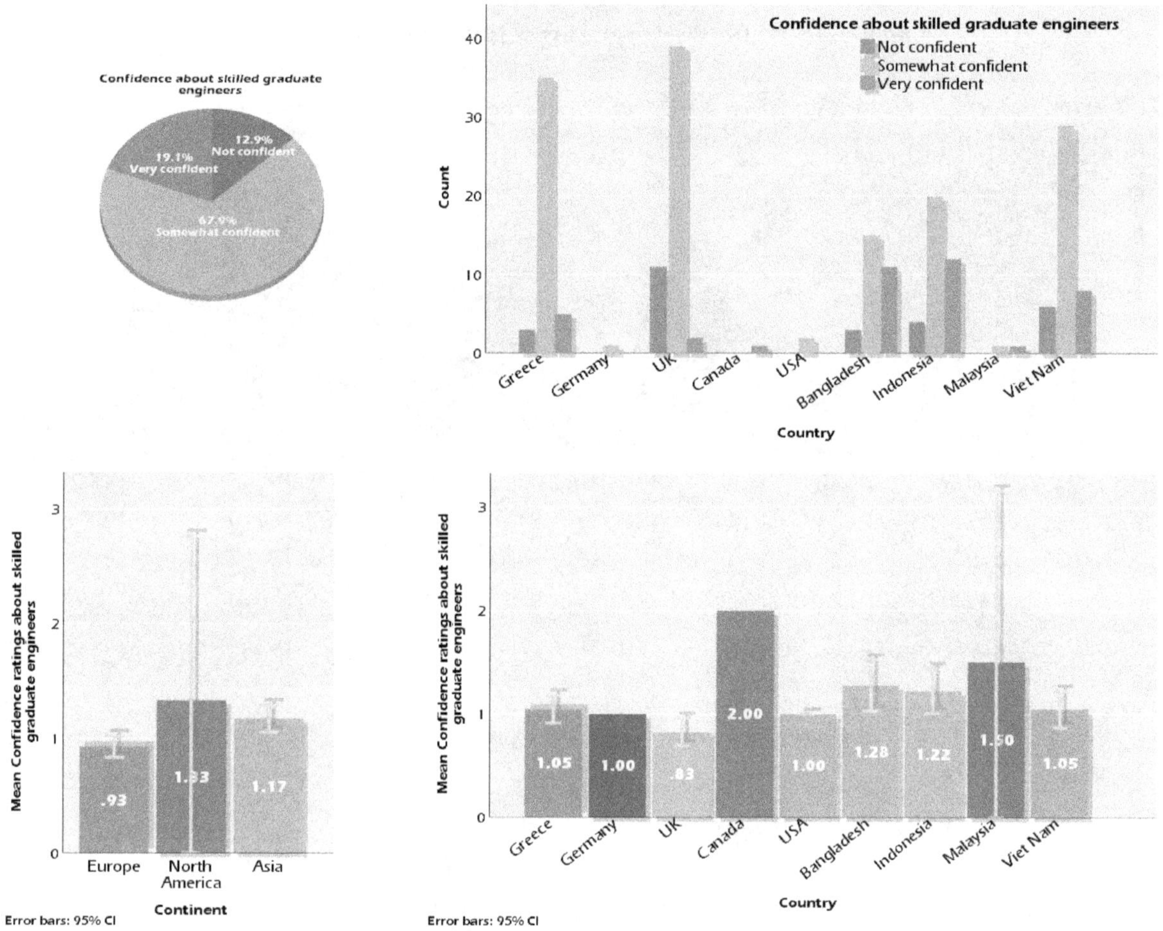

More specifically, Canadian and USA industry appear to find *Career opportunities* and *Opportunities to experience the workplace* as very important features of engineering education, while *Academic Rigour* and *Creativity* are more attractive to European industry. *Accreditation and quality assurance certification* appear to be important across all industries.

Figure 6 illustrates the industry's mean response rating (from 1: for the "Most important" to 10: for the "Least important") concerning 14 attributes (skills, knowledge, experience) of the graduate engineers that (the industry believes that) need to be developed. Therefore, the smaller the bar, the more important the specific attribute.

Figure 6 illustrates the industry's mean response ratings (from 1: for the "Most preferred" to 10: for the "Least preferred") concerning 10 teaching/learning activities that could develop the aforementioned needed attributes (by the industry) of the graduate engineers. Therefore, the smaller the bar, the more preferred the specific activity is. The most preferred activities are *Problem-based learning*, and *Internships and industry placements,* while the least preferred are *Tutorials*, *Seminars* and *Self-directed study*.

Figure 6 illustrates the industry's mean response ratings (from 1: for the "Most needed" to 10: for the "Least needed") concerning 10 topics that graduate engineers needed to have knowledge and under-

Figure 3. Industry responds regarding their confidence of having enough graduate engineers

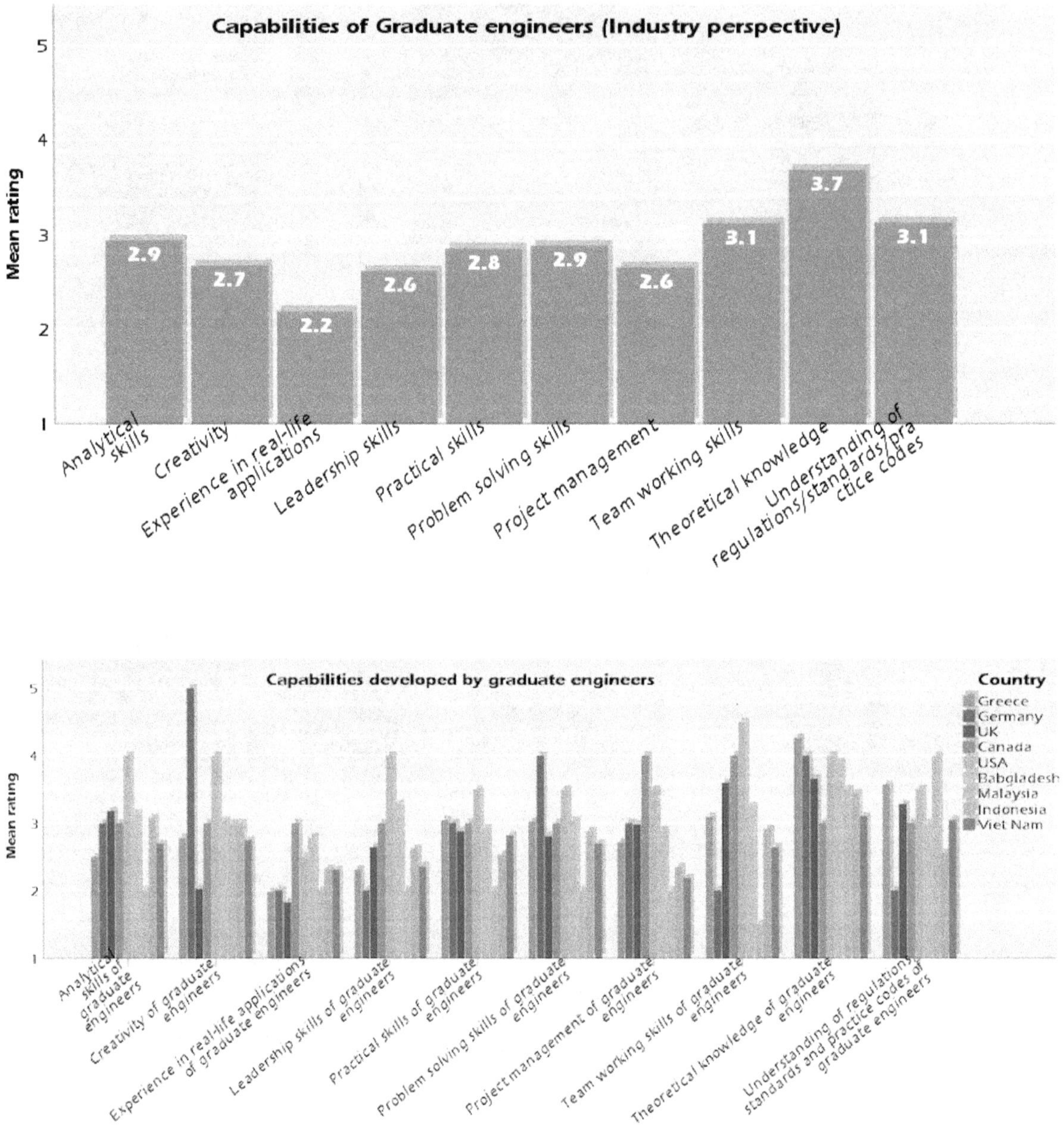

standing. Therefore, the smaller the bar, the most needed the specific topic is. The most needed overall topics are *Professionalism*, *Sustainability*, *and Community needs* while the least needed are *Political factors* and *Culture preservation.*

Figure 4. Means of the Industry's ratings of the 10 capabilities developed by graduate

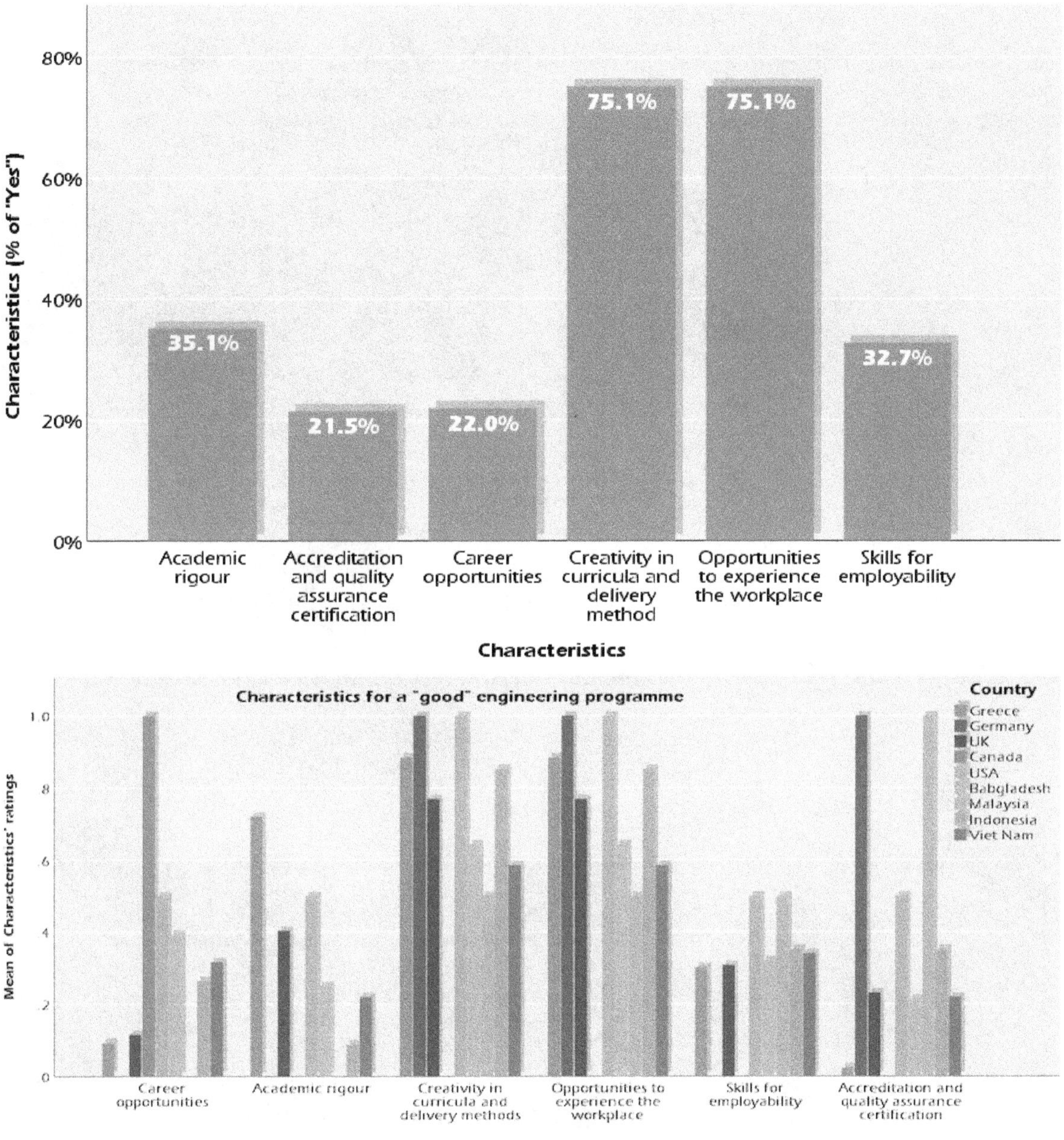

Overall the level of the industry's collaboration/involvement in delivering current engineering programmes rated (by the industry) over 67% as poor and average (Figure 8).

Various types of industry support to HEIs, to help them deliver current engineering degree programmes (to address needed skill challenges) are presented in Figure 9. The means of the response ratings (0: for "No" response and 1: for "Yes" response) for the five support types under study are presented in the middle and bottom clustered bar charts of Fig. 10. Across all countries, the industry support HEIs mainly by offering *internships and/or placements for engineering students* (59.3%), *organising visits for students at the industry* (53.1%) and *delivering invited lectures/talks/seminars/workshops* (49.5%).

Figure 5. The characteristics that the industry believes a "good" engineering programme should have

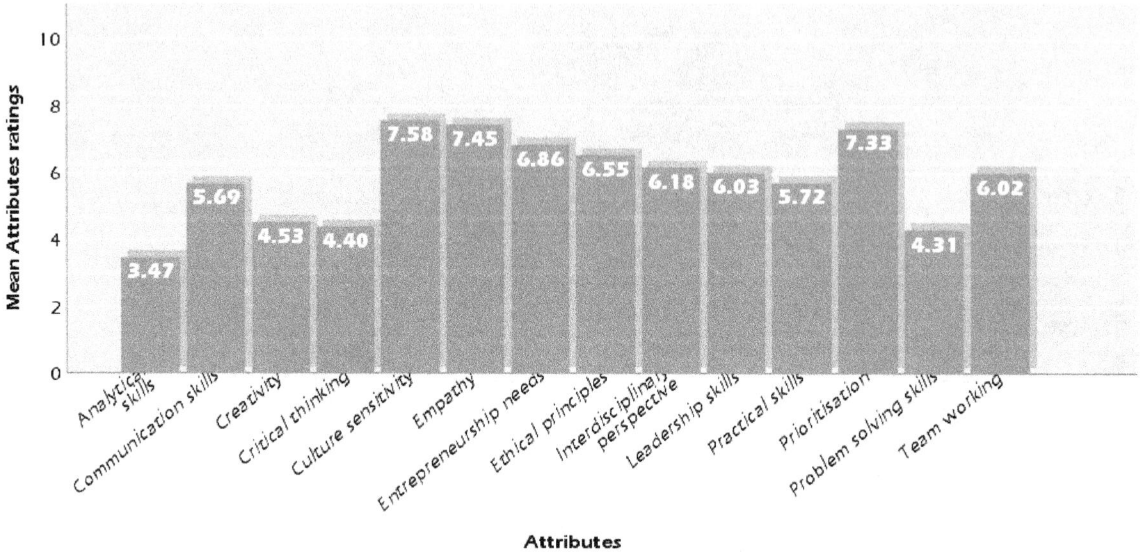

Figure 6. Means of the attributes' ratings that the industry thinks that need to be developed in graduate engineers

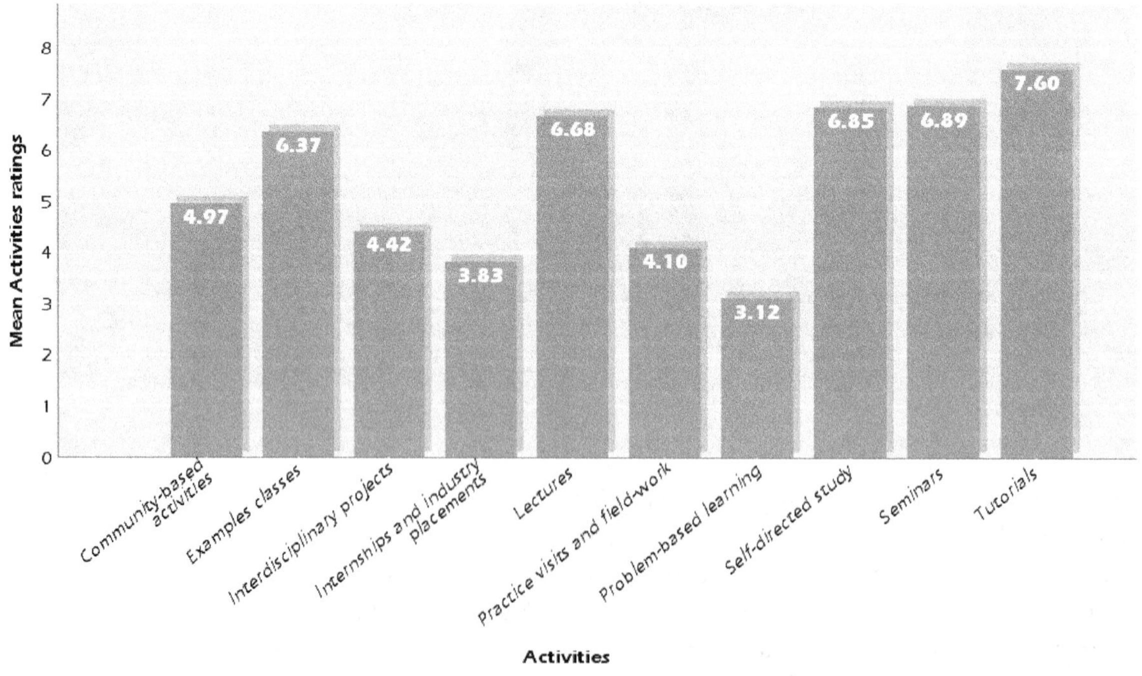

Figure 7. Means of the activities' ratings that the industry thinks that could develop needing attributes for the graduate engineers

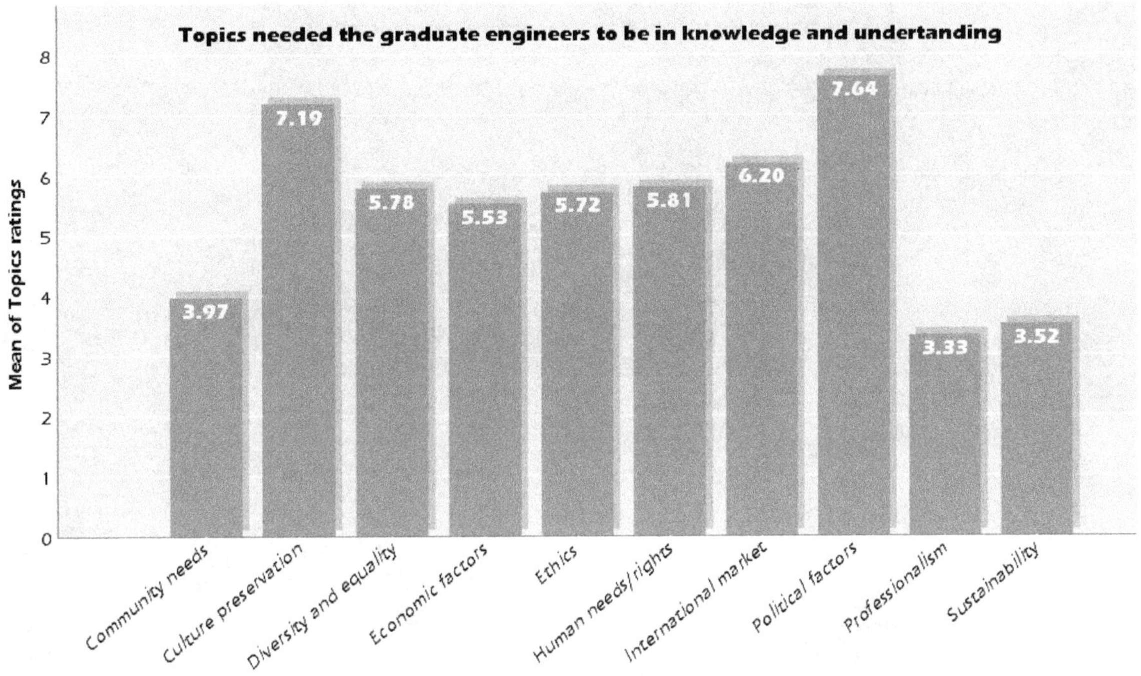

Higher Education Institutions (HEIs) Perception

Figure 10 illustrates the relation of the responses between the HEIs and the Industry regarding the capability of graduate engineers in ten factors (from 1: for "Poor" to 5: "Excellent").

Figure 8. Means of the topics' ratings that the graduate engineers need to have in knowledge and understanding

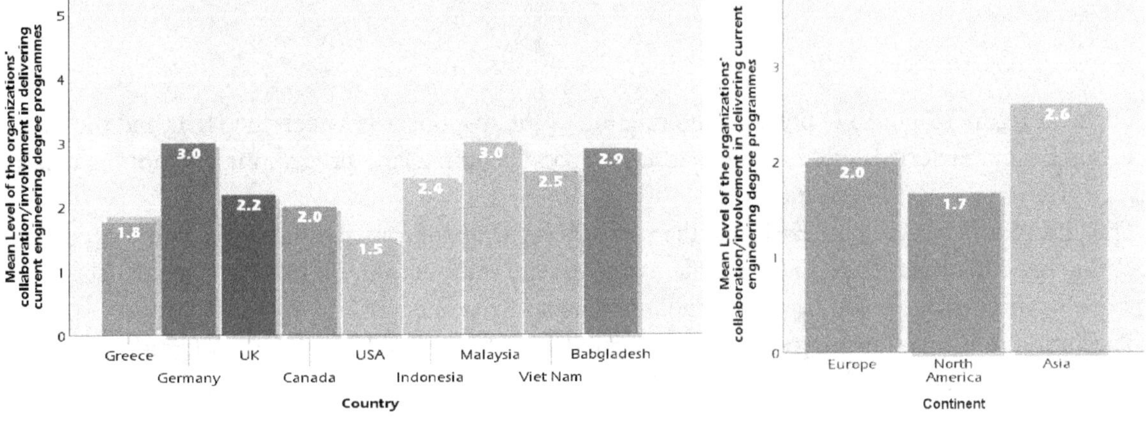

Figure 9. Level of the industry's collaboration/involvement in delivering current engineering programmes

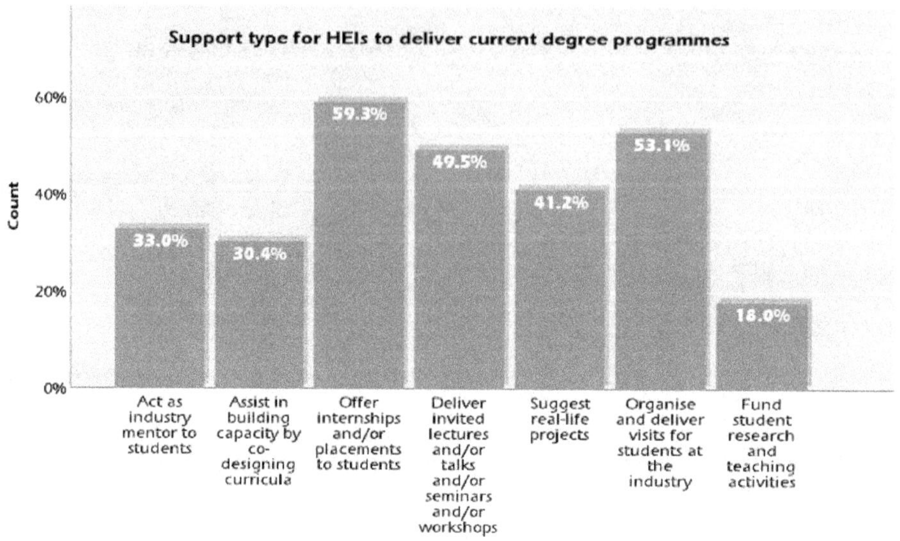

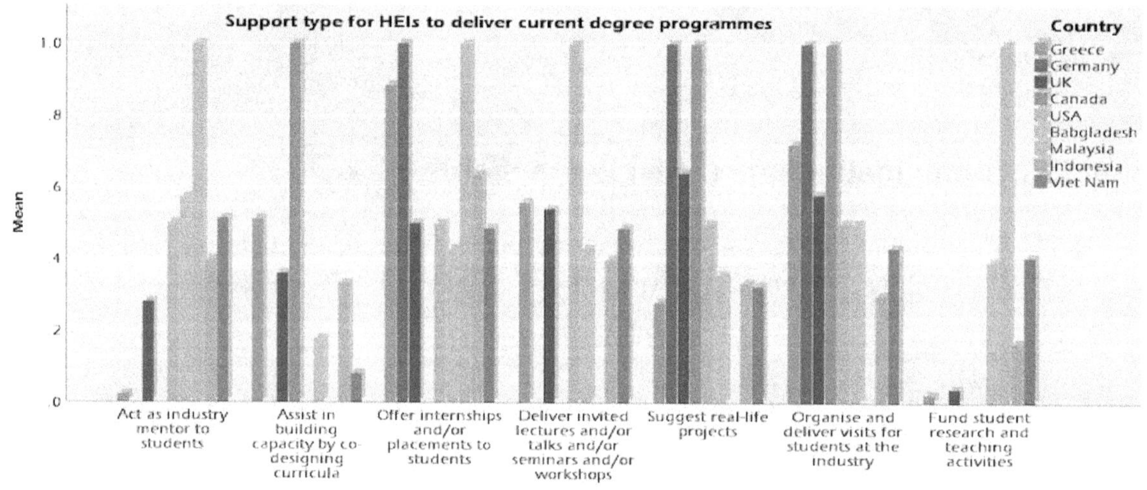

The next figure (Figure 11) presents the relation of the responses between the HEIs and the Industry regarding the characteristics that are needed for a "good" engineering programme, where the mean ratings are from 0: for "No" to 5: for "Yes".

The differences between the HEIs and the Industry regarding the attributes (skills, knowledge, experience) that need to be developed in graduate engineers are presented in Figure 12 as means bars of their response's rating (from 1: for "most needed" to 10: for "least needed").

The various learning/teaching activities that could develop the needed attributes of graduate engineers are presented in Figure 13, both from the HEIs' and Industry's perspective, using means bars of their response's rating (from 1: for "most preferred" to 10: for "least preferred").

Figure 10. Support type for HEIs to deliver current degree programmes

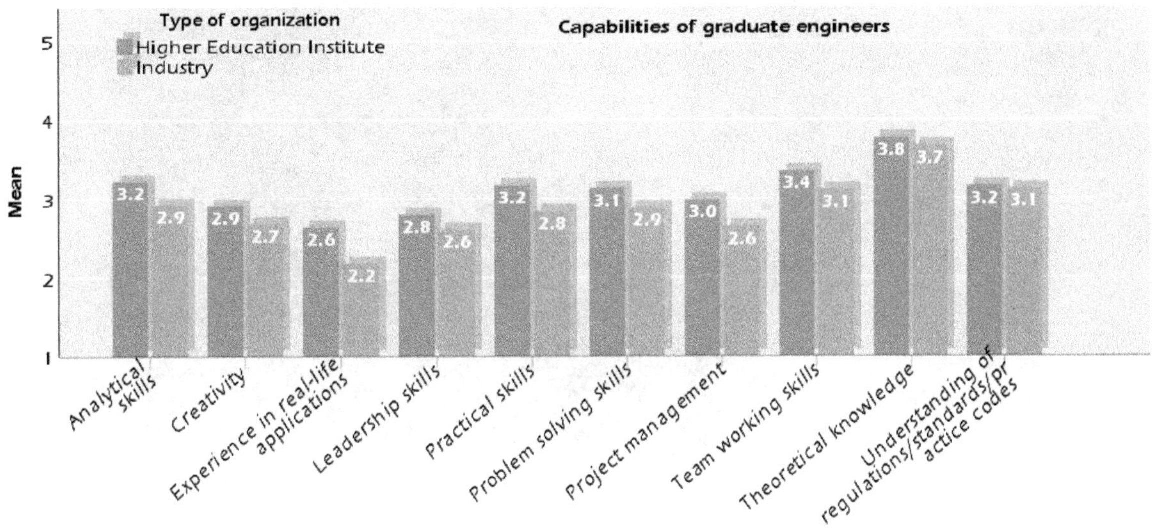

The topics that the graduate engineers need to have knowledge and understanding are presented in Fig. 14, both from the HEIs' and Industry's perspective, using means bars of their response's rating (from 1: for "most needed" to 10: for "least needed").

Regarding the external quality assurance to the HEIs programmes, Figure 15 presents a detailed view.

Figure 11. Industry's support for HEIs to deliver current engineering degree programmes

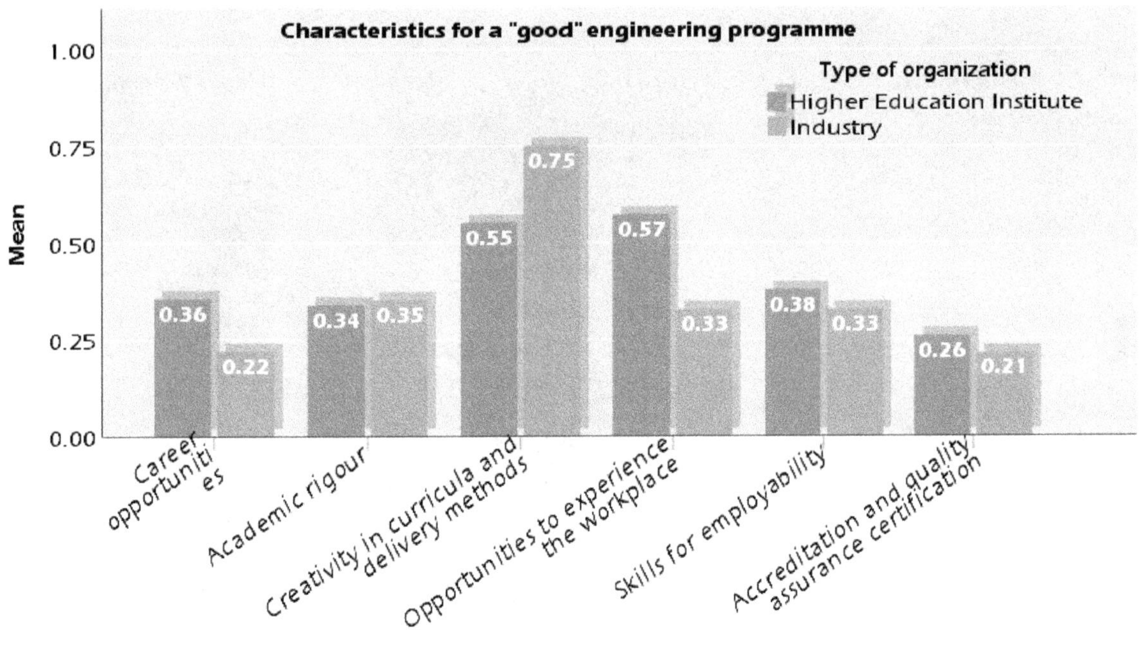

Figure 12. Mean ratings of the graduate engineers' capabilities from the HEIs' and Industry's perspective

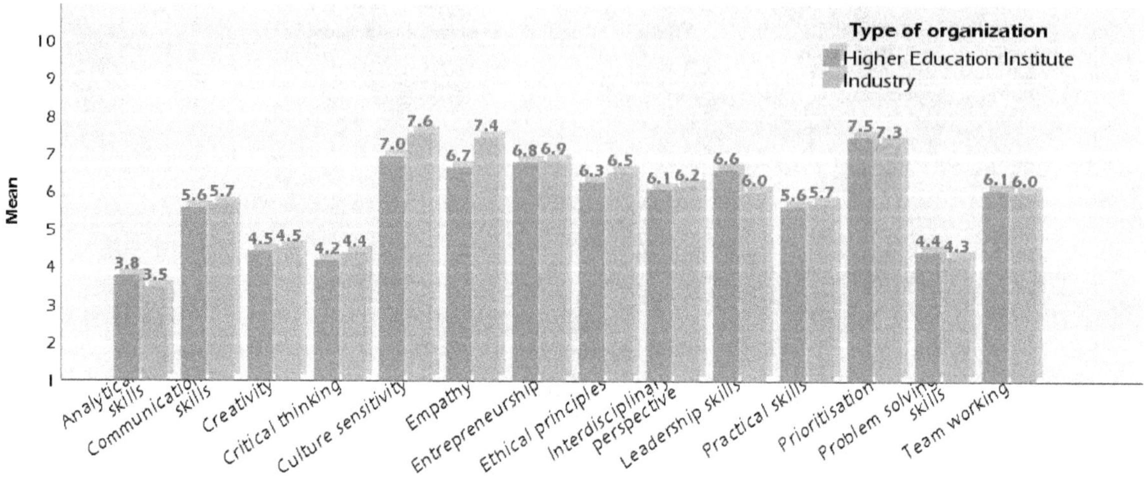

Finally, the responses concerning the existence of a programme related to Humanitarian Engineering, are provided in Figure 18

Quality Assurance Agencies Perception

In our sample, almost 70% of the questioned Quality Assurance (QA) agencies are from Europe (consisted mostly of *Agencies* and *Authorities*) and the rest from Asia (consisted from various types of QA agencies with *Department* type being the most common), and only 1% from USA.

The remit or scope of the European QA agencies is mainly for *Specific learning discipline/field* and for *HEIs,* while Asian QA agencies are more dispersed in their scope. The above differentiation of QA

Figure 13. Mean ratings of the characteristics of a "good" engineering programme from the HEIs' and Industry's perspective

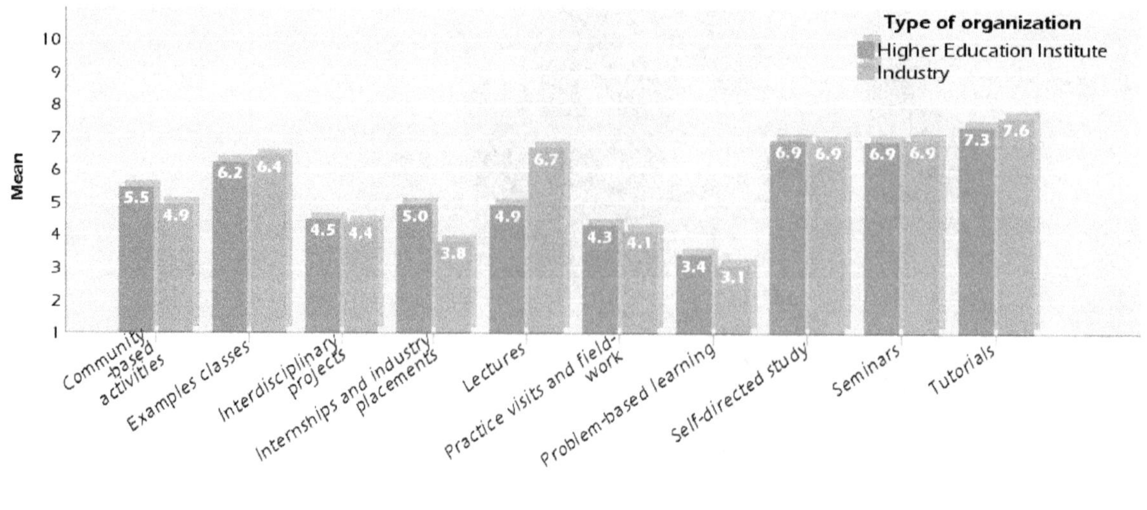

Figure 14. Mean ratings of the attributes needed to be developed by graduate engineers from the HEIs' and Industry's perspective

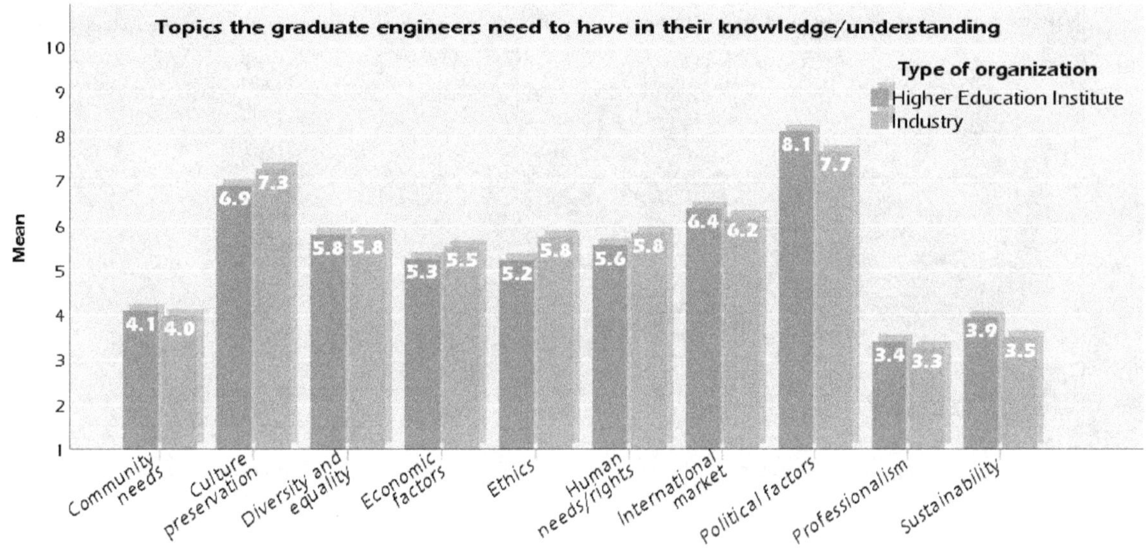

agencies over their scope is statistically true ($p < 0.001$ for the chi-square test for the scope between European and Asian QA agencies).

The importance of various QA activities/resources to the quality review of programmes is presented in Figure 17, where the mean values are used of the response ratings (from 0: for "Unimportance" to 5: for "Essential").

The models/frames of reference used by QA agencies for an external quality review regarding HEIs programmes are presented in Figure. 20, where the mean values of the response ratings were used (0: for "No" and 1: for "Yes").

Figure 15. Mean ratings of the activities that could develop needed attributes of graduate engineers from the HEIs' and Industry's perspective

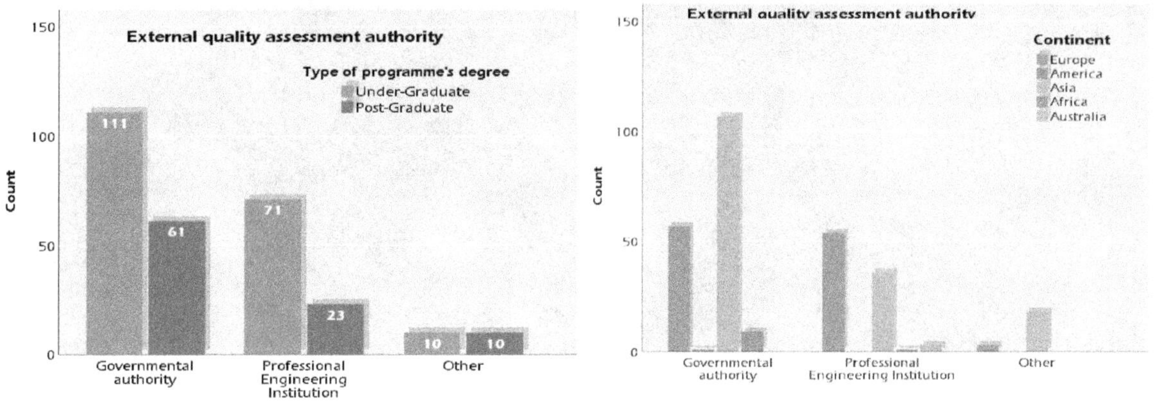

Figure 16. Mean ratings of the topics the graduate engineers needed to have from the HEIs' and Industry's perspective

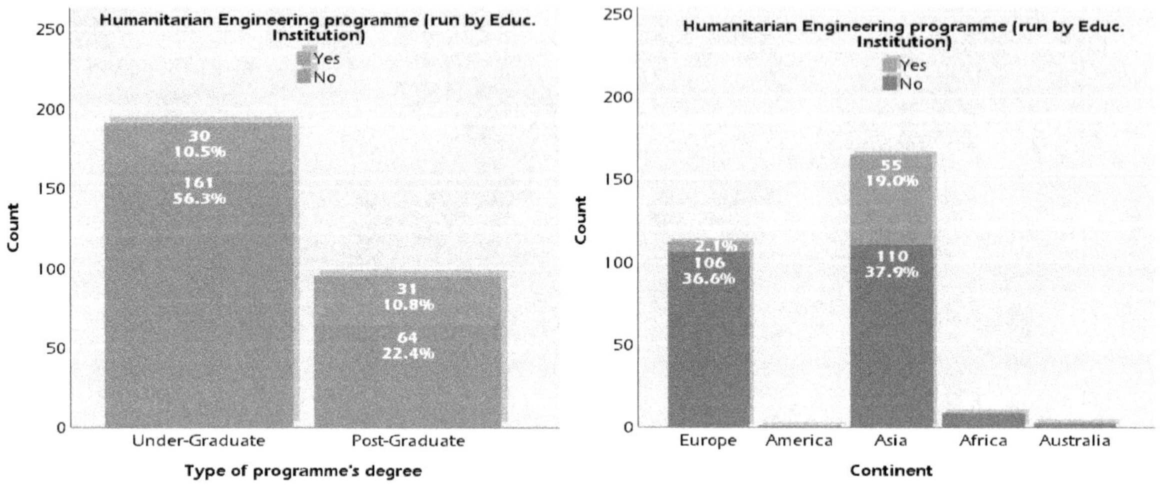

The various problematic areas identified during a programme-level QA external periodical review are presented in Figure 20 via the mean values of the responses ratings (from 1: for "Most problematic" to 10: for "Least problematic").

Finally, Figure 21 offers a detailed view regarding the HEIs most needed QA capacity building areas with direct impact to the delivered programmes, by presenting the mean values of the responses ratings (0: for "No" and 1: for "Yes") for 10 suggested areas.

Figure 17. External quality assessment authorities for the HEIs programmes

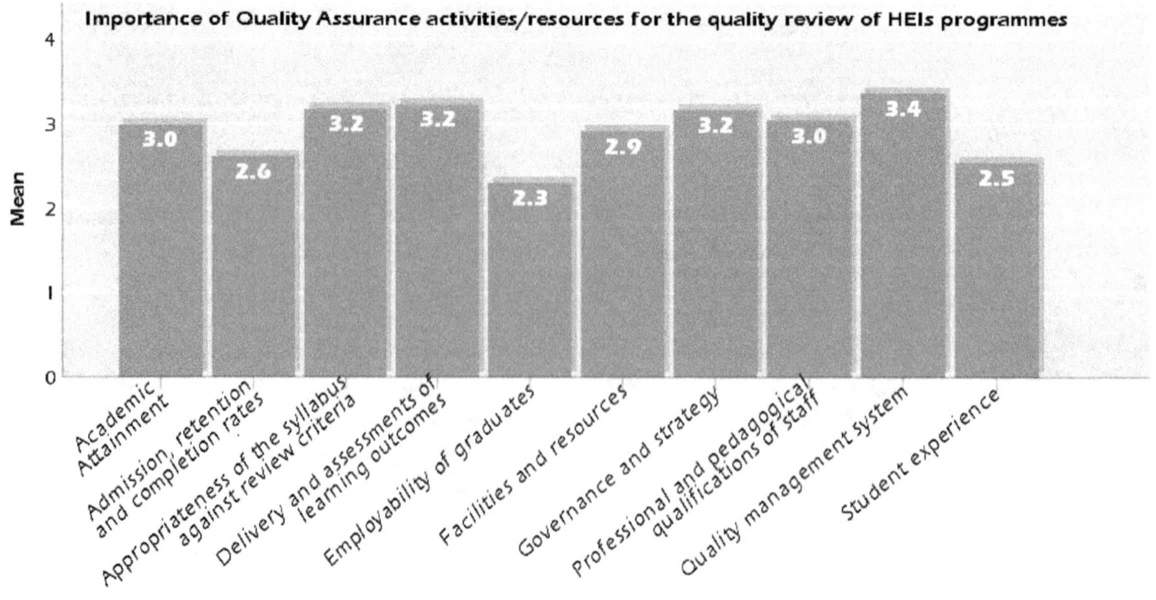

Figure 18. Humanitarian Engineering programmes per type of programme and per continent

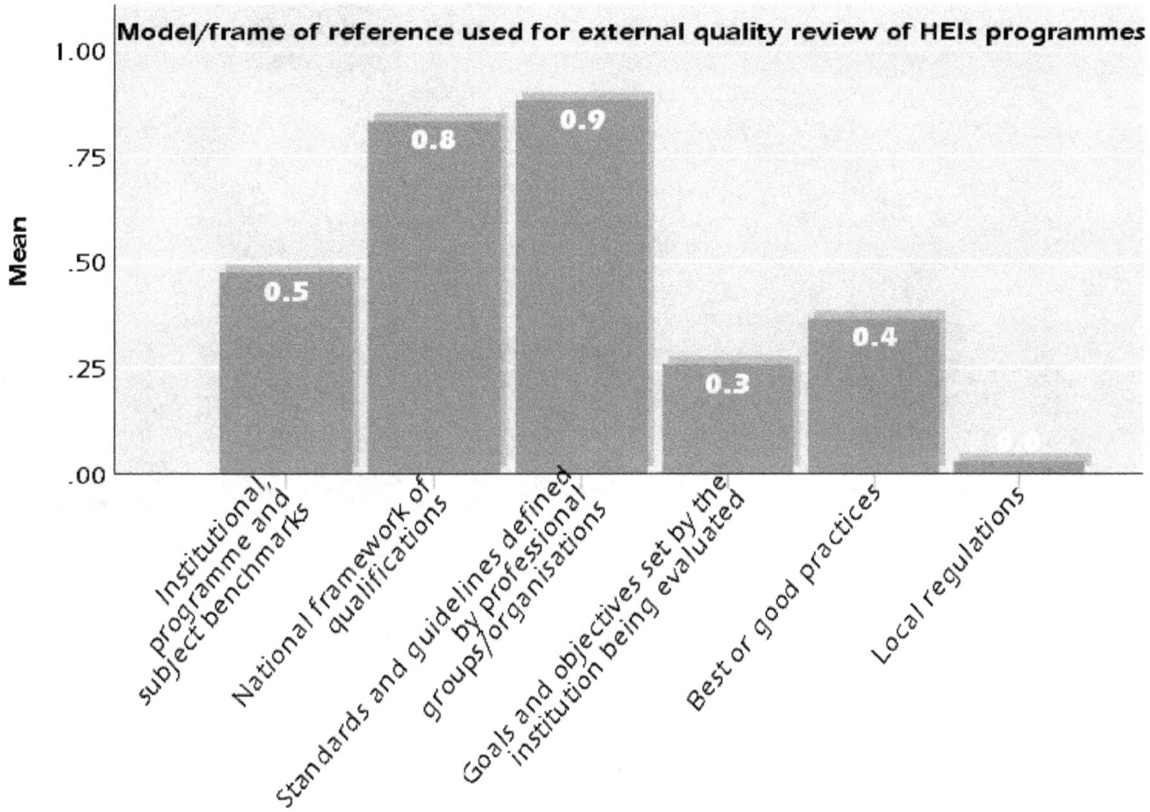

Figure 19. Importance in various QA activities/resources to the quality review of programmes

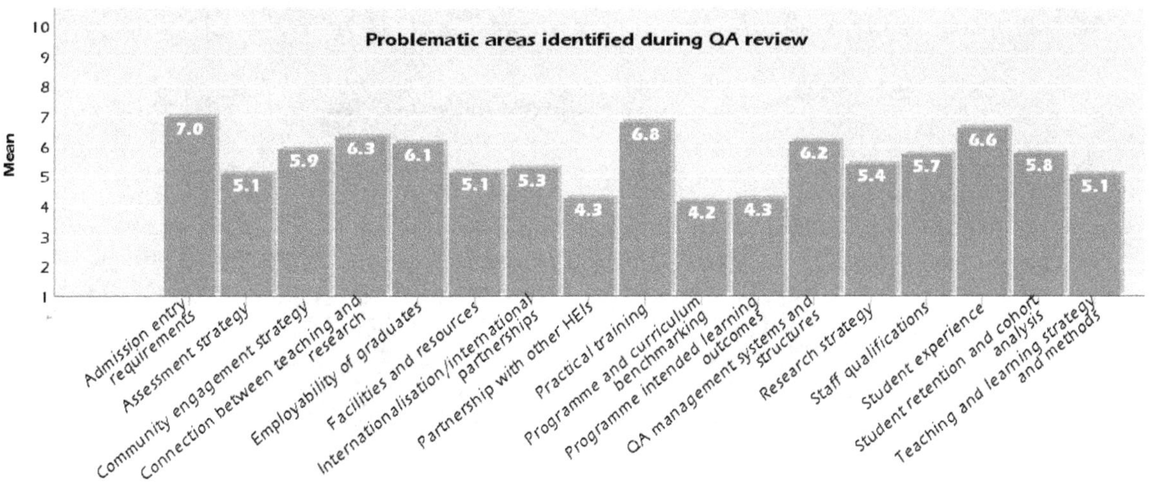

Figure 20. Model/frame of reference used by QA agencies for external quality review of HEIs programmes

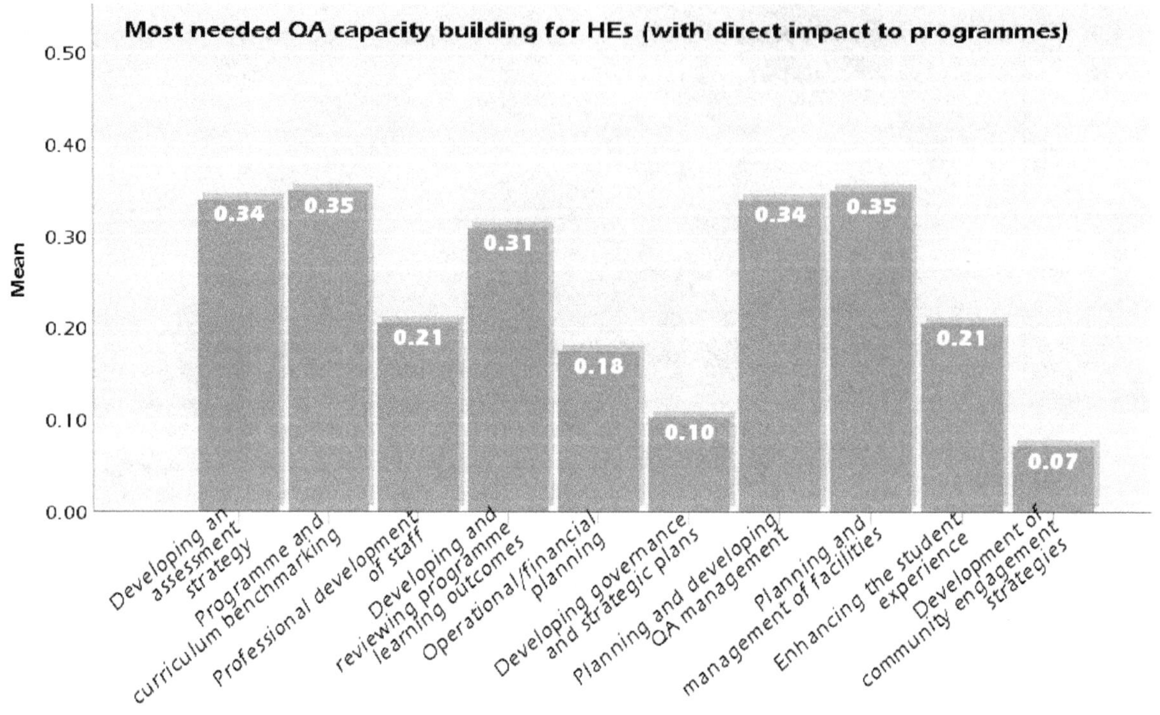

DISCUSSION

This section reveals the different perspectives of the examined organisations across different topics, based on the collected results.

Figure 21. Problematic areas identified during programme-level QA review

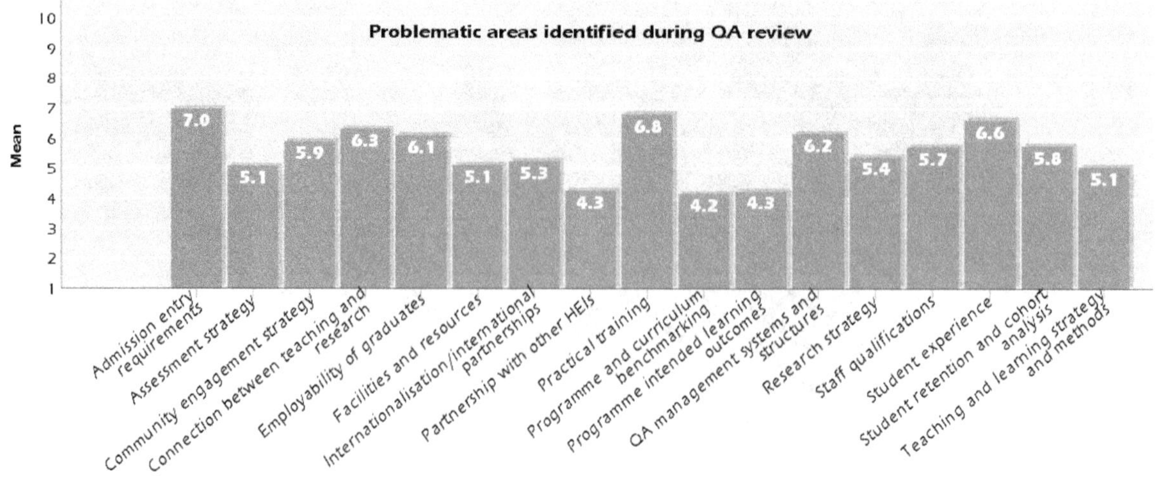

Figure 22. HEIs most needed QA capacity building areas with direct impact to delivered programmes

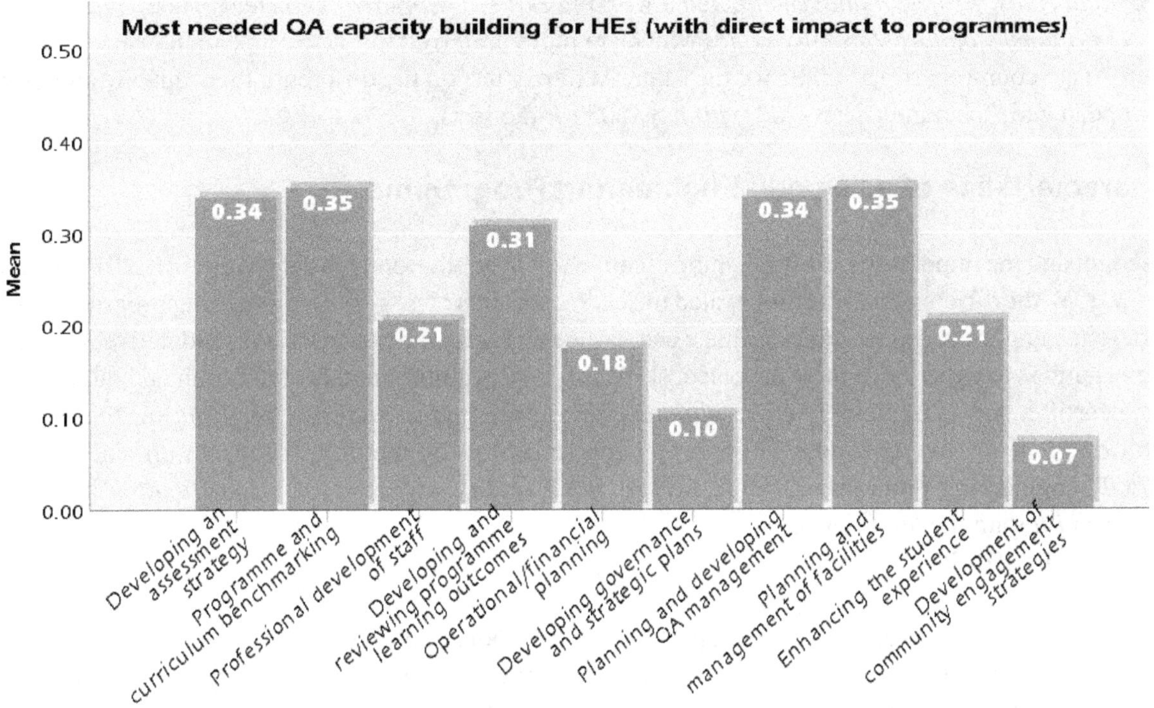

Industry Perception About The Country's Skilled Graduate Engineers

Canada and Malaysia appear to be more confident (on average) among all nine countries. Excluding these two countries due to very few data, it is safe to say that Bangladeshi and Indonesian industries appear to be more confident about their skilled graduate engineers with the UK industries being least confident. The above result is statistically true since all countries have (in a 95% confidence interval) a different probability outlook regarding their confidence about the amount of high-skilled graduate engineers ($p = 0.019 < 0.05$ for the Kruskal-Wallis test). North America seems to be more confident about the amount of high-skilled engineering staff, while Asian industry is more confident than the European one ($p = 0.02 < 0.05$ for the corresponding chi-square test between continents).

The most developed overall capabilities, according to the Industry point of view, are *Theoretical knowledge* followed by *Understanding of regulations/standards/practice codes* while the least developed are *Experience in real-life applications* followed by *Project management*. European industries share the same profile on these capabilities except *Analytical skills* (UK on top), *Creativity* (Germany on top), *Leadership skills* (UK on top), *Problem solving skills* (Germany on top), *Team working skills* (UK on top) and *Theoretical knowledge* (Greece on top) ($p < 0.05$ for the Kruskal-Wallis test of the differentiating capabilities between European industries, and with $p > 0.05$ for the rest). Asian industries share the same profile on capabilities except *Leadership skills* (UK on top), *Problem solving skills*, *Team working skills* and *Theoretical knowledge* (Bangladeshi on top) ($p < 0.05$ for the Kruskal-Wallis test of the differentiating capabilities among Asian industries with $p > 0.05$ for the rest). Distinguishing between the European and Asian industries, they differentiate regarding the capabilities of *Creativity* and *Experi-*

ence in real-life applications which are more preferred in the Asian industry than the European one, and *Practical skills*, *Project management*, *Team working skills*, *Theoretical knowledge*, and *Understanding of regulations/ standards/practice codes* which are more preferred for the European industry rather than their Asian counterparts ($p < 0.05$ for the Mann-Whitney test on the differentiating capabilities between European and Asian industries, and with $p > 0.05$ for the rest).

Characteristics of a "Good" Engineering Programme

Recognising the importance of establishing creative learning environments (Davies *et al.*, 2013; Khatri *et al.*, 2017)), the different organisations rated the characteristics of a "good" engineering programme in six different categories: carrier opportunities; academic rigor; creativity in curricula and delivery methods; opportunities to experience the workplace; skills for employability; and accreditation and quality assurance certification. *Creativity in curricula and delivery method* as well as *Opportunities to experience the workplace* are the most desired/important characteristics by the industry across all countries for a "good" engineering programme, while *Accreditation and quality assurance certification* and *Career opportunities* are the least desired.

More specifically:

1. *Career opportunities* are very important for the Canadian and USA industry, followed by the Bangladeshi, but not so important for all the European countries under study. The Asian industry believe that *Career opportunities* are more important than the European counterparts.
2. *Academic rigor* is more important for Greek industry followed by the USA, but not for the Asian industry. For the European industry, this characteristic is more important than the Asian industry thinks.
3. *Creativity in curricula and delivery methods* are overall important, and most important for Germany and USA industry. This characteristic is also more important for the European industry that the Asian one.
4. *Opportunities to experience the workplace* are most important for the German and USA industry and less important. This characteristic is also more important for the European industry that the Asian one.
5. *Skills and employability* have the same low percentage across all countries, and between European and Asian industry.
6. *Accreditation and quality assurance certification* are important for the German and Malaysian industry. This characteristic is also more important for the Asian industry that the European one.

For the relation of the responses between the HEIs and the, we statistically infer that Industry and HEIs differ in their opinion across all given characteristics except *Academic rigor* and *Skills for employability* ($p = 0.782 > 0.05$ and $p = 247 > 0.05$ for the corresponding chi-square tests regarding these two characteristics between HEIs and Industry, while all the other p-values are < 0.05). In particular, *Career opportunities, Opportunities to experience the workplace* and *Accreditation and quality assurance certificates*, are all rated higher by the HEIs than the Industry while, on the contrary, *Creativity and curricula and deliver methods* are under-rated by the HEIs in relation to the Industry. In a more detailed view, in relation between European and Asian (regarding the HEIs and Industry perspective), there is no difference (between HEIs and Industry) in all five characteristics except *Creativity in curricula and delivery methods* ($p = 0.003 < 0.05$) and *Opportunity to experience the workplace* ($p < 0.001$) among European countries, while there is also no difference in *Academic rigor* ($p = 0.182 > 0.05$), *Skills for employability* ($p = 0.335 > 0.05$) and *Accreditation and quality assurance certification* ($p = 0.184 > 0.05$) among Asian countries. Restricted to HEIs perspective, European HEIs are rating (on average) higher than the Asian counterparts the characteristics of *Academic rigor* and *Creativity in curricula and delivery methods* while the Asian HEIs rating higher than the Europeans the characteristics of *Career opportunities* and *Accreditation and quality assurance certificates* (p-values < 0.05 for the corresponding chi-square tests).

Attributes of the Graduate Engineers That Need To Be Developed

European industries share the same profile except *Analytical skills* (more important by German industry) and *Ethnical principles* (more important by UK's industry) ($p < 0.05$ for the Kruskal-Wallis test of the differentiating attributes among European industries, and with $p > 0.05$ for the rest). The Asian industries share the same profile on capabilities except *Creativity* (more important by Bangladeshi industry), *Critical thinking* (Indonesia's most important) and *Practical skills* (more important by Malaysian industry) ($p < 0.05$ for the Kruskal-Wallis test of the differentiating capabilities among Asian industries, with $p > 0.05$ for the rest). Distinguishing between the European and Asian industry, they differentiate regarding the attributes of *Communication skills, Culture sensitivity, Entrepreneurship, Ethnical principles, Practical skills* and *Prioritisation* which are more important to the Asian rather than the European industry, and *Critical thinking*, *Interdisciplinary perspective*, and *Problem solving skills*, more important to the European rather than the Asian industry ($p < 0.05$ for the Mann-Whitney test on the differentiating capabilities between European and Asian industries, and with $p > 0.05$ for the rest). We statistically infer that HEIs and Industry responded (on average) with the same way regarding all attributes, i.e. HEs and Industry have the same opinion on every attribute (all p-values for the Mann-Whitney tests are > 0.05). In a more detailed view, HEIs and Industry differ in their opinion about these attributes in *Communication skills* ($p = 0.027 < 0.05$ for the corresponding chi-square test) and *Leadership skills* ($p = 0.029 < 0.05$) while all Asian countries have the same opinion for all 14 attributes. Restricted to HEIs perspective, European HEIs prefer (on average) more than the Asian counterparts the attributes (that need to be developed in graduate students) of *Creativity*, *Interdisciplinary perspective* and *Problem-solving skills* while the Asian HEIs prefer more than the Europeans the attributes of *Communication skills*, *Leadership skills* and *Team working* (p-values for the above capabilities are < 0.05 for the corresponding Mann-Whitney tests).

Teaching and Learning Activities To Develop Needed Attributes

European industries share the same profile on the teaching/learning that could develop the aforementioned needed attributes of the graduate students, except for *Example classes*, *Lectures* and *Tutorials* (all are more preferred by German industry) and *Internships and industry placements* (more preferred by Greece industry) ($p < 0.05$ for the Kruskal-Wallis test of the differentiating attributes among European industries, and with $p > 0.05$ for the rest). Asian industries share the same profile on teaching/learning activities except *Problem-based learning* (more preferred by Indonesian industry) and *Self-directed study* (more preferred by Vietnamese industry) ($p < 0.05$ for the Kruskal-Wallis test of the differentiating capabilities among Asian industries under study, with $p > 0.05$ for the rest). Distinguishing between the European and Asian industry regarding these activities, they differentiate in *Examples classes*, *Lectures* and *Practice visits and field-work,* which are more preferred by the Asian industry rather than the European one, and *Interdisciplinary projects*, *Interdisciplinary perspective* and *Problem-based learning*, more preferred by the European industry rather than the Asian counterparts ($p < 0.05$ for the Mann-Whitney test on the differentiating capabilities between European and Asian industries, and with $p > 0.05$ for the rest).

We statistically infer that HEIs and Industry responded (on average) with the same way (and thus they share the same opinion) regarding all the suggested attributes except from *Internships and industry placements* and *Lectures* ($p < 0.001$ for both Mann-Whitney tests concerning the two exceptions while $p > 0.05$ for the rest Mann-Whitney tests). In particular, the HEIs over-rated the internship activity (on average) in relation to the Industry's rating while under-rated the lecturing activity. The bottom clustered bar chart shows the mean response across all continents. In a more detailed view, HEIs and Industry differ in their opinion about these activities in *Community-based activities* ($p = 0.002 < 0.05$ for the corresponding chi-square test), *Lectures* ($p = 0.031 < 0.05$), *Self-directed study* ($p = 0.049 < 0.05$) and *Seminars* ($p = 0.001 < 0.05$) among the European countries as well as in *Internship and industry placements* ($p = 0.004 < 0.05$) and Lectures ($p = 0.001 < 0.05$). Restricted to HEIs perspective, European HEIs prefer (on average) more than the Asian counterparts the activities (that could develop the needed attributes of graduate engineers) of *Interdisciplinary projects*, *Internships and industry placements* and *Problem-based learning* while the Asian HEIs prefer more than the Europeans the activities of *Examples cases*, *Lectures* and *Practice visits and field work* (p-values for the above capabilities are < 0.05 for the corresponding Mann-Whitney tests while for the rest $p < 0.05$).

Ten Topics That Graduate Engineers Need To Know

European industries share the same profile on these topics except *Ethics* (more needed by the German industry) and *Human needs/rights* (more needed by the Greek industry) ($p < 0.05$ for the Kruskal-Wallis test of the differentiating topics among European industries, and with $p > 0.05$ for the rest). Asian industries share the same profile regarding all of these ($p > 0.05$ for the Kruskal-Wallis test for all the topics among Asian industries). Distinguishing between the European and Asian industry regarding these topics, they differentiate in *Diversity and equality* and *Sustainability* which are more needed by the European industry, and *Culture preservation*, *Economic factors*, *Ethics* and *International markets* more needed by the Asian industry ($p < 0.05$ for the Mann-Whitney test on the differentiating topics between European and Asian industries, and with $p > 0.05$ for the rest). We statistically infer that HEIs and Industry share the same opinion regarding all the topics except *Sustainability* ($p = 0.043 < 0.05$ for Mann-Whitney test while $p > 0.05$ for the rest Mann-Whitney tests). More specifically, HEIs over-rated the *Sustainability* (on average) in relation to the Industry's rating. In a more detailed view, HEIs and Industry differ in their opinion about these topics in *Community needs* ($p = 0.028 < 0.05$ for the corresponding chi-square test), *Culture preservation* ($p = 0.027 < 0.05$) and *Sustainability* ($p = 0.01 < 0.05$) among the European countries while all Asian countries have the same ratings with each other. Restricted to HEIs perspective, European HEIs need (on average) more than the Asian counterparts the topics of *Diversity and equality*, *Professionalism* and *Sustainability* while the Asian HEIs need more than the Europeans the topics of *Culture preservation* and *Human needs/rights* (p-values for the above capabilities are < 0.05 for the corresponding Mann-Whitney tests while for the rest $p < 0.05$).

INDUSTRY COLLABORATION WITH HEIS

The Asian industry ranks their involvement more positive than the others. On average, German, Malaysian and Bangladeshi industry have a higher level of involvement while USA has the lower one. In general, the Asian industry collaborating more with the HEIs (in delivering current programmes) with the corresponding USA less. Moreover, the industry across all the Asian countries have the same level of involvement. The above results are statistically confirmed ($p < 0.001$ for the Kruskal-Wallis test for the collaboration ratings). European and Asian industries share the same outlook regarding their supporting type to HEIs except *Assist in building capacity by co-designing curricula*, *Offer internships and/or placements to students* and *Organise and deliver visits for students at the industry* which are more preferred types of support by the European industry, while *Act as industry mentor to students* and *Fund student research and teaching activities* are more preferred support types by the Asian industry.

Capabilities of Graduate Engineers

Ten main capabilities of the graduate engineers were examined, namely: analytical skills; creativity; experience in real life applications; leadership; practical skills; problem solving; project management; team working; theoretical knowledge; understanding of regulations/standards/practice codes. Industry and HEIs differ in their opinion for the graduate engineers' capabilities across all factors except *Theoretical Knowledge* ($p = 0.376 > 0.05$ for the chi-square test regarding the theoretical knowledge of the graduate engineers between HEIs and Industry, while all the other p-values are < 0.05). All capabilities, except *Theoretical Knowledge*, are rated by the HEIs higher than the Industry, and therefore there is a consistent gap between the opinions of HEI's and the Industry, with HEIs seem to over-rating their graduate students. In a more detailed view, in relation to European and Asian HEIs and Industries, there is no difference (between HEIs and Industry) in *Creativity* both for European ($p = 0.056 > 0.05$) and Asian countries ($p = 0.282 > 0.05$). For European countries there is no difference in *Leadership skills* ($p = 0.108 > 0.05$), *Problem solving skills* ($p = 0.087 > 0.05$), *Project management* ($p = 0.229 > 0.05$) and *Team working skills* ($p = 0.577 > 0.05$), and in *Theoretical knowledge* for Asian countries ($p = 0.416 > 0.05$). Restricted to HEIs perspective, European HEIs are rating (on average) higher than the Asian counterparts their graduate engineers' capabilities in *Theoretical knowledge* and *Understanding of regulations/standards/practice codes* while the Asian HEIs rating higher than the Europeans their capabilities in *Creativity*, *Experience in real-life applications* and *Leadership skills*.

Quality Assurance of Graduate Programmes

A difference of the assessment authority is depending upon the programme's type ($p = 0.045 < 0.05$ for the chi-square test between under- and post-graduate programmes) with post-graduate programmes prefer *Governmental authority* for their assessment rather than a *Professional engineering institution*. There is also a strong dependence of the assessment authorities among the European HEIs ($p < 0.001$ for the chi-square test), while there is no dependence of the assessment authority among and Asian HEIs ($p = 0.059 > 0.05$ for the chi-square test). However, Asian HEIs prefer more the *Governmental authority* than the European do and vice-versa (total $p < 0.001$ for the chi-square test).

Moreover, most of the European QA agencies have a review period of 5 years while the Asian ones have 5 years as well as reviews upon request. On average, the frequency of reviewing for Asian

QA agencies are quite higher (every 2.8 years) than the Europeans (every 4.8 years). This frequency difference is also statistically confirmed ($p < 0.001$ for the chi-square test for the review frequencies between European and Asian QA agencies). The formal consequences of both European and Asian QA reviews are mainly on *Approval/Licensing of a programme* with some Asian QA reviews not to have any consequence. However, this concentration around almost one consequence (*Approval/Licensing of a programme*) of the European QA reviews, differentiates overall the formal consequences outlook between European and Asian QA reviews ($p < 0.001$ for the chi-square test for the formal reviewing consequences between European and Asian QA agencies). Overall, 80% of the QA agencies are also responsible for the approval of new programmes and only 20% not. In particular, most of the European QA agencies are responsible for the approval of new programmes while this is not true for their Asian counterparts ($p < 0.001$ for the corresponding chi-square test).

Overall, the most important activity is the *Quality management system* followed close by the *Delivery and assessment of learning outcomes*, *Appropriateness of the syllabus against review criteria* and *Governance and strategy*. European QA agencies differentiate with each other regarding the importance of all the activities/resources except *Academic attainment* and *Student experience*. In particular, UK QA agencies believe (on average) that *Admission, Retention and completion rates* are more important, Italian QA agencies believe mostly for the importance of the *Appropriateness of the syllabus against review criteria* and *Governance and strategy*, *Delivery and assessment of learning outcomes*, *Employability of graduates* and *Facilities and resources*, while the Greek QA agencies in *Governance and strategy*, *Professional and pedagogical qualifications of staff*, and *Quality management system* ($p < 0.05$ for the corresponding Kruskal-Wallis tests is translated as difference regarding the specific activity ratings among European countries, while $p > 0.05$ as equality of the ratings). On the other hand, the Asian QA agencies share the same level of importance for all the activities/resources ($p > 0.05$ for all the Kruskal-Wallis tests between the Asian QA agencies). In comparison between the European and Asian QA agencies, both share the same opinion regarding *Appropriateness of the syllabus against review criteria*, *Delivery and assessment of learning outcomes*, *Governance and strategy*, *Professional and pedagogical qualifications of staff* and *Quality management system* ($p > 0.05$ for the corresponding Kruskal-Wallis tests), while Asian QA agencies believe more in the importance of *Academic attainment*, *Admission, Retention and completion rates*, *Employability of graduates*, *Facilities and resources* and in *Student experience* ($p < 0.05$ for the corresponding Mann-Whitney tests).

The most used model/frame of reference used by QA agencies for an external quality review regarding HEIs programmes is the *Standards and guidelines defined by professional groups/organisations* (mean rating 0.9) followed by the *National framework of qualifications* (mean rating 0.8). Statistically speaking, European QA agencies share the same models/frames of reference except (strongly differentiate) the use of *Institutional programme and subject benchmark*, with UK and Italian QA agencies utilise this model the most, and *Best or good practices*, with Greek QA agencies preferring this model the most ($p < 0.001$ for the chi-square tests for both excluding models, and $p > 0.05$ for the rest models). Note also that for the *National framework of qualifications*, all European QA agencies used this the most while they all do not use at all *Local regulations*. On the other hand, all the Asian QA agencies share the same preference for each model/frames ($p > 0.05$ for the chi-square tests among all Asian QA agencies). Distinguishing between European and Asian QA agencies, they strongly differentiate in their preference in all models/ frames except the *Institutional, programme and subject benchmarks* and the *Best or good practices* ($p > 0.05$ for the chi-square tests of the excluding models and $p < 0.05$ for the rest) with Asian ones preferring more the *Goals and objectives set by the institution being evaluated* than the European ones.

Overall, the most problematic areas during a programme-level QA external periodical review are *Programme and curriculum benchmarking* (mean 4.2) followed by *Programme intended learning outcomes* (mean 4.3) and *Partnership with other HEIs* (mean 4.3), while Admission entry requirements (mean 7.0) and Practical training (6.8) are the least problematic areas. Statistically, the European QA reviews identified the same problematic areas except *Assessment strategy* (most problematic for UK), *Employability of graduates* (most problematic for Greece), *Facilities and resources* (most problematic for Greece), *Programme and curriculum benchmarking* (most problematic for Italy), *Programme intended learning outcomes* (most problematic for UK), *QA management systems and structures* (most problematic for Greece), *Student experience* (most problematic for Greece), and *Teaching and learning strategy and methods* (most problematic for UK/Italy) ($p < 0.05$ for the Kruskal-Wallis tests for the above exceptions and $p > 0.05$ for the rest). The Asian QA reviews, however, identified that each problematic area has the same problematic outlook (rating) for all Asian countries ($p > 0.05$ for all the corresponding Kruskal-Wallis tests). In comparison between the European and Asian QA reviews, the European QA reviews identified to be more problematic than the Asian ones in the following areas: *Partnership with other HEIs* and *Programme and curriculum benchmarking*, while the Asian QA reviews identified to be more problematic than the European ones in *Community engagement strategy*, and in all the other problematic areas European and Asian QA reviews share (on average) the same ratings, i.e. the same problem level ($p < 0.05$ for the Mann-Whitney tests for the differentiating areas and $p > 0.05$ for the rest).

Finally, the most needed QA capacity-building areas with a direct impact to the delivered programmes are *Planning and management of facilities* (mean 0.35) and *Programme and curriculum benchmarking* (mean 0.35) followed by *Planning and developing QA management* (mean 0.34) and *Developing an assessment strategy* (mean 0.35). It is statistically inferred that the European HEIs differentiate regarding their most needed QA capacity-building areas of *Developing an assessment strategy* (most needed in UK), *Programme and curriculum benchmarking* (most needed in Italy), *Professional development of staff* (most needed in UK), *Developing and reviewing programme learning outcomes* (most needed in Italy), *Planning and developing QA management*, *Planning and management of facilities*, and *Enhancing the student experience* (most needed in Greece for the last three) ($p < 0.05$ for the chi-square tests of the differentiating areas and $p > 0.05$ for the rest). In contrast, all the Asian HEIs under study share the same need for each of the QA capacity building areas expect only for *Programme and curriculum benchmarking* (data only for Bangladesh). Distinguishing between Europe and Asia, European HEIs needing more than the Asian the QA capacity building areas like *Programme and curriculum benchmarking*, and *Planning and management of facilities*, while Asian HEIs needing more than the European counterparts the QA capacity building areas like Developing and reviewing programme learning outcomes and *Developing governance and strategic plans* ($p < 0.05$ for the chi-square tests of the differentiating areas and $p > 0.05$ for the rest).

CONCLUSION

Engineers are seen as problem solvers, and engineering has a pivotal role to play in addressing contemporary global challenges. Enabling attributes such as creativity, critical thinking and problem solving to engineering graduates seems to be imperative for enhancing and expanding their engagement to complex, interdisciplinary problems linked to humanitarian challenges. Tackling humanitarian challenges requires knowledgeable and erudite engineers who can handle, combine, transform and create innovative, affordable and sustainable solutions.

This view simultaneously complements and challenges current concepts of an emerging educational movement that, almost without exception, are underpinned by calls for competitive economic growth and technological development.

Within ENHANCE, the optimal setup and systematic design of effective engineering education programmes were explored in alignment with the delivery of engineering graduate attributes for tackling contemporary engineering challenges. The survey explored how intended and desired engineering graduate attributes are aligned with professional, societal and global needs, identified by industry, government/quality assurance agencies and societal needs and how these can be best delivered by deploying effective innovative pedagogies.

Complex engineering problems such as contemporary humanitarian challenges require humanitarian attributes to be enabled to professional engineers through reformed curricula and innovative learning strategies, which once implemented and integrated efficiently in higher engineering education, shall provide students and educators with opportunities to explore interdependencies and links between traditional engineering attributes with skills such as leadership and social influence, international awareness, sustainability, emotional intelligence, professionalism, cultural sensitivity and ethics, and to critically engage with implicit and explicit facets of other disciplinary identities.

Our research demonstrated that interdisciplinary, inter-professional discussions within engineering curricula can be productive, with stakeholders expressing their willingness to support rigorous, less siloed curricula and learning strategies.

ACKNOWLEDGMENT

This study was conducted within the framework of the project ENHANCE – Enabling Humanitarian Attributes for Nurturing Community-based Engineering co-funded by the Erasmus+ Programme, Key Action 2 Cooperation for innovation and the exchange of good practices, Capacity building in higher education, Joint Projects (Project No: 598502-EPP-1-2018-1-UK-EPPKA2-CBHE-JP (2018-2582/001-001). This publication reflects the views only of the authors, and the Commission cannot be held responsible for any use which may be made of the information contained therein.

REFERENCES

Al Bawaba. (n.d.). *BCG Releases "The Future of Education: How Governments Can Help Close the 21st Century Gap" Report*. Available at: https://www.albawaba.com/business/pr/bcg-releases-%E2%80%98-future-education-how-governments-can-help-close-21st-century-gap%E2%80%99-report-

Asian Development Bank. (2015). *Innovations in Knowledge and Learning for Competitive Higher Education in Asia and the Pacific*. Author.

British Council. (2015). *Skills needed: Addressing South Asia's deficit of technical and soft skills Analysing the gap in*. Available at: https://www.britishcouncil.org/sites/default/files/skills_gap_jan2015_print.pdf

Davies, D., Jindal-Snape, D., Collier, C., Digby, R., Hay, P., & Howe, A. (2013). Creative learning environments in education—A systematic literature review. *Thinking Skills and Creativity*, *8*, 80–91. doi:10.1016/j.tsc.2012.07.004

Gill, J., Sharp, R., Mills, J., & Franzway, S. (2008). I still wanna be an engineer! Women, education and the engineering profession. *European Journal of Engineering Education*, *33*(4), 391–402. doi:10.1080/03043790802253459

Kelley, T. R., & Knowles, J. G. (2016). A conceptual framework for integrated STEM education. *International Journal of STEM Education*, *3*(1). Available at: https://stemeducationjournal.springeropen.com/articles/10.1186/s40594-016-0046-z

Khatri, R., Henderson, C., Cole, R., Froyd, J. E., Friedrichsen, D., & Stanford, C. (2017). Characteristics of well-propagated teaching innovations in undergraduate STEM. *International Journal of STEM Education*, *4*(1), 2. doi:10.118640594-017-0056-5

UNESCO. (2014). *Education Systems in ASEAN+6 Countries: A Comparative Analysis of Selected Educational Issues*. Education Policy Research Series Discussion Document No. 5, United Nations Educational, Scientific and Cultural Organization.

Unger, M., & Polt, W. (2017). The Knowledge Triangle between Research, Education and Innovation – A Conceptual Discussion. *Foresight and STI Governance*, *11*(2), 10–26. doi:10.17323/2500-2597.2017.2.10.26

United Nations. (2005). *Hyogo Framework for Action 2005-2015: Building the Resilience of Nations and Communities to Disasters*. A/CONF.206/6.

United Nations. (2015). *Sendai framework for disaster risk reduction 2015–2030*. United Nations Office for Disaster Risk Reduction.

United Nations General Assembly. (2015). *Transforming our world: the 2030 Agenda for Sustainable Development*. A/RES/70/1.

Chapter 9
A Coaching Approach to Foster Scientific Production in Engineering Schools and Research Institutions

Kenza Khomsi
https://orcid.org/0000-0003-3723-5498
General Directorate of Meteorology, Morocco

ABSTRACT

Coaching has been efficient in many aspects: career, teamwork, improving skills, enhancing performance, well-being, and mental health. In higher education settings, studies found that coaching is a promising methodology for building sustainable research careers. It may be an ideal methodology to improve the quality of scientific research work worldwide and could help students and research staff to grow personally and professionally and to thrive in their chosen careers. In the chapter, the author suggests the Mindset-Learning-Knowledge-Sharing (MLKS) model as a new tool for research project coaching. Working through this model is a journey that is similar to the hero's journey of Joseph Campbell. It starts with the call to the adventure and leads to a complete unknown world that facilitates learning, knowledge, and growth and creates sustainable impact on the scientific world. It encourages scientific productivity and achievement and helps unleash the innovation and creativity potential within every researcher.

INTRODUCTION

We use Humanitarian Engineering (capital H and E) to refer to specialized areas of practice involving humanitarian response and community development (Park et al., 2021). Related terms may include Engineering for Developing Communities (Smith, 2019), Global Engineering (Thomas et al., 2021), and Development Engineering (Nilsson et al., 2014). All these concepts aim at emphasizing long-term global community development. For instance, Humanitarian Engineers may focus on implementing solutions that can impact developing countries while considering social–cultural issues (Shekar & Drain, 2016).

DOI: 10.4018/978-1-6684-5619-4.ch009

This highlights the importance of Humanitarian Engineers ability to learn about cultures and engineering practices and the essential role educational systems can play in preparing generations of Humanitarian Engineers. At higher education institutions including Humanitarian Engineering schools, the student prepares his "Who Am I?" answer and build both his identity and motivation to carry on after graduation (Park et al., 2021). An important milestone in an engineer's or any higher education graduate's life is the graduation achievement based, in many cases, on an end of studies research project. When successfully defending the results of this research project, the engineer/graduate confirms his identity and motivation and is, since this valuable moment to him, ready to invade the professional world. Consequently, it is important that educational pathways and curriculums have as main objective to build identity and motivation within the student and contribute to the sustainable and the positive impact he is expected to be delivering to the community afterward. Coaching can fulfill this need and play an integrative role in advancing higher education as a community of learning and teaching (Cruz & Rosemond, 2017).

In this chapter, the author discusses how coaching can help in increasing the motivation, the self-confidence, the performance and the emotional stability of students during their research journey. She will share how a coach can inspire academic staff and research managers to help research students to produce more and of their best, while taking pleasure in their journeys. Coaches can utilize any of the available coaching models when coaching researchers. However, the author believes that the research process needs deep intervention at the individual level to create a solid 'Why', to reach what is unseen and to hear what is unsaid. Drawing on the author's own experience as Masters and Ph.D. degree holder, scientist, engineer, coach, mentor and supervisor, she could propose a new model based on the Hero's journey concept, as defined by Joseph Campbell (Williams, 2019). This chapter presents the new model and discusses in five sections its applicability and positive impact. Section 1 is for an introduction. Section 2 is for the background. In section 3, the author introduces the Hero's journey concept and explains how coaching can change a researcher's life for the better and allow him/her to be a hero. Section 4 presents the new coaching model that the author advocates for scientific research coaching. In Section 5, the author discusses the model's applicability and presents the results of a survey that has assessed the model's impact. In Section 6, future research directions are shared. Finally in Section 7, there is a summary for the chapter's findings and a conclusion.

BACKGROUND

The International Coaching Federation (ICF), defines coaching as "partnering with clients in a thought-provoking and creative process that inspires them to maximize their personal and professional potential. The process of coaching often unlocks previously untapped sources of imagination, productivity and leadership" (ICF, 2021). Either one-to-one (individual) or into a small group context (group), coaching has been efficient in many aspects, such as in: careers; teamwork; improving skills; enhancing performance; well-being; and mental health. Space and Grant (2005) show that coaching in either format (individual or group) appears to assist goal-attainment, improve satisfaction with life, provide a greater sense of control over the environment and open coaches to a greater range of experience, when compared to a no-coaching control group (Spence & Grant, 2005). In high-school settings, studies have found that teachers, who underwent solution-focused coaching, had higher levels of goal attainment, workplace well-being and resilience, and lower levels of stress in comparison with the control group (Devine et al., 2013; Grant et al., 2010). In 2011, Vitae, the body responsible for the support and development of researchers and doctoral students across the UK (https://www.vitae.ac.uk/) concluded that coaching is a promising methodology for building sustainable research careers (Vitae, 2011). More recently, some studies have reference to coaching students in the Higher Education academic framework (Alzen et al., 2021; Molyn, 2020; Sepulveda, 2021; Spencer, 2021), they provide a high level of support for the efficacy of coaching across a wide variety of outcomes ranging from goal attainment to well-being, mental health, and resilience (Atad & Grant, 2021; Howlett et al., 2021). As such, if coaching is as psychology efficient as a change methodology to help people thrive, it would be reasonable to extend it to individuals on research journeys such as students for the award of an engineering degree, Masters, Ph.D., or other qualification or employees in jobs involving research. Coaching may be an ideal methodology to improve the quality and quantity of scientific production worldwide, contribute to resource and time optimization and help researchers and research staff to grow and thrive.

A research project is a scientific endeavor to answer one or more research questions. Çaparlar and Dönmez (2016), define scientific research as "research conducted for the purpose of contributing towards science by the systematic collection, interpretation and evaluation of data in a planned manner. A researcher is the one who conducts this research. The results obtained from a small group through scientific studies are socialized, and new information is revealed with respect to diagnosis, treatment and reliability of applications" (Çaparlar & Dönmez, 2016). Scientific research projects are generally carried out in research institutions, such as engineering technical schools that offer education in industrial arts and applied sciences or universities for Masters' degrees or PhD.

When working on scientific projects, students in research institutions (i.e., researchers) can face many challenges, such as:

- The start is a research subject or topic that guides the identification of the research question. This question is the main idea of the research and exploring it is supposed to create a change in science and impact humanity in a positive way; this is the first challenging step. Formulating the idea and making it walk needs confidence and the ability to create the idea needs strength. It also needs a deep understanding of the current situation in science, with the ideal wanted state and the existing gap. The researcher aim will be to move towards the ideal situation. Moreover, progress by the student needs support, partnership and collaboration. Finding the right idea is about starting once own research, learning what was done in the field of work and being innovative and creative to add to it. A scientific project starts with nothing but an idea and builds a path toward new knowledge.
- During the project's duration, the researcher needs to keep an encouraging degree of motivation and confidence. This can be challenged by limiting beliefs or behaviors from self and surroundings. As such, the researcher can find that he is frequently outside of his comfort zone where his perception is being challenged. Unhelpful thoughts create unhelpful emotions, and threaten the required level of motivation to continue on the appropriate research path. The author believes that this is the main reason why researchers do not successfully complete every project.
- Sharing the new knowledge is challenging as it is about convincing peer review experts that the idea has value and that the researcher has started to really make a change. Science should serve humanity through every new item of knowledge and understanding that it is offering. The researcher is charged with proving that his new contribution is doing so. Usually, the researcher disseminates his findings with the aim to reach decision makers and that's why, the messaging, the method and the approach count in making the dissemination convincing and meaningful.

Nowadays, in the scientific research world, it is generally assumed that students are equipped to somehow work out how to manage their emotions by themselves and will make their projects successful. Scientific supervisors can suggest the research topic and guide the research methodology and the time plan. There is little focus on how researchers can use the understanding of themselves, their thoughts, their emotions, their beliefs, their challenges, their relationships and interactions with their environment (that can support and/or challenge their progress), to keep their motivation and self-confidence, and improve their performance over the project's journey. Research projects may be long and involve rigorous and the highest level of evaluation. Studying for a Ph.D. degree is an example and the attrition rates for higher degree projects are high (Bair & Haworth, 2006; Jaksztat et al., 2021). Statistics show that the attrition rate among PhD candidates can range from 30% to 40% depending on the discipline and the country (Travaglianti et al., 2018). For example 36 to 51% percent of students who begin their doctoral studies in selective colleges and universities in the United States do not persist to graduation (Young et al., 2019). Unhelpful behaviors and believes along the PhD journey might be a cause and challenge the PhD completion. Choosing and sharing the appropriate idea, being open to learning, creating, and sharing a new knowledge can't survive with unhelpful behaviors or believes and then cause the researcher to quit. Attrition from the PhD or any other research project is time and resource consuming and is not serving neither the student nor the stakeholders, including research institutions (Jaksztat et al., 2021). Moreover, some students may be already dealing with personal pressures outside of their research institutions. The author has already coached wives and mothers who are also students taking the PhD journey. With this category of coachees, setting priorities to create the acceptable life-work balance for all stakeholders was the main issue. Thus, thoughts of quitting the project were the best option in many cases. Different researchers analyzing low completion rates argue that it is the unhelpful behaviors and believes, rather than the difficulty of the research itself, which is responsible for the relative high drop-out rates (Kearns, Forbes, et al., 2008; Kearns, Gardiner, et al., 2008; Manathunga, 2005). Sabotaging and self-sabotaging behaviors, including over-committing, procrastination and perfectionism, criticism, social and cultural barriers are all found to have a role to play. The 'personality characteristics' of students contributed to the pressure and lack of success experienced by some (Bakker et al., 2010).

Consequently, there is a need to tailor interventions at the individual researcher/student level. Students and academic staff need more than important policy mandates to flourish as researchers.

A RESEARCH PROJECT: A HERO'S JOURNEY

Introduction to the Hero's Journey Concept

Joseph John Campbell (March 26, 1904 – October 30, 1987) was an American Professor of literature who worked in comparative mythology. In 1949, Campbell released his mythological work, 'The Hero with a Thousand Faces' (https://www.amazon.com/Thousand-Faces-Collected-Joseph-Campbell/dp/1577315936), in which he discusses his theory of the archetypal Hero's journey, termed the monomyth. For the purpose of this work, the word hero is a gender-neutral word and is serving inclusiveness. In Campbell's own words: "A hero ventures forth from the world of common day into a region of supernatural wonder: fabulous forces are there encountered, and a decisive victory is won: the hero comes back from the mysterious adventure with the power to bestow boons on his fellow man". The hero for Campbell is a protagonist whose facing obstacles in pursuit of a goal. It is the archetype who sets out on an adventurous journey, in pursuit of his call for transformation.

According to Christopher Vogler (Vogler, 1999), the monomyth structure has twelve stages as illustrated in Figure 1, and they are:

The Ordinary World

The hero is taking a normal life at the start of the story before the adventure begins. The hero has a background environment with a personal history.

The Call to Adventure

Something shakes up the status quo and the hero is faced with an event, conflict, problem, or challenge either from external pressures or from something rising-up from deep within. The hero is now facing the beginnings of change and the start of an adventure.

The Refusal of the Call

The hero feels the fear of the unknown and initially rejects the call and refuses the adventure. They try to turn away from the adventure because of hesitation, fears, insecurity, or any other number of issues.

Meeting the Mentor

The hero encounters a mentor who can give training, equipment, advice, wisdom, information, or items that enable the hero to be ready for the journey ahead. Our hero now has a source of courage and wisdom.

Crossing the Threshold

The hero decides to leave the ordinary world for the first time by crossing the threshold into a new region or condition with unfamiliar rules and values.

The Tests, Allies, and Enemies

The hero starts to learn the rules of the new special world and endures tests, meets friends, and comes face-to-face with enemies.

The Approach

The hero starts to prepare initial plans to take on conflicts and to resolve them may try different old and new tools, approaches or adopt new ideas.

The Ordeal

The story of the journey goes wrong, and the hero is facing more conflicts. The experience is more difficult and so are the obstacles, some of which may lead to a life-crisis situation. The hero can face death or confront their greatest fear(s). Out of such a critical moment comes a new life.

The Reward

After surviving 'The Ordeal', the hero takes possession of the treasure won by facing death. It may be a physical item or piece of knowledge or wisdom that will help with self-perseveration. Even, a celebration may follow. Yet, the danger and the fear of losing the treasure persists.

The Road Back

The hero is driven to complete the adventure and sees the light at the end of the path. However, our hero may face even more tests and challenges when bringing the treasure home.

The Resurrection

The climax of the journey is where the hero faces a final test. This involves being severely tested once more on the threshold of returning home and the hero uses everything learnt to take on the conflict once and for all.

The Return

The hero brings knowledge and understanding back to the ordinary world and returns home with elements of the treasure that has transformed our hero and has the power to now transform the world.

For the author, the Hero's journey is a journey of faith and growth, taking the traveler from the known to the unknown and back to the known, holding the triumph of a hero. This triumph that could be an added value of knowledge or wisdom, is the transformation, in the being and doing. It becomes then the hero's truth and reality.

The Researcher is a Hero

In the present section, the author looks at the researcher's life through Campbell's lens. This reveals the researcher's journey, challenges and triumphs. The research process is perceived as a journey of multiple scientific steps that has the aim of conducting the research work. Each step is interlinked with the others.

In the literature, there are many scientific research methods and for this work, we consider the 11-stage model, as suggested by Sönmez (2018). These stages are shown in Figure 2, and are as follows:

Manifestation

Is the starting point of the study. It is the only stage that takes place independently from the researcher. It's from the 'manifestation' of the problem that the researcher is called to conduct the research.

Figure 1. The Hero's journey model according to Joseph Campbell

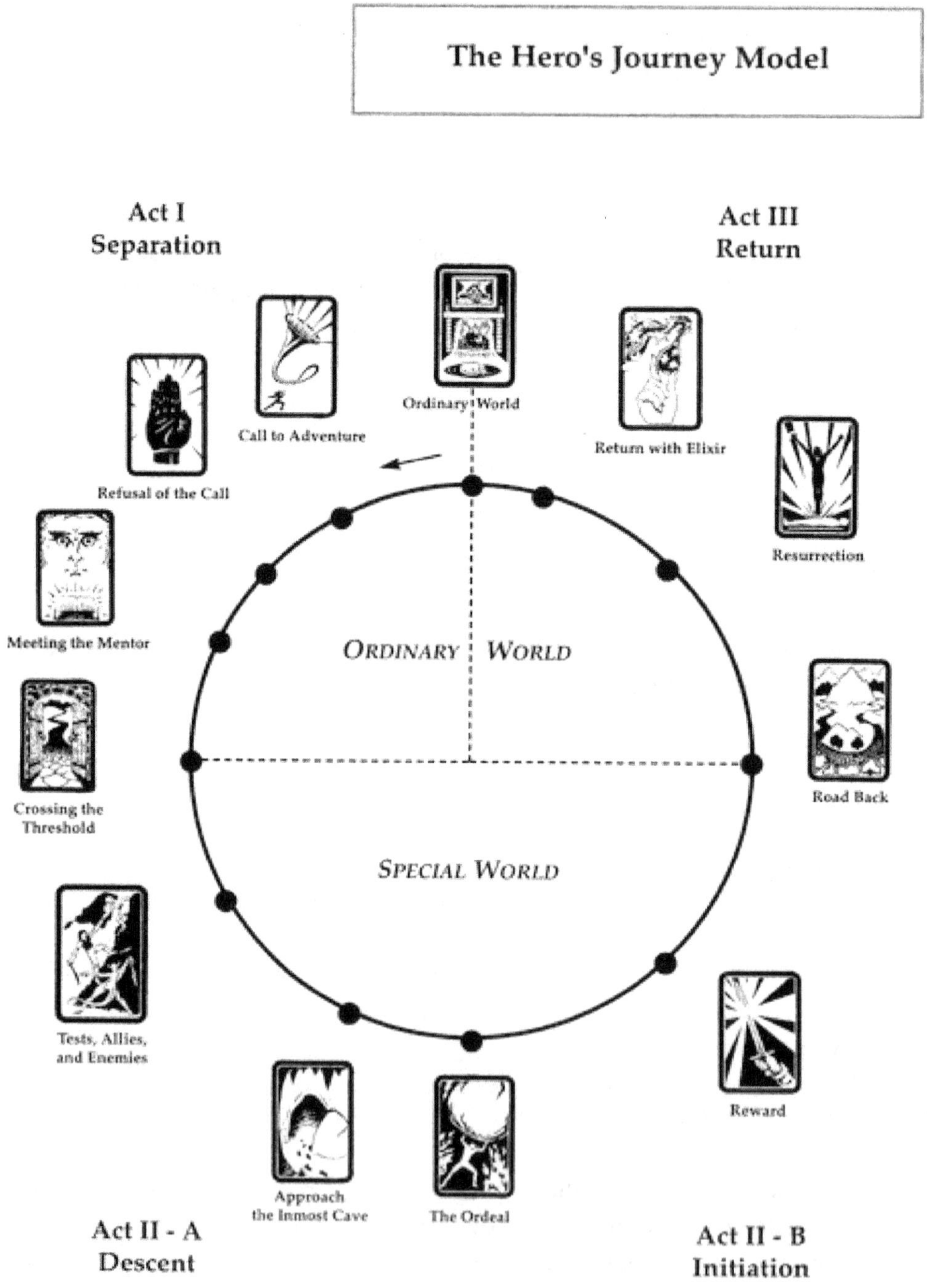

Description

In their co-authored book "the Evolution of Physics", Einstein and his fried Infeld say: "The formulation of a problem is often more essential than its solution" (Brown, 1939). In this step, the researcher describes the problem that has emerged in the 'manifestation' stage. He defines all related features, mainly the location, the environment, the period, the data needed and the stakeholders. The researcher has to use clear language for the 'description' that everyone can understand.

Prediction

Is based on the existing knowledge and reasoning principles and is about finding out what generates the identified problem. This third stage determines the direction of the study and sets the research hypothesis that can be verified or falsified later on.

Investigation

This is the stage of inspection and is the heart of the research work. Of the whole project duration, it takes the longest time and requires the most effort. In this fourth stage, the way can be illuminated, and the methods and the materials can be found. Specifically, by the end, the researcher will know what to use to gather information and produce results, these can be from: observations; surveys; experiments; interviews; archives scanning; statistics; case studies.

Determination

During this stage, the results of the abundant examinations are converted into fewer provisions. The researcher summarizes the articles listed and collects the significant statistical methods, results, and various alternatives obtained in stage 4.

Diagnosis

This concept means naming, personalization, object and duplication, selection, and separation of the maintained information from the previous stage. It can be referred to as a summary stage and it is an act of 'examination' of the information obtained by various information-gathering techniques on the problem or subject being described. Diagnosis signals the end of the previous five stages before the progression to stages 7 to 11.

Verification

This stage is about researching whether something is real or not, proving the truth, and reaching reality by inquiring. The 'Diagnosis' stage requires the 'Verification' of the correctness, which can be done by returning to the earlier stages of the research process, or by conducting multi-faceted analyzes, synthesis, or by comparison between the data from other independent sources or/and data from the summary of the 'Diagnosis' stage.

Treatment

Sönmez (2018) considers that this eighth stage of research is the 'therapy' phase. The researcher implements the application process. The 'treatment' stage is about returning to the normal or the pre-problem conditions, while applying the solutions that are forthcoming from the 'Diagnosis' stage.

Reinforcement

This stage is for strengthening and consolidation. The 'Treatment' stage is not enough, and a kind of rehabilitation is needed. Particularly, when the researcher assesses the long-term impact, it is required to make some additional improvement measures and prevent the appearance of new problems. For example, a study related to a bridge construction needs to reinforce the benefits of the bridge to users by assessing the eco-impact it may have in the long-term and how any detrimental impact can be dealt with.

Progress

Is the penultimate stage about making the treatment and the reinforcement sustainable, and the found results available for use by the next researchers sharing interest in the same topic. If 'Progress' does not occur, it means that the research didn't reach its ultimate aim.

Tracking

When the elements from the research have been 'Progressed' in the previous stage, they are controlled periodically and are recorded in a development tracking form. This processing helps in keeping an eye on there being new problems, complications, and surprises. Thus, the most efficient solutions can be found.

A researcher taking this 11-stage research journey is in many ways taking a Hero's journey. Figure 3. shows how the different steps from both journeys can be linked together, with the Hero's journey on the left-side. The problem manifestation and description stages correspond to the Hero's 'call to the adventure'. In the prediction stage, the researcher, who can be helped by a mentor, or a supervisor, can, based on mentor's knowledge and reasoning principles, be crossing the threshold related to the clarity and the merit of the subject. The investigation stage matches the tests (allies and enemies) in the Hero's journey. The determination and the diagnosis stages of the 11-stage model help the researcher to set the appropriate approach. Verification, treatment, and reinforcement stages help the researcher to find the truth and live the personal ordeal in the Hero's journey. The final stages of progress and tracking is for the researcher to find a way to normalize the new knowledge and understanding, and to return to the normal world.

However, the 11 defined stages of research introduced above has the important omission of the inside world of the researcher. The researcher has been called to move from a comfortable and known situation to an unknown place. He may face pressure to refuse the call and if he finishes by accepting it, he will take the journey looking for 'the truth' and move to where management is needed of doubts, fears, and hesitations to keep the momentum towards gaining new owned knowledge. As such, facing these internal challenges, combined with the external ones, creates, in the researcher, hero status. On the way, the researcher needs to deal with his beliefs, thoughts, emotions, and personal relationships. The rewards he may be receiving for his work on the way as recognition and peer-approval might impact his motivation which could have a big impact on the success of the research project.

Ultimately, coaching can help with the researcher/hero mission. Coaching can ally with advising and supervision to bring out the best of the researcher and facilitate personal and professional transformations towards achieving continuing personal development and, thereby, personal success. According to Alzen (2021), advising focuses on guiding students' degree planning process and typically entails periodic meetings throughout student enrollment. It is about knowledge of policies and curriculum and facilitating student learning and advising on academic technical resources (Alzen et al., 2021). Manathunga and Goozée (2007) define supervision as strategies to develop critical thinking skills including feedback on students' writing, engaging in critical conversations, explicating assessment criteria, and peer-to-peer collaborative learning (Manathunga & Goozée, 2007). Pearson (2009) see supervision as the ongoing process of creating learning opportunities between students and supervisors 'from the ongoing business of doing research' (Pearson et al., 2009). As for coaching, it focuses on developing skills from a holistic perspective, including the co-construction of students' academic and non-academic goals and steps to achieve those goals over a series of frequent coaching sessions during a focused duration of time. Coaching guides the development of academic skills and focuses on motivation and developing self-confidence and self-efficacy (Alzen et al., 2021).

In the next section of the chapter, the author puts forward a simplified coaching model to help the researcher, could be yourself, to succeed when taking the Hero's journey.

THE MLKS COACHING MODEL FOR SCIENTIFIC RESEARCH

Introduction to the MLKS Model

As a scientist, the author believes that improving the quality and quantity of published research has become imperative for many universities worldwide. In the process of dissemination, the resources and time required needs to be optimized during the research process. As a coach, the author believes that both students and academic staff need more than policy mandates to keep-up momentum and motivation, and thereby flourish as researchers. Moreover, the author believes that the research process needs more regular intervention at the student's individual level for the supervision to reach and explore what is unseen, and to hear what is unsaid. Drawing on the author's own experience gained over many years, she is able to propose the **M**indset-**L**earning-**K**nowledge-**S**haring (**MLKS**) coaching model as a new tool for researcher coaching.

The MLKS model is based on four steps that creates a never-ending loop, as shown in Figure 4, and the steps are:

Mindset

The mindset is a collection of thoughts and beliefs that shape our habits. According to Carol Dweck (2015), an American Professor of psychology and author of the popularized book, 'Mindset: The New Psychology of Success', a mindset is a self-perception or 'self-theory' that people hold about themselves. Believing that one is either 'intelligent' or 'unintelligent', a 'good' or 'bad' parent/employee…, are simple examples of a mindset. People can be aware or unaware of their mindsets, but, according to Dweck, this

Figure 2. "11 Steps" Process as a scientific research method

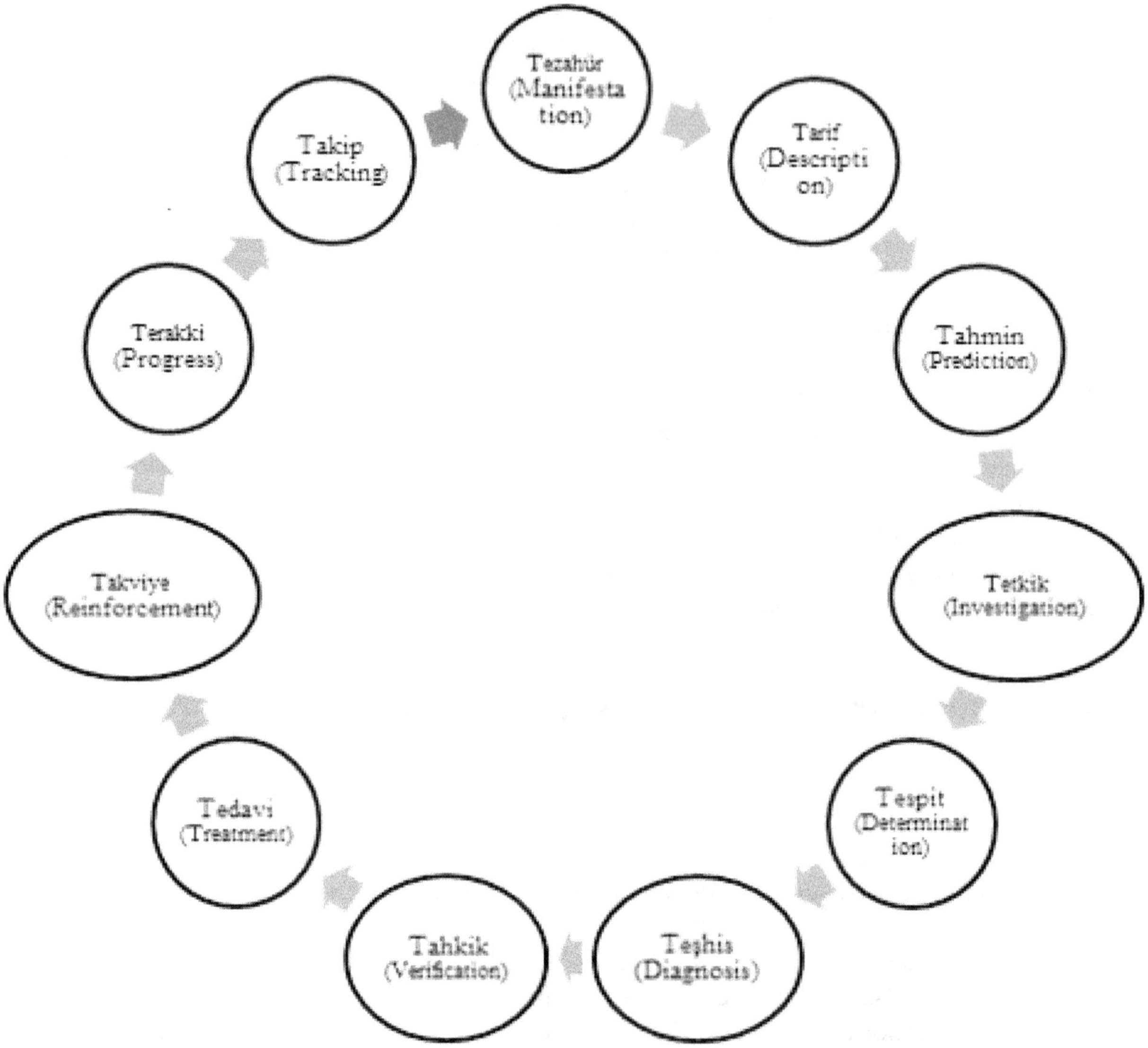

can have a profound effect on learning achievements, skill acquisition, personal relationships, professional success, and many other aspects to a person's life. Preparing the mindset is an essential step before every Hero's journey. It helps in raising self-awareness, understanding and taking advantage of opportunities, and also managing the risks and obstacles on the way. As a starting point of the research project, the coach prepares the student's/researcher's mindset. In a set of coaching exercises, the coach enhances the students' reflection about their life vision and objectives and helps them to see the research project as a part of this vision. The student creates the project vision and understands what is important about leading that project. Next the researcher starts owning the journey and the coach can help the student/researcher to lead a '**V-Step reflection**' (see Figure 5), as follows:

Figure 3. Hero's journey vs researcher's journey

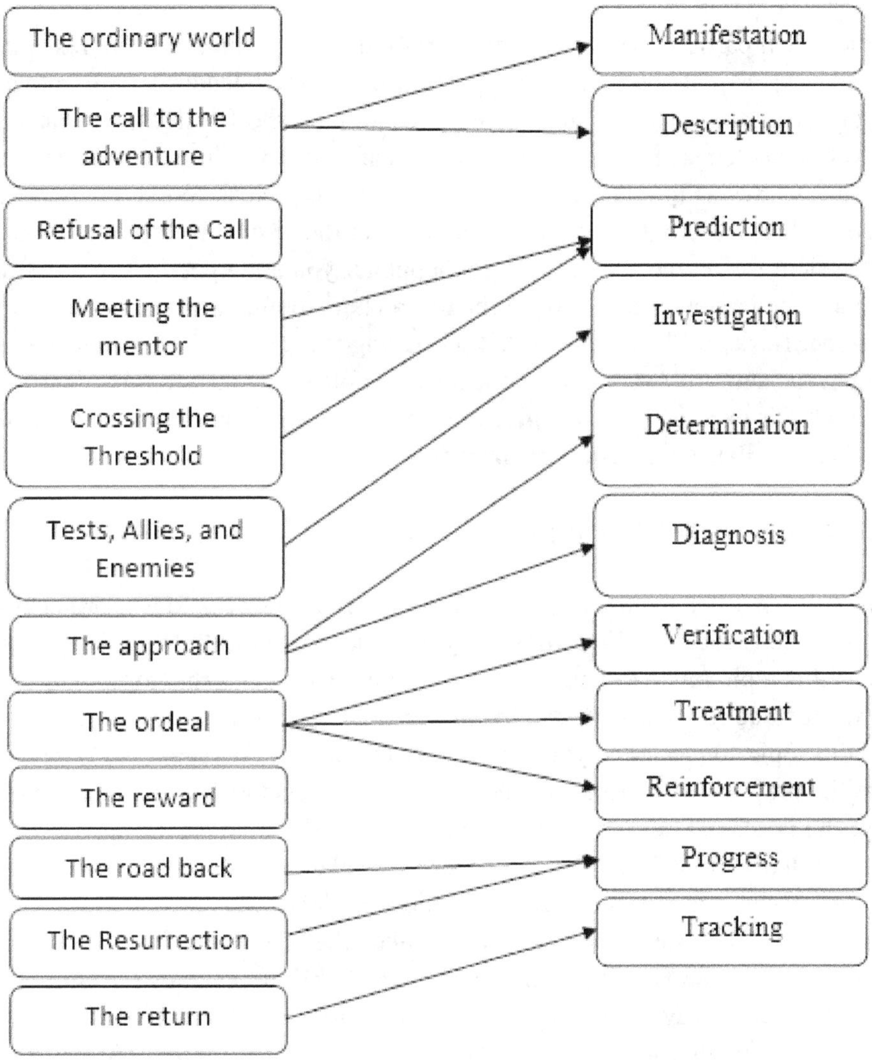

The "\" Step for The Greatest Vision of Life

This is where the coach partners with the researcher to create 'the greatest version of his life' to create the real research project motivation. The coach can ask of the student a set of questions such as: How would you define personal success? How would you like to be remembered?; How would you not like to be remembered?; What value do you contribute to the world and to the other people around you?; etc. At the end of this forward progressing step, the coachee delivers a written declaration of the greatest vision of life. Some students may quit the research journey after setting their 'greatest vision of life'; this is completely acceptable when they discover that the research route is not the most appropriate for them and they decide not to waste their time, and the time and resources of a research institution, in pursuing a research project in which they are not fully invested and which therefore has a high chance of failure.

The "/" Step For The Research Vision

This is where the coach partners with the researcher to allocate meaning to the research journey. The coach can ask a set of questions such as: What does it mean to you to lead research or to have a scientific degree?; What are your strengths that ensure that you can take and finish the journey?; What support do you need?; What challenges may you face?; What can you do with every challenge?; What could hold you back?; etc. Cartesian questions can help here, as well, such as: What will happen if you get the scientific degree?; What will not happen if (when) you get the scientific degree?; What will happen if you don't get the scientific degree?; What will not happen if you don't get the scientific degree?; Future pacing based on a visualization exercise to get the researchers to imagine and visualize their future new life with a new scientific degree is recommended at this stage to connect the coachee to the future desired state. According to the author's experience, coachees are able to connect to their future emotional state when going through this exercise, and this gives meaning to the experience. At the end of this step, the coachee will deliver the **Research vision declaration**.

The "V" Step for the Research Project Vision

This is where the coach partners with the researcher to create a framework for the project topic. The coach can ask a set of questions such as: What topic do you want to work on?; What is interesting about this topic?; How does this topic relate to your research vision?; How does this topic relate to your greatest vision of life?; What value can this topic add to the world?; How is this topic connected to your research vision?; How is this topic connected to your greatest vision of life?; What are the key resources you may need?; Who are the project's stakeholders?; What are their expectations from the project?; etc. At the end of this step, the coachee knows his "Why" and delivers the **Project model** that precisely defines the topic, the research question(s), the key resources and stakeholders for a successful research project. The coach and the coachee can now co-create a 'Project model' template.

The mindset step in the MLKS coaching model takes the researcher from the 'ordinary world' to 'crossing the threshold' in the Hero's journey. In Figure 4 'Mindset' takes the Sun's color of yellow. The author sees Mindset as playing the Sun's role to the Earth by maintaining the Earth's balance and protecting its existence, in other words, Mindset preserves the whole journey. At the Mindset stage students need to develop their self-awareness and to achieve this the coach will recommend attendance at related personal development events/workshops.

Learning

During this step of the MLKS coaching model the researcher is taking the learning path (after coming from the mindset step) with clear life and project visions, and a global idea about the research topic. Decisions about the precise topic will be taken during the learning step. The student needs to start his research project by understanding what is already available within the scientific literature and what, of course, is missing. Identifying the gaps in knowledge and understanding clarifies what new research is needed. The coach will facilitate this process for project description and planning. During this step the student will draw his own map and be prepared to move forward into the 'theoretical' learning. The coach can adopt 'The 3 Project-Pillars' approach by partnering with the student to create/update the 'Project description sheet', the 'Project model', and the 'Project plan'. The coach can suggest templates for these three deliverables whose content is detailed hereafter. Some of the questions to be asked during this learning step are: What is your research topic? (Ask the coachee to formalize the topic clearly in writing); What are the main keywords to describe this topic? What is known and understood nationally and internationally about the topic? (to answer this question the student needs to explore scientific databases); What studies do you think need to be added?; What study can you lead in this context?; What research approaches, methods, and tools could/will you use?; What can hold the progress of your research back?; After this reflection the student will elaborate and revise the 'Project description sheet'. This sheet will clarify the:

1. project's key words;
2. the main objective of the project;
3. the specific objectives of the project;
4. the different phases of the project and their deliverables;
5. the description of the research methodology and the tools that will be adopted.

Then, the student will update his '**Project model**' according to the new knowledge and understanding obtained from carrying out the literature review. The student is supposed to see the results of the scientific work as a product to be delivered to the clients and stakeholders of the project.

Finally, at the end of this step the student will create the '**Project plan**', in which is the clarification of the steps to follow and the deadlines of each milestone, and how it's known that the milestones are realized.

The learning phase takes the researcher through the 'tests, allies, and enemies' and 'the approach' stages in the Hero's journey. In Figure 4, learning takes the color blue that represents water. It refers to rain that is, together with sunlight, needed to make seeds grow, the idea is thereby the seed in the MLKS model.

During the learning step, researchers need to develop curiosity and decision-making skills and so the coach will recommend attendance at related events/trainings.

Knowledge

This step in the research project is the most important, most significant and most meaningful step in the model. The knowledge step is when the researcher creates new knowledge and understanding, which is to be valuable to the scientific community. During the learning step, the student assesses the gaps in the existing knowledge and identifies the need to fill them in. Within the knowledge step, the coachee is fulfilling this need by showing creativity and innovation. He may design new tools to create significant results. This is the step where the student can build self-confidence by creating ownership of the research results. Now, the researcher starts to create his own brand! During the knowledge step the coach helps the student to connect with the 'greatest vision of life' and the 'project's vision'.

Questions that the coach may ask during the knowledge step are: Where have you been, so far, in your research journey?; What new knowledge have you created?; What change has this new knowledge created for yourself and for others?; How is this congruent with the project vision?; How is this congruent with your research vision?; How is this congruent with your greatest vision of life?; How is this honoring your values?; What have you learned about yourself?; Now, as you are here, what would you have done differently to reach this moment?; How do you feel about what you have created?; How do you want to move further?

At the end of this step in the MLKS coaching model, the student/researcher will create his '**Tool Box**'. This box contains the standard research tools to be used and those tools newly created to facilitate the reproduction or the further development of the scientific project. The box may contain manuals, 'how-to-dos' and other guidance related to analyzing the new knowledge being created.

This knowledge step matches 'The Ordeal' and 'The Reward' stages in the Hero's journey, see Figure 1. Creating new knowledge and understanding needs much effort, yet when achieved it is a kind of rebirth offering a valuable treasure to the researcher. In Figure 4 knowledge takes the color of green for herbs. It refers to what appears when seeds receive sunlight and water and new herb plants create the farmer's happiness, in the same way so does the new knowledge and understanding for the student researcher.

Within the knowledge step the coachee needs to build-up his researcher's self-confidence and creativity skills, and so the coach will recommend attendance at related continuing professional events.

Sharing

The researcher must reach the sharing step now holding new knowledge and understanding and be confident about disseminating from the research project added value beyond the literature review, and to him/herself and his/her surroundings. At the sharing step the student has accomplished resolving the research question(s) and needs the results and findings validated and refined. This can be done through a feedback process and is what makes this step important and so valuable. At the sharing step the student will share the knowledge created, reminding, at all times, what has been learnt from the research studies in the learning and knowledge steps. And, again, the coach will facilitate this process. Some questions to ask are: Where have you been so far, in your journey? What new knowledge have you created?; On a scale from 0 to 10 (0 is not confident at all and 10 is very confident) how confident are you about the new knowledge you have created?; How would you share this knowledge with an 8 years old child? ...with your grandparents? ...with your best friend?; How would you share this knowledge with the scientific community?; The coach may also lead a future pacing and visualization exercise making the student see/hear himself sharing and others reacting and then ask: How can you describe yourself while sharing?; How do you feel about you sharing?; How do you want your knowledge to be remembered?; What would be the best way for you to share your knowledge?; What would be challenging in that sharing exercise and the way of sharing?; What efforts do you need to do to overcome the challenges?; What support would you need?; What are the actions you will commit to?; What can hold you back?; How can you hold yourself accountable to that?; etc.

Actually, through the sharing step the researcher is taking 'the road back' to his normal life holding treasure (new knowledge and understanding), yet the student needs to go through 'the resurrection' for the final tests of this new contribution. Participating in conferences, writing and sharing scientific reports, publishing peer-reviewed journal papers (after profound reviews and feedback comments), are all parts of 'the resurrection' stage in the researcher's/Hero's journey, see Figure 1. During the sharing step the researcher will create an action plan containing the SMART goals for sharing, the deadlines and tools needed. SMART stands for: Specific: written in terms of a clear end result, Measurable: easy to assess when they have been achieved, Agreed: jointly agreed by the reviewer and reviewee, Realistic: attainable but challenging, and Timebound: have a deadline. At the end, the researcher is a hero returning safely to the ordinary world holding great pieces of knowledge and wisdom, and ready to take another journey with an upgraded and ready mindset. The author considers that every research project is just the start of another journey. The researcher's reward for a successful project is simply another project, possibly an

Figure 4. MLKS model for coaching

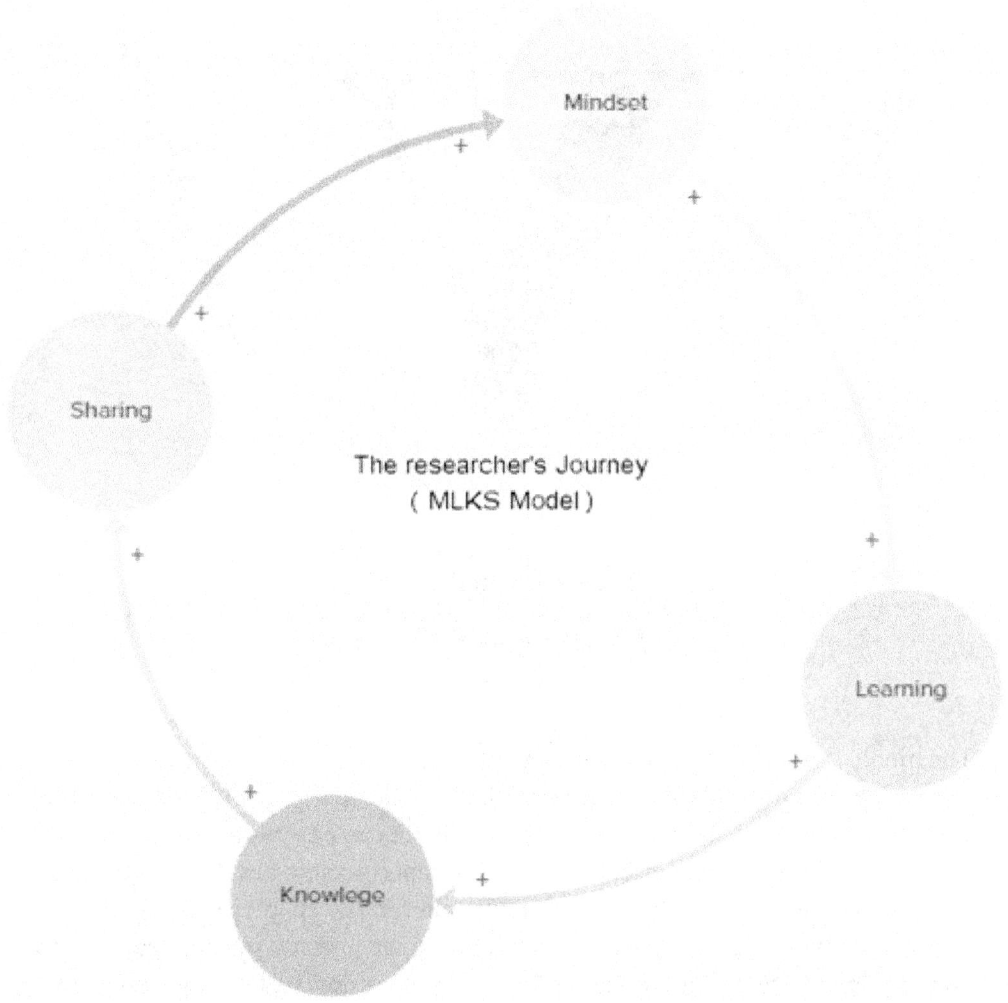

even more challenging one. The difference is that the start of the next project is with increased confidence (the researcher knows he has what it takes to see it through) and increased knowledge (from what he has learned the first time around). Figure 4 attributes the pink color to the sharing step; this refers to 'the life in pink' state of happiness that the researcher can reach when on the 'return' step of the journey.

By the end of this final step the researcher needs to have developed sharing skills and so the coach will recommend attendance at related events. The student needs to be particularly encouraged to learn more about research communication skills, such as: public speaking; storytelling; critical thinking; constructive feedbacks.

Application of the MLKS Model for Scientific Research

Overall, the MLKS coaching model aims at reducing unhealthy behaviours and prepares the students/researchers to approach and complete their research projects successfully and with sound health and

Figure 5. V-Step reflection

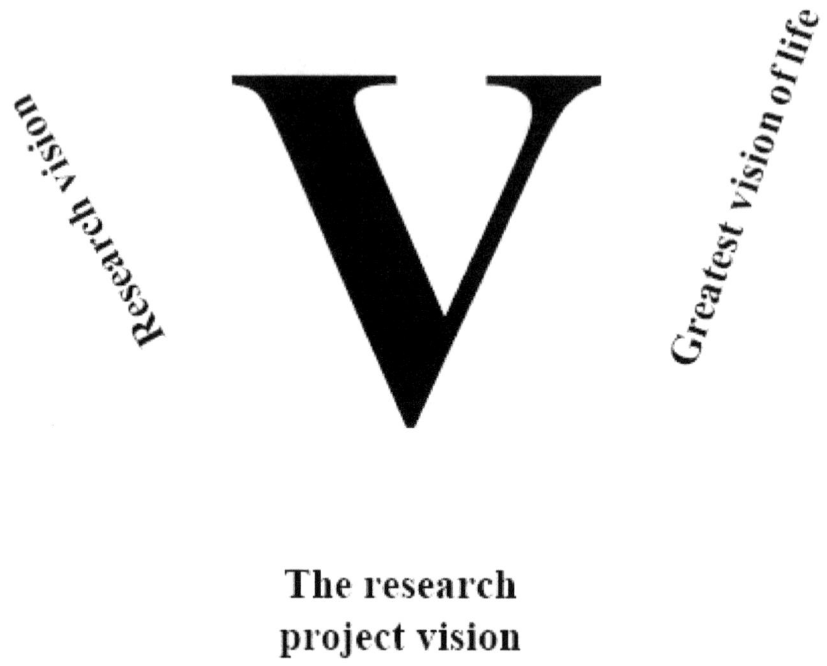

well-being. The model starts from before learning to after sharing, and keeps growth facilitation through continuing motivation. The model is about the coach helping the researcher to find his strong 'Why' and then as Friedrich Nietzsche says, "He who has a why to live for can bear with almost any how". In reality, the MLKS model would be even more effective when it is a partnership between the coaching approach, and the scientific supervision approach as mentioned earlier in the chapter in an academic framework. In this case, the student will be given the freedom to choose the journey starting from the topic through to the approach used for sharing. This can be under the shape of individual or group coaching sessions whose number and frequency are to be decided by way of agreement between the coach and the coachee. Accordingly, the academic institution would decide how to introduce coaching into its research programs.

The author has been using the MLKS model for three years with researchers from different institutions, and one of the main challenges found is that the academic institution may not support coaching as an approach to facilitate their research projects. Actually, even though coaching can be acknowledged as a change-making practice, implementing it in research institutions still needs more time and evidence in order to gain wider acceptance and to standardize the approach in research institutions. So far, the MLKS coaching model is applied, at a reduced scale, between the author (who is coach and supervisor) and his students/mentees or when a client prefers to hire a coach (the author) for his research project. This latter case is quite challenging, as usually the client will have progressed their project, and is basically stuck somewhere along the way when the coaching commences. The MLKS model will then be for the cure rather than prevent, which is acceptable, but is not the main aim of coaching. The coach then needs to take some steps backward to check the mindset and the 'Why' strength of the coachee. If the coaching had been introduced right from the start of the research project, time, energy and researcher's stress would have been saved.

Applicability and Case Studies

The author is both a scientist supervisor and a fully qualified and experienced coach. She has been using the MLKS model over the past two years. During this period, all her students, from different research institutions, undertook the coaching experience for their engineering, Masters or Ph.D. degrees. The starting point was to prepare 'The Mindset'. The coach offers a number of coaching sessions that may be group or individual. This number varies according to the coachees. In some cases, 'The Mindset' coaching had to be intensive when coachees preferred to devote more time to 'the Mindset' stage. The latter usually allows self-awareness insights and is shaping the good start of the research journey. Sessions need to be regular, yet they must respect the coach/coachee agreement.

A preliminary evaluation of the coaching model was realized based on a survey that was distributed to a group of students with and without coaching experience. "Experience" in this context means taking a degree with or without coaching. The idea was to have an experimental group that took coaching in the research journey, and a control group that didn't. In some cases, one participant can have more than one experience (Engineers with PhD or Masters' holders with PhD). The used survey is presented in the Appendix section.

The survey collected 26 responses to assess experiences of students taking engineering undergraduate, Masters and PhD degree programs. The 26 experiences sample is random and depends on the filled surveys received. 53,8% of the 'experiences' benefited from coaching, while 46.2% didn't. Some of the participants had just started their research journey, some were on their way and some had already finished and returned to the ordinary life.

The bar charts in Figure 6 and Figure 7 summarize the survey results for the students without and with coaching experiences, respectively. The survey asks the participants to assess their motivation, self-confidence, pleasure, productivity and performance, stress management and emotional stability during their research journey.

From inspecting the bar charts, we may observe that:

- 92% of the participants who undertook the coaching experience were able to maintain their motivation versus 33% of the participants without coaching experience;
- 100% of coached participants were able to maintain their self-confidence while only 16.6% were able to do this without coaching;
- 100% of the coached researchers found pleasure in their journeys while 41% did without coaching;
- 100% of coached participants agreed they were performant and productive while only 58% consider this to be true when they were without coaching;
- 78.5% of researchers having experienced coaching agree or strongly agree that they were emotionally stable and able to deal with stress, 21.5% were neutral, while only 0.8% said that they were emotionally stable without coaching.

The numbers above show the positive and supportive role that coaching can play in helping the researcher flourish and enjoy the 'Hero's' journey. Some researchers, without coaching, were able to keep their motivation, self-confidence, and pleasure along the way, but when they were asked, they confirmed that some external factors helped with that. They mainly mentioned that the supervisor, the research team, the strength of the topic, and the family all helped them on their journey. The author believes that the research period plays a role as well as these participants were undertaking either Engineering or

Masters' degree but none undertook the PhD. The latter may be more difficult and coaching demanding as it takes a longer period. Ultimately, the coach can help the coachee craft a beautiful experience, using the researcher's inner strength to keep up the momentum. Moreover, it is noteworthy that many of the coachees' research journeys didn't stop after completing the degree, they went on to either take a higher degree journey or chose to be employed in a job involving research. One of the participants shared that the coaching experience not only helped her with her personal life but also, it created in her the motivation to apply for a Masters' degree program and continue her researcher never-ending journey. In addition, all 26 of the participants in the survey were unanimous about recommending coaching, and the MLKS model to their peers.

These findings are in complete agreement with the results from Bettinger and Baker (2014) that focuses on academic coaching using a remote coaching model. Coaches assist students in overcoming academic and non-academic barriers by helping them identify resources and ways to advocate for themselves. The authors found that students who received coaching were 5% more likely to remain enrolled six months from the coaching term and 12% more likely to remain enrolled one year after the end of coaching (Bettinger & Baker, 2014). Another study by Lehan (2018) investigated academic coaching for graduate students in an online degree program. Students met with their academic coaches up to two times per week for a semester. Coaches helped students develop their competence in navigating the institution and supported broader skill development such as time management and how to be a self-directed learner (Lehan et al., 2018). The authors found positive relationships between academic coaching and persistence. They found that students who used coaching services at least once in their first year increased their resilience. Capstick (2019) conducted a study on coaching in urban research university in the southeastern United States. At this institution, the coaches focus on a collaborative relationship to help students focus on personal and professional goals. Coaches help students develop self-awareness, identify strengths, and determine academic plans, all while incorporating students' interests and values. They also help students to identify internal and external barriers to their goals and identify strategies for overcoming those barriers such as study skills, test-taking strategies, and time management (Capstick et al., 2019). The researchers found that coached students earned credits higher than their non-coached counterparts. Additionally, coached students were about 20% more likely to return in the semester after coaching than those who did not receive coaching (Capstick et al., 2019). In addition, many other more recent studies advocate the positive impact of coaching on students' performance in higher education (Hakro & Mathew, 2020; Ismatova, 2021; Sepulveda, 2021). Although this study confirms that coaching may add value to scientific journeys, it has some limitations related to the sampling technic that considered all the research experiences (Engineering, Masters and PhD) to be similar even though they differ in the period and the context. Also, coaching was realized by only one coach (the author) who suggests the model, differences in coaching styles may influence the results. In a study designed to understand the practices that make coaching a distinct service, Sepulveda (2021) found that coaching is made up of six practices: coaching beliefs; coaching skills; coaching meetings using a conversational framework; coaching over multiple meetings; training, growth, and development as a coach; and the coaching role in the context of higher education (Sepulveda, 2021). Further, Sepulveda (2021) found that the consistent combination of the six together is what makes coaching distinct. Consequently, the MLKS model is a tool that every coach can add to the research projects coaching toolbox, yet all the six above mentioned coaching practices would be considered.

Figure 6. The survey's results for participants with NO COACHING experience

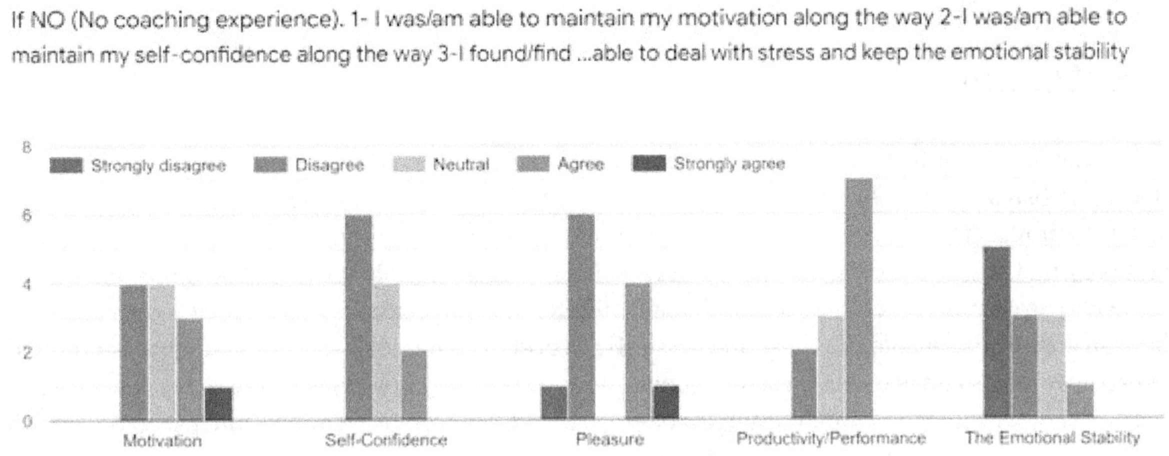

Figure 7. The survey's results for participants with COACHING experience

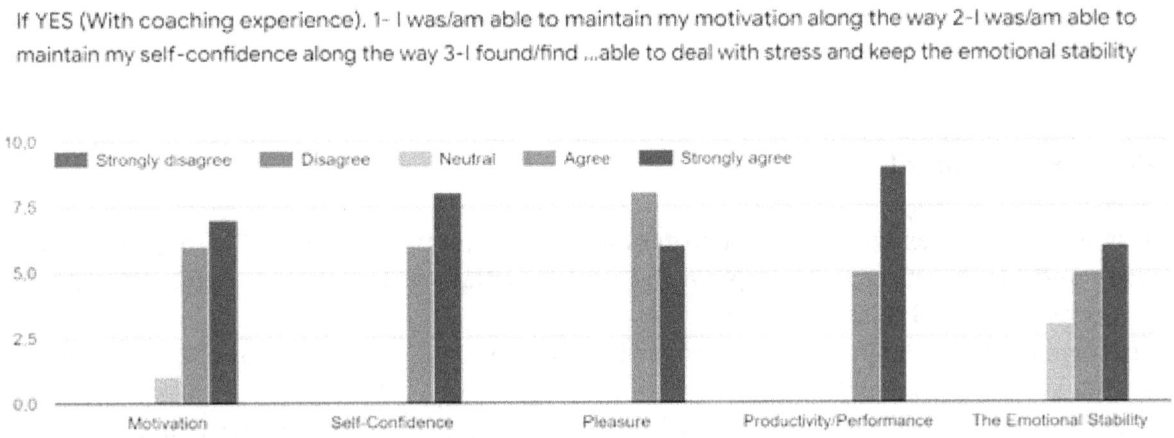

FUTURE RESEARCH DIRECTIONS

At the end of this research journey, the author was able to define a new coaching model serving researchers. The MLKS model is coachee focused. It starts with the mindset and facilitates coping with the surroundings. It is a transformational model and can enable change in a constantly changing environment. The author will carry on experiencing the model and assessing the outcomes for further improvement. He will explore how the nature of the experience (Engineering, Masters' degree and PhD) can influence the impact from coaching. Also, the author intends to investigate how the research-coaching model can impact students' personal lives and help them thrive in other areas outside of studies and/or careers.

Implementing partnerships with universities that will be ready to implement coaching tools in their research strategies is a promising path for the author as well.

CONCLUSION

In this chapter, the author, a coach and a scientist supervisor, suggests the 'Mindset-Learning-Knowledge-Sharing (MLKS) as an effective coaching model to coach student researchers to help them to make the best of their research projects. Working with the MLKS model and giving meaning to every step of it, is a journey that is similar to the Hero's journey of Joseph Campbell. It starts with a call to the adventure and leads to a completely unknown World that facilitates learning, knowledge, understanding and growth, that can create sustainable impact in the scientific World. The model encourages scientific productions and achievements and helps unleash the innovation and creativity potential within every researcher. This is absolutely serving the research strategies of engineering schools, universities and research institutions. It assists in the optimization of time and resource used in completing successful research projects.

The survey with 26 students that led to an assessment of the model considered an experimental group with coaching and a control group without coaching. The participants were asked about their motivation, self-confidence, pleasure in the journey, performance and productivity and emotional stability. The results from the survey indicate that the MLKS model was successful in improving students' motivation, their ability to manage their time with lower levels of stress, and improved their ability to progress. Through using the suggested coaching model, the student's journey in not only more productive but is found to be more enjoyable and pleasant as well.

ACKNOWLEDGMENT

The author wishes to express her profound gratitude to the Mindfreeks community (https://www.mindfreeks.com/) for their mentorship program and especially Bernard Kates who mentored the author for his work on the MLKS coaching model.

The author thanks Adijat Olanike Akinyemi for her coaching sessions. Dear colleague, you made this work and paper possible!

This research received no specific grant from any funding agency in the public, commercial, or not-for-profit sectors.

REFERENCES

Alzen, J. L., Burkhardt, A., Diaz-Bilello, E., Elder, E., Sepulveda, A., Blankenheim, A., & Board, L. (2021). Academic Coaching and its Relationship to Student Performance, Retention, and Credit Completion. *Innovative Higher Education*, *46*(5), 539–563. Advance online publication. doi:10.100710755-021-09554-w

Atad, O. I., & Grant, A. M. (2021). Evidence-based coaching as a supplement to traditional lectures: Impact on undergraduates' goal attainment and measures of mental well-being. *International Journal of Mentoring and Coaching in Education, 10*(3), 249–266. Advance online publication. doi:10.1108/IJMCE-05-2020-0024

Bair, C. R., & Haworth, J. G. (2006). Doctoral Student Attrition and Persistence: A Meta-Synthesis of Research. In Higher Education: Handbook of Theory and Research. doi:10.1007/1-4020-2456-8_11

Bakker, A. B., Boyd, C. M., Dollard, M., Gillespie, N., Winefield, A. H., & Stough, C. (2010). The role of personality in the job demands-resources model: A study of Australian academic staff. *Career Development International, 15*(7), 622–636. Advance online publication. doi:10.1108/13620431011094050

Bettinger, E. P., & Baker, R. B. (2014). The Effects of Student Coaching: An Evaluation of a Randomized Experiment in Student Advising. *Educational Evaluation and Policy Analysis, 36*(1), 3–19. Advance online publication. doi:10.3102/0162373713500523

Brown, G. B. (1939). The Evolution of Physics: The Growth of Ideas from the Early Concepts to Relativity and Quanta. By Albert Einstein and Leopold Infeld (Cambridge: at the University Press, 1938. Pp. x + 319. Price 8s. 6d.). *Philosophy (London, England), 14*(54), 242. Advance online publication. doi:10.1017/S0031819100011670

Çaparlar, C. Ö., & Dönmez, A. (2016). What is scientific research and how can it be done? Turk Anesteziyoloji ve Reanimasyon Dernegi Dergisi. doi:10.5152/TJAR.2016.34711

Capstick, M. K., Harrell-Williams, L. M., Cockrum, C. D., & West, S. L. (2019). Exploring the Effectiveness of Academic Coaching for Academically At-Risk College Students. *Innovative Higher Education, 44*(3), 219–231. Advance online publication. doi:10.100710755-019-9459-1

Cruz, L., & Rosemond, L. (2017). Coaching Academia: The Integration of Coaching, Educational Development, and the Culture of Higher Education. *Journal on Excellence in College Teaching.*

Devine, M., Meyers, R., & Houssemand, C. (2013). How can Coaching Make a Positive Impact Within Educational Settings? *Procedia: Social and Behavioral Sciences, 93*, 1382–1389. Advance online publication. doi:10.1016/j.sbspro.2013.10.048

Frankl, V. E. (1997). Man's Search for Meaning. *Language.* doi:10.1037/a0019212

Grant, A. M., Green, L. S., & Rynsaardt, J. (2010). Developmental coaching for high school teachers: Executive coaching goes to school. *Consulting Psychology Journal.* doi:10.1037/a0019212

Hakro, A. N., & Mathew, P. (2020). Coaching and mentoring in higher education institutions: A case study in Oman. *International Journal of Mentoring and Coaching in Education, 9*(3), 307–322. Advance online publication. doi:10.1108/IJMCE-05-2019-0060

Howlett, M. A., McWilliams, M. A., Rademacher, K., O'Neill, J. C., Maitland, T. L., Abels, K., Demetriou, C., & Panter, A. T. (2021). Investigating the Effects of Academic Coaching on College Students' Metacognition. *Innovative Higher Education, 46*(2), 189–204. Advance online publication. doi:10.100710755-020-09533-7

ICF. (2021). *What is Coaching?* https://coachingfederation.org/about

Ismatova, M. S. (2021). Development of coaching technologies in the educational space of the university. *Current Research Journal of Pedagogics*. doi:10.37547/pedagogics-crjp-02-05-25

Jakstat, S., Neugebauer, M., & Brandt, G. (2021). Back out or hang on? An event history analysis of withdrawal from doctoral education in Germany. *Higher Education*, *82*(5), 937–958. Advance online publication. doi:10.100710734-021-00683-x

Kearns, H., Forbes, A., Gardiner, M., & Marshall, K. (2008). When a high distinction isn't good enough: A review of perfectionism and self-handicapping. *Australian Educational Researcher*.

Kearns, H., Gardiner, M., & Marshall, K. (2008). Innovation in PhD completion: The hardy shall succeed (and be happy!). *Higher Education Research & Development*, *27*(1), 77–89. Advance online publication. doi:10.1080/07294360701658781

Lehan, T. J., Hussey, H. D., & Shriner, M. (2018). The influence of academic coaching on persistence in online graduate students. *Mentoring & Tutoring*, *26*(3), 289–304. Advance online publication. doi: 10.1080/13611267.2018.1511949

Manathunga, C. (2005). Early warning signs in postgraduate research education: A different approach to ensuring timely completions. Teaching in Higher Education. doi:10.1080/1356251042000337963

Manathunga, C., & Goozée, J. (2007). Challenging the dual assumption of the "always/already" autonomous student and effective supervisor. *Teaching in Higher Education*, *12*(3), 309–322. Advance online publication. doi:10.1080/13562510701278658

Molyn, J. (2020). *The Role and Effectiveness of Coaching in Increasing Self-Efficacy and Employability Efforts of Higher Education Students*. doi:10.29007/294n

Nilsson, L., Madon, T., & Sastry, S. S. (2014). Toward a new field of Development Engineering: Linking technology design to the demands of the poor. *Procedia Engineering*, *78*, 3–9. Advance online publication. doi:10.1016/j.proeng.2014.07.032

Park, J. J., Park, M., & Smith, J. (2021). *Engineering students' concepts of humanitarian engineering and their identity development as humanitarian engineers*. Sustainability. doi:10.3390u13168845

Pearson, M., Cowan, A., & Liston, A. (2009). PhD education in science. In Changing practices of doctoral education (pp. 100–112). Academic Press.

Sepulveda, A. (2021). Coaching college students to thrive: Exploring coaching practices in higher education. *Dissertation Abstracts International. B, The Sciences and Engineering*.

Shekar, A., & Drain, A. (2016). Community engineering: Raising awareness, skills and knowledge to contribute towards sustainable development. *International Journal of Mechanical Engineering Education*, *44*(4), 272–283. Advance online publication. doi:10.1177/0306419016668966

Smith, J. (2019). Impacts of a humanitarian engineering education pathway on student learning and graduate outcomes. *International Journal for Service Learning in Engineering, Humanitarian Engineering and Social Entrepreneurship*. doi:10.24908/ijsle.v14i1.12555

Sönmez, S. (2018). *"11 steps" process as a research method.* Universal Journal of Educational Research. doi:10.13189/ujer.2018.061125

Spence, G. B., & Grant, A. M. (2005). Individual and group life coaching: Initial findings from a randomised, controlled trial. In Evidence-based coaching, Vol 1: Theory, research and practice from the behavioural sciences. Academic Press.

Spencer, D. (2021). Understanding the Coaching Experiences of Non-traditional Students in Higher Education in the UK. *International Journal of Evidence Based Coaching and Mentoring*. Advance online publication. doi:10.24384/00v3-nm61

Thomas, E., Salvinelli, C., Harper, J., MacDonald, L., Klees, R., Platais, G., Javernick-Will, A., & Linden, K. (2021). A Body of Knowledge and Pedagogy for Global Engineering. *International Journal for Service Learning in Engineering, Humanitarian Engineering and Social Entrepreneurship*. doi:10.24908/ijsle.v16i1.14483

Travaglianti, F., Babic, A., & Hansez, I. (2018). Relationships between employment quality and intention to quit: Focus on PhD candidates as traditional workers. *Studies in Continuing Education, 40*(1), 115–131. Advance online publication. doi:10.1080/0158037X.2017.1396448

Vitae. (2011). *Coaching for research in UK higher education institutions: A review.* https://www.vitae.ac.uk/vitae-publications/reports/coaching-report-2012-vitae.pdf

Vogler, C. (1999). The Stages of the Hero's Journey. *Myth and the Movies: Discovering the Mythic Structure of 50 Unforgettable Films,* 300.

Williams, C. (2019). The Hero's Journey: A Mudmap for Change. *Journal of Humanistic Psychology, 59*(4), 522–539. Advance online publication. doi:10.1177/0022167817705499

Young, S. N., Vanwye, W. R., Schafer, M. A., Robertson, T. A., & Poore, A. V. (2019). Factors Affecting PhD Student Success. *International Journal of Exercise Science*. PMID:30761191

KEY TERMS AND DEFINITIONS

Coaching: Is a partnership between the coach and the client to help the client to get where he wants.

Hero's Journey: A journey that a person takes to reach his ultimate goal. It is an adventure that will make him a hero after overcoming all the challenges and getting the rewards on the way.

Higher Education: Education at engineering schools and universities to get engineering, master's or PhD certifications.

Mental Health: Emotional and psychological health and social well-being.

MLKS Model: A coaching client focused model that can be used in coaching researchers aiming at engineering, master's, or PhD certifications. It can also be used within scientific institutions outside the academic context.

Scientific Research: A research conducted in research institutions for the purpose of contributing to the worldwide scientific progress.

Scientific Research Coach: An individual who helps a scientific researcher to reach the research goals while maintaining a good mental health.

Scientist Supervisor: An experienced scientist who uses his knowledge and experience to advise and guide the student, taking a scientific research journey, to develop his scientific learning, knowledge, performance, and productivity.

Chapter 10
What Vietnamese Employers and Educators Want in Impactful Engineering Degree Programmes

Thuy Tien Nguyen
https://orcid.org/0000-0002-2132-9599
Ho Chi Minh City University of Transport, Vietnam

Anh Pham Thi
https://orcid.org/0000-0001-8510-1625
Ho Chi Minh City University of Transport, Vietnam

Tuan Anh Nguyen
https://orcid.org/0000-0003-4696-9418
Ho Chi Minh City University of Transport, Vietnam

Luu Mai
Ho Chi Minh City University of Transport, Vietnam

J. Toby Mottram
University of Warwick, UK

Georgia Kremmyda
University of Warwick, UK

Yiannis Koumpouros
https://orcid.org/0000-0001-6912-5475
University of West Attica, Greece

Angelos Georgoulas
https://orcid.org/0000-0001-8845-3781
University of West Attica, Greece

Dong Doan Van
Ho Chi Minh City University of Transport, Vietnam

ABSTRACT

This chapter presents results and analysis from the ENHANCE project survey for the views of employers and educators towards the impactful improvement of engineering degree programs in Vietnam. This building capacity project aims to find ways of strengthening the employability of engineering graduates by improving the quality of engineering programs. The survey was carried out by inviting respondents from industry (employers) and from higher education (educators) to answer online questionnaires. The number of respondents was 116, of which 69 are educators and 47 are employers. This chapter presents results that encompass the following three topics: university graduate attributes, capabilities,

DOI: 10.4018/978-1-6684-5619-4.ch010

knowledge, and understanding; the teaching and learning methods used in universities; and the future of engineering programs in Vietnam.

INTRODUCTION AND VIETNAMESE CONTEXT

The engineering profession in Vietnam is for versatile and impactful careers that create a legacy for society and for the lives of everyone (Arunkumar et al., 2018). To support their businesses, employers are demanding engineering graduates be better trained and better educated. Within the Higher Education (HE) sector, today's agenda in Vietnam focuses on engineering graduate employability and the development of measurable outcomes for higher quality in graduates' attributes and skills. Because of these society-driven HE changes, there is an unstoppable push for engineering programmes to go far beyond the traditional core teaching of fundamental principles in science and technology. This is reflected in the Vietnamese university system, which has three pillars to the content of graduate programmes. These unpinning pillars are knowledge; abilities and skills; and personal qualities and attitudes. To successfully achieve the required learning outcomes from embedding these pillars into engineering programmes Vietnamese HE must identify the desired attributes of its graduates. It is observed that these graduate attributes are delivered primarily through traditional or online lecturers and/or stand-alone specialized laboratory-based practical.

It is generally accepted that engineering graduates need to be prepared for the increasing use of advanced and appropriate technologies in their workplaces. In 2007 Patil and Codner (Patil & Codner, 2007) distinguish hard, soft and global skill sets as essential components of engineering students' competencies. of Hard skills are fundamental knowledge; expertise in engineering subjects; engineering design and problem solving skills; project management skills, etc. Under the soft skills are listed general knowledge; communication skills; managerial and organization skills; negotiation and interpersonal communication skills ; ethics, empathy; leadership and listening; financial management and budgeting; and safety and sustainability; The global skills set encompasses understanding global, political and societal issues; cross and multi-cultural issues; international labour market and workplace imperatives; engineering solutions/applications in a global context, etc.. For Vietnamese students to effectively be able to apply soft and global skill sets in the engineering workplace we observed that education and training requires the introduction of new graduate attributes and non-traditional teaching and learning methods, which are not familiar within current Vietnamese HE. These same Vietnamese students are well-known historically to cope well with education scoping the hard skills by way of the teaching of fundamental principles covering core science and technology subjects that are assessed by traditional paper examinations.

In recent years there has been an upsurge in the attention in Vietnamese universities for programme leaders to map and develop a new set of workable graduate attributes. The underlying factors that have influenced the growing importance of these attributes, include education being a lifelong process, a focus on graduate employability and the development of outcome measures to demonstrate the overall quality of HE throughout the country (Bath et al., 2004). Most Vietnamese universities prepare their engineering curricula in accordance with the frameworks and guidelines of the well-known engineering accreditation bodies, such as Accreditation Board for Engineering and Technology (ABET) in North America and ASEAN University Network (AUN). Departments and faculties with engineering programmes implement these accreditation requirements to define, monitor, and articulate essential graduate attributes. Accordingly, Vietnamese universities have recently addressed the importance of properly including employability skills in their graduate programmes. This is achieved through courses with learning outcomes linked to assessing development and applications of skills. It is noteworthy that in 2016 the Vietnamese government issued the National Qualifications Framework which embeds the concept of learning outcomes to increase quality of HE with respect to graduates' employability skills.

The authors are partners in the Erasmus+ project *Enabling Humanitarian Attributes for Nurturing Community-based Engineering*, which is known be the abbreviation ENHANCE. This project under No: 598502-EEP-1-2018-1-UK-EPPKA2-CBHE-JP (2018-2582/001-001) was to have a duration of 36 months starting in November 2018. Because travelling was suspended from March 2020 owing to the Coronavirus (COViD-19) pandemic the project completion was postponed to the end of May 2023. As an integral part to the project's workplan an online survey (pre-COVID) was run in partner countries (namely, Bangladesh, Greece, Indonesia, United Kingdom, and Vietnam) from 16 July to 18 November 2019. Its questions and responses focused on interdisciplinary teaching and learning, skills and attributes that graduate engineers may need to permit them to tackle complex interdisciplinary, inter-professional problems, which can involve humanitarian engineering challenges. The first part of the ENHANCE project survey has common background questions and the second part is with bespoke questions for either HE (educator) or industry (employer) respondents.

For background context, the ENHANCE project (https://warwick.ac.uk/fac/sci/eng/enhance/.) has European partners from UK (The University of Warwick as project lead) and Greece (University of West Attica), that are working with eight universities in the three South Asian partner countries of Bangladesh, Indonesia, and Vietnam. There are three universities in Indonesia (Gadjah Mada University, Institute Technology Bandung and Universitas Brawidjaja), two in Bangladesh (Bangladesh University of Engineering and Technology (BUET) and University Dhaka) and two in Vietnam (Ho Chi Minh University of Transport (UT) and Ho Chi Minh City University of Technology (HCMUT)) (Kremmyda et al., 2020). The partners from the Ho Chi Minh University of Transport carried out the survey and its analysis in Vietnam.

The ENHANCE project is to nurture advancement of education and professional attributes (skills) in the field of community-based engineering that serve unsupported communities in collaboratively identifying problems and defining solutions that are sustainable. ENHANCE focuses on initiating sustained educational changes by modernising study programmes to address inclusive, interdisciplinary, problem-based, experiential curricula in programmes in South Asian partner countries.

In summary, the overall aim of this multi-partner ERASMUS+ project is to strengthen and build capacity in engineering programmes in the three South Asian countries by:

- introducing modern, innovative pedagogical approaches;
- identifying and tackling inherent barriers to quality, knowledge organization, curriculum content, and hierarchy; and,
- ensuring advancement of emerging skills to increase accountability, raise the quality of engineering graduates for employability, raise the quality and consistency of humanitarian services and to improve in-country capacity to operate safely.

In this paper, the authors are presenting the results from 116 responses to the 2019 ENHANCE project online survey sent in by contributors in Vietnam. 69 of these responses were received from educators (lecturers on engineering programs for undergraduate or/and postgraduate degrees), and another 47 responses are from employers (industry-based), which represent the opinions of Vietnamese companies and organizations employing graduate engineers. In what follows we use the survey analysis to elaborate on the established competency gaps between current graduate attributes and expectations of industry. Herein, we are using the terminology 'industry' to encompass, in a single word, all types of employers who require Vietnamese engineering graduates. The paper identifies ways in which engineering graduate attributes may be aligned to the preferred skills linked to employability. To address the issue of variability of taught attributes in Vietnamese HE programmes the paper presents results from the survey concerning how Vietnam universities taught, assessed, and reported employability skills. This clarifies mismatches between skills that are developed during existing degree programmes and those other skills needed in the workplace for graduates to be successful professional engineers.

- Based on the survey's structure, the authors have divided the paper into three sections: Section 2 for attributes of graduate engineering students from the point of view of educators and industry
- Section 3 for teaching and assessment methods on the quality of engineering programmes; and,
- Section 4 with proposals for future engineering programmes.

In the framework of the ENHANCE project the online survey was run concurrently in the other four partner countries of the UK, Greece, Bangladesh, and Indonesia. This enabled the authors of this paper to compare selective survey results from the other partner countries in South-East Asia with the survey findings from the 116 respondents in Vietnam in methods of teaching and learning that lecturer using and quantify the percentage of their contribution to the delivery of their program.

Attributes of Graduate Engineering Students in Vietnam

This section presents the survey's analysis for employers and educators' opinions regarding three the main pillars of graduate skills associated with employability, namely: 1) attributes, 2) capabilities, and 3) knowledge and understanding.

Graduate attributes are defined in several publications, with the definition in Bowden *et al.* (Bowden et al., 2000) relevant to the ENHANCE project. This definition says that they possess: "*…the qualities, skills and understandings a university community agrees its students would desirably develop during their time at the institution and, consequently, shape the contribution they are able to make to their profession and as a citizen*". It is to be noted that the term 'graduate attributes' can also be referred to by other descriptors such as: generic skills, graduate qualities, generic attributes, and graduate capabilities (Nair et al., 2009). This lack of a single, well-recognized descriptor (and definition) can create a certain ambiguity in discussion and dissemination because it is observed that these descriptors have been used interchangeably and can have ambiguous meanings if unfamiliar.

Presented in Table 1 are the ENHANCE project survey results for graduate attribute gaps between HE and the expectation of industry. In column (2) of the table there is, in alphabetic order, a list of 14 graduate attributes (skill sets) that had been ranked as 'most important' within the framework of the objectives of the ENHANCE project. There might well be other graduate attributes that could have been prioritized, yet the the project partners decided to limit to 14 the number surveyed. Column (1) in Table 1 allocates an integer number to each of the 14 attributes, which in column (2) have conveniently been ordered in alphabetical order. To answer this survey-question a respondent was asked to rank 10 of the 14 attributes, with 1 being the most important and 10 the least important. Column (3) presents

Table 1. Rating of the importance of graduate attributes by respondents from higher education and industry

No.	Graduate attribute	Higher Education (*HE*)	Industry (*Id*)	*D* (%)	*(HE+Id)/2*
(1)	**(2)**	**(3)**	**(4)**	**(5)**	**(6)**
1	Analytical Skills	**3.88**[1]	**3.81**[1]	1.82	3.85
2	Communication Skills	**4.54**[3]	4.78	-5.15	4.66
3	Creativity	4.58	**4.39**[2]	4.24	4.49
4	Critical Thinking	**4.41**[2]	5.68	-25.2	5.05
5	Culture Sensitivity	8.05	8.29	-2.94	8.17
6	Empathy	7.60	6.82	10.8	7.21
7	Entrepreneurship	7.05	6.13	14.0	6.59
8	Ethical principles	6.19	4.82	24.9	5.51
9	Interdisciplinary Perspective	6.77	7.63	-11.9	7.20
10	Leadership Skills	6.77	6.00	12.1	6.39
11	Practical Skills	4.65	**4.77**[3]	-2.55	4.71
12	Prioritization	7.75	6.22	-21.9	6.99
13	Problem Solving	5.00	5.47	-8.98	5.24
14	Team Working	5.42	5.77	-6.26	5.60

Note: $D(\%) = \dfrac{HE - Id}{0.5 \times (HE + Id)} \times 100\%$

for Higher Education (variable *HE*) the mean ranking scores for the 14 attributes based on 69 educator survey returns, and the numbers (in bold font) with superscript numbers 1, 2 or 3 are to highlight the attributes with the three lowest scores (i.e., they are ranked **most important**). The equivalent survey results for Industry (variable *Id*) are similarly reported in Column (4) using responses from 47 surveys.

Reported in column (5) are the percentage differences, *D*, calculated using the formula given in the Note at the bottom of Table 1. These percentage differences are determined based on the mean of the scores in columns (3) and (4). The overall mean values for the 14 attributes in columns (3) and (4) are 5.90 for *HE* and 5.76 for *Id*. These means should be the same at 5.5 when every survey return ranks ten attributes with integer numbers 1 to 10. The higher means show that surveys were not always completed to the instructions. Although there are 22 more survey respondents from HE than from industry, the lower number of 47 for industry is deemed sufficient for the authors to treat the two populations as being sufficiently representative of their sector so that a direct unweighted comparison between the mean ranking scores can be made with confidence. For additional comparison column (6) shows the average of the mean HE and Id scores from respectively column 3 and 4 ((HE+Id)/2).

To commence a discussion on the survey results presented in Table 1 the authors first considered the mean ranking scores for the three most important attributes. The lower the score listed in the table the higher is an attribute's importance to Vietnamese engineering graduates. Based on the fourteen attributes *HE* ranks the first three attributes in order of importance to be: Analytical Skills (3.88 is the lowest score); Critical Thinking (4.41); and Communication Skills (4.54). It is noteworthy that because Creativity with 4.58 and Practical Skills with 4.65 have ranking mean scores within 0.07 of Communication Skills (4.54) there's evidence to say that these three attributes are of equal importance. The sixth ranked attribute is Problem Solving with a score of 5.00, and with an increase in the mean of 0.35 this attribute can be said to belong in the less important HE group that includes nine attributes with scores between 5.00 and 8.05 (the latter has the second highest value in Table 1 for a least important ranking). In line with these findings, over the recent years, the education sector in Vietnam has directed higher education institutions to actively innovate training content, curricula, and teaching methods in the direction of a competency approach, enhancing application and practical skills. Training institutes link training with the labour market, proactively grasp the needs of enterprises or develop training programmes with the participation of enterprises to ensure the quality of their training output (Bui Thi Hong Dung, 2019).

Applying the same evaluation methodology to the Industry (*Id*) survey results in column (4) it is seen that employers rank their first three attributes in order of importance to be: Analytical Skills (3.81), Creativity (4.39); and Practical Skills (4.77). It is observed for *Id* that because Communications with 4.78 and Ethical Principles with 4.82 have ranking scores within 0.03 of Practical Skills there's a case to be made that each of these three attributes are of equal importance to industry. The sixth industry ranked attribute is Problem Solving at 5.47, and with an increase in the mean score of 0.65 this attribute belongs in the lesser important group with the remaining eight attributes, whose scores are from 5.68 to 8.29 (the latter is highest values in Table 1 for the least important ranking).

Though employers' needs vary from country to country, and among industries, analytical skills, creativity, practical capabilities, and skills in teamwork, communication and problem-solving are generally desired by the majority of employers. According to an employers' assessment, most graduates can't immediately start doing professional jobs, students can't plan their own study to perfect themselves at work and they lack the necessary soft skills to serve the assigned work. Accordingly, in order to provide training related to the needs of enterprises, universities must cooperate with enterprises in building training program objectives and content based on what society and learners need (Pham Thi Hang, 2021).

Given the high-level of commonality between the means of the two groups the average score values (column (6)), given by $(HE + Id)/2$ is used to find a joint ranking of attribute importance. This results in the highest importance being Analytical Skills (3.85), which is for the core and fundamental principles underpinning engineering programmes. This is not a surprising finding given the prominence of this attribute in existing engineering programmes.

In combination with Analytical Skills (table 1, column (6)) Creativity at 4.49, Communication skills at 4.66 and Practical Skills at 4.71 belong to the top four attributes of highest importance having a combined mean score < 5.0. The fifth ranked attribute is Critical Thinking with a combined score of 5.05. Whereas educators rank Critical Thinking as second most important (*HE* score is 4.41), which suggests they find that this attribute should have improved delivery, industry does not share this opinion and ranks Critical Thinking seventh with a significantly higher mean score of 5.68. From knowledge of their business activities employers rank Ethical Principles fourth with a score of 4.82, indicating this attribute lacks relevant content in existing curricula. HE on the other hand, returns a score for Ethical Principles of 6.19 (and rank this attribute eighth), which suggests that educators believe that the existing offer in teaching and learning is satisfactorily scoping this attribute in support of graduate employability. One important finding from the Vietnamese survey analysis is that HE and industry need to hold conversations to discuss and agree on the importance in teaching and learning of the two attributes of Critical Thinking and Ethical Principles.

This evaluation of the survey results next targets a gap analysis using the scores in the range 3.8 to 6.0 and the variable D for percentage differences, reported in column (5) of Table 1. Figure 1 plots the mean scores for these eight attributes with the solid circle symbol for *HE* and the solid square symbol for

Figure 1. Comparison of the eight mean ranking scores from 3.8 to 6.0 for the most important student graduate attributes

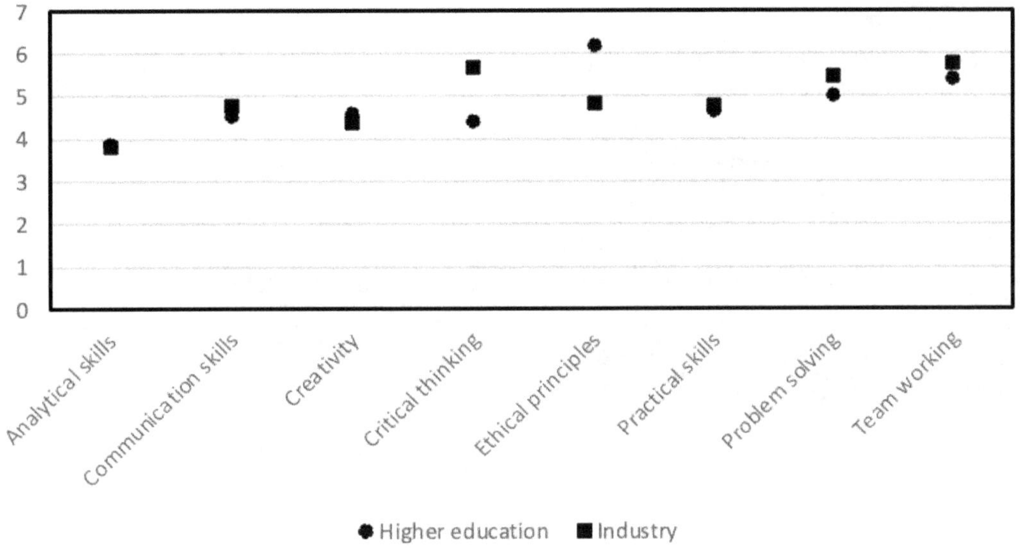

Id. In addition, the bar chart in Figure 2 exposes the magnitudes of differences by plotting the absolute values of D for the eight attributes. The percentage differences between HE and industry are found to be from 2 to 25%. With percentages of 9, 25 and 25 (to nearest percentage) the three attributes with the

Figure 2. Absolute differences as a percentage (D) between higher education and industry for the most important student graduate attributes receiving a mean ranking score in range 3.8 to 6.0

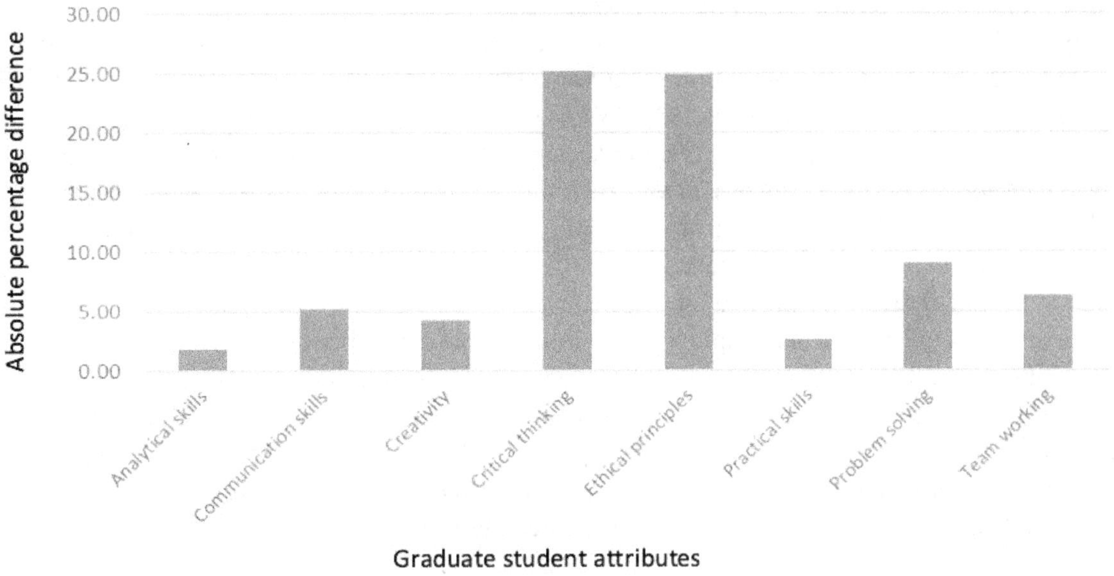

Figure 3. Confidence levels rated by 47 employer respondents to the survey question: 'do Vietnamese engineering graduates have the skills to fill high-skilled job vacancies?

greatest divergence are Problem Solving, Ethical Principles and Critical Thinking. On the other hand, the three attributes of Analytical Skills, Practical Skills and Creativity have percentage differences < 5%, which from the gap analysis are deemed not significant. Another finding from the survey analysis is that both Vietnamese sectors consider the relevance and importance of these attributes in the same way.

Another outcome from the ENHANCE project survey is the finding that in the opinion of employers Vietnamese engineering graduates neither have the depth or breadth in Analytical Skills (ranked first) and Problem-Solving Skills (ranked sixth) as required in the future. Information from the survey's free-text questions further shows that employers also desire graduates to possess higher levels of application skills, which is inherently linked to the attribute Practical Skills (ranked third). Vietnamese industry knows from experience that such graduate skills are essential for working efficiently, and for business successes with improvements in creativity, innovation, and productivity.

Survey results presented in Table 1 and Figures 1 and 2 show consistency with a finding from another question in the industry's bespoke set of survey questions. The question is: "*As we move towards an economy driven by the Fourth Industrial Revolution (also known as Industry 4.0), there are clear implications for the engineering sector and its skills needs. How confident are you that there are enough engineering graduates in your country with the skills to fill high-skilled job vacancies?*" The bar chart in Figure 3 presents this question's results from 47 employer respondents. 12% said they are 'not confident' and 66% only thought they are 'somewhat confident' that Vietnamese engineering graduates possess the necessary skills to fill their high-skilled job vacancies.

The survey subsequently invited employers to categorize the capability of graduate engineers by way of 10 capabilities and rank them as either Poor, Average, Good, Very Good or Excellent. The results are presented in Table 2 as percentages rounded to the nearest integer except for the Excellent percentages. These are rounded up or down to one decimal. Column (2) of the table lists capabilities in alphabetical order with their ordering number in column (1). Our evaluation clusters the capabilities as perceived by Employers and Educators into one of the following three classifications:

Table 2. Rating by employers and educators of 10 capabilities as perceived in graduates of engineering programmes

		Poor and Average		Good and very good		Excellent	
	Capability	Industry	HE	Industry	HE	Industry	HE
		%	%	%	%	%	%
(1)	(2)	(3)	(4)	(5)	(6)	(7)	(8)
1	Analytical Skills	35	12	63	86	2.5	2.3
2	Creativity	51	37	44	58	4.9	4.7
3	Experience in Real-Life Applications	59	29	41	69	0.0	2.4
4	Leadership Skills	61	47	39	51	0.0	2.3
5	Practical Skills	30	19	70	77	0.0	4.7
6	Problem Solving Skills	41	16	59	81	0.0	2.3
7	Project Management	68.	44	32	56	0.0	0.0
8	Team Working Skills	46	30	54	63	0.0	7.0
9	Theoretical Knowledge	17	5	78	81	4.9	14
10	Understanding of Regulations, Standards and Codes of Practice	32	26	63	72	4.9	2.3

- Average and Poor Good;
- Very Good; and
- Excellent.

For each capability the three percentages across the row reported in columns (3), (5) and (7) for Industry or columns (4), (6) and (8) for Higher Education should add up to 100%.

The percentage values listed in column (3) show us that the five capabilities Creativity, Experience in Real-Life Applications, Leadership Skills and Project Management were judged as Poor or Average by more than 50% of industry respondents. This judgment may be interpreted as an appeal from employers to educators in HE to improve on these capabilities in the curricula. Figure 4 illustrates the percentages for the ten capabilities in Table 2 with the abscissa axis numbering corresponding to the capability numbers in column (1) of the table. Secondly, we can see from the data in Figure 4 that educators consistently assess the capabilities as less poor/average. This is true for Analytical Skills with a 23% difference, Experience in Real-Life Applications showing 30% difference, Problem-Solving Skills with 25% difference, and lastly Project Management having 24% difference. These four capabilities are judged with a difference > 20%. It could be argued that employers possess the most realistic understanding on the skill needs of graduates entering the workforce. As a consequence of this observation the authors want to see an increase in the number of employer influencers on committees preparing new or revised engineering curricula because this will undoubtedly benefit the graduates' capabilities and their employability.

Thirdly, we find that employers rated four of the capabilities (>60%) as Good and very good. These are Analytical Skills, Practical Skills, Theoretical Knowledge, and Understanding of Codes and Practices. With respect to the relative importance of these four capabilities they agreed more or less with

Figure 4. Percentages for capabilities (1 until 10) assessed Poor and Average with higher education values given by the circle solid symbols and industry values given by the square solid symbols

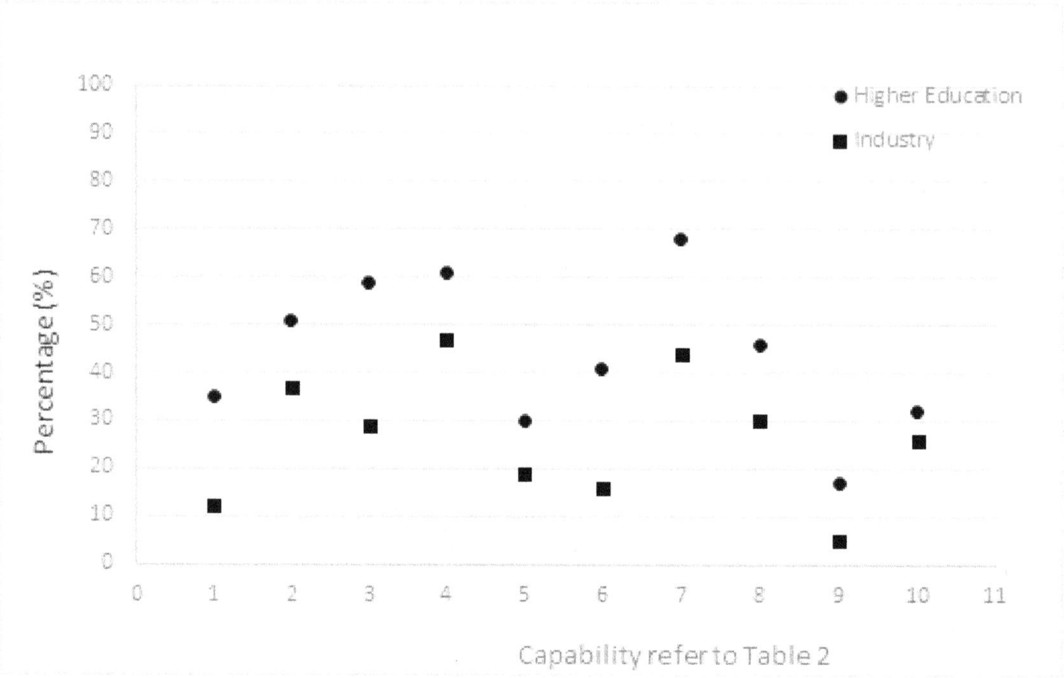

the educators. The outcome that not less than 35% of employers find that graduates have a poor/average capability with respect to Analytical Skills (and of highest importance), must be an incentive in HE to implement actions for an improvement.

The abscissa numbers are the capability numbering in column (1) of Table 2.

Another survey question scoping topics related to the soft and global skill sets in engineering education is: *Rate from 1 to 10 the need for graduate engineers to have knowledge and understanding of the following topics: (1) Community Needs; (2) Culture Preservation; (3) Diversity and Equality (Gender, Sex, Age, Ethnicity, etc.); (4) Economic Factors; (5) Ethics; (6) Human Needs and Human Rights; (7) International Market; (8) Political factors; (9) Professionalism; (10) Sustainability; (11) Other (please specify)*". This question was answered by all 116 respondents with the results presented in Table 3. The respondents were asked to score the most needed topic by 1 and the least needed of the ten topics by 10. Column (1) gives the topic a number based on alphabetical ordering of their names in column (2). Columns (3) and (4) present the HE survey results with mean values in column (3) having the same meaning as the equivalent values in column (3) of Table 1, and column (4) is for their rank order. The columns (5) and (6) present the results from the industry respondents.

As can be observed from inspecting the results in Table 3 both Industry and HE agree by-and-large on the order of importance of the 10 topics for soft and global skill sets. Both are assigning highest importance (first and second ranking) to students receiving curriculum exposure to Community Needs (scores of 3.43 and 4.08) and Professionalism (scores of 4.45 and 3.51). Of lower, yet significance importance, both sets of survey results rank third and fourth Economic Factors (scores of 4.46 and 4.82)

Table 3. Survey results for the importance of topics of knowledge and understanding for engineering graduates

No.	Topics of knowledge and understanding	Mean HE	HE ranking	Mean Industry	Industry ranking
(1)	(2)	(3)	(4)	(5)	(6)
1	Community Needs	3.43	1	4.08	2
2	Culture Preservation	5.81	7	6.38	8
3	Diversity and Equality (gender, sex, age, ethnicity, etc.)	5.95	8	7.05	10
4	Economic Factors	4.64	3	4.82	4
5	Ethics	5.74	6	5.65	6
6	Human Needs and Human Rights	5.62	5	6.13	7
7	International Market	6.55	9	5.46	5
8	Political Factors	7.12	10	6.95	9
9	Professionalism	4.45	2	3.51	1
10	Sustainability	5.38	4	4.79	3

and Sustainability (scores of 5.38 and 4.79). Again demonstrating commonality, HE and Industry rank Ethics and Human Needs in the range of fifth to seventh.

Employers give low importance to students having received knowledge and understanding to the important topics of Diversity and Equality and Political Factors. It would be interesting, in particular for these topics to find out to what degree these rankings are specific to Vietnamese educators and industry or are ranked similarly by their counterparts in other countries. Employers are found to consider the topic International Markets to be of greater importance (ranked fifth) than by educators (ranking it ninth). Educators in Vietnam either judge this topic to be outside of their engineering programmes or consider it to be integrated into the knowledge of Economic Factors, which is ranked in the top four of the 10 skill sets (see Table 3).

For a clear overview the topic ranking given in table 3 is plotted in figure 5.

To complete this section, we collate the three most important attributes from Table 1 and three most important topics from Table 3. From an industry position they are Analytical Skills, Creativity and Practical Skills (from Table 1) and Professionalism, Community Needs and Sustainability (from Table 3). These can be compared with the survey results from HE which have Analytical Skills, Critical Thinking, and Communication Skills from Table 1 and Community Needs, Professionalism and Economic Factors from Table 3. Encouragingly we find that there is a high degree of similarity in the priorities of the two survey groups. Possible major gaps in the priorities for teaching and learning on undergraduate/postgraduate programmes can be with Critical Thinking, which is ranked higher by HE and Ethical Principles, which is ranked higher by the employers. With regards to the perceived capabilities of engineering students on graduation we observe from the results in Table 2 that Vietnamese HE assesses graduates to be consistently more capable than does industry.

Employers' responses to a question on the confidence they have in the skills of graduates (refer to Figure 3) show that their confidence is low. When we now consider the perceived shortcomings in graduates' capabilities both sets of survey results agree with respect to weaknesses in project management and leadership skills. Moreover, employers would like their graduated engineers to have more education

Figure 5. Mean values from Table 3 for topics of knowledge and understanding (with abscissa numbering according to column (1) of the table)
Note: Values assigned by higher education are indicated by circle symbols and by industry by square symbols.

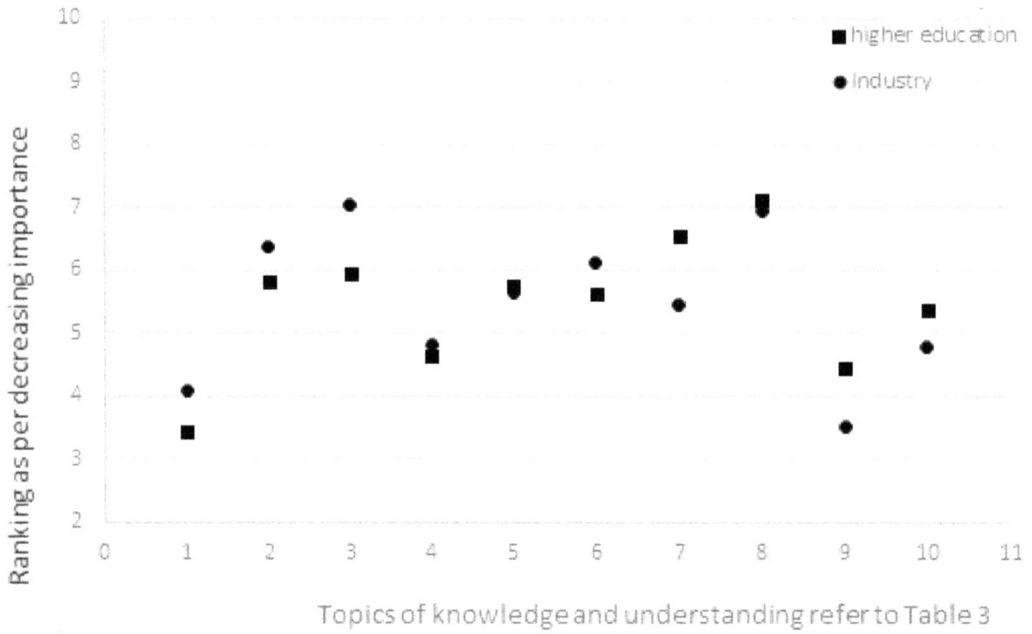

covering analytical skills and real-life experience(s). To conclude on the main outcomes from this section of the paper, the authors are encouraged that HE in Vietnam is providing much of what industry expects from graduate engineers, and following the ENHANCE project survey results, we strongly recommend that there is dialogue and collaboration so that industry's priorities are aligned with what educators advance in their engineering programmes.

IMPACT OF TEACHING AND ASSESSMENT METHODS IN VIETNAM

This section presents survey analyses of teaching and learning and of performance assessment methods that are current in Vietnamese engineering programmes. One of the bespoke survey questions in the HE survey is "*In the list below identify which methods of teaching and learning you are using and quantify the percentage of their contribution to the delivery of your program (total = 100%)?*".

For teaching and learning methods Table 4 presents the survey results expressed as a percentage of delivery time over degree programmes. Column (1) allocates an integer number to 14 methods with their descriptors listed in column (2). The maximum values from the 69 Higher Education (HE) surveys are reported in column (3), together with mean values in column (4). The summation of the 14 mean values equals 100% and to highlight their relative importance this is shown by the bar chart in Figure 6.

These ENHANCE project survey results can lead us to the following observations. With a mean at 32%, which is three times greater than the next highest mean value, and with the highest maximum at 70%, method No. (4) for traditional Lectures is by far in way the most applied approach for the delivery

Table 4. Maximum and mean percentages of teaching time spent on different methods for teaching and learning on engineering programmes

No.	Method for teaching and learning	Maximum (%)	Mean (%)
(1)	(2)	(3)	(4)
1	Game-Based	20	4.3
2	Flipped Classroom	40	4.4
3	Journal Club	12	0.9
4	Lectures	70	31.9
5	Personalised Learning	30	7.8
6	Practical Activities (Laboratories, field visits, etc.)	30	8.3
7	Problem-Based Learning	30	9.0
8	Project-Based Learning	70	9.9
9	Student as Researcher	12	2.7
10	Student-Centred Learning	60	7.0
11	Technologically Enhanced Learning (e.g., e-learning)	20	3.3
12	Tutorials (or seminars)	40	5.8
13	Work-based learning	30	4.3
14	Others (please specify)	16	0.4

of teaching and learning. This expected finding is vividly highlighted by the tallest bar in Figure 6. Next, with means of up to 10% are Problem-Based Learning and Project-Based Learning. These two are followed in the rankings by Practical Activities for laboratories, field visits, etc., with a mean percentage just over 8%. Practicals, of all types, form a core teaching activity in the delivery of engineering programmes. The remaining 10 methods accounts for about 42% (or on average four percent each) of the delivery time and a for a wide variety of methods. From amongst these ten, a middle group, namely Personalized Learning, Student-Centred Learning, and Tutorials represent over 20% of delivery time, having means, respectively, of 7.8%, 7.0% and 5.8%. The remaining seven methods are less important taking minor time percentages.

It is of interest that column (3) in Table 4 shows us that the maximum time recorded is usually, and unsurprisingly, much higher than the average time. This finding does hold, yet at a more modest degree for Lecturing (mean 31.9%, maximum 70%). Despite the solid role of lecturing it is noteworthy to record that the minimum time spent can be 0%, indicating that on certain programmes lecturing has no place. Although the evidence is hidden from how the survey results have been analysed there is every likelihood that when there are no or few lectures this method has been replaced by Project-Based Learning, which has a maximum at 70%. Given enticing thoughts on how newer programmes are structured for their delivery time column (3) in Table 3 informs us that maximum percentages are relatively high at 60% for Student-Centred Learning and at 40% for Flipped Classroom and Tutorials; these maximums are to be compared with means of 7.0%, 4.4% and 5.8%, respectively. Another modern method with an impact on at least one programme is Game-Based learning, which has a maximum at 20%, yet a mean barely exceeding 4%.

Figure 6. Mean percentages of delivery time over degree programmes from 14 methods of teaching and learning currently employed on Vietnamese engineering programmes

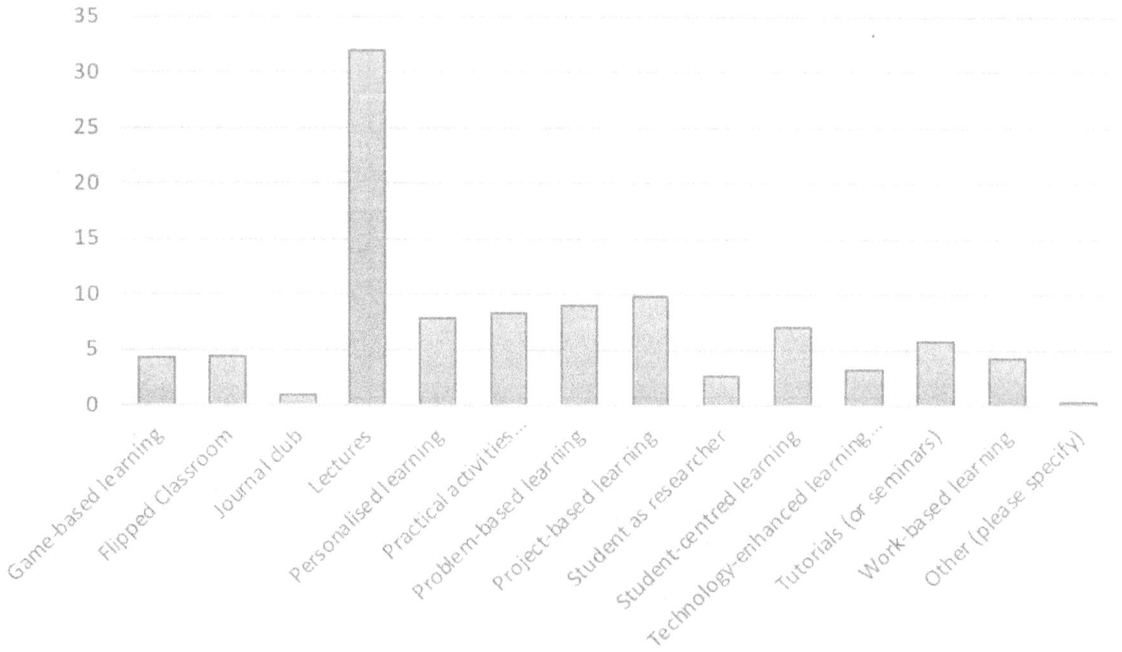

As expected the face-to-face traditional lecture method is the most common approach for instructional delivery in most countries in the world, especially when programmes have curricula dominated by traditional teaching of fundamental engineering principles for analytical skills (Arunkumar et al., 2018) (Akor et al., 2019). Vietnam, HE programmes are no exception. By introducing into this discussion, the ENHANCE project survey results from the countries There are established that lecturing contributes 41.22% in Bangladesh, 33.67% in Indonesia, 36% on average to the total teaching delivery methods in Asia countries.

It is understood that the Lecture method offers several pedagogical advantages, this is because they:

- present material not otherwise available to students by another person who can be familiar with the students;
- present substantial amounts of information in an effective way;
- be specifically organized to meet the needs of particular student audiences;
- appeal to those who learn by listening (Bonwell, 1996) (Cashin, 1985).

Moreover, a fifth advantage is that experienced lecturers can use lecturing to communicate the intrinsic interest of a subject through their enthusiasm and (localized) subject expertise.

Yet is well-known that lecturing has disadvantages that makes it not suitable for the acquisition of every skill set that Vietnamese employers are seeking in graduate engineers. Some of these disadvantages are discussed by Bonwell and Cashin (Bonwell, 1996) (Cashin, 1985), and can be summarized as:

- lectures are not well suited for teaching complex, abstract material;
- lecturing presumes that all students learn at the same pace and are at the same level of understanding;
- lecturing requires effective speakers (who should have empathy with the needs of the students);
- lectures are not suited for delivering material in support of learning higher orders of thinking, such as for applications, analysis, synthesis or evaluation, for teaching physical skills and for influencing professional attitudes or values.

Moreover, it is known that a fifth weakness is that information tends to be forgotten quickly when students are passively learning in the lecture room environment. A sixth disadvantage is that the lecture method emphasizes learning by listening (with watching), which is a challenge for students who prefer to educate themselves by way of other methods that deliver teaching and learning (refer to the other 13 methods in Table 4).

Highlighted from the ENHANCE project survey results in Table 4 is that teaching and learning methods in Vietnamese universities are diverse, with important contributions for a mean percentage of 35.0% from Project-Based Learning, Problem-Based Learning, Practical Activities and Personalised Learning. Different delivery methods are associated with different skills or student attributes that engineering students possess on graduation. Embracing the survey results there can be an aspiration that, even accounting for the diversity between employers and educators, they collaborate to find common ground for programme developments. As a consequence, confidence in graduate skills would grow and we shall find in, say, five years a positive change to the current poor situation, as shown in Figure 3. Furthermore, our survey finds that employers are seeking professional graduates with improved capabilities to be creative, having sound experiences of real-life applications, and possessing strong leadership and project management abilities. It is important to note that these attributes are not likely to be enhanced or developed if lecturing is the method of teaching and learning.

Clearly, methods of assessment play a key role in ensuring the level and quality of student learning and for ranking the knowledge, skills, and behaviour of graduates. In order to guarantee that on graduation Vietnamese engineering students have career skills, alternative teaching methods are advocated, such as Project based learning and Problem-Based Learning. The approach to assessing students is changing accordingly to (Akor et al., 2019), (Kipper & Ruutmann, 2012) and (McDowell et al., 2004), with formative assessment methods now given more prominence than summative assessment methods (Subheesh & Sethy, 2018). In Vietnam, owing to the lack of consistency in their application, there is considerable scope for changing to more innovative methods of assessment. Currently, the government regulations on the percentage composition of marks in degree programmes creates local difficulties for universities to make radical changes in teaching delivery. Throughout Vietnam the convention in 2022 is to employ a marking framework based on 10% for diligence, 20-30% for mid-term, and 60-70% for end-of-term assessments. Adding further restrictions to what Vietnamese educators in HE can do to assess their engineering students these internal marking frameworks are often fixed by government regulations for specific degree subjects.

To understand which methods of assessment are used the survey included the following question, "*In the list below please identify which methods of assessment you are using and quantify the percentage of their contribution to the delivery of your programme (total = 100%): (1) Artefact, (2) Blog; (3) Essay; (4) Group Oral Presentation; (5) Group Portfolio; (6) Individual Oral Presentation; (7) Individual Portfolio; (8) Journal Article; (9) Laboratory Report; (10) Online Tests (including Quizzes); (11) Peer Assessment; (12) Poster Presentation; (13) Reflective Essay; (14) Written Examination; (15) Other (please specify)*".

According to this question the educator has 14 different assessment methods to choose from, plus a fifteenth, should the programme use another assessment method. Evaluation of answers using the 79 HE surveys show that the prominent four methods of assessment are by Written Examinations at 36%, Group Oral Presentations at 17%, Essays at 11%, and Individual Oral Presentations at 9%. Combined these four traditional methods of assessment account for 73% to the total contribution. Because programmes will change, and new programmes start the authors expect that, if an ENHANCE project survey is repeated in 2024 (five years after the original survey), the total contribution from the other 10 assessment methods will be higher than at 27% in 2019.

To conclude this section of the paper, we have found that university teaching of engineering students is dominated, in terms of delivery time, by lecturing, with a complementary role for problem-based learning, project-based learning and all forms of practical activities, including internships. The internship program is usually designed in all engineering training programs as 2 credits, placed in the last semester to support students in fact-finding, serving the implementation of their graduation thesis. In addition, the study program, may include 1 to 2 credits for site visits or field trips to the industry for students to visit and learn related to their field of study. Because the latter method is ideal to expose students to real-life experiences (refer to Table 1) the time allocated to this method of delivery has to be increased, from its current 8.3% contribution, to say between 12 and 15%. Because of government regulations assessment methods are remaining mostly traditional. The results of the ENHANCE project offers support from industry that educators can be empowered to introduce new educational goals and assessment methods.

In view of the societal challenges that Vietnamese engineering graduates must cope with lecturing needs to be supplemented by appropriate delivery methods that will stimulate and develop analytical skills, creativity, and skills in project management and leadership.

FUTURE ENGINEERING PROGRAMMES IN VIETNAM

One finding from the discussion in Section 2 is that employers in Vietnam need graduate engineers with improved capabilities regarding analytical skills, creativity, experience in real-life applications, and skills in leadership and project management. These graduates should also be professionals with open eyes to community needs. What needs to be done to bring about improvements in graduate capabilities? Common to the delivery of many of the desired capabilities for skill sets, as presented in Sections 2 and 3, is that students need increased exposure to the World and its demands, outside the university campus environment. Relevant to the future of engineering programmes is that ASCE (ASCE, 2004) has established that a lack of cooperation between industry and HE in delivering degree programmes elicits the danger of undesirable outcomes, to the detriment of all stakeholders. The authors are aware that to increase the desired exposure that will build capabilities in graduate employability requires pro-active participation between employers, educators and students (i.e. all the stakeholders).

How could universities interact and mutually collaborate with industry so that employers can contribute to the successful development and delivery of engineering programmes? Interaction and collaboration can take various forms, such as industry providing high-quality internships and relevant societal and industrial problems (in the framework of problem-oriented studies). In addition, industry should be prepared to take part in the final assessment at the end of students' studies. Furthermore, universities would welcome employers that are willing to work closely with educators to strengthen both programme content and methods for delivery of teaching and learning.

The survey has confirmed the need for educators to adapt degree programmes, e.g. by replacing traditional lecturing with problem-oriented studies (laboratory work, or project or problem-based learning activities). Possibilities could be to reserve laboratory space for more laboratories and to allocate teaching funds for field work (in student groups) undertaken in the framework of problems proposed by industry or other societal actors (e.g. NGOs).

From the students perspective there already exists an understanding that Vietnamese students highly appreciate working on problem-oriented studies. This enthusiasm stems from the fact that these activates open students to real-life engineering problems, which cater for the needs of local and global communities, industry and the Vietnamese society at large.

Teaching students to be creative as engineers, which is a key graduate attribute, will certainly be stimulated by an increased exposure to real societal problems. Engineers need to be creative in how they go about solving problems they are tasked to work on. This is especially so in 2022 because Vietnam is experiencing a rapid societal transformation. Seventeen years ago, Cropley *et al.* (Cropley & Cropley, 2005) reviewed findings on fostering creativity in engineering education within the United States, and concluded that there was little support for engaging creative students. Two years later Sternberg (Sternberg, 2007) outlined the three learning activities that promote the habit of students/graduates working creatively. First, he said that students must have the opportunity to engage in exercises whose solutions require creativity. This learning opportunity must be woven, holistically, throughout degree programmes in an integrated and mutually reinforcing manner. Second, this author posited that students must receive positive encouragement from facilitators as they engage in learning tasks requiring them to give creative solutions. Third, Sternberg said that students must be rewarded with marks when they achieve outcomes that demonstrate creativity. In the changes towards a stronger interaction between engineering programmes and societal practice the recommendations regarding creativity should be taken to heart; it must be prominent in teaching and learning activities and be supported by assessments of study outcomes.

Given existing regulations in Vietnam the necessary pedagogical changes, highlighted by the ENHANCE project survey, are likely to require structural adjustments in Vietnam's university system. Here, not only individual educators and their departments and faculties must play a role, but for necessary changes (meeting the needs of all stakeholders) to take root, they must be embraced by higher administrative levels. Without such high level adoption, we can say that radical modifications of assessment methods in degree programmes (linked to innovative methods of teaching and learning) are not going to be approved.

For future engineering programmes in Vietnam the authors propose three approaches to assure a better alignment of graduate attributes based on the survey results discussed in this paper, and they are that:

- Curricula need to integrate more problem-oriented learning with traditional instructions to develop graduates with practical skills.
- Engineering educators and employers need to work together so that more teaching and learning activities are dedicated to graduate attributes that support student employability.
- To satisfy the industry's need for graduates having enhanced competency standards for professional and personal attributes, knowledge and skills, engineering programmes require a higher education alignment to workforce needs through curricula content with more practical experience, and employer-educator collaboration delivering new teaching and learning activities.

CONCLUSION

The world has entered the era of the fourth generation of industrial transformation, with information and communication technologies driving changes and their impacts are felt in every sector of Vietnamese society. Undergraduate/postgraduate higher education is one of the many sectors that are severely impacted, meaning that higher education educators need to update their curricula and modify their teaching and learning, and their assessment methods to be impactful in meeting the continuing changing requirements of graduate engineers entering the industry workforce. Continuous change will need to be flexible and innovative, and appropriate to cope with development trends in technology and communication. To ensure that advances in teaching and learning activities are aligned with the set of graduate attributes, as presented in this paper by way of a ENHANCE project survey, the authors advocate that there must be stronger and more effective collaboration between employers, educators and student engineers.

Engineering jobs require of Vietnamese graduates stronger attributes and capabilities (skill sets) in a variety of domains. The ENHANCE project survey has shown that in 2019 employers were only moderately confident that engineering graduates could satisfy the requirements of their job vacancies. While they found that graduates should possess analytical skills, creativity and practical skills, their opinion was that graduates needed to have more exposure during degree programmes to real-life experience(s) and be better equipped in terms of project management and leadership skills. In addition, graduates should have open eyes to the needs of the communities they serve. By way of the survey also showed' that the opinions of educators did not differ much from those of employers; yet, educators valued graduate skill sets generally higher than did the employers.

Amongst a list of 14 teaching and learning methods the survey confirmed that lecturing takes up most delivery time in engineering programmes. Modern methods of delivery such as project-based learning, problem-based learning, practical activities or personal-based learning have their place, but their share of programme delivery time needs to be enhanced in view of present societal needs (for soft and global topics), and the employability of graduates. This important conclusion is evident from the survey's results.

The survey informs us that assessment of performance in engineering programmes is usually dominated by written examinations. This is not consistent with the views on the importance of graduate skill sets. The authors thereby recommend that innovative assessment methods need to be aligned with the goals and teaching and learning methods employed in future engineering programmes. Another recommendation is that educators need to adjust their curricula, teaching and learning, and assessment methods according to societal needs. Among these needs, those required by Vietnamese industry and the development trends in technology and communication are preeminent for the future content of state-of-the-art engineering programmes.

Finally, the authors strongly recommend to all Vietnamese stakeholders that for the country to have impactful engineering degree programmes they must implement the recommendations for changes (refer to Sections 2 to 4) that are evidenced by the analysis of the ENHANCE project survey results of 2019.

ACKNOWLEDGMENT

The ENHANCE project survey and its analysis were undertaken within the framework of the project ENHANCE - *Enabling Humanitarian Attributes for Nurturing Community*-based Engineering (project No: 598502-EPP-1-2018-1-UK-EPPKA2-CBHE-JP (2018-2582 /001-001), funded by the Erasmus +

KA2 Cooperation for innovation and the exchange of good practices – Capacity building in the field of Higher Education. This publication reflects the views only of the authors, and the Commission cannot be held responsible for any use which may be made of the information contained therein.

REFERENCES

Akor, T. S., Bin Subari, K., Binti Jambari, H., Bin Noordin, M. K., & Onyilo, I. R. (2019). Engineering and related programs' teaching methods in Nigeria. *International Journal of Recent Technology and Engineering*, *8*(2), 1279–1282. doi:10.35940/ijrte.B1915.078219

Arunkumar, S., Sasikala, S., & Kavitha, K. (2018). Towards enhancing engineering education through innovative practices in teaching learning. *International Journal of Engineering and Advanced Technology*, *8*(2), 153–159.

ASCE. (2004). *Civil Engineering Body of Knowledge for the 21st Century*. American Society of Civil Engineers.

Bath, D., Smith, C., Stein, S., & Swann, R. (2004). Beyond mapping and embedding graduate attributes: Bringing together quality assurance and action learning to create a validated and living curriculum. *Higher Education Research & Development*, *23*(3), 313–328. doi:10.1080/0729436042000235427

Bonwell, C. C. (1996). Enhancing the lecture: Revitalizing a traditional format. In T. E. Sutherland & C. C. Bonwell (Eds.), *Using active learning in college classes: A range of options for faculty. New Directions for Teaching and Learning No. 67*. doi:10.1002/tl.37219966706

Bowden, J., Hart, G., King, B., Trigwell, K., & Watts, O. (2000). *Generic Capabilities of ATN University Graduates*. Final Report to DETYA, Australian Technologies Network Teaching and Learning Committee, Melbourne, Australia.

Bui Thi Hong Dung. (2019). Improve the quality of higher education - Nâng cao chất lượng giáo dục đại học. *Figures and Event Journal*. https://consosukien.vn/nang-cao-chat-luong-giao-duc-dai-hoc.htmhọc

Cashin, W. E. (1985). *Improving lectures. Idea Paper No. 14*. Kansas State University, Center for Faculty Evaluation and Development.

Cropley, D. H., & Cropley, A. J. (2005). Engineering creativity: A systems concept of functional creativity. In J. C. Kaufman & J. Baer (Eds.), *Faces of the muse: How people think, work and act creatively in diverse domains* (pp. 169–185). Lawrence Erlbaum.

Kipper, H., & Ruutmann, T. (2012). Teaching for understanding in engineering education. *2012 15th International Conference on Interactive Collaborative Learning, 3*(1), 55–63. 10.1109/ICL.2012.6402148

Kremmyda, G., Georgoulas, A., & Koumpouros, Y. (2020). Demystifying Humanitarian Engineering: A comparative study on perceptions in UK and Asia. *2020 IEEE Global Engineering Education Conference (EDUCON 2020) 27-30 April 2020*, 1009–1014.

McDowell, L., White, S., & Davis, H. C. (2004). Changing assessment practice in engineering: How can understanding lecturer perspectives help? *European Journal of Engineering Education*, *29*(2), 173–181. doi:10.1080/03043790310001633151

Nair, C. S., Patil, A., & Mertova, P. (2009). Re-engineering graduate skills - A case study. *European Journal of Engineering Education*, *34*(2), 131–139. doi:10.1080/03043790902829281

Patil, A., & Codner, G. (2007). Accreditation of engineering education: Review, observations and proposal for global accreditation. *European Journal of Engineering Education*, *32*(6), 639–651. doi:10.1080/03043790701520594

Pham Thi Hang. (2021). University and enterprise cooperation in the current context - Hợp tác đại học và doanh nghiệp trong bối cảnh hiện nay. *State Management Review*.

Sternberg, R. J. (2007). Creativity as a habit. In A.-G. Tan (Ed.), *Creativity: A handbook for teachers* (pp. 3–25). World Scientific. doi:10.1142/9789812770868_0001

Subheesh, N. P., & Sethy, S. S. (2018). Assessment and evaluation practices in engineering education: A global perspective. *3rd International Conference of the Portuguese Society for Engineering Education, CISPEE 2018*, 1–5. 10.1109/CISPEE.2018.8593451

Chapter 11
The Engineering Curriculum of Higher Education in Indonesia Revisited:
The Perspective of Higher Education Providers, Quality Assurance Agencies, and Industry

Intan Supraba
Universitas Gadjah Mada, Indonesia

Surjono
https://orcid.org/0000-0002-3024-5397
Universitas Brawijaya, Indonesia

Budi Hasiholan
Institut Teknologi Bandung, Indonesia

Indra Perdana
Universitas Gadjah Mada, Indonesia

Irwanda Laory
University of Warwick, UK

Angelos Georgoulas
https://orcid.org/0000-0001-8845-3781
University of West Attica, Greece

Yiannis Koumpouros
https://orcid.org/0000-0001-6912-5475
University of West Attica, Greece

Georgia Kremmyda
University of Warwick, UK

ABSTRACT

Over the years humanitarian issues have increased in many parts of the world including Indonesia. Such issues drive the need to evaluate Indonesia's engineering higher education curriculum. Aiming to provide a constructive contribution to inherent humanitarian aspects in the engineering education curriculum, this study revisits the teaching and learning programs in each curriculum of three public universities in Indonesia (i.e., ITB, UB, and UGM). The survey was conducted on 112 respondents from higher education providers and quality assurance agency and 43 respondents from the industry. Results show that about half of postgraduate and a quarter of undergraduate programmes have already run humanitarian engineering programs. According to the industry, undergraduates have better average scores in creativity

DOI: 10.4018/978-1-6684-5619-4.ch011

and leadership, while postgraduates have higher average scores in other parameters such as analytical skills, experience in real-life applications, practical skills, and teamwork skills.

INTRODUCTION

There are two aspects underlying the need to review the engineering curriculum of higher education in Indonesia. The first is the need for higher education to focus more on axiology than ontology and epistemology. This is motivated by various humanitarian problems due to natural and man-made disasters. Located on the Pacific Ring of Fire (an area with high tectonic activity), Indonesia is faced with the constant risk of volcanic eruptions, earthquakes, landslides, floods, and tsunamis (Siagian, Purhadi, Suhartono, & Ritonga, 2014). Those natural disasters inflict serious damage which causing economic losses. Economic losses from natural disasters have been increasing in recent decades especially in disaster prone areas. Future natural disaster losses are expected to increase due to climate change (Botzen, Deschenes, and Sanders, 2019). Hence, both, natural disasters and poverty alleviation are still the main target that must be resolved, as stated in the Sustainable Development Goals (SDGs). In this context, axiology becomes more contributing than epistemology and ontology.

This article discusses the technical higher education curriculum, especially for three universities in Indonesia: Universitas Brawijaya (UB), Universitas Gadjah Mada (UGM), and Bandung Institute of Technology (ITB). Those three universities are Indonesia partner universities in Erasmus+ Enhance Project (https://warwick.ac.uk/fac/sci/eng/enhance/). Thus, best practices of implementation higher education curriculum in those universities are being analysed. Due to uncertain future, the higher education curriculum must pay attention to the importance of cultivating value-conscious personality traits, one's motivation, and the ability to self-determination through understanding existential life forms (Kiryakova, Olkhovaya, Melekesov, & Presnov, 2015). The second aspect is related to the competitiveness of Indonesian universities internationally especially with regards to global ranking. In terms of ranking, Indonesian universities are generally still ranked lower than universities in Malaysia and Singapore in the regional context, especially within the Association of Southeast Asian Nations (ASEAN) (Volodymyr, 2018). In Asia, UGM is ranked 57 followed by ITB ranked 62 and UB is still ranked 271-280 (QS Asia University Rankings, 2021). Curriculum development does not directly affect the main indicators of ranking but indirectly affects publications, academic reputation, and employer reputation (Albashiry, Voogt, & Pieters, 2015).

There are several significant aspects of curriculum development in higher education. The credit system in higher education in Indonesia is rigid regarding the number of credits for undergraduate and postgraduate programmes. The distribution of credits for each course and the number of courses in the curriculum design influence the average contact and self-study time. The contact time would influence learning outcomes (Mohamed, 2006). Most undergraduate programmes in Indonesia have less than 10 hours per week, while postgraduate programmes have contact time between 10 to 20 hours per week. Methods of teaching and learning are also important issues. Problem-based learning, blended learning, and student-cantered learning are already popular in Indonesian higher education institutions (Zainuddin & Keumala, 2018). However, methods such as flipped classroom, journal club, game-based learning, and other recent approaches are not very popular yet in Indonesia (Farida, Alba, Kurniawan, & Zainuddin, 2019). Another recent issue is related to an effective method of teaching and learning during the Covid-19 pandemic. Covid-19 has strengthened distance learning methods in Indonesia in terms of its system, regulation, and communication technologies in higher education institutions. The initiation of the Merdeka Belajar Kampus Merdeka Programme ('learning freedom': students' right to take courses from other universities, institutions, and/or industries) was also started in the pandemic period (Suryaman, 2020). Based on these backgrounds, this paper aims to describe the potential and problems of community-based engineering education curriculum in the three participated universities (i.e., ITB, UGM and UB), to analyse the existing engineering higher education teaching-learning programmes from the point of view of the three stakeholders, and to introduce new collaborative design of engineering education program in Indonesia, based on the results of the survey.

In Indonesia, higher education is based on the national standard and is viewed as political. The standard was prepared by teams that were primarily experts from universities, teaching universities, and the Ministry of Education (Burrill, Lappan, & Gonulates, 2015). The current Indonesia's government policy assigned to the higher education institutions is aimed to support achieving a new level of economic development by 2025 through human resource development. Volodymyr (2018) argued that the standard was nationwide, and the development process was top-down. Higher education in Indonesia is regulated in the Higher Education Act No 12 of 2012 (The Republic of Indonesia, 2012), while its standard is formulated in the Regulation of Minister of Education and Culture No. 3 of 2020. The ministerial regulation comprises standards of a) graduate competencies, b) learning contents, c) learning process, d) learning assessment, e) teachers and academic staff, f) learning facilities and utilities, g) management, and h) finance. Curriculum development and evaluation must refer to this regulation. (The Ministry of Education and Culture, 2020). Such regulation is necessary and required to ensure the quality of higher education institutions in Indonesia.

One of the indicators to measure the quality of universities in Indonesia is the adoption of world university rank published by Quacquarelli Symonds (QS), Times Higher Education (THE), and other prominent institutions. In general, research-related indicators in Indonesian universities need to be improved by establishing re-orientation and adaptation of policies and developing more research collaboration to increase prestigious Indonesian universities world ranks (Burrill et al., 2015). The low achievement in world rankings is explained by the current lack of institutional autonomy, students' critical thinking skills, and limited teaching methodologies, impacting the internationalization and scientific outcomes of tertiary institutions (Volodymyr, 2018). Therefore, the Government of Indonesia should be more open to the role of foreign educational institutions and private sectors to enable educational policies to meet political, economic goals (Council on Foreign Affairs, 2016). At the national level, the Indonesian Higher Education General Directorate classify universities in Indonesia into several clusters due to their performance (Herlina, 2020). The quality of higher education in Indonesia can also be measured from its adaptability to the speed of technological change. Educational institutions, teachers, and educators must adjust to Generation Z students' current environment, needs, and expectations (Moore, Jones, & Frazier, 2017). These adjustment and update are also required to provide specific skills and knowledge which are needed to address current humanitarian issues related to the risk of natural and man-made disasters as well as climate change.

Humanitarian engineering is a complex and integrated endeavour requiring both technical and social aspects to be addressed simultaneously. The involvement of community members is the key to the sustainability of the activity, and in the end, it is the community that benefits from the solutions proposed by the engineers (Mazzurco & Jesiek, 2017). Since community engagement is significant, a trans-disciplinary approach that includes social sciences and humanities perspectives into engineering activities and projects is crucial. In humanitarian engineering education, Tran disciplinarity is required to sustain engineering curriculums (Tejedor, Segalàs, & Rosas-Casals, 2018). Tejedor et al. (2015) propose 4 (four) varieties of transdisciplinary, i.e., soft, hard, inclusive, and reflexive. These four varieties are essential to adjust knowledge integration and stakeholder's collaboration due to circumstances change (Tejedor et al., 2018).

Humanitarian engineering education programmes actively address goals and targets of UN SDGs related to humanitarian action and development. Many curriculum approaches use project-based learning, related to SDGs goals, through service common. The core of humanitarian engineering programme approaches, and methods developed by countries and institutions is to improve educational outcomes and ethical development practices. Three approaches are given to this, building on individual institutions, national discussions, and professional recognition (Smith, Tran, & Compston, 2020). Humanitarian engineering education is also essential to shaping global engineers with ethically driven social competencies (Shekar & Tunnicliffe, 2018).

Humanitarian engineering education programmes actively address goals and targets of UN SDGs related to humanitarian action and development. Many curriculum approaches use project-based learning, related to SDGs goals, through service common. The core of humanitarian engineering programme approaches, and methods developed by countries and institutions is to improve educational outcomes and ethical development practices. Three approaches are given to this, building on individual institutions, national discussions, and professional recognition (Smith, Tran, & Compston, 2020). Humanitarian engineering education is also essential to shaping global engineers with ethically driven social competencies (Shekar & Tunnicliffe, 2018).

Curriculum development consists of planning, constructing, implementing, and evaluating learning opportunities intended to produce desired learning outcomes and realize a high-quality, relevant curriculum (Albashiry et al., 2015). At the programme level, curriculum development is commonly conducted through a systematic approach that is divided into 5 (five) steps: a) need assessments, b) formulate the goals and learning outcomes (LO), c) identify resources, d) develop educational strategies and implement the curriculum, and 5) evaluate and modify the curriculum. Formulation of the programme's learning outcomes is a crucial activity in curriculum development. The learning outcomes' formulation corresponds to programme content selection, educational strategies that are influenced by student performance and perceptions on an assessment, and alignment of education methods with assessment methods (van Woezik, Koksma, Reuzel, Jaarsma, & Jan van der Wilt, 2020). The process shows a sequential process. However, curriculum development is an interactive and dynamic process. Students are central to this process (students centred learning). Students' needs and prior experiences are significant factors for teachers and instructors to develop educational strategies that meet different learning styles (Brodsky & Newman, 2010).

Current teaching and learning methods have developed significantly with the support of technology. Many educators have moved from the traditional approaches, i.e., lectures and tutorials, to internet assisted teaching methods such as flipped classroom, blended learning, game-based learning, and technology-enhanced learning (Hwang & Chen, 2016; Souza, Veado, Moreira, Figueiredo, & Costa, 2018). The new method, such as the blended learning method, is designed to improve the traditional learning model and e-learning practice (Farida et al., 2019). Integrating technology is a complex process of educational change, particularly in developing countries, and the extent of technological applications used in schools is still varied. The degree teachers integrate the technology is still a question (Spector, Elen, & Bishop, 2014); (Fraillon, Ainley, Schulz, Friedman, & Gebhardt, 2014). The use of technology also requires adequate standardization in learning outcomes and the classification of technology-assisted methods, while the standardization is still lacking (Souza et al., 2018). However, recent technologies have provided new opportunities to enhance learning and student engagement.

COVID-19 pandemics boosts the opportunity and the use of technology. Distance learning and blended learning methods that showed slow progress prior to the COVID-19 are currently used as the primary teaching delivery methods. Due to the characters of educational programmes, many uses blended learning, while the others use total distance (online) learning. The pandemic has forced higher education worldwide to use internet assisted technology. Various learning management systems are currently used. LMS Blackboard (https://www.blackboard.com) became the LMS with the largest user in the world, 41%, followed by Moodle (https://moodle.org/) as much as 23%, Desire2learn (https://www.d2l.com/) as much as 11% and the rest is V class rooming (www.vclassrooming.com) (Krasnova & Demeshko, 2015). Concepts such as distance learning, and open universities, are now becoming more widely used for teaching and learning. However, due to the nature of the subject domain, the teaching of Science, Technology, and Engineering are still relatively having problems when using total online approaches (Potkonjak et al., 2016).

The research presented in this paper is based on a survey in 2019 to three stakeholders of higher education: higher education providers, quality assurance agencies, and industry in Indonesia. One of the targets in this study is engineering curriculum development for the aforementioned three universities in Indonesia i.e., UB, ITB, and UGM. Curriculum development includes credit systems, teaching methods, assessment methods, student feedback, quality assurance, engineer's attributes, and (humanitarian) needs assessment.

METHODOLOGY

Data collection was conducted through a survey which targeted respondents from three key stakeholders including Indonesian higher education institutions, industrial partners, and quality assurance sector. The sample, which is categorized into three different groups, consists of (1) 99 respondents from the Higher Education Institutions/HEIs, (2) 45 respondents from various Industries, and 13 respondents from Quality Assurance/QA authorities. Each group was given a set of questions that were specific to the group. However, for comparison purposes, some questions were asked to both the HEIs and Industry. Questions for higher education institutions were addressed to collect information related to existing teaching delivery, learning assessment, student involvement, and external quality assurance. In addition, capabilities of graduate engineers including skills, knowledge, and experience as well as the importance of humanitarian engineering activity for capacity development were also captured. On the other hand, questions for respondents from industrial sectors focused on several aspect of skills and the capability of new engineering graduate needed in industry. The respondents were inquired to rate the capability of graduate engineers in the country and mention the attributes (i.e., skills, knowledge, and experience) required to improve in the graduate engineers as well as the learning method needed. Questions prepared for respondents from quality insurance sectors aimed at gathering information related to how quality assurance for higher education is ensured within a country and its main requirements. The program-level response of universities to improve an educational programme following quality assurance review was also captured.

The survey was conducted under a digital platform prepared by Erasmus+ Enhance team through which all data were digitally recorded. Information about the survey was widely spread through any kind of electronic and social media to reach as many respondents as possible in the country. Number of respondents was always monitored as a guidance to get a proportional number and distribution of participating people in the survey. The data collected from the three groups of stakeholders were then statistically analysed. The results are presented and discussed in the following sections.

FINDINGS AND DISCUSSION

The findings from the survey are presented in the following manner, starting with the current practices at Higher Education Institutions, then the perceived capabilities of graduate engineers from Higher Education Institutions' perspectives and Industries' perspectives, and finally, the practice of Quality Assurance in Higher Education.

Current Practice at Higher Education Institutions

This section is aimed at exploring the current practices among the Indonesia HEIs in delivering graduate engineering education including the study time, emerging teaching and learning activities, as well as methods for assessment and capturing student's feedback. Almost 70% of the respondents from HEIs teach in Undergraduate Programme and from which almost 75% of them are in Applied Sciences (mostly Engineering) field, while the rest teach in Postgraduate Programme. The following findings reflect the current practice at Higher Education Institutions, particularly in the Undergraduate Programme.

Contact Time, Expected Self-Study Time, and Weeks of Teaching

Figure 1. Average contact time and expected average self-study time per student per week, and weeks of teaching

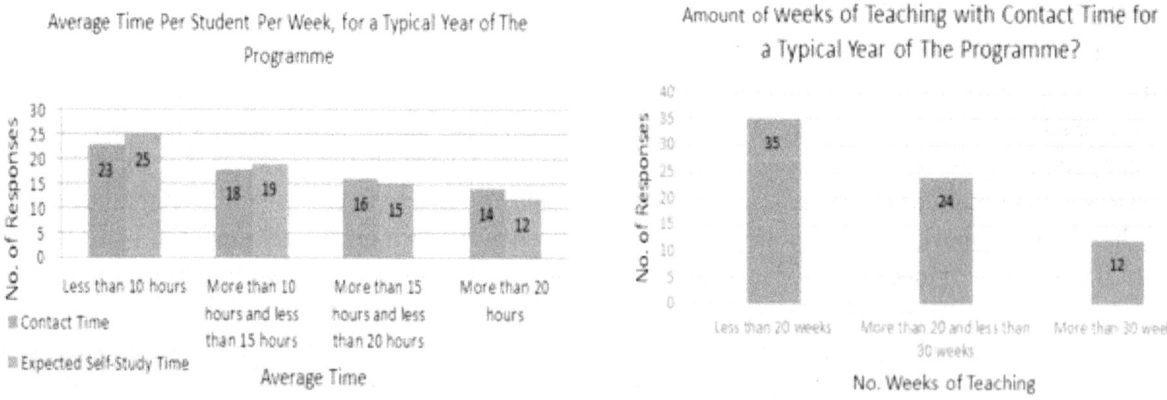

Students Learning are generally in the form of (1) contact time through lectures, tutorials, seminars, laboratory activities, fieldwork, etc., and (2) self-study time. A study by Goodhew (2010) showed that based on detailed surveys of teaching methods, typically, engineering and science students in the majority of the universities have two types of study time: contact hours and private study time. Figure 1 presents the results for three questions regarding the contact time, average self-study time per week as well as the number of academic weeks per year. It shows that most undergraduate programmes have more than 10 hours of contact time (e.g., lectures, tutorial, seminars, laboratory, and fieldwork). Only 23 out of 71 respondents have less than 10 hours student contact time per week, while the least number of respondents (i.e., 14 out of 71) have more than 20 hours student contact time in their programmes.

Based on Minister of Education and Culture's Decree No. 3/2020 on "National Standard for Higher Education", for lectures and tutorials, one credit is defined as 50 minutes per week in the form of Face-to-face learning in the classroom, another 60 minutes per week in the form of structured assignment, and another 60 minutes per week for expected self-study time, or in total about 170 minutes per week of learning activities. For each semester, students can take up to maximum 24 credits. If one course is 2 credits, then students can take up to 12 number of courses. Based on the given definition of 1 credit, for 12 courses which each course has 2 credits, students will have about 20 hours (12 courses x 50 minutes per credit x 2 credits per course) face to face learning per week. This is similar to the average student contact time in the UK, Australia, and the USA, as students in engineering and science typically have about 20 scheduled 'contact hours 'per week – time (Goodhew, 2010). By using the same approach, about 10 to 11.60 hours per week are expected to be spent as self-study time. Some surveys reveal that students devote about 7 hours per week to non-contact study in their first year, rising to about 11 in their final year (Goodhew, 2010).

Results on Self-Study Time

Regarding the number of weeks of teaching, Figure 1 shows that "less than 20 weeks" become the highest response (35 respondents), for 16 weeks of teaching per semester is a common practice and defined within the aforementioned decree. The number of weeks in a semester in Indonesia conforms typical number worldwide that is generally 14 – 18 weeks. In Indonesia, normally, these 16 weeks consist of 14 weeks of teaching and 2 weeks for the examination period (midterm and final). Table 1 shows that a student normally spends 8 to 9 hours per day, or 48 to 54 hours per week. Therefore, a normal load for a student is between 16 to 19 credit (6 days learning activities). The significant issues are therefore not related to the number of weeks, the number of credits, or the number of hours/minutes in each credit, since the government regulation related to teaching standards is rigid and uniform nationwide (see table 1). The more relevant issues are related to teaching-learning methods and techniques used by the lecturers.

Table 1. Student learning load per credit/week

<table>
<tr><th>Lecture/response/tutorial</th><th>Structured assignment</th><th>Self-study time</th></tr>
<tr><td>50 minutes/week/semester</td><td>60 minutes/week/semester</td><td>60 minutes/week/semester</td></tr>
<tr><td rowspan="2">Seminar/similar forms</td><td>Face-to-face</td><td>Self-study</td></tr>
<tr><td>100 minutes/week/semester</td><td>70 minutes/week/semester</td></tr>
<tr><td colspan="2">practicum, studio, workshop, field, and work practice, research, design or development, student exchange, internship, entrepreneurship, and/or community service,</td><td>170 minutes/week/semester</td></tr>
</table>

Based on that concept, the number of credits for different level of education of higher education in Indonesia is described in Table 2. With an assumption that a normal learning time per day is between 8 to 9 hours, the minimum credits for undergraduate programmes are 144 credits, while for postgraduate programmes are 36 credits. Thus, if an undergraduate student takes undergraduate thesis as 5 credits, students community service as 3 credits, and practical work as 2 credits, then the student must take at least 67 courses (2 credits/course) to fulfil the minimum requirements to graduate.

Table 2. Minimum credits for higher education programmes in Indonesia

No	Programme	Minimum credits	Maximum credits
1	D1	36	40
2	D2	72	80
3	D3	108	120
4	Undergraduate	144	160
5	Profession	36	40
6	Postgraduate (master programme)	72	80

Methods of Teaching and Learning

There were 14 methods of teaching and learning presented in the questionnaire, namely (1) Game-based Learning, (2) Flipped Classroom, (3) Journal Club, (4) Lectures, (5) Personalised Learning, (6) Practical Activities (Laboratories, Fieldwork, Site visits, etc.), (7) Problem-based Learning, (8) Project-based Learning, (9) Student as Researcher, (10) Student-centered Learning, (11) Technology-enhanced Learning (including e-Learning), (12) Tutorials (or Seminars), (13) Work-based Learning, and (14) Others.

As far as the teaching/learning methods the HEIS use as contribution percentages to the delivery of their programmes, as shown in Figure 2, lectures are the most common method, followed by Practical Activities and Problem-based methods. Following them, another 3 methods of teaching that worth noting

Figure 2. Percentage of contribution on methods of teaching and learning in delivering the programme

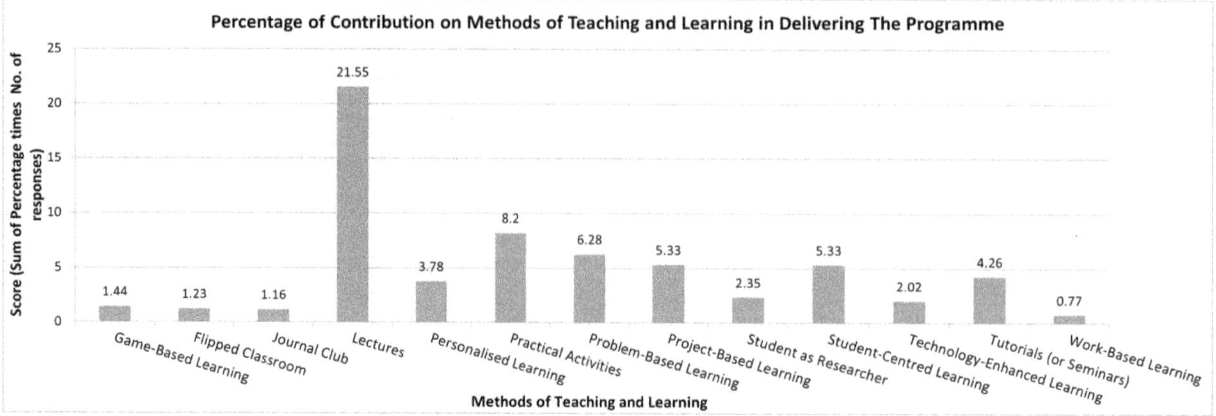

are Project-based Learning, Student-centred Learning, and Tutorial (or Seminars). These results are in line with the outcome of a study conducted by Goodhew (2010). The study highlighted that the Lecture method appears to be efficient in transmitting information, and particularly direct presence to the student to define the syllabus and thus the content to be examined; but very dependable on the lecturer as well as low achievement in encouraging deep learning and almost all research indicates that a listener can only give proper (effective) attention to a speaker for about 15 minutes, therefore it should be coupled with one or more of the many techniques available for maximizing the effectiveness. Problem-based learning is said to be promoting the students' efforts in learning (to discover and understand) and engagement; but very dependable on the problems, therefore requires well-planned preparations.

Methods of Assessment

Figure 3. Percentage of contribution on methods of assessment in delivering the programme

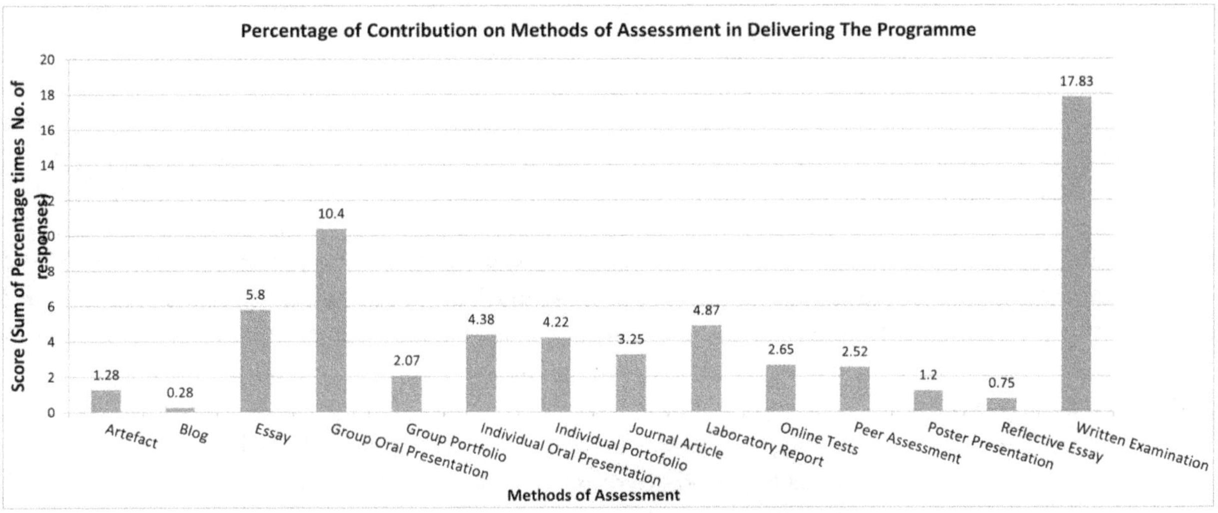

There were 15 methods of assessment being asked in the questionnaire, namely (1) Artefact, (2) Blog, (3) Essay, (4) Group oral presentation, (5) Group portfolio, (6) Individual oral presentation, (7) Individual portfolio, (8) Journal article, (9) Laboratory report, (10) Online tests (including quizzes), (11) Peer assessment, (12) Poster presentation, (13) Reflective essay, (14) Written examination, and (15) Others. Regarding the assessment methods the HEIs use as contribution percentages to the delivery of their programmes, Figure 3 highlights that Written Examination is the most common method, followed by Group oral presentation and essay. Following them, another 3 methods of teaching that worth noting are Laboratory report, Individual oral presentation, and Individual portfolio. Planning the assessment techniques for the module should be the second activity in the constructive alignment process, after defining the intended learning outcomes and before considering teaching and learning approaches (Goodhew, 2010). The motive in setting the assessment should be clear, whether to allow the students to demonstrate competence, or to differentiate the smart from the average, or to give credit for further non-directed study (Goodhew, 2010). In surveys of student satisfaction in the UK in the years 2005-2010, one of the most consistent problems identified was that of inadequate feedback on assessment.

A straightforward attitude would be that students deserve feedback on all work which they submit, and they need it in time to act on it before they make the same errors again.

Methods for Capturing Student Voice on the Content, Quality, and Experience

Figure 4. Methods for capturing student voice on the content, quality, and experience

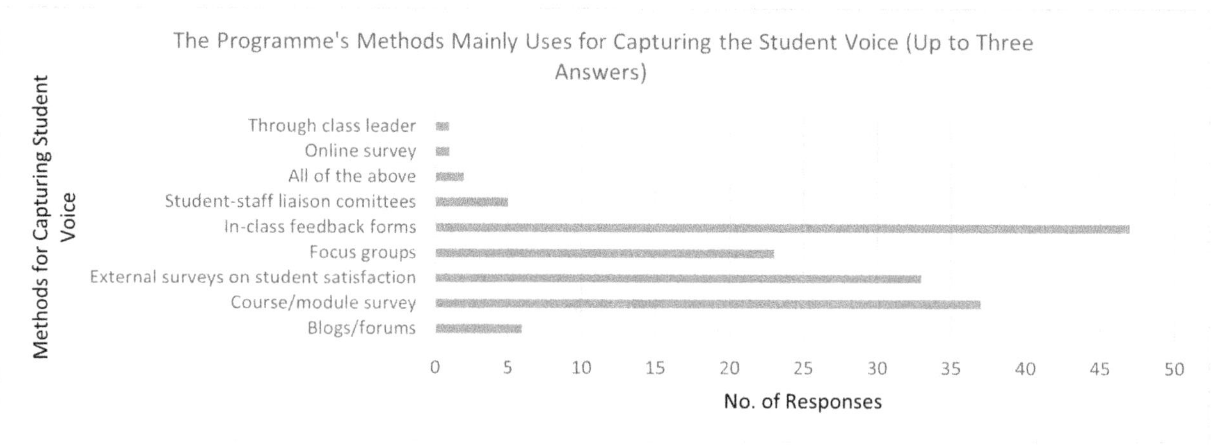

As far as the students' feedback is concerned, Figure 4 presents methods that are mainly used by the HEIs to capture the student voice on the content, quality, and experience of their programmes. In-class feedback forms and course/module survey are by far the most used methods of obtaining feedback from students, followed by external surveys on students' satisfaction. The least used methods are online survey and feedback via class leader. With regards to the process of updating the programme, several methods were presented, namely (1) students co-create curricula with academic staff (with 2 responses), (2) students provide feedback which leads to update/ refresh (with 45 responses), (3) both of the above (1 and 2) with 11 responses, and (4) students get information from alumni (with 1 response). It is clear that "students provide feedback which leads to update/ refresh" is the most common method on how students involve in the process of refreshing or updating the programme of parts of it.

Graduate Engineers

This part is aimed at exploring (1) to what extent engineering curricula offered by the Indonesia HEIs reflect the industrial, community, national engineering needs; (2) the challenges faced in integrating industrial, community, national engineering needs in engineering education; and (3) the contribution of industry in the delivery of engineering education. For comparison purposes, in this part, the same questions were asked to both the HEIs group and Industries group.

Capability of Graduate Engineers

Figure 5 illustrates the mean ratings (from 1: for Poor to 5: Excellent) of the responses from HEIs and Industries on the perceived capabilities of graduate engineers in Indonesia. It is statistically inferred

Figure 5. The perceived capabilities of graduate engineers in Indonesia, from HEIs and Industries' perspectives

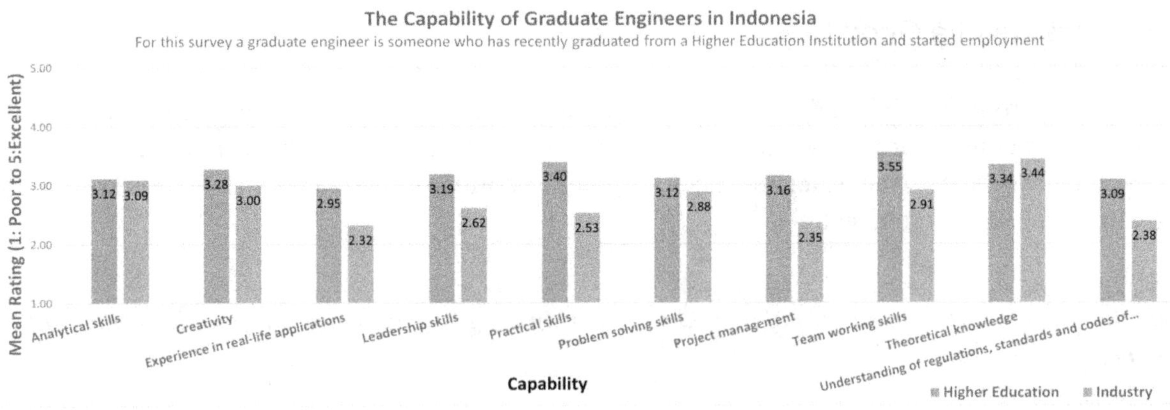

that HEIs and Industries are indifferent in their opinion for the graduate engineers' capabilities in four out of ten factors (p-values are > 0.05 for the Kruskal-Wallis test, while all the other p-values are < 0.05), i.e., analytical skills, creativity, problem-solving skills, and theoretical knowledge. In particular, except for theoretical knowledge, all capabilities are rated by the HEIs higher than that by the industries. HEIs rate team working skills as the strongest capabilities of their graduates, followed by practical skills, theoretical knowledge, and creativity, while Industries rate theoretical knowledge as the strongest capabilities, followed by analytical skills, creativity, and team working skills. HEIs rate experience in real-life applications as the weakest capabilities, followed by understanding of regulations, standards, and code of practice, while Industries rate experience in real-life applications as the weakest capabilities (same as HEIs), followed by project management and understanding of regulations, standards, and code of practice (same as HEIs). There is a notable gap between the opinions of HEIs and Industries, particularly practical skills (the largest gap) and followed by project management, while experience in real-life applications and team working skills are also worth noting.

Most of the respondents from HEI mentioned that graduate engineers have team working skills, practical skills, and creativity. Meanwhile, the majority of respondents from Industry mentioned that graduate engineers have theoretical knowledge, analytical skills, and creativity (see Figure 5). These results reflect the real situation that often occurs where despite the fact that practical skills are part of the learning outcome in the learning process in HEIs, most practitioners in industry still think that this is the major attribute that is lacking from the graduates. For example, one of the comments received by a UGM student during his/her internship programme in industry is that he/she did not have sufficient practical skills. On another occasion, during a meeting with Civil and Environmental Engineering Department (CEED) UGM alumni who are directors of several government contractors, similar feedback was also received. According to the alumni, the majority of graduates from CEED UGM do not have good communication skill and have more theoretical knowledge than practical skills. Such feedback from industry is not unique for UGM and can be observed in other HEIs as well. This is the main reason for the Ministry of higher education to take the initiative and introduce a new programme, Merdeka Belajar Kampus Merdeka (MBKM). This new programme provides the student with opportunities to explore and learn more from industry. CEED UGM has revised its curriculum and the current curriculum has been adjusted to accommodate MBKM.

Characteristics of a Good Engineering Programme

The responses from the HEIs and the Industry regarding the characteristics that are needed for a good engineering programme are shown in Figure 4. It is inferred that HEIs and Industries are different in their opinion across all given characteristics except for opportunities to experience the workplace and supportive learning environment.

Figure 6. Expected Characteristics of a good engineering program in Indonesia, from HEIs and industries' perspective

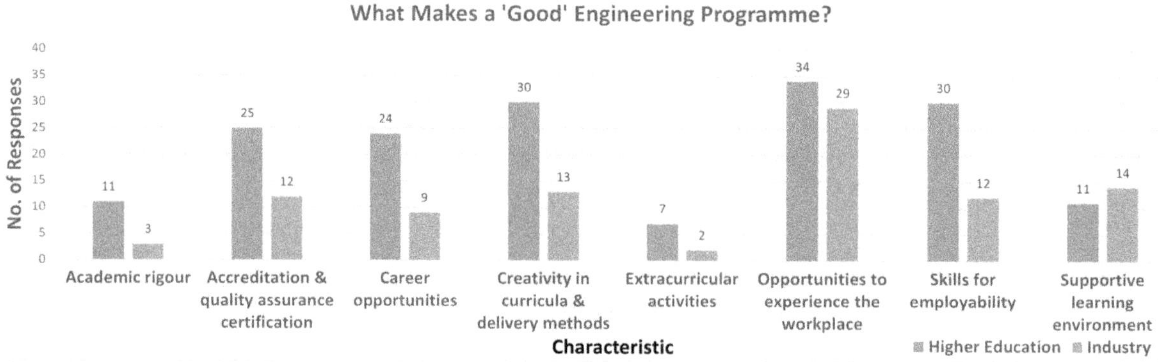

All characteristics, except for supportive learning environment, are rated by the HEIs higher than the industries. HEIs rate opportunities to experience the workplace as the highest characteristic, followed by creativity in curricula and delivery methods which tie in responses with skills for employability. On the other hand, industries rate opportunities to experience the workplace as the highest characteristics, followed by supportive learning environment and creativity in curricula and delivery methods. HEIs and Industries rate the lowest characteristics indifferently with extracurricular activities, followed by academic rigour. There is a notable gap between the opinions of HEIs and Industries, particularly skills for employability (the largest gap) and followed by creativity in curricula and delivery methods and career opportunities. Both HEI and industry agreed that the most important factor to create a good engineering programme is opportunities to experience the workplace (see Figure 6). The opportunities to experience the workplace is by doing the internship programme as explained before.

Attributes Needed to be Developed by Graduate Engineers

The means of response's rating (from 1: for "most important" to 10: for "least important") from the HEIs and the Industry regarding the attributes (skills, knowledge, experience) that need to be developed in graduate engineers are presented in Figure 7. Worth to note that lower means are more important and vice versa higher means are less important. The means of response's rating (from 1: for "most important"

Figure 7. The Attributes needed to be developed by graduate engineers, from HEIs' and industries' perspective

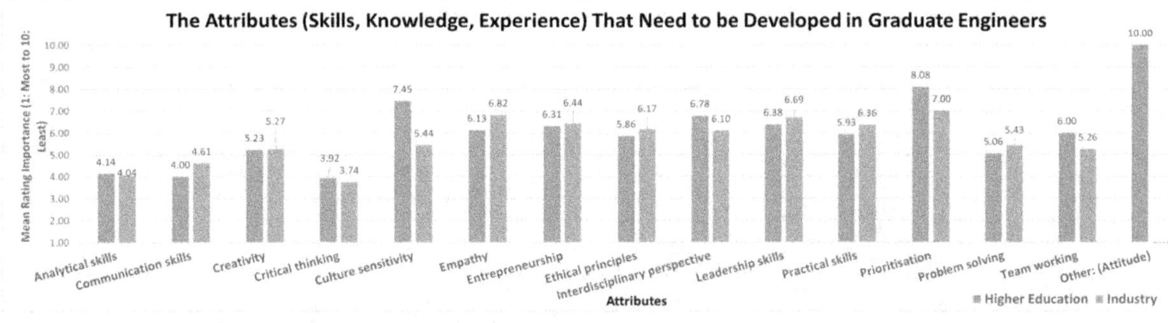

to 10: for "least important") from the HEIs and the Industry regarding the attributes (skills, knowledge, experience) that need to be developed in graduate engineers are presented in Figure 5. Worth to note that lower means are more important and vice versa higher means are less important. It is inferred that HEIs and Industries are indifferent in their opinion for the attributes needed to be developed by graduate engineers across all attributes except "culture sensitivity". In particular, eight out of four teen attributes (more than half), are rated by the HEIs lower (more important) than the industries, i.e., "communication skills", "creativity", "empathy", "entrepreneurship", "ethical principal", "leaderships skills", "practical skills", and "problem solving". The lowest (most important) attributes that need to be developed, is "Critical thinking", followed by "Analytical skills" and "communication skills", while the highest (least important) attributes is "prioritisation", followed by "leadership skills" and "empathy".

HEIs people said that the most important attributes to be developed in graduate engineers is attitude (see Figure 7). This result seems relevant to concern that is always raised by faculty members at CEED UGM about integrity. New regulations are always developed to make sure that students will not be cheating during examination. One example, once a student proved cheating during exam, then he/she will not pass all the subjects that he/she took in that semester. It is believed that if students got used to cheat, they will tend to cheat when later they work such as doing corruption.

The Activities That Could Develop Needed Attributes of Graduate Engineers

The various learning/teaching activities that could develop the needed attributes of graduate engineers are presented in Figure 8, both from the HEIs' and Industry's perspective, using means bars of their response's rating (from 1: for "most preferred" to 10: for "least preferred"). It is statistically inferred that HEIs and Industry responded (on average) the same way (and thus they share the same opinion) regarding all the suggested attributes except "lectures". In particular, all activities, except from "lectures", "practice visits and fieldwork", and "tutorials", are rated by the HEIs higher (less preferred) than the industries. The lowest (most preferred) activities that could develop the needed attributes, is "problem-based learning", followed by "practice visits and fieldwork" and "internships and industry placements", while the highest (least preferred) activities are "seminars", followed by "tutorials" and "self-directed study".

Figure 8. The activities that could develop needed attributes of graduate engineers, from HEIs' and Industries' perspective

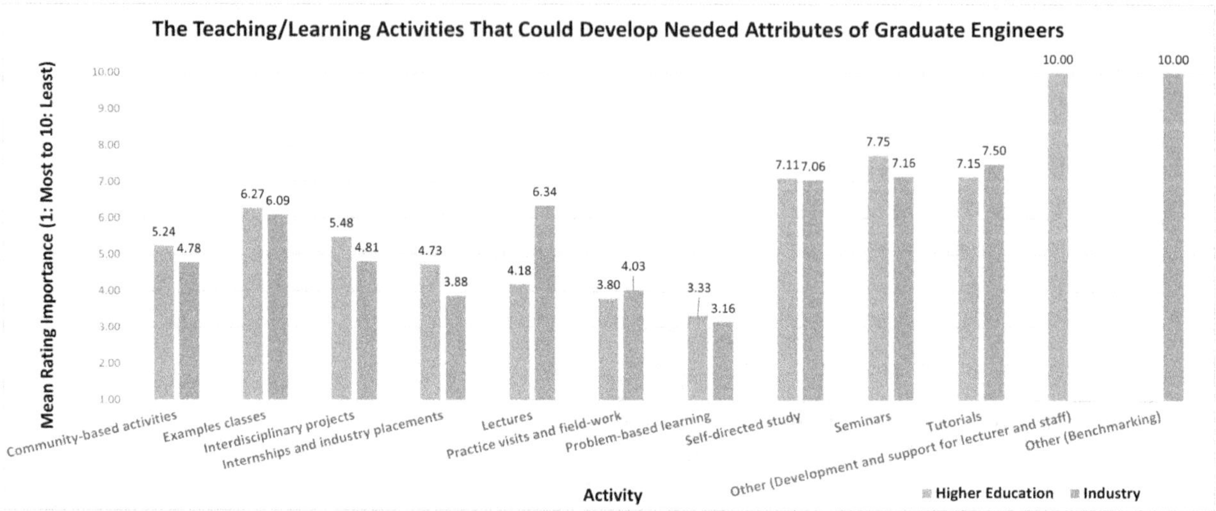

The Topics the Graduate Engineers Needed to Have

Figure 9 presents the topics that the graduate engineers need to have knowledge and understanding, from HEIs' and Industries' perspective, using means bars of their response's rating (from 1: for "most needed" to 10: for "least needed"). It is statistically inferred that HEIs and Industry responded (on average) indifferently (and thus they share the same opinion) on all the topics except "Ethics" (p-values are < 0.05 for the Kruskal-Wallis test, while all the other p-values are > 0.05). In particular, six out of ten topics, are rated by the HEIs lower (more needed) than the industries, i.e., "community needs", "culture preservation", "diversity and equality", "human needs and human rights", "professionalism", and "sustainability". The lowest (most needed) topics needed to have been "professionalism", followed by "sustainability" and "community needs", while the highest (least needed) topics is "political factors", followed by "culture preservation" and "diversity and equality".

Figure 9. The topics the graduate engineers needed to have, from HEIs' and Industries' perspective

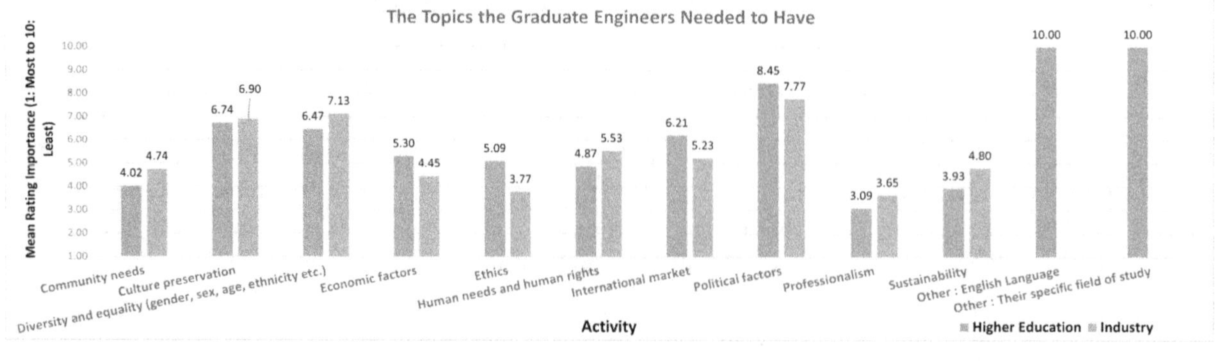

Industry people thought that the most important knowledge needed by graduate engineers is political factors (see Figure 9). This results truly reflects the real situation in Indonesia where many decisions often affected by politics.

THE PRACTICE OF QUALITY ASSURANCE IN HIGHER EDUCATION

This part is aimed at exploring the implications of Quality Assurance requirements in the development of engineering curricula. Almost 75% of respondents from the Higher Education Institutions answer that Governmental Authority (e.g., Ministry of Education, National Accreditation Agency) provides external Quality Assurance to their programme, while 22% choose Professional Engineering Institutions. Based on the survey to Quality Assurance Agency, there were 13 respondents who fill up the questionnaire

Figure 10. Frequency of periodical external quality assurance review

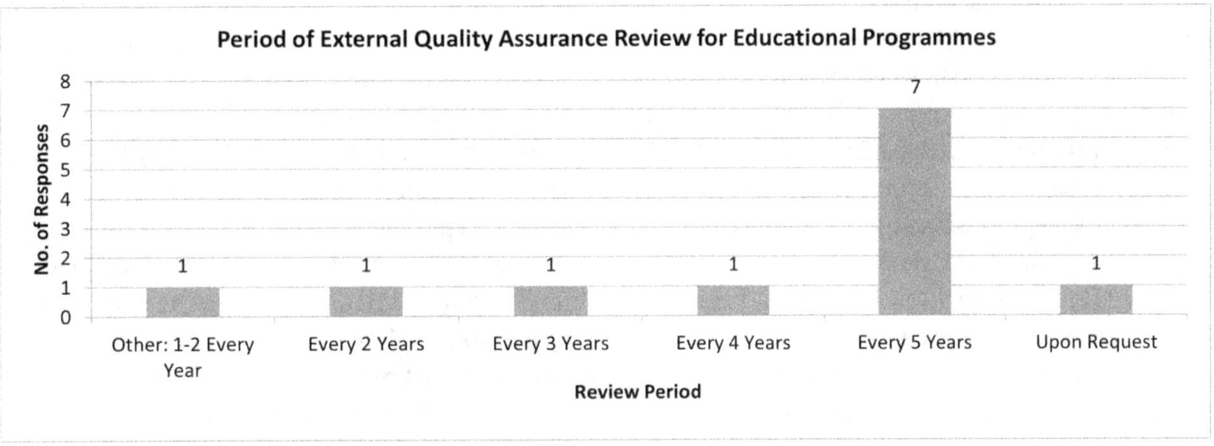

Figure 11. The importance rating of Quality Assurance Resources

Figure 12. Reference to review external quality

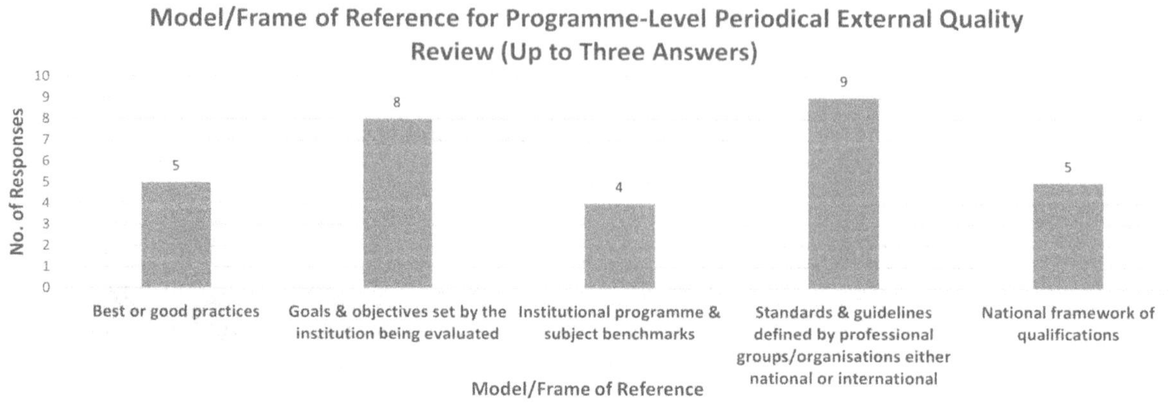

which 9 respondents were the assessor for Higher Education Institutions. Those assessors evaluate the curriculum in Higher Education Institutions every 5 years, and mostly approve it (see Figure 10).

The review results showed that academic attainment, facilities and resources, and quality management system are the most important factors to assess the quality of a study program (see Figure 11).

Standards and guidelines defined by professional groups/ organisations, and goals set by the institutions are used as the reference to review the quality of department/study program (see Figure 12).

This is acceptable because each study programs apply for a specific accreditation program. For example, Department of Civil and Environmental Engineering at Universitas Gadjah Mada (CEED UGM) and Civil Engineering Study Program at Institut Teknologi Bandung (ITB) are accredited by international accreditation institution i.e., Accreditation Board for Engineering and Technology (ABET) from United States of America (USA). ABET requires 25% of curriculum consist of basic sciences subjects such as Mathematics, Basic Physics, Chemistry, and the final project should be related to design for example the students are able to design reservoir, highway, bridge, building, etc. Hence, CEED UGM have revised the curriculum accordingly to make sure that the new curriculum has consisted 25% of basic science

Figure 13. Panel membership

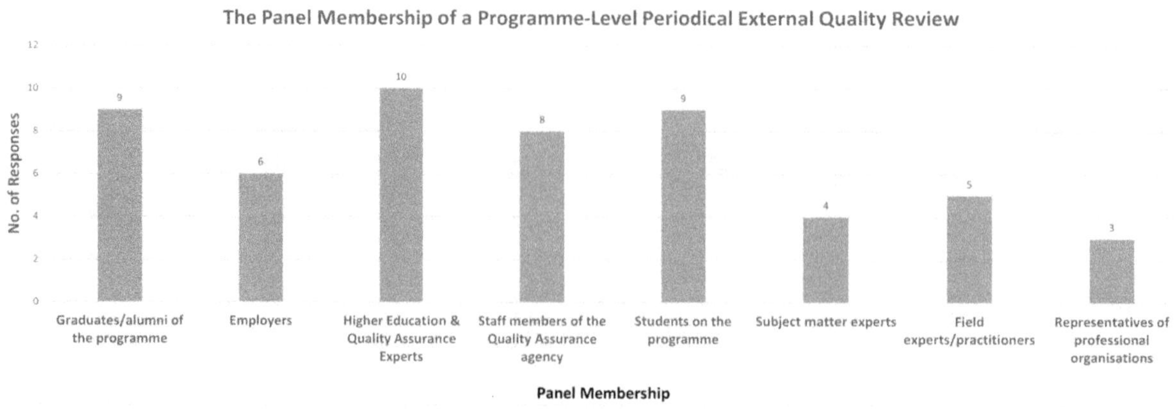

Figure 14. The most problematic area affects the quality of study program

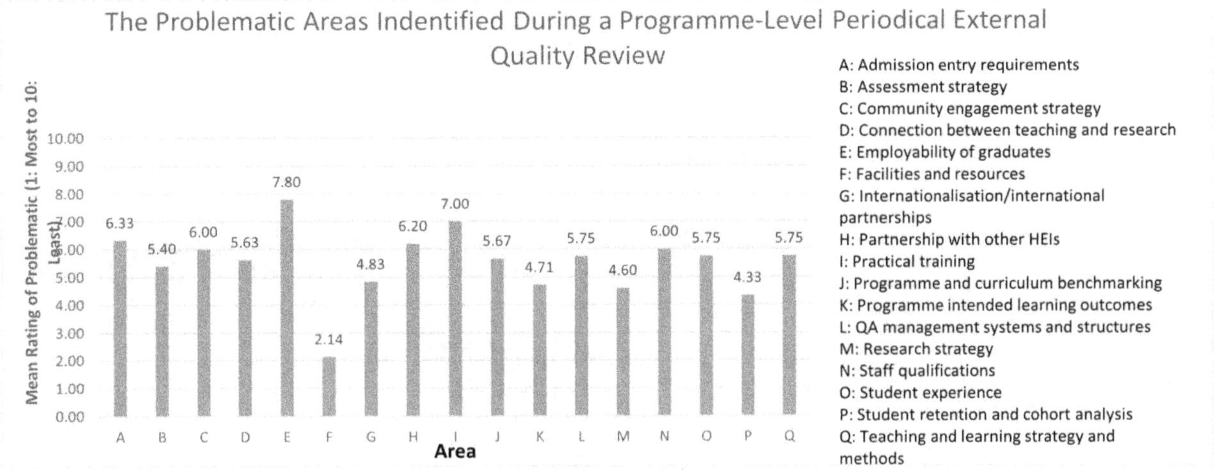

subjects, so one of the department mission is to conduct research activities for supporting education program especially in the design level (Visi dan Misi – Departemen Teknik Sipil dan Lingkungan (ugm.ac.id)) while some of the student outcomes are graduates will be able to apply basic sciences in civil engineering field and be able to design system and infrastructure in Civil Engineering area by considering economic, environment, health, and safety aspects (Sarjana Teknik Sipil – Departemen Teknik Sipil dan Lingkungan (ugm.ac.id)).

Hence, the assessor evaluates whether that department/study program has truly achieved the goals that already defined by reviewing the achievement of student outcomes objectively. To make sure that it is quite objective, while making the assessment, the members of quality assurance agency will interview the faculty members, the students, and the graduates. This is in line with the survey results (see Figure 13).

The most problematic area identified during the external quality review is facilities and resources (see Figure 14).

The results reflect the actual condition in many Higher Education Institutions (HEI) in Indonesia that do not have adequate laboratories facilities to support advance research activities. For example, Sanitation Engineering Laboratory at CEED UGM do not have CO_2 measuring device or supercomputer being used to simulate global climate model. Hence, so far there is no research related to air pollution or climate change. Many faculty members aware about this situation and hope that in the future, there is a policy to give adequate budget in the HEI to conduct advance research.

Survey result showed that QA capacity building that is most needed by HEI in Indonesia is program to developing and reviewing learning outcomes (see Figure 15). This is aligned with the discussion within faculty members. The HEIs have determined the student outcomes for each subject. However, there is a question whether those graduates have truly achieved the learning outcomes or not. Right now, it is only measured by the grades. And the grading system is determined by the HEIs itself. It may not reflect the actual competency of graduates.

The stakeholders like industries may have different opinions about the quality of graduates. Recently there is a news mentioned that industries do not satisfy with the competency of graduates from HEIs. Indeed, they complain that the curriculum in HEIs is 5 years behind the industry needs (Kemendikbud: Lulusan Kampus Tertinggal 5 Tahun dari Industri (cnnindonesia.com)).

Figure 15. The required quality assurance capacity building

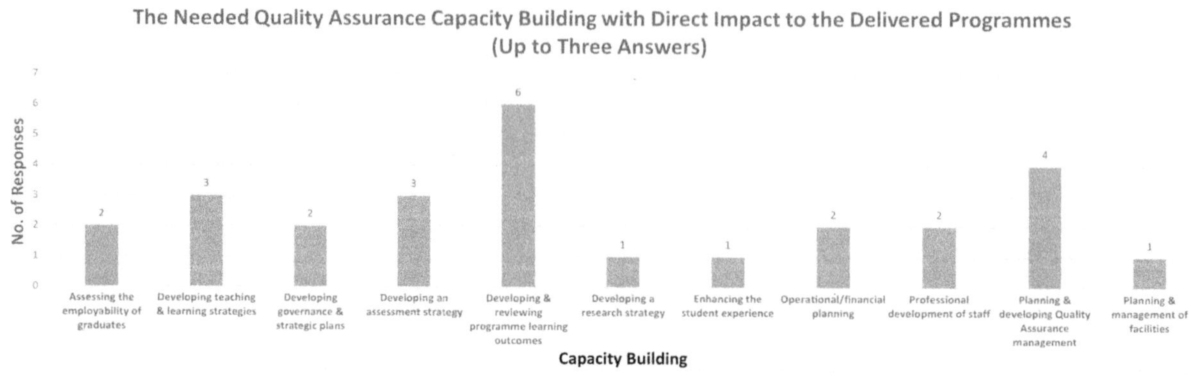

Hence, it is important to develop tools to measure learning outcomes objectively.

IMPLICATIONS

Following the survey conducted in this study, UGM and UB re-visited their curriculum and developed an action plan to further incorporate the humanitarian attributes in their undergraduate and post graduate programmes. Regarding the curriculum, this study provided valuable input for the development of the master's degree Program of Urban and Regional Planning (MPURP) - Universitas Brawijaya (UB) that have been completed in 2021. The survey also highlighted some important points to be improved in the implementation of the curriculum: 1) precision associated with learning time and the designed credit system; 2) the use of various more innovative teaching methods to complement the traditional one, lecture method; 3) synchronization of assessment method with learning outcomes of the course based on the Outcome Based Education (OBE) approach; and 4) addressing the weaknesses of the capability of the graduates engineers based on the perception of the industry, particularly experiences in real life application, project management, and understanding of regulation, standards and codes of practices. These weaknesses can be addressed via the MBKM programme, launched by the Ministry of education and Culture in 2021. The response from industry partners in this programme have been shown to be very positive, confirming the outcome of the survey conducted in this study.

Civil and Environmental Engineering Department Faculty of Engineering Universitas Gadjah Mada (CEED FoE UGM) revisited its 2016 curriculum into 2021 curriculum to implement the MBKM programme, in which students have the options to learn in another study program within university, to learn in another university, and to do internship in industry up to 20 credits. In 2021, one student joined student exchange for 1 semester at Waterloo University, Canada and many other students joined internship programmes at government construction companies. The curriculums of all study programs in CEED were being discussed with the stakeholders such as Ministy of Public Works and Housing, Ministry of National Development Planning of Indonesia, and Government Construction Companies. The revised curriculum also discussed with advisory board of CEED FoE UGM to get feedback of things to be improved. The advisory board includes Director General of Water Resources of Ministy of Public Works and Housing, Commissioner and Director of Human Resources of Wijaya Karya Construction Company (Advisory Board DTSL FT UGM – Departemen Teknik Sipil dan Lingkungan).

Not only in curriculum development, the outcome of the survey is also used to develop teaching delivery methods in both universities. The survey shows that traditional teaching method (lecture method) is still the most popular in Indonesia. MPURP - UB, in response to the result of the survey and the implementation of OBE approach, insists lecturers to use recent innovative methods such as flipped learning and student-centred learning, while problem-based learning and practical activities, which are relevant to respond the industry's feedback, are already implemented in the Department of Urban and Regional Planning (DURP) - UB.

Covid-19 pandemic has a significant role in the acceleration of adjusted blended learning method, hybrid method, and total online learning management system in UB. Lecturers that were formerly encouraged to use blended learning method, are currently familiar with the system/virtual learning management (VLM) and find some benefits of using that method. Other popular platform, i.e., Google classroom is also used by several study programs in UB. Two most common platforms for communication and discussion are Zoom and Meet. During 2021 to first half semester in 2022 all forms of teaching – learning in MPURP-UB are conducted online, except for project based learning courses, i.e., urban, and regional planning studios. The site survey and observations are conducted by only several members of the groups who are living in or close to the city where the case study was taken. In UGM, all of lectures during the pandemic are conducted in the online basis. All of exams are also conducted online using zoom. As practicums are cancelled, then the laboratory staffs were asked to make videos. The students watched the videos before the class and have discussion with the lecturers during the class. All of field trips are cancelled and student community service are also conducted in the online basis. During student community service the students made videos, booklets, and leaflets to explain about their programs to the community in the villages.

CONCLUSION

Higher education institutions shall adapt to the ever-evolving education agenda to accommodate for the needs of industry and the society. Traditional approaches of teaching are required modernisation in the current era hence academic staff are expected to enhance their teaching and learning practice using innovative and modern pedagogies supported by digital technologies including the use of social media as needed. Networking and collaboration with all stakeholders are also required to make sure that the students have adequate exposure to the life of work during their studies. Lessons learnt from the survey recommend that the three Indonesian universities (UB, UGM, ITB) would benefit from adjusting their curricula of both undergraduate and postgraduate programmes. The new curricula for both undergraduate and postgraduate programs shall improve students' opportunities to experience real-life applications, develop and practice skills for employability, and optimise their understanding of regulations, standards, and codes of professional practice.

ACKNOWLEDGMENT

This study was conducted within the framework of the project ENHANCE – Enabling Humanitarian Attributes for Nurturing Community-based Engineering co-funded by the Erasmus+ Programme, Key Action 2 Cooperation for innovation and the exchange of good practices, Capacity building in higher education, Joint Projects (Project No: 598502-EPP-1-2018-1-UK-EPPKA2-CBHE-JP (2018-2582/001-001). This publication reflects the views only of the authors, and the Commission cannot be held responsible for any use which may be made of the information contained therein.

REFERENCES

Albashiry, N., Voogt, J., & Pieters, J. (2015). Improving curriculum development practices in a technical vocational community college: Examining effects of a professional development arrangement for middle managers. *Curriculum Journal*, *26*(3), 1–27. doi:10.1080/09585176.2015.1040041

Botzen, W. J. W., Deschenes, O., & Sanders, M. (2019). The Economic Impacts of Natural Disasters: A Review of Models and Empirical Studies. *Review of Environmental Economics and Policy, 13*(2). doi:10.1093/reep/rez004

Brodsky, D., & Newman, L. (2010). Educational Perspectives: A Systematic Approach to Curriculum Development. *NeoReviews*, *12*(1), e2–e7. doi:10.1542/neo.12-1-e2

Burrill, G., Lappan, G., & Gonulates, F. (2015). Curriculum and the Role of Research. In S. J. Cho (Ed.), *The Proceedings of the 12th International Congress on Mathematical Education* (pp. 247–263). 10.1007/978-3-319-12688-3_17

Council on Foreign Affairs. (2016). *Indonesia's Education Gap*. Retrieved from Asia Unbound website: https://www.cfr.org/blog/indonesias-education-gap

Farida, R., Alba, A., Kurniawan, R., & Zainuddin, Z. (2019). The development of flipped learning model with Bloom's Taxonomy in teaching the Indonesian Political System course. *Kwangsan, 7*(2), 104–122.

Fraillon, J., Ainley, J., Schulz, W., Friedman, T., & Gebhardt, E. (2014). *Preparing for Life in a Digital Age* (1st ed.). doi:10.1007/978-3-319-14222-7

Goodhew, P. (2010). *Teaching Engineering All you need to know about engineering education but were afraid to ask*. UK Centre for Materials Education.

Herlina, N. (2020). *Direktorat Jenderal Pendidikan Tinggi Umumkan Klasterisasi Perguruan Tinggi Indonesia tahun 2020*. Retrieved September 20, 1BC, from Kemendikbud website: https://dikti.kemdikbud.go.id/kabar-dikti/kabar/direktorat-jenderal-pendidikan-tinggi-umumkan-klasterisasi-perguruan-tinggi-indonesia-tahun-2020/

Hwang, G.-J., & Chen, C.-H. (2016). Influences of an inquiry-based ubiquitous gaming design on students' learning achievements, motivation, behavioral patterns, and tendency towards critical thinking and problem solving: Inquiry-based ubiquitous game for students' learning. *British Journal of Educational Technology*, *48*(4), 950–971. Advance online publication. doi:10.1111/bjet.12464

Kiryakova, A., Olkhovaya, T., Melekesov, G., & Presnov, A. (2015). The Axiological Approach to the Analysis of the Problems of Modern University Education. *Mediterranean Journal of Social Sciences*, *6*. Advance online publication. doi:10.5901/mjss.2015.v6n2s3p22

Krasnova, T., & Demeshko, M. (2015). Tutor-mediated Support in Blended Learning. *Procedia: Social and Behavioral Sciences*, *166*, 404–408. doi:10.1016/j.sbspro.2014.12.544

Mazzurco, A., & Jesiek, B. (2017). Five Guiding Principles to Enhance Community Participation in Humanitarian Engineering Projects. *Journal of Humanitarian Engineering*, *5*(2). Advance online publication. doi:10.36479/jhe.v5i2.80

Ministry of Education and Culture. (2014). Minister Decree No. 49/ 2014 on National Standard of Higher Education, Indonesia.

Mohamed, Z. (2006). *Managing Student Learning Time for Effective Learning*. Enhancing Human Capital Development Through Quality Technical Education.

Moore, K., Jones, C., & Frazier, R. S. (2017). Engineering Education For Generation Z. *American Journal of Engineering Education*, *8*(2), 111–125. doi:10.19030/ajee.v8i2.10067

Potkonjak, V., Gardner, M., Callaghan, V., Mattila, P., Guetl, C., Petrović, V. M., & Jovanović, K. (2016). Virtual laboratories for education in science, technology, and engineering: A review. *Computers & Education*, *95*, 309–327. doi:10.1016/j.compedu.2016.02.002

QS Asia University Rankings. (2021). https://www.topuniversities.com/university-rankings/asian-university-rankings/2021

Shekar, A., & Tunnicliffe, M. (2018). An introductory course with a humanitarian engineering context. *Proceedings of the 13th International CDIO Conference*, 137–148. Retrieved from https://prism.ucalgary.ca/handle/1880/52101

Siagian, T. H., Purhadi, P., Suhartono, S., & Ritonga, H. (2014). Social vulnerability to natural hazards in Indonesia: Driving factors and policy implications. *Natural Hazards*, *70*(2), 1603–1617. doi:10.100711069-013-0888-3

Smith, J., Tran, A. L. H., & Compston, P. (2020). Review of humanitarian action and development engineering education programmes. *European Journal of Engineering Education*, *45*(2), 249–272. doi:10.1080/03043797.2019.1623179

Souza, M. R. A., Veado, L., Moreira, R. T., Figueiredo, E., & Costa, H. (2018). A systematic mapping study on game-related methods for software engineering education. *Information and Software Technology*, *95*, 201–218. doi:10.1016/j.infsof.2017.09.014

Spector, J., Merrill, M.D., Elen, J., & Bishop, M.J. (2014). Handbook of Research on Educational Communications and Technology: Fourth edition. *Technology, Knowledge and Learning, 20*(1), 123–128. doi:10.1007/s10758-014-9231-7

Suryaman, M. (2020). Orientasi Pengembangan Kurikulum Merdeka Belajar. *Seminar Nasional Pendidikan Bahasa Dan Sastra, 1*(1).

Tejedor, G., Segalàs, J., & Rosas-Casals, M. (2018). Transdisciplinarity in higher education for sustainability: How discourses are approached in engineering education. *Journal of Cleaner Production, 175*, 29–37. doi:10.1016/j.jclepro.2017.11.085

The Ministry of Education and Culture. *Peraturan Menteri Pendidikan dan Kebudayaan (The Regulation of Ministry of Education and Culture) No. 3 of 2020.*, Pub. L. No. Permen No 3/2020 (2020).

The Republic of Indonesia. (2012). *Undang-Undang Nomor 12 Tahun 2012 tentang Pendidikan Tinggi (Higher Education Act)*. Indonesian Act.

van Woezik, T., Koksma, J., Reuzel, R., Jaarsma, D., & Jan van der Wilt, G. (2020). How to encourage a lifelong learner? The complex relation between learning strategies and assessment in a medical curriculum. *Assessment & Evaluation in Higher Education, 45*(4), 513–526. doi:10.1080/02602938.2019.1667954

Volodymyr, K. (2018). Indonesias Higher Education: Context, Policy, And Perspective. *Asian Journal of Contemporary Education, 2*(2), 159–172. doi:10.18488/journal.137.2018.22.159.172

Zainuddin, Z., & Keumala, C. M. (2018). Blended Learning Method within Indonesian Higher Education Institutions. *Jurnal Pendidikan Humaniora, 6*, 69–77.

Chapter 12
Promoting Humanitarian Engineering Approaches for Earthquake-Resilient Housing in Indonesia

Krishna Suryanto Pribadi
Institut Teknologi Bandung, Indonesia

Muhamad Abduh
https://orcid.org/0000-0001-6926-6665
Institut Teknologi Bandung, Indonesia

Patria Kusumaningrum
Institut Teknologi Bandung, Indonesia

Budi Hasiholan
Institut Teknologi Bandung, Indonesia

Reini D. Wirahadikusumah
Institut Teknologi Bandung, Indonesia

Roi Milyardi
Institut Teknologi Bandung, Indonesia

ABSTRACT

Indonesia is an earthquake-prone country. In many earthquakes, the housing sector suffers the most damage. Repairing and rebuilding houses is often the most expensive part of post-disaster recovery, placing a burden on the government. Standards and guidelines for earthquake-resilient houses exist, yet houses continue to be damaged in earthquake events. Although Indonesia has numerous engineers, both experienced and newly qualified, only a few are interested in pursuing and identifying affordable solutions to the housing problems caused by earthquakes that are sustainable and resilient. Moreover, people are unaware of and unable to implement the proposed and promoted solutions. With many people living in earthquake-prone areas with little access to affordable and appropriate technology, the challenge facing Indonesia today is to produce more engineers who have the level of sensitivity, empathy, and passion required to address this great humanitarian need. The promotion of humanitarian engineering to provide appropriate solutions must start early on in higher engineering education.

DOI: 10.4018/978-1-6684-5619-4.ch012

INTRODUCTION

This chapter discusses the humanitarian issues faced by a large proportion of low- and middle-income people in Indonesia who are at risk of earthquake hazards because of their lack of accessibility to earthquake-resilient housing. Humanitarian engineers are needed to provide services to these people, bringing their knowledge and affordable technology to the construction industry and strengthening houses to improve their earthquake resistance and resilience. This chapter is introduced by presenting the context in terms of the archipelago's seismic hazards, the typology, the performance, and the vulnerability of the housing sector; this is then followed by a discussion on the technology available and approaches to providing safer housing. The latter part of the chapter focuses on the role of humanitarian engineers in providing services to improve access to more affordable earthquake-resilient technology for the people affected and describes how a new generation of humanitarian engineers is being promoted through education for young engineers, both in training and in service. The terminology of earthquake-resilient house is used throughout the discussion in this chapter to refer to the characteristics of a house having adequate robustness to resist the damaging impact of a design earthquake and can be quickly put back in use with minimum effort.

BACKGROUND

The Indonesian archipelago lies above a complex tectonic zone, where the Eurasian, Indo-Australian, Philippine and Pacific Plates meet, resulting in significant seismic hazards across the country. In Indonesia in the last 30 years (1991–2020), the Emergency Events Database reveals that earthquakes have caused damage valued at a total of US$18,468,041,000 (adjusted) and resulted in a total death toll of 183,748, with 164,963 injured and a reconstruction cost of US$1,726,563,000 (adjusted) (Centre for Research on the Epidemiology of Disasters, 2022). These earthquake hazards have become a major humanitarian challenge considering that earthquake disasters in the past have caused damage to the housing sector, mostly affecting the low- and medium-income population. Based on the 2019 national statistical data (BPS, 2020), from a total of approximately 67.5 million residential housing units in Indonesia, 31 million are located in high seismic hazard zones and another 21 million in medium seismic hazard zones. In 2022, BNPB (the National Disaster Management Agency) reported that in the last 10 years, 406,000 houses were damaged by earthquakes, also resulting in a large number of fatalities and injuries (BNPB, 2022). Most of these houses are built by local craftsmen and masons without the input of architects and/or engineers. Despite the availability of manuals and guidelines for earthquake-resilient house construction, which are produced by numerous organisations including relevant government agencies, many of the houses are built without adherence to the minimum requirements for earthquake resistance. Various factors have contributed to this situation, including a lack of earthquake awareness, limited understanding of available and affordable technology for earthquake resistance and limited economic capacity to procure appropriate building materials, such as construction timber, cement, and steel. Although in the past, traditional vernacular houses using locally available timber proved to be resilient to earthquake hazards, the current preferred materials for house construction are masonry and reinforced concrete (RC). Different types of bricks are used as the main material for load bearing as well as infill walls, which are of non-uniform quality and far from the available standards. A few reasons for this trend are that timber has become more expensive and difficult to obtain, carpentry skills are becoming less commonly available and there is a social perception that masonry houses are an indicator of social and economic progress. Unfortunately, the adoption of the more modern technology of masonry and RC is not complemented by an understanding of the stricter requirements for earthquake-resilient houses. In Indonesia, most of the local craftsmen and masons are not formally trained but have learned their trades from their seniors through an informal apprenticeship period at various construction sites. Good practices in earthquake-resilient masonry and RC houses are generally not properly communicated during this process, and therefore, most craftsmen and masons are not aware of the various recommendations set out in the guidelines and manuals for simple earthquake-resilient houses. As the homeowners are mostly of low- to medium-income, they cannot afford to obtain the services of engineers to provide advice on improving the resistance of their houses, and government engineers from the relevant public sector are limited in number and lack the capacity to provide adequate assistance to ensure the safety of houses built by local craftsmen. Most engineers are not interested in dealing with simple housing problems and design; they are more interested in providing services to more sophisticated clients with high-rise buildings and large infrastructure projects.

In this regard, the emergence of humanitarian engineers brings increased opportunities to tackle this major humanitarian challenge by mobilising their creativity, critical thinking, knowledgeability and problem-solving capacity to create innovative, affordable and sustainable solutions to address the earthquake vulnerability of ordinary peoples' houses. It is believed that humanitarian engineers, sensitive to the needs of low-income households, are more committed to engaging with the complex interdisciplinary problem of enhancing safety in relation to earthquake hazards. These engineers require a good understanding of the socio-cultural and economic context of the situation rather than just the traditional technical engineering approach. Therefore, promoting humanitarian engineering education is a key issue for engineering education in Indonesia to produce more engineers with enhanced passion and empathy towards people in need and who are not only interested in solving technical problems and achieving economic objectives when serving clients.

EARTHQUAKE HAZARDS IN INDONESIA AND THE IMPACT OF PAST EARTHQUAKES ON THE HOUSING SECTOR

Indonesia is prone to earthquakes because it is situated at the meeting point of three major tectonic plates, the Indo-Australian, Sunda–Eurasian and Philippine–Pacific Plates. Figure 1 illustrates the relative movement of these plates (Research and Development Agency, Ministry of Public Works and Housing [PUPR], 2017). The Sunda–Eurasian Plate moves in a south-easterly direction, and the Indo-Australian Plate moves northward, where it meets the Sunda–Eurasian Plate, creating a subduction zone known as the Sumatra–Java (Sunda) deep submarine trench, under the Sunda–Eurasian Plate in the southern part of Sumatra and Java. The Sunda trench, where most earthquakes in Indonesia occur, lies approximately 300 km from the west coast of Sumatra and south of Java. The Philippine–Pacific Plate moves southwestward under the Sunda–Eurasian Plate, creating a subduction zone, which forms volcanoes in the central part of Indonesia. In eastern Indonesia, the Australian and Pacific Plates collide, forming mountainous Papua, with Puncak Jaya as Indonesia's highest peak. The subduction zones may slip along thrust faults, causing megathrust earthquakes at the bottom of the ocean trenches, which may generate tsunamis that have a more devastating effect than the earthquake itself, as demonstrated by the M9.0 Aceh–Andaman

Figure 1. Tectonic map of Indonesia

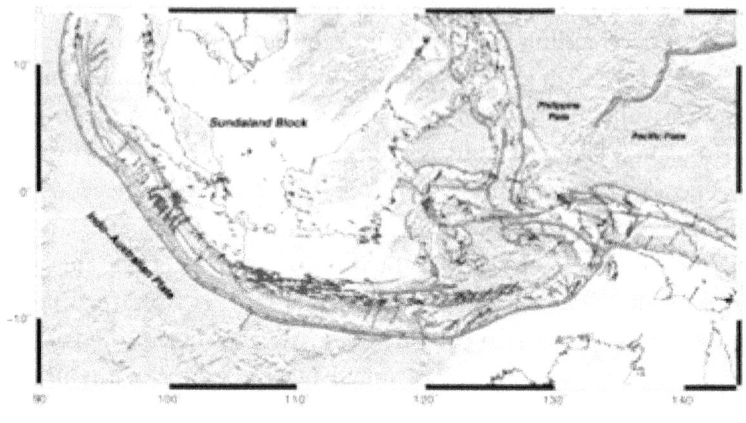

earthquake on 26 December 2004. Other than subduction zones at the bottom of the ocean, Indonesia also has at least 295 active faults on shore, which may trigger earthquakes when they slip.

To better understand earthquake hazards as part of earthquake risk reduction efforts, Indonesia has been developing a seismic hazard map since 1966 (Irsyam et al., 2017). The first seismic hazard map was introduced in the 1966 Indonesian Concrete Code (PBI), referring to the 1962 Geophysical Notes No. 2, which describe three earthquake zones. In 1970, Irian Jaya was added to the map, and the revised map was stipulated in the 1970 Indonesian Loading Code, the first code that required a seismic load design. In 1978, the map was refined with the inclusion of six earthquake zones and was stipulated in the 1983 Indonesian Earthquake Resistant Building Design Code (PPTGIUG), which was based on a New Zealand code that included a Maximum Considered Earthquake (MCE) response spectrum with a return period of 225 years (20% probability of exceedance [PE] in 50 years) and incorporated structural ductility, sway mechanisms and capacity designs. The 1983 building code (PPTGIUG), which was based on the 1978 seismic map, was updated twice, once in 1987 as the 1987 Earthquake Resistant Building and House Design Code (SKBI 1.3.53.1987) and then in 1989 as the Procedure for the Design of Earthquake Resistance for Houses and Buildings (SNI 03-1726-1989). In 2002, the seismic map was updated to take into account the average value of seismic accelerations resulting from four different studies conducted by national experts, which focused on the seismic activities of subduction zones (Irsyam et al., 2017). There was also a shift of reference, as the updated SNI 03-1726-2002 building code was based on the 1997 US Uniform Building Code and International Building Code, with the MCE's spectral acceleration derived from a return period of 500 years (10% PE in 50 years). National research on earthquake geohazards continued to be conducted comprehensively, and in 2010, the seismic hazard map was changed substantially by incorporating data from the high magnitude earthquakes that occurred between 2004 to 2010 and seismic activity data from 81 active faults, using attenuation model analysis to obtain seismic acceleration. The map was incorporated into the Procedure for the Design of Earthquake Resistance for Buildings and Non-building Structures (SNI 1726-2012), referring to the American Society of Civil Engineers (ASCE)/Structural Engineering Institute (SEI) 7-10 standard, which derive the spectral acceleration of the Risk-targeted Maximum Considered Earthquake from a return period of 2,500 years (2% PE in 50 years). Housing was included in the code under risk category II. Newly discovered seismic sources incorporating no less than 295 active faults led to a further update of the seismic maps in 2017.

Figure 2. Number of houses in disaster risk areas in 2021

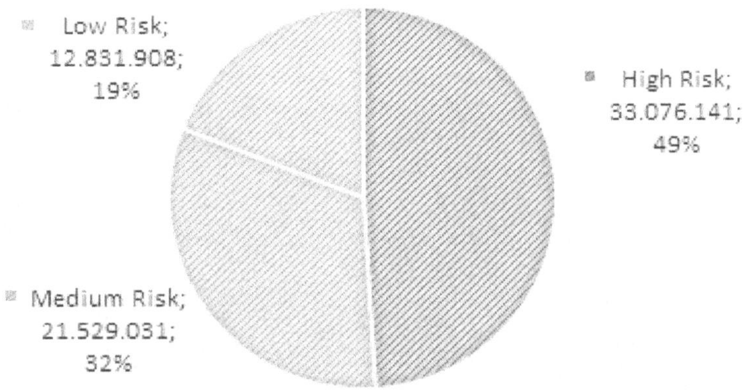

This map remains in use and has been included in the updated SNI 1726-2019, which requires more stringent confinements in accordance with ASCE/SEI 7-16.

As of 2021, based on the latest seismic hazard map, 49% of the residential housing stock is in high seismic areas and 32% in medium-risk areas (Fig. 2). According to the records, 13 strong earthquakes with magnitudes of M6.0 and above occurred between 2004 and 2021, resulting in considerable damage to residential housing and causing fatalities to the building occupants, as plotted in Figure 3.

In Indonesia, 93% of residential houses are non-engineered (Winarno et al., 2010) and are typically 'designed and built' by local craftsmen and masons based on traditional knowledge shared from generation to generation and using locally available materials without interventions from engineers or architects. Most are improperly regulated by (local) governments; hence, they rarely follow available (official or non-official) guidelines for earthquake-resilient residential housing. This may be attributed to a lack of awareness of earthquake hazards in the area, knowledge of how to build affordable earthquake-resilient houses or, most importantly, economic constraints. Non-engineered houses in Indonesia can be classified into two categories: lightweight houses made of timber with light roofs (constituting 9% of houses) and heavyweight houses made from either unconfined masonry (pure masonry) or confined masonry (masonry walls with confining elements made from RC columns and beams and sometimes from timber), either with light metal roofing or heavy roof tiles (constituting 84% of houses) (Winarno et al., 2010). The latter category is the main cause of casualties during earthquake disasters, as these houses are extremely vulnerable and have a much higher probability of collapse as a result of the failure of unreinforced masonry walls, the failure of frames in confined masonry houses caused by inadequate beams and columns supporting heavy roofs (Fig. 4), the failure of foundation tie beams or soil liquefaction when subjected to a strong earthquake.

Most of the structural system of engineered houses in Indonesia, estimated at about 7% of housing number, can be categorized as masonry-infilled RC frame. They are prone to collapse in earthquake due to beam-column joint shear failure or soft storey failure caused by the strong-beam weak-column mechanism. These mechanisms are still often found as the cause of structural failure in both confined masonry and masonry-infilled frame-type houses (Pribadi et al., 2008). To prevent these masonry-infilled RC frames from structural collapse, the design should be in line with SNI 1726-2019, while failure caused

Figure 3. Earthquake impact on (a) casualties and (b) damage to residential houses, 2000–2021 (BNPB, 2021b)

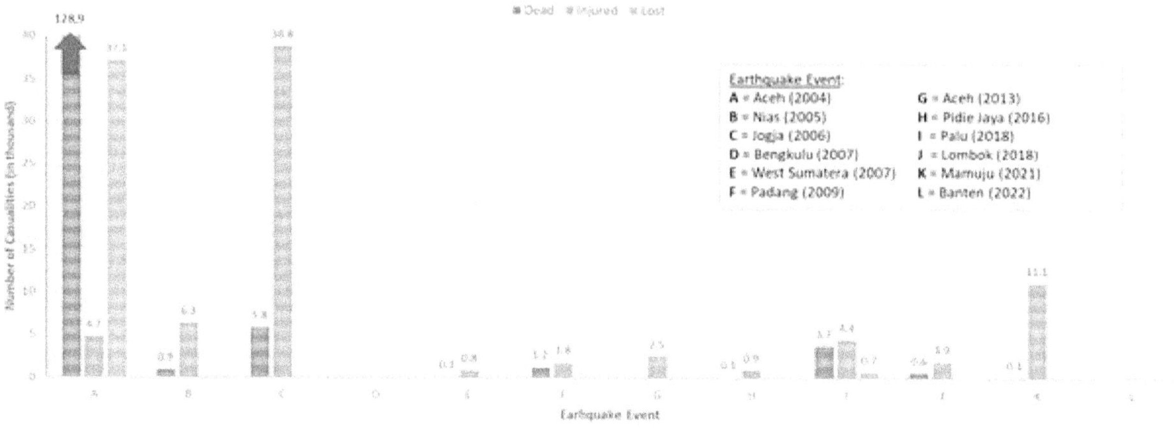

(a)

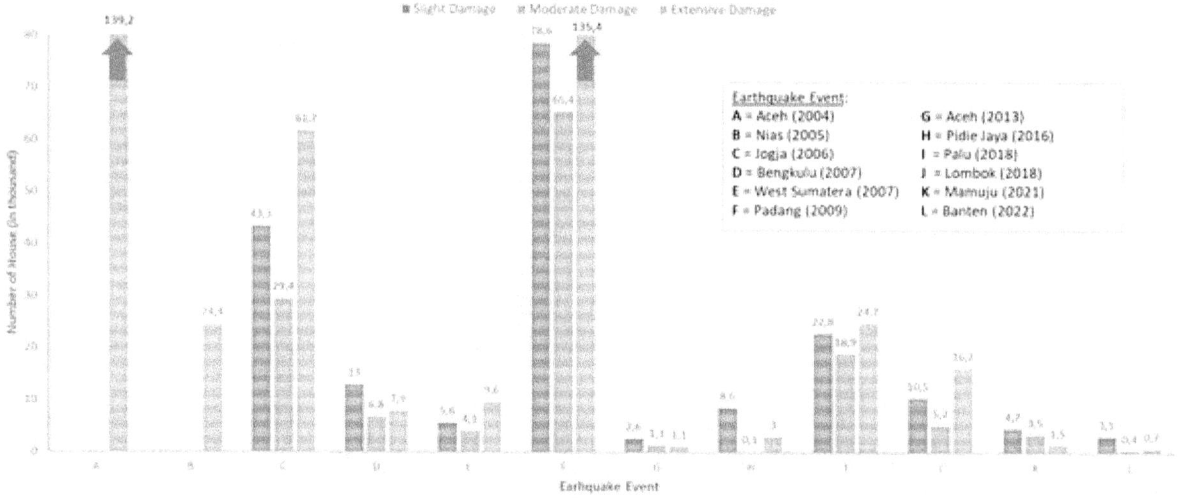

(b)

by dynamic amplification in soft soils and liquefaction may be prevented by proper soil investigation and the geohazard mapping of the respective region.

Housing reconstruction plays a key role in post-earthquake recovery. This is mainly a government-led operation, although in more recent earthquakes, community groups and/or non-governmental organisations (NGOs) have participated in the process. The process begins with a housing sector damage assessment, which divides the damage into three categories: lightly damaged (*rusak ringan*, RR), medium damaged (*rusak sedang*, RS) and heavily damaged (*rusak berat*, RB). Houses considered RR or RS can be repaired and sometimes need strengthening and/or retrofitting, whereas RB houses are usually rebuilt according to the design guidelines for earthquake-resilient houses. However, the implementation of housing reconstruction projects often faces major challenges related to community involvement, project manage-

Figure 4. Progressive failure of conventional masonry caused by the (a) failure of unreinforced masonry walls (2009 Tasik earthquake) and (b) liquefaction (2018 Palu earthquake)

(a) (b)

ment, institutional establishments and their coordination, project administration, technical design, and quality assurance (Pribadi et al., 2020). These challenges must be addressed properly for the housing reconstruction project to be successful.

The involvement of the affected community has been shown to provide a better participatory approach, increasing awareness and knowledge of earthquake-resilient house-building practices (Sadiqi et al., 2017), enabling people to rebuild and intensify their living, at first, through safer housing construction activities. With proper guidelines and expert supervision, their engagement during the reconstruction process may further enhance their knowledge and skill, enabling them to construct their own earthquake-resilient homes from the time of 'built back better' onward and increase their disaster risk awareness,

Figure 5. Masonry house in Biak, Papua (Photos by Pribadi, 2006)

Figure 6. Humanitarian engineering challenges for earthquake-resilient housing in Indonesia"

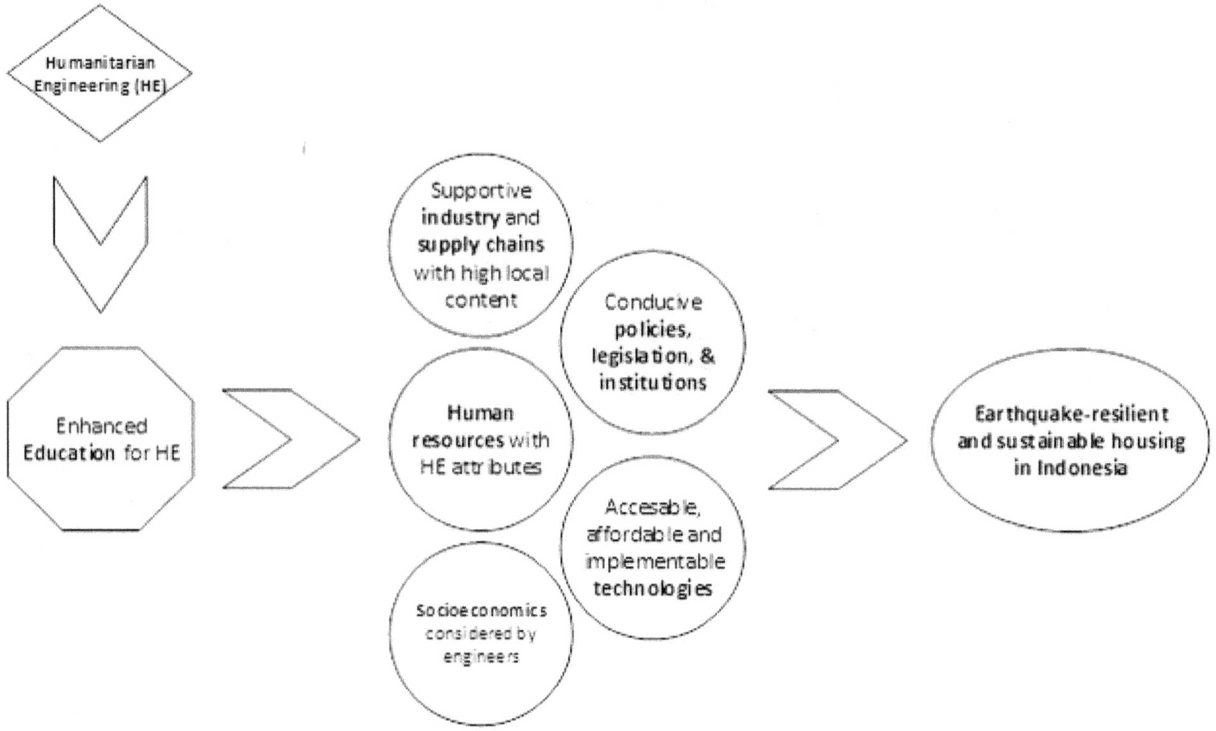

which will remain with them throughout their lives (Abe et al., 2018; Pribadi et al., 2020). However, long-term monitoring is required to evaluate the impact of the applied approaches on the community. Humanitarian engineers are needed for raising awareness of earthquake hazards in the affected communities and strengthening their knowledge of building practices so that they understand the need to construct earthquake-resilient houses.

CURRENT PRACTICE IN HOUSE BUILDING

Types of House Construction

Indonesia, with its archipelagic nature, is home to more than 1,300 ethnic groups. Each of the ethnic groups has its own distinct form of aesthetic traditional vernacular houses, known as Rumah Adat, which are built according to the beliefs and values of the local community. With the help of local craftsmen, traditional vernacular houses are built by the local community using natural materials available locally, such as timber, bamboo or coconut trunks, employing palm leaves, *alang-alang* grass or rice straw to form a roof cover. In the past, many traditional vernacular houses were lightweight in nature and proved to be earthquake-resilient structures (Pribadi et al., 2009). Many traditional houses in Sumatra, Java, Bali and Nusa Tenggara have distinctive characteristics, such as wooden columns supported by natural stone blocks at their base in the form of '*umpak batu*' or stone pedestal, with no fixed connection between them. This support system worked similarly to a base isolation system, absorbing the ground shock caused by earthquakes through the shear friction between the wooden columns and their respective supporting stones (Suwantara & Rusli, 2013). A more advanced system of 'umpak batu' was observed in the Nias vernacular houses, with each supporting stone block having a hemispherical basin, which corresponded to the shape of the column base, allowing it to rotate freely when subjected to earthquake motion.

People in modern society tend to build new houses with masonry walls and/or RC, with heavy roofs made of clay tiles. One reason is that masonry walls and clay-tiled roofs provide effective thermal comfort in the tropical environment of Indonesia. Another reason may be that wooden houses, including vernacular houses, have been stigmatised by classifying them as semi-permanent houses and people with better economic situation tend to opt for masonry wall houses which are classified as permanent houses by the government. Pure masonry structures have been present in Indonesia since the era of the Sriwijaya Kingdom in the seventh to twelfth centuries; the Muara Takus temple site provides evidence of this (Ministry of Education and Culture of Indonesia, 2022). The Trowulan site from the thirteenth-century Majapahit Kingdom era also provides evidence of this type of construction (Kuswanto, 2007). Later, under the Dutch colonial administration, knowledge of pure masonry structures with double wythes and clay-tiled roofing was introduced to high-class society, with large numbers of masonry structures being built in Indonesia. Because of the stiffness of double-wythe masonry walls, these structures were observed to be elastic during earthquakes of low-to-medium intensity. However, once the yield capacity of the structure was reached, it abruptly failed. Severe damage caused by strong earthquakes to renowned double-wythe masonry structures was reported in Batavia (now Jakarta); notably, the Mohr Observatory, a six-storey masonry structure, was damaged by a strong M7 earthquake in 1780 (National Geographic Indonesia, 2015). Later, after the independence of Indonesia in 1945, more single-wythe pure masonry structures were built because of the low economic conditions during that period. This change lowered the yield capacity substantially, increasing the vulnerability of such structures to earthquakes. Confined masonry houses with adequate confining RC beams and columns framing the masonry walls have become popular as typical houses in urban areas since the 1970s. Communities in rural areas have also reconstructed their vernacular wooden houses as confined masonry structures. As the economy has grown since the 1990s, RC infilled frames have been employed as the typical structural system for luxury homes of three stories or more, albeit in limited cases in more developed urban areas.

Houses built using confined masonry and RC infilled frames need to be properly designed and constructed as engineered buildings because they are heavy by nature, and their performance during an earthquake relies on the ductility of the structure, especially the beam-column joints. The government and NGOs have already developed some practical instructions and construction guidelines on how to build earthquake-resilient houses. However, these guidelines have not been well disseminated and are still far from being implemented properly. Most house construction in Indonesia is considered to be non-engineered, built by local people and craftsmen using local traditional building knowledge and customs passed down from generation to generation and employing traditional local and locally available materials (Arya et al., 1986). As there is no engineer and/or architect intervention during the construction process and because of the limited funds available, poor attention to detail is generally observed. This may further increase the vulnerability of houses to earthquakes.

One of the key factors of an earthquake-resilient house is the quality of the building materials, such as clay bricks, concrete and mortar, and reinforced steel bars. The clay brick industry in Indonesia, compared with that of other countries in Asia, such as Vietnam, India and Nepal, has received less attention in terms of technology development to reduce carbon dioxide emissions and produce bricks of standardised quality (Heierli & Maithel, 2008). There is no government enforcement of brick manufacturers and house builders to control the quality of products. Brick manufacturers in Indonesia are typically home industries; hence, the quality of products available on the market varies since no standard is employed. In this case, the national government, represented by the Ministry of Industry, should step in to enforce product quality in accordance with the national industry standard (SNI), as has been done, for example, in Vietnam. The local government may also take part in controlling the domestic industry. However, if this is done, the price of bricks may increase. Likewise, concrete and mortar cast in-situ by the owner or contractor, depending on the project delivery method, results in diverse quality, depending on the water to cement ratio applied and the quality of aggregate available in the respective area. Only cement and steel reinforcing bars that are quality controlled by the manufacturers meet specific requirements. Nonetheless, steel reinforcing bars that do not meet the SNI requirements are still traded and circulated within the domestic market. As they are cheaper than standardised rebar, many homeowners prefer these bars. Because of its archipelagic nature, the availability of aggregate in some parts of Indonesia is limited, especially in coastal areas and small islands and/or areas away from active volcanoes such as Kalimantan. In coastal areas, only sand with a fine rounded grain and uniform gradation containing chloride salt and sulphate with unfavourable properties is available for concrete. The situation is similar for the availability of cement and reinforcing bars. In remote areas, such as Biak in Papua, unfired cement bricks are observed to have low compressive strength, and therefore low shear strength, because they are typically made using a 1 to 10 cement to sand ratio. Confined masonry houses are constructed by using locally available wooden beams and columns (Fig. 5). In addition, because of the limited knowledge of locals and craftsmen, wooden beam and columns connection uses only nails without notches and pegs, resulting in poorly confined masonry with low earthquake resilience. Moreover, the wood preservation and lamination technologies adopted by domestic industry are rather costly, and they are difficult to find nationally because these technologies are mostly for export; they are therefore not preferred by homeowners. Structural wood readily available on the market is untreated, and Indonesia's high humidity makes it susceptible to termites.

Another key factor in earthquake resilience in housing is the roof, which should be light. In the past, vernacular houses used wooden roofs covered with palm leaves, *alang-alang* grass or rice straw. As hardwood is now hard to get and has become more expensive and because of the high humidity, heavy rainfall and strong winds, although much used in the past, wooden roofs are rarely used now. Currently, the commonly adopted roof system is made of cold-formed light-gauge steel covered with clay tiles. Cold-formed steel is thin and brittle. When covered with heavy clay tiles which increase the building mass, it does not provide sufficient lateral stiffness.

When subjected to an earthquake, the well-built confined masonry structure may remain intact, but the roof may fail prematurely because of its weight, harming the house occupants. In some parts of Indonesia, such as Sumatra, Kalimantan and Sulawesi, corrugated iron sheets, a technology introduced by the Dutch, are used to cover roofs; they are used more than burnt clay roofs, possibly because of local beliefs. Recently, along with material technology development, roof covers using galvalume and zinc-alum have been developed, offering a better option for a lightweight roof cover.

House Construction Delivery Method

Several project delivery methods for house construction have been observed. For example, in rural areas, the most utilised project delivery methods are the following:

Figure 7. National education curriculum framework for earthquake-resilient and sustainable housing

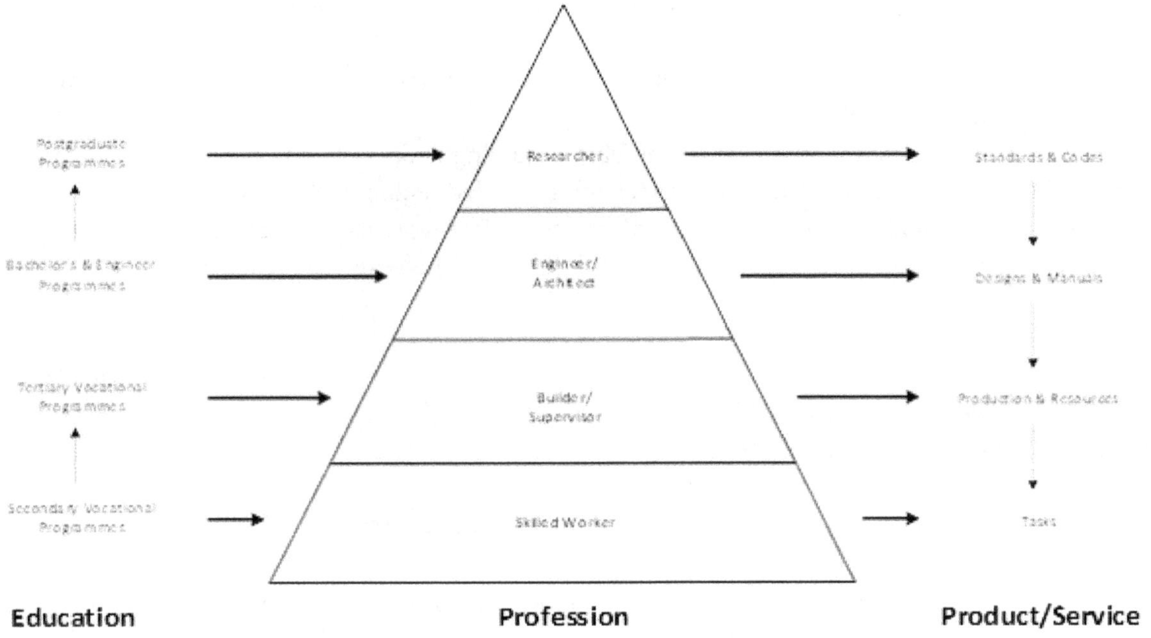

1. Owner built, with or without help from relatives and neighbours in the community.
2. Owner managed, mobilising local tradespeople and craftsmen (masons and carpenters).
3. Small-scale contractors, mostly informal, appointed by the house owner to build the house.
4. Small-scale contractors, either formal or informal, appointed by the house owner to build the house.
5. In some cases, for those who have the financial capacity, house owners may appoint a design–build contractor, providing architectural input to the design of the house while the construction is carried out by craftsmen or skilled workers under the supervision of the contractor.
6. In real estate housing schemes, a developer (private or public sector) builds houses on parcel lots obtained through a land subdivision to be sold to homeowners. The developer often involves an architect and/or engineers in the design of the houses and local infrastructure. The houses are built either by small-scale or medium-scale contractors appointed by the developer, depending on the size of the land subdivision.

In rural areas, people mostly build their houses using methods 1, 2 or 3, whereas in urban areas, methods 3 to 6 are more popular. For post-disaster recovery house reconstruction, either a community-based approach or contractor-based approach is used by the government. In these approaches, the government usually requires the involvement of engineers to supervise the construction.

The construction quality of houses depends on whether there is supervision by qualified personnel (engineers) or not, but for methods 1 to 3, supervision by qualified personnel rarely occurs. However, the quality of the house construction is more likely to be determined by the field operators, such as the craftsmen (masons and carpenters), and depends on whether they have a good understanding of earthquake-resilient house construction practices. Unfortunately, few local craftsmen and masons in Indonesia have received formal training in house building and construction skills, learning their trade from their seniors through informal apprenticeships at different construction sites. If their senior mentors do not understand the technical skills required to build earthquake-resilient masonry and RC houses, apprentices cannot learn this skill, and most of them are unaware of the various practices recommended by the guidelines and manuals on earthquake resilience for simple houses. This situation contributes considerably to the vulnerability of houses.

Therefore, it is expected that the role of humanitarian engineers is to bridge the gap in knowledge regarding the implementation of building codes with detailed technical specifications, which may be difficult for the community to understand, and promote affordable technology in the construction of earthquake-resilient houses. This can be done within local public sector agencies or through NGO initiatives.

PROGRESS IN EARTHQUAKE-RESILIENT HOUSING TECHNOLOGY IN INDONESIA

Guidelines on Earthquake-Resilient House Construction

Initiatives promoting earthquake-resilient houses in Indonesia through guidelines, as far as can be traced through online resources, started in 1993 with the *Guideline for Constructing an Earthquake-Resistant Building* (Department of Public Works, 1993), provided by PUPR. This guideline provides guidance (1) on how to conduct basic planning for earthquake-resilient buildings (particularly houses), including layouts and the placement of walls and openings, roofing and foundations, and (2) on how to construct bamboo-framed and bamboo-walled houses, (3) wooden houses, (4) steel-framed and masonry-walled houses, (5) masonry houses, (6) hollow concrete block houses with RC and (7) RC houses. In 2006, *Technical Guidelines for Earthquake-Resistant Houses and Building* (Department of Public Works, 2006) were issued, containing various guidelines on healthy houses (masonry, half-masonry half-timber, timber houses on stilts and landed timber houses). These technical guidelines, which are still widely used to date, include the construction of (1) new houses made of (a) wood, (b) RC and (c) a steel frame; (2) single-story buildings that are made of (a) wood and (b) RC, as well as multi-story RC buildings; and (3) methods for repairing and retrofitting houses.

Another guideline for constructing new houses, provided by Boen (2009a) and distributed by the United Nations Centre for Regional Development, is entitled *Constructing Seismic-Resistant Masonry Houses in Indonesia*. This guideline focuses on confined masonry houses and provides details on the construction stages of safer masonry houses, including (1) general requirements and the layout of houses, (2) basic utilities, (3) principles of seismic-resistant house construction, (4) building materials, (5) preparation works, (6) foundation works, (6) reinforcing bar works, (7) concreting works, (8) bricklaying works and roofing. In 2009, Boen also provided *The Minimum Requirements for Safer Houses*, particularly for confined masonry houses (Boen et al., 2009b). Prior to that, in 2006, Boen provided a series outlining the minimum requirements for earthquake-resilient houses, covering (1) one-brick-thick walled masonry house, (2) one-brick-thick walled masonry house with RC, (3) wooden house and (4) masonry house with timber confinement (Boen, 2006a, 2006b, 2006c, 2006d).

Apart from these government or domestically provided technical guidelines, in practice, particularly for internationally funded human settlement rebuilding projects in disaster-affected areas, international guidelines are referred to. In issuance date order, some of the international guidelines include but are not limited to (1) *Earthquake-Resistant Confined Masonry Construction* by the National Information Center of Earthquake Engineering (Brzev & Mitra, 2007), (2) the *Seismic Design Guide for Low-Rise Confined Masonry Buildings* by the Earthquake Engineering Research Institute (EERI) (Confined Masonry Network, 2011), *Guidelines for Earthquake-Resistant Non-Engineered Construction* by the United Nations Educational, Scientific, and Cultural Organization (Arya et al., 2013), and the *Guide Book (For Builders) for Building Earthquake-Resistant Houses in Confined Masonry* by EERI (Architects CCR et al., 2018).

Development of New Technology for House Building

In responding to the need for new and safer houses in disaster-affected areas, the government, through PUPR's Research and Development Agency, launched a knockdown modular (precast) technique for single-story houses, known as RISHA, in 2006 (following the 2004 Aceh and 2005 Nias earthquakes), with the introduction limited to Aceh and Nias reconstruction projects (Pinhome, 2021; Waluyo, 2021). In 2013, it was reported that RISHA had been implemented in a wider area (Pinhome, 2021). Not earlier than 2017, the *Practical Guide on Requirements of Earthquake-Resistant Buildings for RISHA* was launched for wider implementation (Ministry of Public Works and Housing, 2017c). This guideline promotes the second generation of RISHA, known as RUSPIN. Around the same time, the government (Ministry of Public Works and Housing, 2017b, 2017d) launched another two guidelines, the *Practical Guide on Requirements for Earthquake-Resistant Buildings* for (1) wooden buildings and (2) confined masonry buildings.

As for the existing house structures, either damaged or intact, the options are to repair (damaged houses) or retrofit (intact or lightly damaged houses) the structure, with several guidelines being available. The government included the repair and retrofitting guidance in the 2006 technical guidelines (Department of Public Works, 2006). In 2010, a guideline was provided by Boen, in cooperation with the University of Andalas, University of Indonesia, Institut Teknologi Bandung (ITB), Islamic University of Indonesia (UII) and University of Gadjah Mada (UGM), entitled *Ways to Repair an Earthquake-Damaged Simple Structure* (Boen et al., 2010). In 2012, as part of a Japan International Cooperation Agency sponsored project, the same group provided another guideline entitled *Repairing and Retrofitting Simple Masonry Structures* (Boen et al., 2012). Moreover, Boen provided guidelines on building earthquake-resilient masonry houses with ferrocement layers in 2015 and 2019 (Boen, 2019) and with fibrocement (fibre-reinforced cementitious [FRC] composite) (Boen, 2021a) and textile fibre composite (TRC) layers in 2021 (Boen, 2021b).

For the past 20 years, at least 22 earthquake-resilient housing systems of numerous structural types have been approved by PUPR and built en masse as part of earthquake post-disaster recovery projects. The dome house, initiated by Domes for the World (DFTW), is one peculiar type of shell structure made from Styrofoam without foundations as part of the reconstruction project following the 2006 Yogyakarta earthquake (UFO Dome House Indonesia, 2022). Barrataga (Sarwidi, 2020) and RIKO (Pradoto et al., 2022), classified as RC confined masonry houses, have incorporated sufficient confining rebars at the beam-column joints. Barrataga, constructed by UII Yogyakarta, has the special feature of sand layers beneath the foundations, which absorb the ground shock. RCI (Ridha et al., 2021) by PT Raswari is an RC house cast on-site. The RC walls, reinforced with single-layer wire mesh, are cast together with the beams and columns by incorporating a reusable moulding system. RISHA, further developed as RUSPIN (Ministry of Public Works and Housing, 2017b), and RAPI, developed by PT Indocement, which falls into the RC precast house category, utilise precast RC panels, with the structural components framing the lightweight mortar brick wall. The readily available cold-formed normal-strength steel frame system may use either lightweight board as the wall cover, such as Bale Lestari or RAISA, or a relatively heavy infilled wall of lightweight concrete, such as DOMU, RISDA or RICAG (Ridha et al., 2021). However, the cold-formed lightweight steel frame systems use a lightweight wall cover, either in the form of expanded polystyrene sandwich panels, such as RISTA or RISMA, steel modular panels, such as MESRA or RISBARI, or calciboard, such as RISBA and RITA (Ridha et al., 2021). Among other lightweight steel frame systems, RITA has a distinctive base isolator system using springs installed at the columns' base. An option for a more sustainable and environmentally friendly wooden house is provided by the RIKA system developed by PUPR, using laminated veneer lumber technology. KUMAC and CONWOOD panel houses (Ridha et al., 2021) have load-bearing walls developed by connecting the panels using adhesive glue and bolted connections, respectively. A more sophisticated earthquake-resilient and fire-resistant house developed by BPPT is BALE KOHANA (BRIN, 2019), which is a modular prefabricated polymer composite house equipped with rubber bearings as the base isolator.

Prefabricated systems, using materials such as RC, galvanised steel, cold-formed steel, or timber, may shorten the duration of construction, ensure the quality of components through monitored quality control and quality assurance by manufacturers, reduce on-site material waste (thereby being more cost-effective) and provide a safer working platform. However, on-site installation requires skilled labourers to construct the specified system and good quality control by a knowledgeable manufacturer. In the implementation of post-disaster reconstruction projects, detailed supervision and monitoring and on-site quality control to improve the buildability of earthquake-resilient houses are difficult to do in practice, as on-site facilitators are limited in number and technical knowledge (Pribadi et al., 2020). This causes further problems in terms of quality, such as (i) non-compliance with the structural technical requirements, including lower material strength and the poor workmanship of beam-column joint reinforcements and steel connections, and (ii) poor workmanship of the non-structural components, such as lightweight cement bricklaying.

The technology of the described earthquake-resilient housing is, however, anecdotal in practice and has emerged only during disaster recovery, mostly with financial support from the government and NGOs. This technology has not been adopted by the community as an option when they build houses routinely because of the significant technology gap and high dependency on the industry. Also, some of these technologies may not be compatible with the natural conditions and cultural beliefs of the community. Therefore, the community tends to use the old unsafe technology of non-engineered masonry houses with the clay bricks they are familiar with. To be widely adopted, such technology requires immense industrial support in terms of price, making it affordable for the community, and supply, to be continuously available within the community, and government support in terms of law enforcement in relation to the building of residential housing. Innovations in new building materials and technology also need to consider the economic capacity of the community. Building materials should not be expensive and must be readily available, especially in remote areas.

In conducting disaster risk reduction, PUPR's Directorate General of Human Settlements and Housing launched a self-subsistent housing renovation programme aimed at retrofitting existing vulnerable houses. However, the programme deviated from the main goal, which was to reduce the structural vulnerability of houses, because it was influenced by social perceptions within the community that people considered only the physical appearance of the house without considering its safety. The repairs conducted tended to be cosmetic, such as tiling the floors and changing half-height brick walls into full-height brick walls, rather than a structural retrofit, thus retaining the vulnerability of the house.

With the rapid development of science and technology, new innovations in earthquake-resilient construction materials have been developed by many researchers aiming to

- Use by-products or waste materials from the industry as much as possible as new building materials to reduce the excessive use of natural resources, such as coal fly ash, Sidoarjo mud (Nuruddin et al., 2010), steel slag, nickel slag and other pozzolanic materials.
- Introduce and implement high-performance, lightweight materials to reduce the weight of buildings and houses, such as cold-formed steel, zyncalume and galvalume roofing sheets, lightweight concrete bricks, autoclaved aerated concrete (PCA, 2019), laminated timber (Naturally:wood, 2022), laminated bamboo, bamboo beehives and bamboo zephyr.
- Strengthen commonly used building materials and components to enhance the building capacity, such as ferrocement (Boen, 2019), engineered cementitious composite (Boen, 2021a; Li, 2008) and textile fibre composite (Boen, 2021b).

Innovations in new building materials also need to consider the economic capacity of the community. They must not be expensive and should be readily available, especially in remote areas.

House Retrofitting Technology

In 2006, the government provided technical guidelines incorporating repairing and retrofitting techniques for existing house structures (Department of Public Works, 2006). The guidelines cover the following: (1) retrofitting techniques for (a) one-brick-thick masonry structures by adding (i) horizontal rebar to brick layers, (ii) wire mesh to brick layers and (iii) strips of iron sheeting to brick layers, (b) RC structures, such as (i) strength enhancement and (ii) ductility enhancement; and (2) repairing techniques for (a) cracked or crushed masonry walls with (i) grouting and covered with nailed wire mesh and plaster and (ii) anchors in every 10 layers of brick, (b) rafters displaced from their mountings, (c) intersections between beams and columns, (d) various types of damage to columns, and (e) the meeting of continuous walls and columns and between two side walls at corners. More elaborate guidance on repairing and retrofitting techniques was also provided by Boen et al. (2010) and a team of five higher education institutions in 2010, in which some of the techniques are more advanced versions of those included in the 2006 technical guidelines.

In addition, Boen has explored ferrocement strengthening as a retrofitting technique for masonry houses, issuing guidelines in 2015 and 2019. Moreover, guidelines for ferrocement and TRC strengthening as retrofitting techniques were issued in 2021. Ferrocement strengthening can be applied to both confined and unconfined masonry houses. The technique consists of applying a layer of wall-nailed wire mesh to both sides of a masonry wall, particularly those prone to damage or that are weak in critical areas, and then covering it with plaster. As for FRC strengthening (for a new structure), this technique uses a mixture of fibre (either E-glass, basalt, steel, polyvinyl alcohol or polypropylene), cement and fine aggregates as plaster, which is combined with a strand of four wires acting as a shear connector that is installed in every 20 cm of brick layers. TRC is similar to FRC but uses fibre mesh (either polyparafenilenbenzobisoxazolo, aramid or aragrid, basalt, carbon or glass) instead of wire mesh.

DISSEMINATION OF TECHNOLOGY

Methods and initiatives for promoting earthquake-resilient housing technology within communities have been introduced using various means and by various actors; this includes raising awareness of technical guidelines and technologies. The 2006 technical guidelines for RISHA, RUSPIN, RIKO and RIKA, developed by the government, have been disseminated at every level of government (from central to local government and across related agencies) and reached elements of the community (academia, NGOs, builders, engineers, and homeowners). Most of the dissemination materials are already in the form of booklets and practical guides that contain illustrations as well as details on the raw materials, tools and equipment required and a breakdown of costs to facilitate budget calculations. Some of the dissemination materials are available online for ease of access. It is crucial that the community apply these guidelines and technologies as soon as possible to reduce the vulnerability of their homes when earthquakes strike. Most of the technologies have been introduced and implemented in response to earthquake events. RISHA, for instance, was developed in 2004 and implemented initially only in the Aceh and Nias reconstruction projects. Subsequently, in the reconstruction projects following the 2018 Palu and Lombok earthquakes, a total of approximately 18 models (in Lombok) of earthquake-resilient houses were proposed by the government, academia and industries and then recommended by central and provincial governments (West Nusa Tenggara Provincial Government, 2019). However, the various building models caused confusion in the affected communities, who were faced with many options but had little awareness or understanding of them, resulting in delays in the reconstruction projects (Pribadi et al., 2020). For the technical guidelines and technologies to be properly implemented, knowledge transfer (counselling, training, technical assistance, and competency building) should be conducted, particularly to those who plan and supervise (engineers) as well as execute (builders) the new buildings or who perform the repairs and retrofitting. PUPR initiated a housing makeover programme in 2014 that supposedly repaired or retrofitted vulnerable houses (Ministry of Public Works and Housing, 2017a). In 2018, the government issued its National Standard of Work Competence, particularly aimed at RISHA, which established a national qualification, the delivery of professional education and training, competence tests and professional certification (Ministry of Manpower, 2018). Because of the region's vast area and limited government personnel, the involvement of community elements (academia and NGOs) in the dissemination, as well as the technical assistance provided for implementation, is imperative. Academia is involved through their respective community outreach activities that take place on an annual or per-event basis, whereas NGOs tend to be involved on a per-event or project basis. It is worth noting that methods for building new houses, as well as repairing and retrofitting existing houses, should be simple, affordable and replicable.

With all these new technologies, it is important to understand that the capacity of local communities to receive and adopt the technology is limited, resulting in a 'business as usual' mindset within the community in terms of building or renovating their houses. Retrofitting is mostly unaffordable for low-income communities, both with regard to technology access and economic capacity. These are then areas that humanitarian engineers would intervene in, improving the capacity of local craftsmen and masons to enable them to absorb the skills required to build safer houses using available and affordable technology.

HUMANITARIAN ENGINEERING CHALLENGES IN IMPLEMENTING EARTHQUAKE-RESILIENT HOUSING IN INDONESIA

Challenges for Implementing Earthquake-Resilient Housing

Humanitarian engineering, by its definition, is designed to directly improve the well-being of under-served populations (Mitcham & Muñoz, 2010). In practice, it can be viewed as an effort to integrate engineering with the sciences, social sciences, and arts to invent, create, design, develop or improve technologies, processes and tools that promote the well-being of communities facing major humanitarian challenges. Even though the humanitarian challenges may differ in their complexity from one location to another, humanitarian engineering always aims to improve the well-being, welfare and comfort of individuals and communities in disadvantaged contexts (Kremmyda et al., 2020).

As stated by the International Federation of the Red Cross and Red Crescent Societies (IFRC, 2010), Asia is the world's most disaster-prone region, with countries suffering from insufficient implementation capacity and strategies. Based on a previous study in four countries, the UK, Vietnam, Indonesia and Bangladesh, earthquake disasters are categorised as Indonesia-specific humanitarian challenges (Kremmyda et al., 2020), and using the IFRC's 2020 Strategy, saving lives, protecting livelihoods and strengthening recovery from earthquake disasters are the main themes for Indonesian engineers to focus on (IFRC, 2010). Pribadi et al. (2021) recognised, based on lessons learned from major earthquake disasters in Indonesia over 30 years, that a significant contribution of Indonesian engineers would be to implement earthquake-resilient technologies in the housing sector. However, based on existing capabilities, the challenges to making earthquake-resilient housing a reality in Indonesia are considerable because they are related to the following important multifaceted and multidisciplinary aspects:

1. **Socioeconomics**. This aspect relates to how the majority of Indonesians perceive their socioeconomic status and express it in their daily lives, including their housing. It is common for Indonesians to build permanent houses, such as brick and concrete houses, instead of traditional houses to show their improved socioeconomic status; however, they are also not willing to spend much money on construction.
2. **Technology**. This aspect is related to the availability of safe housing technologies and knowledge that are easily accessed and implemented by the majority of Indonesians. Although the traditional and proven knowledge and technologies used to build earthquake-resilient houses, based on local wisdom, are no longer acknowledged, new and more modern knowledge and technologies are not available to most Indonesians. However, since housing can be categorised as non-engineered construction, limited efforts are being made in terms of engineering and its practices to apply the knowledge and technologies.
3. **Industry**. This aspect involves the supply chains that support housing construction and technologies, such as standardised, local and affordable construction materials and methods. Even though there are some standardisations for construction materials related to housing, such as concrete, the industries and their supply chains cannot provide these materials as they are supposed to, and enforcement is weak; many technologies are supported by the informal sector.
4. **Human Resources**. This aspect includes the availability of all related professions to support earthquake-resilient housing, including skilled workers to stack bricks, builders to manage construction resources, supervisors to oversee the works in the field, architects, and engineers to design

the houses, researchers to produce knowledge and technologies and policymakers. It seems that earthquake-resilient housing technologies, as well as the construction sector, are not of interest to most human resource professionals for economic reasons.

5. **Policy, Legislation, and Institutions**. Effective implementation of earthquake-resilient housing needs to be supported by relevant, synchronised, and strong policies, legislation and institutions related to the previous four aspects. Housing is mostly provided by the private sector, and the current government and its policies for establishing earthquake-resilient houses cannot access this sector effectively.

The Need for Humanitarian Engineering for Implementation

The emergence of humanitarian engineers provides greater opportunities for Indonesia to address the humanitarian challenge of earthquake disasters and the implementation of earthquake-resilient housing by mobilising engineers' creativity, critical thinking, knowledge, and problem-solving skills to create innovative, affordable, and sustainable solutions to the earthquake vulnerability of ordinary peoples' houses. It is believed that humanitarian engineers who are sensitive to the needs of low-income communities are more committed to engaging with the complex interdisciplinary problem of enhancing safety in relation to earthquake hazards. They require a good understanding of the socio-cultural and economic context of the situation rather than just adopting a traditional technical engineering approach. Humanitarian engineers are needed to provide services to these people, enhancing all aspects of building and strengthening houses for improved earthquake resistance and resilience (Fig. 6).

As a response to reports by the Asian Development Bank (2015 and 2018) that the main constraint in higher education for capacity building in mitigating disasters is the traditional delivery, Kremmyda et al. (2020) introduced a taxonomy of humanitarian attributes to upskill professional engineers, suggesting that the taxonomy should be implemented and integrated efficiently in higher engineering education using modern and innovative delivery. The taxonomy is organised into three components of attributes: (a) knowledge: the theory, technical detail and information required to successfully perform a task; (b) skills: the practical application of knowledge required to successfully undertake a task; (c) behaviours: the mindsets, attitudes or approaches required for competence.

Kremmyda et al. (2020) reported that in Indonesia, several humanitarian challenges are discussed in the current engineering curricula of certain modules, field courses and final projects. However, stronger emphasis should be placed on these challenges throughout the curricula across all engineering programmes, regardless of the disciplinary content. They suggest that the delivery methods need to be modernised and refreshed for the expected learning outcomes to be effectively met; problem-based interdisciplinary learning supported by enhanced technology is a critical enabler. Therefore, promoting humanitarian engineering education is a key issue for engineering education in Indonesia to produce more engineers with enhanced passion and empathy towards people in need and who are not only interested in solving technical problems to achieve solely economic objectives when serving their clients.

Figure 8. Before, during and after the construction of an earthquake-resilient school by the 2011 SCEP group

STRATEGY FOR PROMOTING AND DEVELOPING HUMANITARIAN ENGINEERING IN INDONESIA

Capacity-Building Strategy in National Education for Construction Industries

Capacity building to enhance the education of Indonesian human resource professionals for earthquake-resilient housing cannot be achieved through just one programme or level of education, such as only in the engineering programme; a more comprehensive and integrated solution is required to support the Indonesian construction industry, as depicted in Figure 7.

Abduh et.al. (2022) reported that training and education for the construction industry are skewed towards professional programmes in engineering and architecture rather than the more practical education of technicians at the diploma level or skilled workers at the vocational level, leading to a situation in which many construction labourers are unskilled or have received no training. Simultaneously, the government's certification and registration system has failed to improve the competency of construction personnel at professional, technical and skilled worker levels; an estimated 7 million workers out of a total of 8.3 million may be working informally or casually even though the Construction Services Law requires all workers to be registered (Pribadi & Chan, 2022).

Moreover, Abduh (2020) notes that the number of vocational institutions in the civil engineering and construction fields is limited and exhibits a decreasing trend. Moreover, a minority of high school graduates are interested in studying further in vocational higher education institutions, with the majority more interested in applying for engineering degrees because higher education engineering graduates receive higher salaries. However, some vocational higher education institutions are developing engineering programmes that focus not just on building but also on helping the engineer design.

The informal sector adds a further set of challenges but also provides the opportunity to develop foremen into managers and entrepreneurs. Workers are reluctant to seek certification, as the certificates do not lead to higher wages. Surprisingly, the government has forged ahead with a one-year postgraduate professional engineering programme to improve the skills of engineers but has done little to promote the upskilling of construction labourers. Industry players argue that they cannot afford to invest in training, and they pass this responsibility to the government, which seems content with the limited number of construction programmes at polytechnics (Pribadi & Chan, 2022).

Enhancement Strategy for Architecture and Engineering Programmes in Higher Education

In higher education, according to Kremmyda et al. (2020), weaknesses in the skills of graduate engineers in Indonesia, Bangladesh and Vietnam are mainly related to collaboration, networking, agility and adaptability, effective oral and written communication and exposure to real-life problems and applications. The skills are distinguished from technical or hard skills, which the graduates of these three Asian countries appear to be confident with. As soft skills are applicable and useful in various contexts, graduates have expressed their appreciation that these skills can be developed or transferred between professions and disciplines. With regards to interdisciplinary interprofessional awareness, the study demonstrated that (1) graduates seem to be unable to recognise the relationship between their own discipline and other disciplines and (2) graduates do not value the contributions of multiple technical and non-technical perspectives to a given interdisciplinary problem.

To provide the knowledge and practices required by the Indonesian construction industry, higher education should focus on the gaps identified. For higher education institutions focused on civil engineering and architecture, the design abilities of students should include the capability to design products and understand the design process. These design abilities should be demonstrated by students during their study within a capstone course project design. Many traditional subjects related to construction should also be strengthened through greater learning experiences. Changing the delivery method is needed nowadays with the focus on how students learn. The lecturer should facilitate the students learning through a student-centred approach, such as problem-based learning and simulation (Abduh, 2020).

Based on the taxonomy of attributes for humanitarian engineers and innovative methods of delivery, some enhancements have been implemented at ITB, Indonesia, to promote humanitarian engineering in higher education through different methods, for example, in the form of curricular, intracurricular and extracurricular programmes:

Enhancement of Existing Courses (Curricular Programmes)

The taxonomy of attributes for humanitarian engineers could be used to enhance student outcomes of relevant existing courses. This strategy is considered the easiest to implement, considering that the control is on the lecturer's side. However, for courses that cover several different engineering or architecture programmes, such as university-wide mandatory courses, coordination between programmes and lecturers can be another challenge. Nevertheless, the enhancement of students' outcomes could also be achieved through innovative delivery methods and assessments. The contents related to humanitarian engineering (e.g. the definition, challenges, attributes and methods) should be introduced at the beginning of the course or in related sessions as well as in the context of Indonesia (i.e. natural disasters, earthquakes and the need for resilient housing). At ITB, this strategy has been implemented in university-wide mandatory courses, such as Introduction to Engineering and Design and Civic Education, and in compulsory civil engineering programme courses, such as Construction Management, by introducing the concept of sustainable development in the construction industry. Moreover, for the postgraduate programme in civil engineering at ITB, this enhancement strategy strengthens existing courses, such as Introduction to Infrastructure Management and Disaster Mitigation in Infrastructure, leading to the development of a new course, Rural Infrastructure Development.

Enhancement of the Summer Course Programme (Intracurricular)

Since the enhancement strategy in the curricular programme only covers a limited number of students from a specific discipline, civil engineering, another strategy has been introduced to reach more students from other disciplines, such as architecture, other engineering fields and the sciences, by establishing a short summer course. This programme also provides an opportunity for civil engineering students to collaborate with other disciplines. Because of the limited time of the programme, a specific theme is selected to focus on, such as earthquake-resilient housing, while still providing a broad understanding of humanitarian engineering principles. Lecturers have been invited from academia and industry through several international networks. In 2021, the course was held online and called the International Virtual Course, and it had 55 participants (18% international, 18% ITB students, 2% professionals and 62% Indonesian students). Different innovative delivery methods have been introduced in this course, such as problem-based and project-based methods. At the end of the course, students present their group projects using a case study of engineers' roles in mitigating earthquake disasters resulting from local geological faults for schools and housing in the surrounding areas.

Enhancement of the Student Community Engagement Programme (Intra and Extracurricular)

Student community engagement programmes are commonly held in Indonesia's universities and can be either intracurricular or extracurricular. This community service programme was introduced in 1971 by the Ministry of National Education and Culture, and it then became part of the national agenda as the student community service or engagement programme, Kuliah Kerja Nyata (KKN). At ITB, KKN is held annually as an intracurricular programme, and many other programmes, called student community engagement programmes (SCEPs), are held as extracurricular programmes. The enhancement strategy strengthens existing programmes with humanitarian engineering attributes and principles. In the last two years, ITB has collaborated with two other universities, UGM and Brawijaya University, to hold this enhanced KKN. Students from these three universities from different disciplines collaborate to solve the humanitarian challenges faced by selected communities. However, ITB's civil engineering programme has been conducting the SCEP programme for 30 years, supported by its student association. SCEP aims to provide students with the experience of real infrastructure challenges by providing rural communities with the needed infrastructure and simulating a real project-based management environment. SCEP is based on the stages of the infrastructure life cycle: initiation, planning, design, implementation and usage. At each stage, the involvement of students is maximised, except for tasks that need professional interventions, such as design approvals, which are done by certified lecturers, and skilled construction work, which is performed by skilled workers or contractors. SCEP provides the graduates with the necessary competences and hard and soft skills required by civil engineers through a programme that is delivered using an experiential and project-based learning approach (Abduh & Putri, 2022). Two projects related to providing earthquake-resilient non-engineered buildings were conducted in 2011 (three classrooms) and 2019 (informal playgroup). Figure 8 is taken from the 2011 SCEP project.

In addition to the above enhancement strategies, it is worth mentioning here that the following enhancement strategies, which are related to the enhancement of delivery methods for the existing programme, creating a new collaborative course and establishing a brand-new multidisciplinary programme, will be introduced at ITB:

Excursion to Vulnerable and Disaster-Prone Areas

As an effective delivery method, an excursion to a vulnerable or disaster-prone area is encouraged for any courses related to civil engineering and architecture. This activity will expose students to the reality of national and local capacities regarding earthquake-resilient housing by observing the vulnerabilities existing at the location and the practices of related stakeholders.

Establishment of a New Collaborative Course on Non-Engineered Buildings

Housing is categorised as non-engineered construction, but few civil engineers are interested in this subject, and architects are not well-equipped with the relevant technology to provide earthquake-resilient housing. This course is therefore needed. This is a collaborative course between the civil engineering and architecture programmes, although students from other relevant disciplines are welcome to participate.

Establishment of a Multidisciplinary Disaster Mitigation Postgraduate Programme

The previous two strategies focus on students' learning objectives, but education should also be supported by research and its application. There is also a need for disaster mitigation professionals. Therefore, a degree and multidisciplinary programme on disaster mitigation is required. However, there should be a specialisation that is structured to address non-engineered buildings.

CONCLUSION

A major humanitarian challenge for Indonesia is to provide earthquake-resilient housing, considering that earthquake disasters in the past have caused substantial damage to the housing sector, which primarily affects the low- and medium-income population. The available knowledge, technologies, codes, standards and practices related to housing, which is categorised as non-engineered construction, is limited and seems to be neglected by Indonesian architects and engineers, exacerbating the complexity of the problem. This problem has multifaceted and multidisciplinary aspects, including socioeconomics, knowledge and technology, industry, human resources, policy, legislation and institutions. As most housing construction in Indonesia is performed by the local informal construction industry and involves unregistered builders and craftsmen, the challenge is how to develop knowledge and skills in earthquake-resilient house construction within this informal sector to ensure that non-engineered house constructions incorporate engineering principles already provided by various agencies and organisations.

The emergence of humanitarian engineering provides opportunities to address this challenge by producing engineers with enhanced passion and empathy towards people in need and who are capable of creating innovative, affordable and sustainable solutions to the problem of earthquake vulnerability in ordinary peoples' houses. Strategies to promote humanitarian engineering in Indonesian engineering education, such as those implemented at ITB, include enhancements to existing engineering courses and programmes as well as the establishment of new courses if required.

The enhancement of existing courses may include (1) adding new content, such as humanitarian engineering, earthquake engineering and non-engineered buildings, (2) revising student outcomes to improve soft-skill competencies such as ethics and empathy, and (3) introducing innovative delivery methods, such as problem-based, project-based, collaborative (multidisciplinary) and experiential (excursion and community engagement) learning. The existing programmes could be improved by revising curricula to accommodate humanitarian engineering attributes as well as introducing new courses, such as disaster mitigation, rural infrastructure development and a collaborative course for architecture and civil engineering students related to earthquake-resilient housing. The establishment of a new multidisciplinary programme related to earthquake disaster mitigation is necessary because the teaching programme should also be supported by solid research activities.

REFERENCES

Abduh, M. (2020). *Advancing the construction engineering and management in Indonesia.* Construction Plus Asia. https://www.constructionplusasia.com/id/advancing-the-construction-engineering-and-management-in-indonesia/

Abduh, M., Pribadi, K. S., Soemardi, B. W., & Chan, T.-K. (2022). Education, training, and certification. In T.-K. Chan & K. S. Pribadi (Eds.), *Construction in Indonesia: Looking back and moving forward* (p. 12). Routledge. https://doi.org/https://doi.org/10.1201/9781003149866

Abduh, M., & Putri, A. T. (2022). Capacity building in sustainable infrastructure through student's community engagement projects Indonesia. *Proceedings of the 9th Civil Engineering Conference in the Asian Region (CECAR).*

Abe, M., Ochiai, C., & Okazaki, K. (2018). Is post-disaster housing reconstruction with participatory method effective to increasing people's awareness for disaster prevention? *Procedia Engineering*, *212*, 411–418. doi:10.1016/j.proeng.2018.01.053

Arya, A. S., Boen, T., & Ishiyama, Y. (2013). *Guidelines for earthquake resistant non-engineered construction*. UNESCO.

Arya, A. S., Boen, T., Ishiyama, Y., Martemianov, A. I., Meli, R., Scawthorn, C., Neumann, J. V., & Xian, Y. Y. (1986). *Guidelines for earthquake resistant non-engineered construction*. The International Association for Earthquake Engineering.

BNPB. (2021a). *Indonesia disaster risk index in 2020*. BNBP. http://mpbi.info/wp-content/uploads/2021/07/BUKU-IRBI-2020-K
P.pdf

BNPB. (2021b). *Casualties and damage by disaster*. BNBP. https://dibi.bnpb.go.id/kbencana

BNPB. (2022). *Indonesian disaster information data*. BNBP. https://dibi.bnpb.go.id/xdibi

Boen, T. (2006a). *The minimum requirements for earthquake-resistant house: Masonry house with timber confinement*. https://teddyboen.com/Poster_Membangun_Rumah_Tahan_Gempa.htm
l

Boen, T. (2006b). *The minimum requirements for earthquake-resistant house: One brick thick-walled masonry house*. https://teddyboen.com/Poster_Membangun_Rumah_Tahan_Gempa.htm
l

Boen, T. (2006c). *The minimum requirements for earthquake-resistant house: One brick thick-walled masonry house with reinforced concrete*. https://teddyboen.com/Poster_Membangun_Rumah_Tahan_Gempa.htm
l

Boen, T. (2006d). *The minimum requirements for earthquake-resistant house: Wooden house and several variations*. https://teddyboen.com/Poster_Membangun_Rumah_Tahan_Gempa.htm
l

Boen, T. (2009a). *Constructing seismic-resistant masonry houses in Indonesia*. United Nations Centre for Regional Development. http://www.confinedmasonry.org/wp-content/uploads/2018/11/masonryhousesindonesiaguide.pdf

Boen, T. (2019). *Building earthquake-resistant masonry house with ferro-cement layers*. https://teddyboen.com/Membangun_Rumah_Tembokan_Tahan_Gempa_dengan_Balutan_Lapisan_Ferosemen.html

Boen, T. (2021a). *Building earthquake-resistant masonry house with fibrocement (fibre reinforced cementitious/FRC) layers*. https://teddyboen.com/Membangun_Rumah_Tembokan_Tahan_Gempa_dengan_TRC_FRC.html

Boen, T. (2021b). *Building Earthquake-resistant Masonry House with Textile Fibre Composite/TRC Layers*. https://teddyboen.com/Membangun_Rumah_Tembokan_Tahan_Gempa_dengan_TRC_FRC.html

Boen, T., Firmansjah, J., Ismail, F. A., Kusumastuti, D., Lase, Y., Pribadi, K. S., Sarwidi, S., Satyarno, I., & Widodo, L. (2010). *Ways to repair an earthquake-damaged simple structure*. Australia-Indonesia Facility for Disaster Reduction and Indonesian National Board for Disaster Management.

Boen, T., Saputra, A., Ismail, F. A., Pribadi, K. S., Satyarno, I., & Widodo, L. (2012). *The minimum requirements for safer houses*. Japan International Cooperation Agency.

Boen, T., Suprobo, P., Sarwidi, S., Pribadi, K. S., Arkani, M. I., Satyarno, I., & Saputra, A. (2009b). *The minimum requirements for safer houses*. Directorate General of Housing and Settlement and Japan International Cooperation Agency. https://teddyboen.com/Persyaratan_Pokok_Rumah_yang_Lebih_Aman.html

BPS. (2020). *Housing and Settlement Statistics 2019*. BPS. https://www.bps.go.id/publication/2020/08/31/6a9e70d6154fde75499239e6/statistik-perumahan-dan-permukiman-2019.html

BPS. (2021). *Percentage of Households by Province and Widest Non-Bamboo/Other Walls (Percent), 2019-2021*. BPS. https://www.bps.go.id/indicator/29/851/1/persentase-rumah-tangga-menurut-provinsi-dan-dinding-terluas-bukan-bambu-lainnya.html

BRIN. (2019). *Bale Kohana, earthquake resistant residential technology innovation*. http://pusyantek.bppt.go.id/id/posts/artikel/bale-kohana-inovasi-teknologi-hunian-tahan-gempa

Brzev, S., & Mitra, K. (2007). *Earthquake-resistant confined masonry construction*. National Information Center of Earthquake Engineering, Indian Institute of Technology Kanpur.

Carlevaro, N., Roux-Fouillet, G., & Schacher, T. (2018). *Guide book (for builders) for building earthquake-resistant houses in confined masonry*. Swiss Agency for Development and Cooperation and Earthquake Engineering Research Institute. https://www.shacc.ch/documents/files/library/Manual/1806-CM-masons-guide_eng_SMALL.pdf

Centre for Research on the Epidemiology of Disasters. (2022). *Earthquake disaster data in Indonesia 1991–2020*. EM-DAT, The International Disaster Database. https://www.emdat.be/

Confined Masonry Network. (2011). *Seismic design guide for low-rise confined masonry buildings*. Earthquake Engineering Research Institute.

Department of Public Works. (1993). *The Guideline of Constructing an Earthquake-Resistant Building - Decree of Directorate General of Housing and Settlement No. 111/KPTS/CK/1993*. Directorate General of Human Settlement.

Department of Public Works. (2006). *Technical Guideline for Earthquake-Resistant Houses and Building*. Directorate General of Housing and Settlement.

Heierli, U., & Maithel, S. (2008). Brick by brick: The Herculean task of cleaning up the Asian brick industry. In P. Osborn (Ed.), *Swiss Agency for Development and Cooperation* (1st ed.). Swiss Agency for Development and Cooperation.

IFRC. (2010). *Disasters in Asia: The case for legal preparedness*. https://disasterlaw.ifrc.org/sites/default/files/media/disaster_law/2020-09/Disasters in Asia_EN 2010.pdf

Irsyam, M., Asrurifak, M., Mikail, R., Sabaruddin, A., & Faisal, L. (2017). 2017 Indonesia earthquake source and map update. One Day Seminar on Disaster in Commemorating the 72nd Indonesian Independence Day, 28 August 2017, Jakarta, Indonesia.

Kremmyda, G., Georgoulas, A., & Koumpouros, Y. (2020). Demystifying humanitarian engineering: A comparative study on perceptions in UK and Asia. *2020 IEEE Global Engineering Education Conference (EDUCON 2020)*, 1009–1014.

Kremmyda, G., Georgoulas, A., Koumpouros, Y., & Mottram, J. T. (2020). Taxonomy of engineering attributes for tackling humanitarian challenges. *Proceedings of the 2020 IEEE Global Engineering Education Conference (EDUCON)*, 1003–1008.

Kuswanto. (2007). *Report on reviewing findings of brick structure in Nglinguk Hamlet, Trowulan Village, Trowulan District, Mojokerto Regency.* BP3Jatim.

Li, V. C. (2008). Engineered cementitious composites (ECC) material, structural, and durability performance. In E. Nawy (Ed.), *Concrete construction engineering handbook* (2nd ed.). CRC Press.

Ministry of Education and Culture. (2022). *Muara Takus Temple*. https://petabudaya.belajar.kemdikbud.go.id/Repositorys/muara_takus/

Ministry of Manpower. (2018). *Indonesian national standard of work competence, category: Construction, Main group: Building construction, Field: RISHA building, Decree of Minister of Manpower No. 221/ 2018*. Ministry of Manpower.

Ministry of Public Works and Housing. (2017a). *Ministry of Public Works and Housing renovated 1,707 houses in Prabumulih*. https://pu.go.id/berita/kementerian-pupr-bedah-1-707-rumah-di-prabumulih

Ministry of Public Works and Housing. (2017b). *Practical guide on requirement of earthquake-resistant building for confined masonry building*. Ministry of Public Works and Housing.

Ministry of Public Works and Housing. (2017c). *Practical guide on requirement of earthquake-resistant building for RISHA*. Ministry of Public Works and Housing.

Ministry of Public Works and Housing. (2017d). *Practical guide on requirement of earthquake-resistant building for wooden building*. Ministry of Public Works and Housing.

Mitcham, C., & Muñoz, D. (2010). *Humanitarian Engineering* (Vol. 5). Morgan & Claypool.

National Geographic Indonesia. (2015). *Tracing the site of the oldest observatory in Indonesia*. National Geographic. https://nationalgeographic.grid.id/read/13297532/melacak-tapak-observatorium-tertua-di-indonesia?page=all

Naturally:wood. (2022). *Cross-laminated timber (CLT)*. Naturally Wood. https://www.naturallywood.com/products/cross-laminated-timber/

Nuruddin, M. F., Bayuaji, R., Masilamani, M. B., & Biyanto, T. R. (2010). Sidoarjo mud: A potential cement replacement material. *Civil Engineering Dimension*, *12*(1), 18–22.

PCA. (2019). *Autoclaved Aerated Concrete*. PCA. https://www.cement.org/cement-concrete/paving/buildings-structures/concrete-homes/building-systems-for-every-need/autoclaved-aerated-concrete

Pinhome. (2021). *RISHA (Rumah Instan Sehat Sederhana)*. https://www.pinhome.id/kamus-istilah-properti/risha/

Pradoto, R., Oktavianus, A., Pribadi, K. S., Rasmawan, I. M. A. B., & Wulandari, L. D. (2022). Palu housing reconstruction process: Reviewing and learning after the 2018 earthquake. Sustainable Infrastructure and Built Environment (SIBE-2022), Bandung, Indonesia.

Pribadi, K. S., Abduh, M., Wirahadikusumah, R. D., Hanifa, N. R., Irsyam, M., Kusumaningrum, P., & Puri, E. (2021). Learning from past earthquake disasters: The need for knowledge management system to enhance infrastructure resilience in Indonesia. *International Journal of Disaster Risk Reduction*, *64*, 102424. https://doi.org/https://doi.org/10.1016/j.ijdrr.2021.102424

Pribadi, K. S., & Chan, T.-K. (2022). Construction workers and the informal sector. In T.-K. Chan & K. S. Pribadi (Eds.), *Construction in Indonesia: Looking back and moving forward* (p. 18). https://doi.org/https://doi.org/10.1201/9781003149866

Pribadi, K. S., Hidayat, B., Triyadi, S., & Harapan, A. (2009). Indigenous knowledge on house building system in West Sumatra and Southern-West Jawa, Indonesia. In R. Shaw, A. Sharma, & Y. Takeuchi (Eds.), *Indigenous knowledge and disaster risk reduction*. Nova Science.

Pribadi, K. S., Kusumastuti, D., & Rildova. (2008). *Learning from recent Indonesian earthquakes: An overview to improve structural performance*. The 14th World Conference on Earthquake Engineering, Beijing, China.

Pribadi, K. S., Pradoto, R., Hanafi, E., & Rasmawan, I. M. (2020). Lombok earthquake, one year later: Housing sector recovery. *E3S Web of Conferences, 156*, 1012. https://doi.org/ doi:10.1051/e3sconf/202015601012

Research and Development Agency PUPR. (2017). 2017 *Indonesia earthquake source and hazard map*. Ministry of Public Works and Housing.

Ridha, R., Rahmawaty, A. A., & Santoso, H. (2021). Strategy to accelerate post-earthquake rehabilitation and reconstruction through earthquake resistant house zoning in North Lombok Regency. *Planoearth National Seminar, 2*, 33–41.

Sadiqi, Z., Trigunarsyah, B., & Coffey, V. (2017). A framework for community participation in post-disaster housing reconstruction projects: A case of Afghanistan. *International Journal of Project Management*, *35*(5), 900–912. https://doi.org/https://doi.org/10.1016/j.ijproman.2016.11.008

Sarwidi, S. (2020). *Barrataga*. http://barrataga.com/

Suwantara, I. K., & Rusli, R. (2013). Experimental study of shear-friction connection of lontar pillars (Borassus flabellifer) with stone in Ammu Hawu traditional house (NTT). *Jurnal Teknik Sipil*, *20*(2), 79. doi:10.5614/jts.2013.20.2.1

UFO Dome House Indonesia. (2022). *Dome house*. https://www.domehouse.id/

Waluyo, D. T. (2021). *RISHA instant technology builds disaster victim's house*. https://infopublik.id/kategori/sorot-ekonomi-bisnis/537996/risha-teknologi-instan-membangun-rumah-korban-bencana?show=

Winarno, S., Griffith, A., & Stephenson, P. (2010). Reducing earthquake risk to non-engineered buildings in Indonesia. *Construction Innovation*, *10*(4), 421–434. doi:10.1108/14714171011083588

Chapter 13
Examining the Quality Assurance in Engineering Education in Bangladesh

G. M. Tarekul Islam
Bangladesh University of Engineering and Technology, Bangladesh

Georgia Kremmyda
University of Warwick, UK

Angelos Georgoulas
https://orcid.org/0000-0001-8845-3781
University of West Attica, Greece

Yiannis Koumpouros
https://orcid.org/0000-0001-6912-5475
University of West Attica, Greece

Syed Labib Ul Islam
https://orcid.org/0000-0002-5113-6122
Bangladesh University of Engineering and Technology, Bangladesh

A. K. M. Saiful Islam
Bangladesh University of Engineering and Technology, Bangladesh

Sujit Kumar Bala
Bangladesh University of Engineering and Technology, Bangladesh

Mohammad Shoeb
Dhaka University, Bangladesh

ABSTRACT

Quality assurance in higher education is a global practice. However, the quality assurance aspect in the engineering educational institutions of Bangladesh is still at the inception level. It is crucial to monitor, assess, and improve the system to make the system effective in the context of Bangladesh. This study highlights the present practice and the perception of the professionals involved with the agencies as well as the way forward to further improve the quality assurance aspects. A questionnaire survey was conducted to assess the current status of quality assurance practices in the educational institutions of Bangladesh. It is found that there is a limited scope of quality assurance in engineering education in Bangladesh. It is also revealed that there appears to be a lack of direction as there are a number of ideas for quality assurance that have a similar level of support among the respondents. There is a long way to go for the quality assurance set-up in Bangladesh to mature and be more effective.

DOI: 10.4018/978-1-6684-5619-4.ch013

INTRODUCTION

The quality assurance aspect of engineering education is highly challenging and crucial in the modern era (Chowdhury et al., 2013). Under the guidance of the University Grants Commission (UGC), each higher education institution in Bangladesh has set up the Institutional Quality Assurance Cell (IQAC) to ensure the quality of higher education. Higher education institutions can achieve quality assurance in engineering education in many ways. For instance, accreditation by national and global agencies can help ensure the quality of engineering education. The Washington Accord, signed in 1989, is a multi-lateral agreement between bodies responsible for accreditation or recognition of tertiary-level engineering qualifications within their jurisdictions who have chosen to work collectively to assist the mobility of professional engineers. Ali et al. (2011) discussed the essential requirement of engineering accreditation under the Washington Accord and its benefits. They uncovered that a minimum fulfillment of the Washington Accord could render an excellent opportunity for third-world countries to provide quality engineering education. It is worth noting that quality assurance in engineering education is a continuous process and must be evaluated and updated with changing circumstances. Li and Lei (2015) studied the quality assurance mechanisms in engineering education with a case study of Purdue University in the United States of America. They found that the concept of continuous quality improvement was not well implemented in the practice of quality assurance.

As the world faces crises on multiple fronts, engineers are at the forefront of tackling the problems alongside professionals from other sectors. Disasters induced by climate change and the adaptation and mitigation of climate change are the most prominent problems, and engineers play a crucial role in solving these problems (Milovanovic et al., 2022). There is an increasing demand for sufficiently skilled engineers who can work toward sustainable development while addressing social and humanitarian needs (Smith et al., 2019). This is a significant shift from the traditional education methods that are still widely applied all across the world, particularly in developing countries (Ahmed et al., 2016).

The quality of the education that graduate engineering students have access to is also a point of major concern as the largely theoretical knowledge may not be sufficient to enter the professional field (Daun et al., 2016; Faizi & Umar, 2021). Engineering graduates need to have interdisciplinary knowledge in order to provide sustainable solutions to emerging problems (Hadgraft & Kolmos, 2020). Underpinning these transformational changes, the role of quality assurance and accreditation cannot be ignored (Chowdhury et al., 2013; Manzoor, 2017).

The engineering education system in Bangladesh possesses a similar curriculum to many other countries, yet it is not mutually recognized due to non-compliance issues related to quality assurance (Alam et al., 2016). In recent years there has been some progress in this regard. However, the mechanism set for quality assurance in Bangladesh is yet to mature, where the formal arrangements for governance and regulation are insufficient and backdated (Alam, 2020). Moreover, the lack of a robust quality assurance framework hinders progress, as Rahnuma (2020) highlights.

With this background, this paper aims to gain insight into the quality assurance agencies and their roles in Bangladesh in relation to engineering education. This is particularly important as the quality assurance aspect in the engineering educational institutions of Bangladesh is still at the inception level, and it is crucial to monitor, assess, and improve the system to make the system effective in the context of Bangladesh.

This study highlights the present practice and the perception of the professionals involved with the agencies, as well as the way forward to further improve the quality assurance aspects. This paper can help the policymakers and professionals engaged with education quality assurance to develop a robust periodic quality assurance framework and implementation mechanism for engineering education in Bangladesh.

METHODOLOGY

The study was conducted as part of Enabling Humanitarian Attributes for Nurturing Community-based Engineering (ENHANCE) project funded by the European Union (EU). A number of Asian and European countries were part of this project, where large-scale data collection was carried out. Within the framework of ENHANCE and aiming primarily to introduce relevant and appropriate innovations in graduate engineering education in the Partner Higher Education Institutions (HEIs), an online survey was developed and circulated among the project partners and beyond at the international level. The survey was taken by 734 respondents spanning from three distinctive sectors; Higher Education, Industry and Quality Assurance. Participation in our online survey was completely voluntary. Anyone could choose to withdraw participation at any time without giving a reason by contacting one of the research team. The survey followed the ethics and deontology rules applied in the EU. Ethics approval was granted by the University of Warwick, Biomedical and Scientific Research Ethics Sub-Committee.

The online survey was addressed to three target groups; representatives from Higher Education Institutions, Quality Assurance Agencies and Industry. Each target group was asked to respond to a set of questions. The average time to complete the online questionnaire was around 11 minutes. The sample consisted of 734 respondents in total; 221 respondents from Industry, 403 respondents from Higher Education Institutions (HEIs), and 110 respondents from Quality Assurance Agencies (QAA), mainly from UK, Greece, Bangladesh, Indonesia and Vietnam.

The respondents were purposively selected so that they were affiliated with different universities in the engineering faculties, professionals from the engineering industry sector, or engineering education quality assurance agencies.

The respondents were requested to complete a questionnaire through an online medium, which had three diverging blocks. The three blocks were designed to represent the Higher Education Institutions, Industries, and Quality Assurance Agencies. The specific questions were of both closed and open-ended types to enable extracting the maximum possible insight.

For this study focusing on Bangladesh, only the responses from the Quality Assurance Agencies of Bangladesh were selected for carrying out a simple statistical analysis. The valid, complete responses were considered for this purpose following data cleaning and sorting. The questions requiring the respondents to rank different options were analyzed based on the mean rank values assigned by all the valid responses.

A limitation of this study is the fact that the Quality Assurance of engineering education in Bangladesh is a comparatively new sector, and the availability of sufficiently experienced experts in this area in the context of the country is limited in number. However, the perception-based study in this regard with the current professionals engaged in Quality Assurance issues is very important to gauge the current situation and the way forward to develop the framework of quality assurance reviews in engineering education in Bangladesh.

Organizational Denomination

Quality Assurance in higher education being a comparatively new domain in Bangladesh with a limited scope, a total of 8 responses could be collected for this study. 42.86% of the respondents mentioned the denomination of their organization is *Department*, while 28.57% were from *Agency* denomination, as shown in Figure 1. In addition, there were denominations of *Wing*, *Research*, and *Corporation* among the respondents as well.

Figure 1. Organisational denomination

Scope of Quality Assurance Agencies in Bangladesh

In terms of the remit or scope of the Quality Assurance Agencies, 25% of the respondents mentioned their scope falls within *A specific discipline or field of learning*, *Higher Education Institutions*, and *Research Units/Centers* (Figure 2). In addition, two respondents mentioned about *Agricultural Development Department* and the *Implementation of research* in this regard. Specifically, the *Irrigation sector in Bangladesh* and *Civil Construction Work* were highlighted.

Figure 2. Scope of quality assurance agencies in Bangladesh

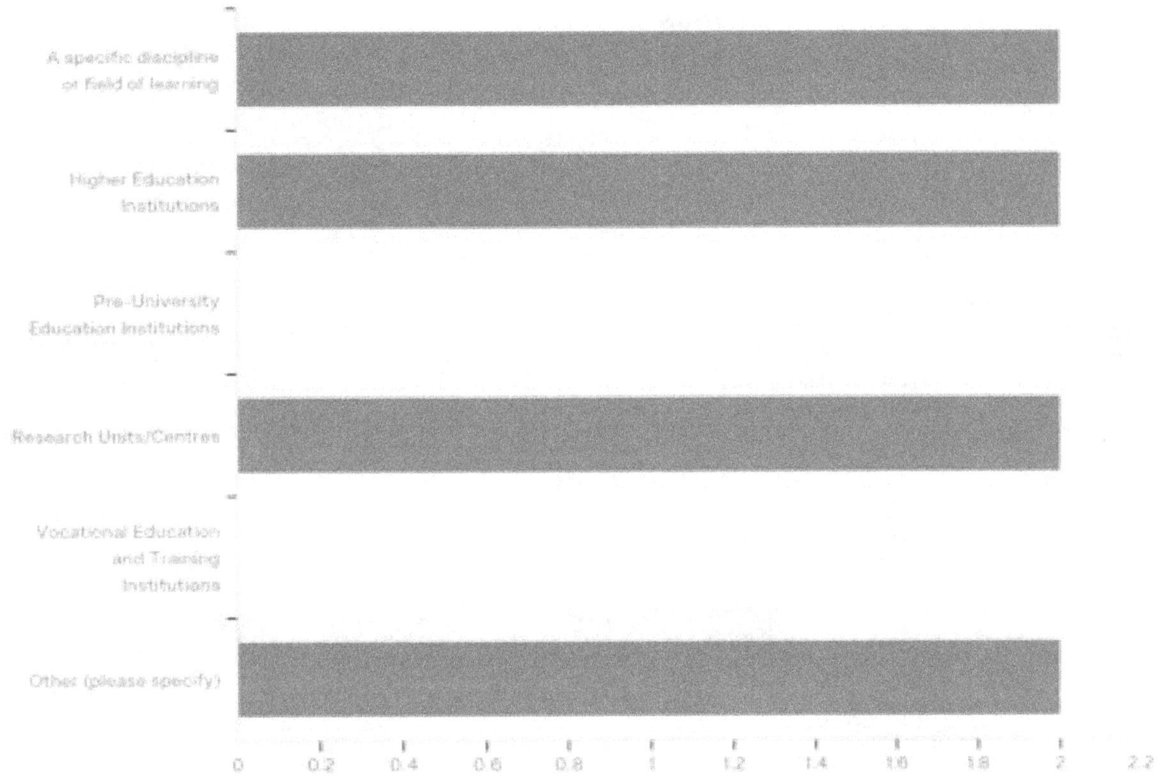

Frequency of External Quality Assurance Review

Figure 3 shows the frequency of periodical external quality assurance reviews for educational programs. 75% of the respondents mentioned that their review is done *Upon request*, while 12.50% mentioned they do it *Every 5 years*. Besides, another respondent mentioned that their organization carries out a review every year.

Consequences of External Quality Assurance Review

As for the consequences of external quality assurance review for Higher Education Institutions, 50% of the respondents mentioned that there is *No formal consequence* (Figure 4). *Approval/Licensing of a program* and *Funding/Withdrawal of funding to a program* each were highlighted by 25% of the respondents. This indicates that there is not a lot of meaningful consequence of external reviews for educational institutions.

Importance of Quality Assurance Activities

The respondents rated the importance of periodical quality review as shown in Figure 5. In their perception, the quality review is essential for *Governance and strategy* and the *Professional and pedagogical qualifications of staff* are essential, as each was highlighted by 50% of them. However, the most consensus

Figure 3. Frequency of external quality assurance review

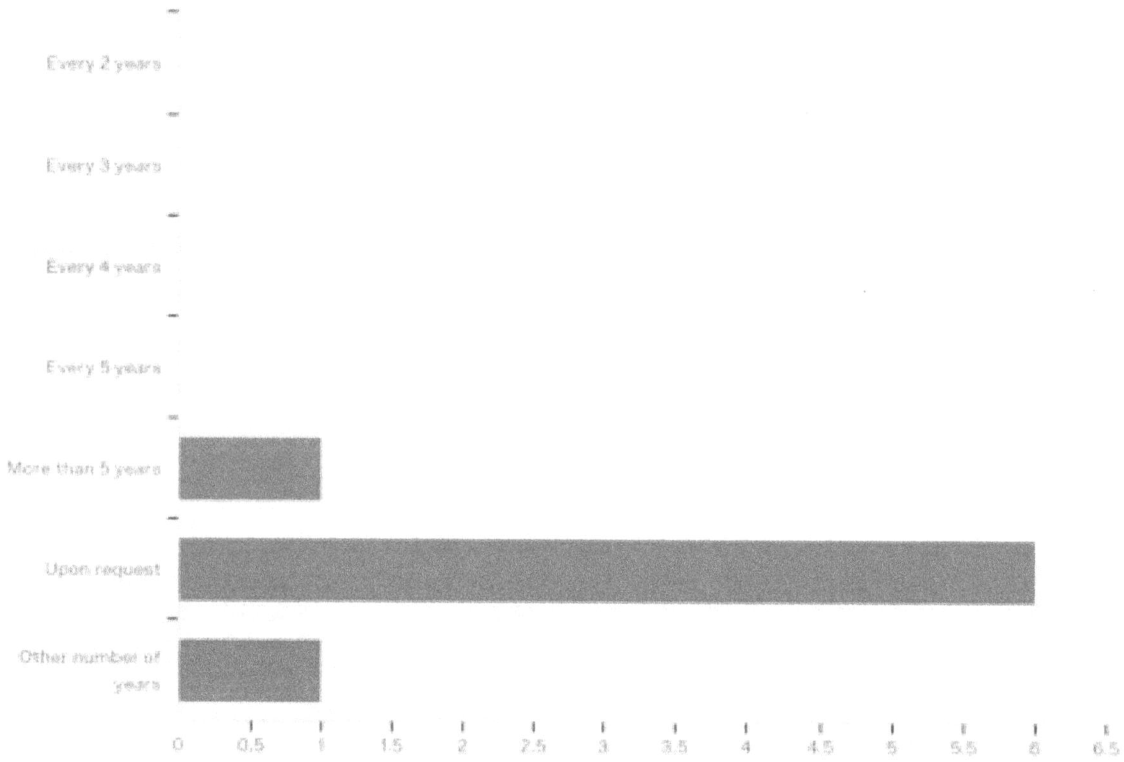

Figure 4. Formal consequences of external quality assurance review

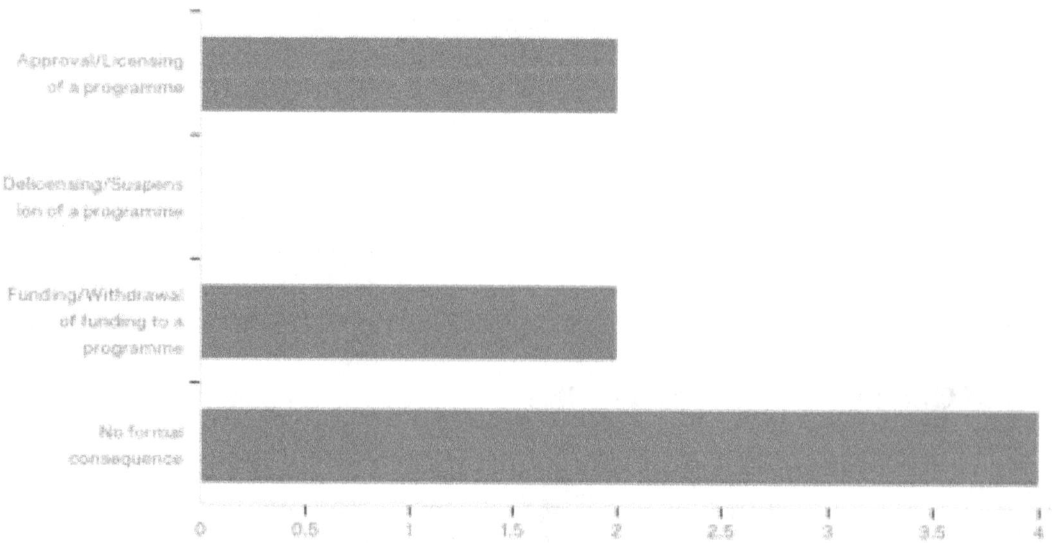

Figure 5. Importance of quality assurance activities

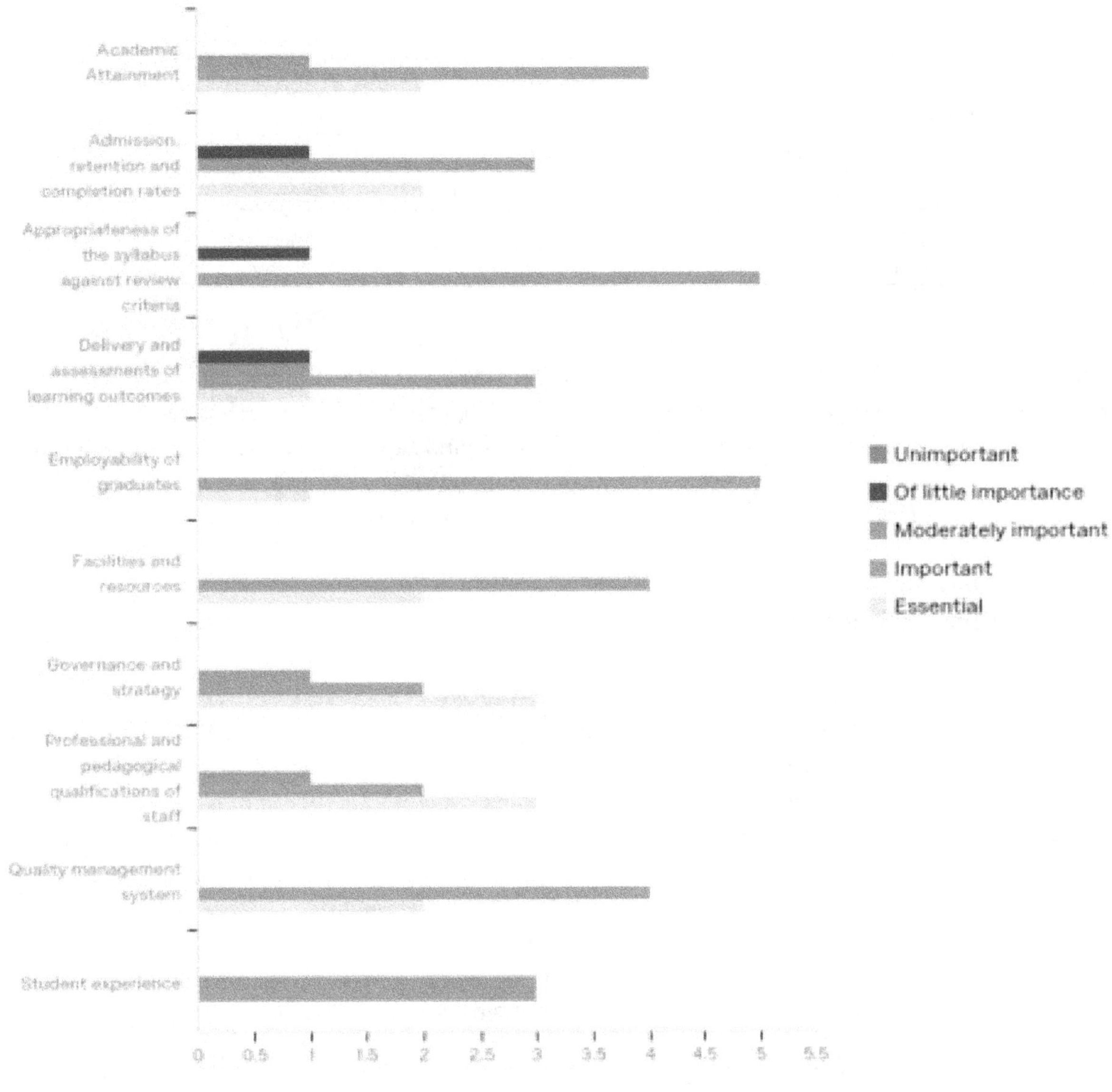

was found in the *Appropriateness of the syllabus against review criteria* and *Employability of graduates*, each of which was ranked *Important* by 83.33% of the respondents, while 66.67% marked *Facilities and resources* as *Important*. On the other hand, *Admission, retention and completion rates*, *Appropriateness of the syllabus against review criteria*, and *Delivery and assessments of learning outcomes* – each being ranked *Of little importance* by 16.67% of the respondents.

Models Used in Bangladesh for Quality Assurance Review

Table 1 shows the perception regarding the models/frames of reference used for the program-level periodical external quality review. The most used model was found to be *Standards and guidelines defined by*

Table 1. Models used for external quality reviews

Sl. No.	Models Used for Reviews	Percentage
1	Standards and guidelines defined by professional groups/organizations either national or international	36.36%
2	Goals and objectives set by the institution being evaluated	27.27%
3	Institutional, program and subject benchmarks	18.18%
4	Best or good practices	9.09%
5	Local regulations	9.09%
6	National framework of qualifications	0.00%

professional groups/organizations either national or international with 36.36% of the responses, closely followed by *Goals and objectives set by the institution being evaluated* with 27.27%. *Institutional, program and subject benchmarks* were highlighted by 18.18%, and *Best or good practices*, as well as *Local regulations*, were mentioned by 9.09% of the respondents. However, nobody mentioned any *National framework of qualifications* which is reflective of the lack of national guidelines in the aspect of quality assurance for educational institutions.

Responsible Authorities for New Programs

Table 2 shows the responsibility for the approval of new programs, and it is revealed that 50% of the respondents mentioned that their quality assurance agency is responsible for the approval of new programs, while the rest of the 50% mentioned they were not. It was also found that 66.67% of the respondents mentioned *The HEI is solely responsible (self-regulation)* while 33.33% mentioned *Ministry* is responsible for the approval of new programs (Table 3). This indicates that most higher education institutions approve new programs themselves – being self-regulated and, in some cases, overseen by the Ministry. The role of external agencies was found to be very limited in this regard.

Table 2. Quality assurance agency's responsibility of approving new programs

Sl. No.	QA Agency Responsibility	Percentage	Number of Responses
1	Yes	50.00%	3
2	No	50.00%	3

Panel Membership and Problematic Areas for External Quality Review

The panel membership taxonomy was found to be composed of a diverse range of backgrounds (Figure 6). 21.05% of the respondents highlighted the presence of *Field experts/practitioners* and *Representatives of professional organizations*. 15.79% mentioned *Employers* and *Higher Education and Quality Assurance Experts*. The most problematic areas for periodical external quality review were also ranked between 1 to 10, with 1 being the most problematic and 10 being the least (Table 4). It was found that

Table 3. Authority responsible for approving new programs

Sl. No.	Responsible Authority	Percentage	Number of Responses
1	Ministry	33.33%	1
2	The HEI is solely responsible (self-regulation)	66.67%	2

Community engagement strategy and *Staff qualifications* were the most problematic – each scoring a mean of 2.00. They were followed by *Research strategy* (2.67), *Practical training* (4.00), *QA management systems and structures* (4.00), *Connection between teaching and research* (4.33), and *Teaching and learning strategy and methods* (4.50). On the other hand, *Student experience* was identified as the least problematic area, with a mean score of 10.00, followed by *Admission entry requirements* (9.50). This highlights that research and training are prioritized by quality assurance agencies while the student experience is overlooked. This needs to be changed as the involvement of students can be crucial for effective quality assurance and quality enhancement (Elassy, 2013; Elassy, 2015).

Required Capacities for Quality Assurance

Figure 7 shows the quality assurance capacity that is most needed by Higher Education Institutions. The most important aspects were identified to be *Operational/financial planning* and *Program and curriculum benchmarking* each with 16.67% of the responses. The remaining qualities received similar attention with *Developing an assessment strategy, Developing and reviewing program learning outcomes, Developing a research strategy, Developing governance and strategic plans, Development of community engagement strategies, Developing teaching and learning strategies, Planning and developing Quality Assurance management*, and *Planning and management of facilities* all receiving 8.33% of the responses. These findings illustrate that the quality assurance aspects are very much in the inception phase in Bangladesh, and more initiatives are required to establish a pragmatic setup that is applicable and beneficial for the quality of students and HEIs in Bangladesh.

CONCLUSION

The importance of quality engineering education is evident, as engineering solutions are required to address emerging challenges around the world. However, the quality assurance framework is a new concept for Bangladesh. In conducting this study, a limited number of responses were collected, which is indicative of the limited scope of quality assurance in engineering education in Bangladesh. From the discussion of the findings, it is also revealed that there appears to be a lack of direction as there are a number of ideas for quality assurance, which have a similar level of support among the respondents. This is down to the fewer experts available and the lack of a country-wide homogeneous framework for quality assurance.

There are several aspects revealed in this study where improvements can be made. The majority of the respondents mentioned that their agency conducts external quality reviews *upon request*, instead of on a regular frequency. There also appears to be a lack of *formal consequences* while some respondents highlighted that the licensing and funding can be on the line in some instances. *Governance and*

Figure 6. Panel membership of external quality review

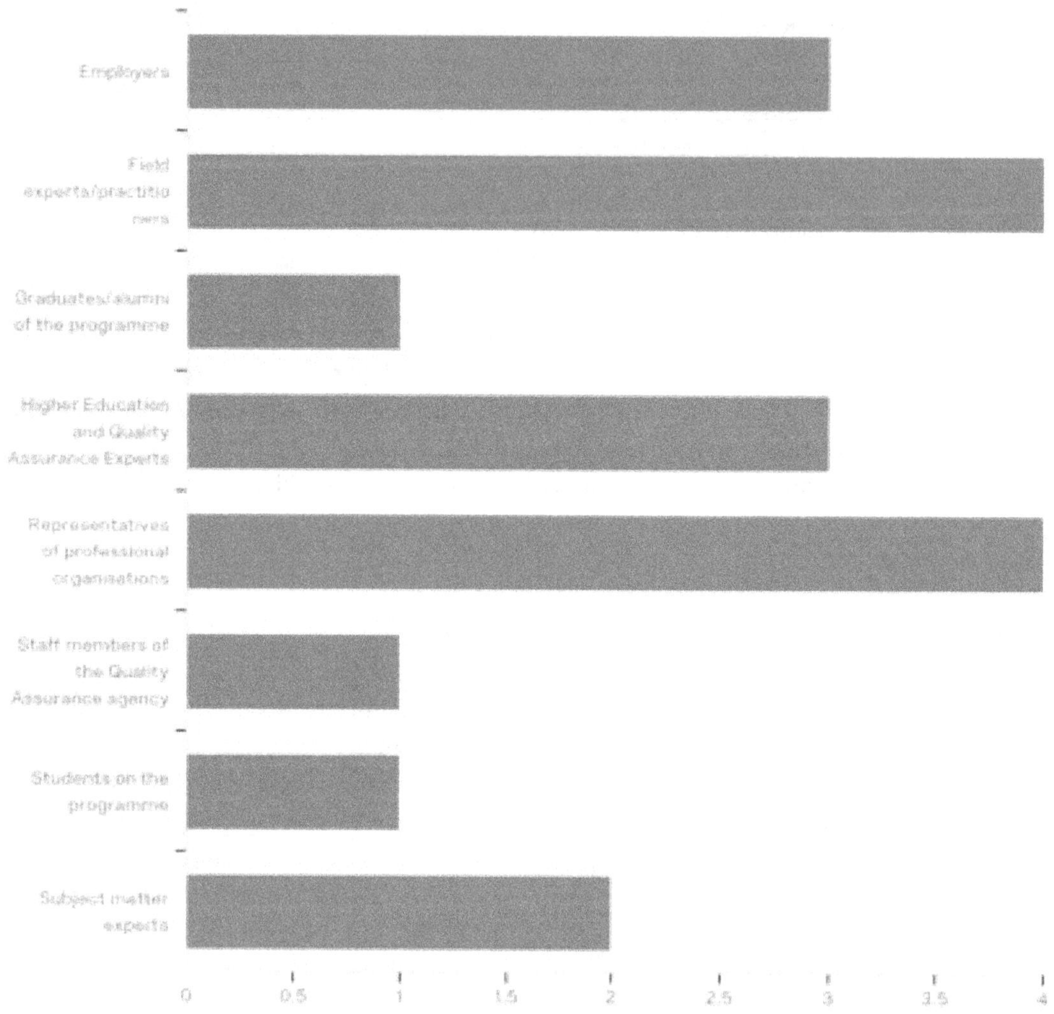

strategy and *Professional and pedagogical qualifications of staff* are prioritized by the current quality assurance setup, while *delivery and assessments of learning outcomes* are deemed to be of comparatively less importance. Furthermore, a number of different quality assurance models are found to be used in Bangladesh, with *standards and guidelines defined by professional groups/organizations either national or international* and *Goals and objectives set by the institution being evaluated.* On the other hand, half of the respondents mentioned the Quality Assurance Agencies were responsible for approving new programs, while the other half mentioned they were not. Those who responded that the Quality Assurance agencies were not responsible, mentioned the Ministry or the self-regulated HEIs were solely responsible for approving new programs. These reflect the lack of clarity in the quality assurance sector at this stage in Bangladesh to regulate engineering education effectively.

The external quality assurance review panel membership was found to be composed of professionals from varied background, and along with the *community engagement strategy*, *staff qualifications* was found to be one of the most problematic areas for quality assurance. To improve the situation of quality

Table 4. Problematic areas identified during external quality reviews

Sl. No.	Problematic Areas	Mean Value by the Respondents (Closer to '1' is 'More Problematic')
1	Community engagement strategy	2.00
2	Staff qualifications	2.00
3	Research strategy	2.67
4	Practical training	4.00
5	QA management systems and structures	4.00
6	Connection between teaching and research	4.33
7	Teaching and learning strategy and methods	4.50
8	Facilities and resources	5.25
9	Assessment strategy	6.33
10	Partnership with other HEIs	6.50
11	Internationalization/international partnerships	6.75
12	Program and curriculum benchmarking	7.00
13	Program intended learning outcomes	7.50
14	Employability of graduates	8.50
15	Admission entry requirements	9.50
16	Student experience	10.00

assurance in Bangladesh, the financial and operational aspects were prioritized by the respondents. Equally prioritized was the benchmarking of the programs. This reveals the need for standardizing engineering programs and ensuring an effective governance framework for the operation of the quality assurance cells.

There is a long way to go for the quality assurance setup in Bangladesh to mature and be more effective. However, the findings through studies similar to this one will be crucial to gain insight from the experience of the professionals involved to build upon the existing quality assurance setup for engineering education.

ACKNOWLEDGMENT

The study has been undertaken in the framework of the project ENHANCE - ENabling Humanitarian Attributes for Nurturing Community-based Engineering (project No: 598502-EPP-1-2018-1-UK-EP-PKA2-CBHE-JP (2018-2582/001-001), funded by the Erasmus + KA2 Cooperation for innovation and the exchange of good practices – Capacity building in the field of Higher Education. This publication reflects the views only of the authors, and the Commission cannot be held responsible for any use which may be made of the information contained therein.

Figure 7. Most needed capacities for quality assurance

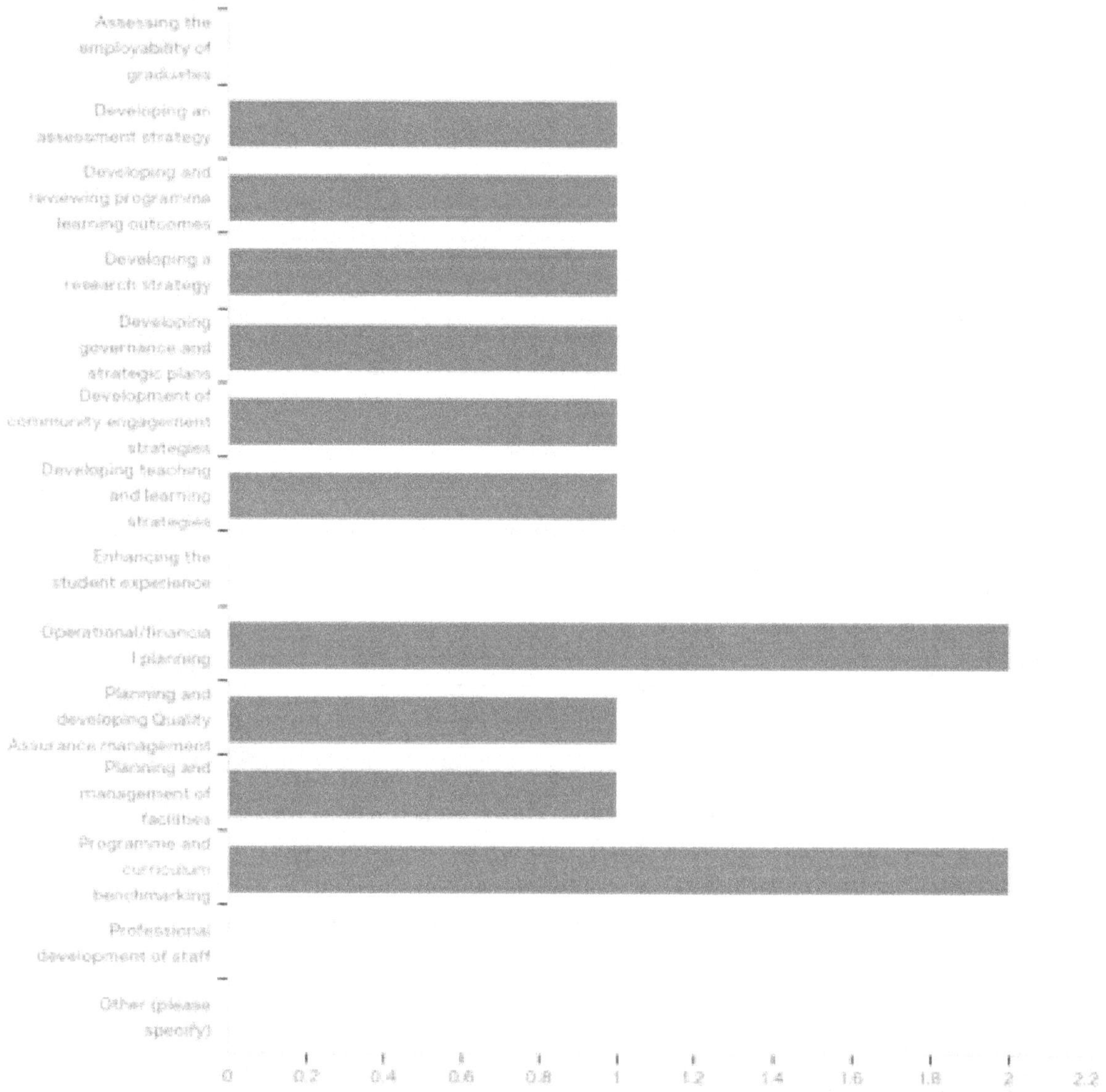

REFERENCES

Ahmed, A., Haque, M. M., Rahman, A., & Rahman, A. (2016). *Engineering Education in Bangladesh: Implementation of Online Delivery Approach. In Proceedings of International Conference on Engineering Education and Research*. Western Sydney University. http://handle.westernsydney.edu.au:8081/1959.7/uws:39681

Alam, F., Sarkar, R., La Brooy, R., & Chowdhury, H. (2016). Engineering Education in 21st Century. *AIP Conference Proceedings*, *1754*, 020002. doi:10.1063/1.4958344

Alam, G. M. (2020). Quality Assurance for Private Universities in Bangladesh: A Quest for Specialised Institutional Governance, Management and Regulatory Mechanism. *International Journal of Comparative Education and Development*, *22*(1), 1–15. doi:10.1108/IJCED-01-2019-0006

Ali. (2021). Quality Assurance in Engineering Education: Accreditation and Its Global Influence. *Proceedings of the 11th Annual International Conference on Industrial Engineering and Operations Management.*

Chowdhury, H., Alam, F., Biswas, S. K., Islam, M. T., & Islam, A. K. M. S. (2013). Quality Assurance and Accreditation of Engineering Education in Bangladesh. *Procedia Engineering*, *56*, 864–869. doi:10.1016/j.proeng.2013.03.208

Daun, M., Salmon, A., Weyer, T., Pohl, K., & Tenbergen, B. (2016). Project-Based Learning with Examples from Industry in University Courses: An Experience Report from an Undergraduate Requirements Engineering Course. *CSEET IEEE 29th International Conference.* 10.1109/CSEET.2016.15

Elassy, N. (2013). A Model of Student Involvement in the Quality Assurance System at Institutional Level. *Quality Assurance in Education*, *21*(2), 162–198. doi:10.1108/09684881311310692

Elassy, N. (2015). The Concepts of Quality, Quality Assurance and Quality Enhancement. *Quality Assurance in Education*, *23*(3), 250–261. doi:10.1108/QAE-11-2012-0046

Faizi, J., & Umar, M. S. (2021). A Conceptual Framework for Software Engineering Education: Project Based Learning Approach Integrated with Industrial Collaboration. *International Journal of Education and Management Engineering*, *11*(5), 46–53. doi:10.5815/ijeme.2021.05.05

Hadgraft, R. G., & Kolmos, A. (2020). Emerging Learning Environments in Engineering Education. *Australasian Journal of Engineering Education*, *25*(1), 3–16. doi:10.1080/22054952.2020.1713522

Li, M., & Lei, Q. (2015). On Quality Assurance Mechanisms in Engineering Education: A Case Study of Purdue University. In *International Conference on Social Science, Education Management and Sports Education* (pp. 494-497). Atlantis Press. 10.2991semse-15.2015.126

Manzoor, A., Aziz, H., Jahanzaib, M., Wasim, A., & Hussain, S. (2017). Transformational Model for Engineering Education from Co0ntent-based to Outcome-based Education. *International Journal of Continuing Engineering Education and Life Long Learning*, *27*(4), 266-286.

Milovanovic, J., Shealy, T., & Godwin, A. (2022). Senior Engineering Students in the USA Carry Misconceptions about Climate Change: Implication for Engineering Education. *Journal of Cleaner Production*, *345*, 131129. doi:10.1016/j.jclepro.2022.131129

Rahnuma, N. (2020). The Bangladeshi Higher Education Quality Assurance Framework: A Pathway for Transformation. *Quality in Higher Education*, *26*(1), 14–31. doi:10.1080/13538322.2020.1729309

Smith, J., Tran, A. L. H., & Compston, P. (2019). Review of Humanitarian Action and Development Engineering Programmes. *European Journal of Engineering Education*, *45*(2), 249–272. doi:10.1080/03043797.2019.1623179

Chapter 14
Humanitarian and Soft Engineering for Disaster Mitigation in Coastal Bangladesh

Syed Labib Ul Islam
https://orcid.org/0000-0002-5113-6122
University of Birmingham, UK

G. M. Tarekul Islam
Bangladesh University of Engineering and Technology, Bangladesh

ABSTRACT

Developing countries like Bangladesh are facing multidimensional challenges due to climate change. The traditional engineering sector needs to evolve to contribute to the solution of the emerging problems by incorporating social and technical solutions. Integration of humanitarian and soft engineering and proper project formulation and implementation can go a long way in addressing the challenges. Harmonised and clear policies to this end are of high importance, along with political will and accountability in the public investments. Efficient utilisation of resources are needed through objective and proper prioritisation of projects – free from external influence or interference. Geopolitical cooperation between countries, coordination between the government and non-government sectors, as well as within the departments of the government is crucial for the success of humanitarian and soft engineering in resource-constrained countries.

INTRODUCTION

Humanitarian Engineering is based on flexible context-based solutions, which is of crucial importance for adaptation to climate change (Johnson, 2021). The inclusion of Humanitarian Engineering elements into various sustainability, disaster mitigation, post-disaster recovery, and poverty alleviation projects

DOI: 10.4018/978-1-6684-5619-4.ch014

has demonstrated success in the past few decades (Younger, 2018). This requires interdisciplinary knowledge, but there is a significant shortage of technical and soft skills in the professional sectors at present (White et al., 2021).

After the World Wars, there was a need for rapid recovery of the damaged infrastructure. This prompted the use of concrete and steel at a massive scale (Ghafoori & Dutta, 1995). Consequently, the use of flexible nature-based solutions or soft engineering techniques was often overlooked. The unsustainability of exclusively hard engineering measures came to the attention of the wider society later on – particularly with the onset of the impacts of climate change. The engineering education, practices, as well as policies had largely not been specifically designed to counter the issues such as climate change and sustainability. This significantly contributed to the present situation. Facing increasing climate change-induced natural disasters, Humanitarian Engineering issues have become more important than ever for developing disaster resilience across the society to avoid a humanitarian crisis in the post-disaster periods.

In this chapter, the authors review the soft engineering approaches concerning Humanitarian Engineering and the practicability of Humanitarian Engineering in resource-constrained developing countries such as Bangladesh. The purpose of this chapter is to collate and interpret the recent works in this field concerning developing countries and climate change. The authors also aim to share their understanding of the present situation from relevant professional experience. Finally, in this chapter, the authors aim to provide recommendations for further development of Humanitarian and Soft Engineering for disaster mitigation in the coastal area as well as for overall development in the country.

BACKGROUND

Climate change has gradually been creating a paradigm shift in the approach of development and engineering as the world becomes more aware of the importance of sustainability and adaptation. The engineering sector at the academic and professional level needs to evolve and tackle the emerging challenges by incorporating sustainable strategies. The gap between technical and social elements needs to be reduced and paying closer attention to the social elements influenced by engineering projects is essential to address both social and engineering concerns.

Humanitarian Engineering provides a unique premise for assuring social justice through engineering works (Leydens & Lucena, 2014). It promotes socio-technical thinking, as the humanitarian engineering projects are inherently interdisciplinary – working at the interface of efficient engineering and sustainable development of the community as a whole (Mazzurco & Daniel, 2020). Humanitarian engineering education and relevant projects have been on the rise in the past few decades to impart social benefits from engineering projects (Smith et al., 2017). Engineering education and ethics play a crucial part in equipping students with sufficient technical and social skills to practice Humanitarian Engineering effectively in the professional sectors (Martin et al., 2021).

The water sector is prime with opportunities for Humanitarian Engineering interventions, as access to safe usable water is becoming a precarious issue in many developing countries due to the increasing impacts of climate change. This is a focus of Sustainable Development Goal 6, aiming to provide a safe water supply and sanitation to everyone. Lack of safe water has been spreading water-borne diseases especially in the economically vulnerable classes of society. Salinity intrusion is contaminating the groundwater resources – especially in the coastal areas of developing countries such as Bangladesh (Rakib et al., 2019; Chen & Mueller, 2018). Due to the complexity prevalent in such issues, humanitarian engineering needs to address social, economic, and political dimensions of the problems if a sustainable solution is to be offered (Ali et al., 2008). Engineers Without Borders have been working on several projects in the water and sanitation sector of Peru, Panama, and Nicaragua (Montgomery et al., 2020; Budny, 2018; Walters et al., 2017; Miller, 2017). The Peru and Panama projects focused on water treatment and reliable water distribution systems along with educating the community. The Peru Project focused on rainwater harvesting, along with a simple purification process to supply safe water to the community for daily consumption. The arsenic education program prompted community engagement in conjunction with the involvement of the local authority. The alternative water source relieved the community from the reliance on arsenic-contaminated groundwater, and the community engagement raised awareness about the maintenance of the water distribution system and local water resources (Engineers Without Borders, 2021a). Similarly, in the Panama Project, the focus was also on water storage and supply to reduce the spread of waterborne diseases. Water distribution networks with communal taps demonstrated lower coliform counts compared to the traditional method of water storage. The project collaborated with the community to build up awareness and educate the community to equip them with the knowledge required to manage the water system on their own (Engineers Without Borders, 2021b). Subsequently, in Nicaragua, they are designing projects for sanitation and water supply considering the long-term sustainability issues as the local community currently has to walk long distances to collect water and the sanitary hygiene system is outdated and vulnerable to floods (Engineers Without Borders, 2021c; Engineers Without Borders, 2021d). Projects in Kenya have addressed the water supply issues in the Lake Bogoria Basin where the Nyamilu community water project secured the availability of safe drinking water and clean water for household consumption (John, 2013). The project had large-scale community engagement to overcome various engineering challenges, and the piped water supply along with an increased number of boreholes has contributed to minimizing the waterborne diseases that previously were rampant among the community. This paves the way for sustainable development in the community.

The importance of ecosystem services is increasingly being recognized with the Asian and South Asian region being a hotspot for ecosystem restoration and reservation. This can prove to be an important tool for Humanitarian Engineering due to the interdependency between the ecosystem and local human communities. These solutions are often recognized by the public due to the increasing disaster risks in the backdrop of climate change, and they offer multiple benefits to society (Bhardwaj and Gupta, 2021). Himalayan ecosystem goods and services are crucial for the survival of the upstream and downstream populations, where soft intervention strategies such as ecological restoration and conservation of forests are the only viable options (Bhatt et al., 2020). This approach along with community engagement has proven to be essential for underpinning a wide range of climate change related issues such as the management of wetlands, forests, water, soil, and livelihoods along with physical infrastructure planning (Pulhin & Tapia, 2021; Tenzin & Norbu, 2020). There is a need to increase the institutional capacities and further evidence-based research in this regard, but from the available literature on this issue, both rural and urban landscapes can greatly benefit from the ecosystem or nature-based solutions (Frantzeskaki et al., 2019).

Innovative non-structural interventions have already proven to be effective in disaster-prone developing countries like Bangladesh (Schismenos et al., 2021a; Schismenos et al., 2021b; Azad et al., 2020). Community-based and ecosystem-based adaptation techniques along with early warning systems have been identified as the key components for climate resilience (Filho et al., 2018). These measures improve the social situation by helping to alleviate poverty in addition to performing the engineering functions. The success of such projects depends on interdisciplinary involvement and community participation. Besides, studies have found that practical involvement with the fieldwork of such projects can encourage students to cross the disciplinary boundaries, and impart them with the necessary technical and soft skills to prepare them for professional life (Volger et al., 2018).

HUMANITARIAN AND SOFT ENGINEERING

The Constraint of Resources

The constraint of resources has been an ever-present factor throughout human history. Every country has to consider the demand and limitation of resources. No matter how good the intention is, any initiative can be stunted when it comes to the question of resources. This is particularly true for developing countries. The constraint of resources comes in many different forms. There are obvious constraints in finances and natural resources. There are also constraints in land and sufficiently skilled manpower. The political visions and intentions of a country, along with transparency and accountability often dictate resource-related issues. This holds not only at a macro level formulating national policies to develop and utilise resources, but also at a human level down to every single stakeholder. The effect of this is wide-ranging and has precipitous long-term impacts. The policy and governance issues will ultimately determine the success or failure of Humanitarian Engineering and Soft Engineering initiatives.

Humanitarian Engineering Perspective

The developing countries face a wide range of problems, which often have a contradictory nature. For example, a socioeconomically vulnerable but agriculturally highly productive region may benefit from a new infrastructure project. However, that same project may take up a large chunk of the agricultural land, which can create a significant shortfall in the expected agricultural yield from that region creating a risk in national food security. Finding sustainable and balanced solutions to the problems is often highly complicated for policymakers. The tradeoffs to be made are highly sensitive and finding a consensus is difficult, to say the least.

The majority of the poverty-stricken people are living in developing countries (Mamun & Ullah, 2020). Breaking the cycle of poverty can often seem impossible for people living in dire situations, with limited access to education, healthcare, social and financial security, and basic amenities. Climate change has been exacerbating this situation with unpredictable extreme weather events. Lack of safe water availability due to extreme droughts and extreme floods have become commonplace in recent decades with increasing frequency and intensity (Dodd et al., 2020; Li & Zha, 2020; Maxwell et al., 2019). In this reality, ensuring reliable and safe water supply, transportation network, and food relief are the major concerns in the aftermath of natural disasters where Humanitarian Engineering can play a prominent role.

Natural disasters can drastically alter the priorities of the affected people in the immediate post-disaster recovery period and long term. Understanding these parameters is of paramount importance for the practicing engineers who are tasked with designing and implementing development projects. Foresight is required for this purpose so that the humanitarian issues related to the project are a priority – not an afterthought. Reactionary actions simply lead to repeated expensive recovery programs if humanitarian elements are not integrated with the rebuilding effort.

The interface between social and political elements, and the technical or engineering elements must overlap for sustainable development. The engineering education system has to evolve to account for the emerging challenges in developing countries. Breaking away from rigid curriculum teaching with an inherently unbalanced focus on hard engineering measures while taking a naïve approach to social, economic, and political parameters is a necessity. There is a significant shortage of skills to tackle these issues in the professional engineering as well as engineering education sector, and the development of engineers with the necessary skillsets needs to be an immediate priority.

Soft Engineering Perspective

Traditional engineering in a sense was soft engineering with a focus on nature-based solutions. However, the necessity of rapid infrastructure development and economic growth, as well as the rapid rebuilding of countries after endless cycles of wars, created the unbalanced focus on hard engineering measures that appear to provide a sense of security. In the process of doing so, several important aspects appeared to have been overlooked. The most glaring of these overlooked aspects is the environment. In addition, this drove the inequitable development efforts where the countries with limited or depleted resources had to either lag or use unsustainable means in an attempt to catch up with the other wealthier nations. However, the actual costs of pursuing solely hard engineering measures became clearer with the onset of climate change impacts.

As natural disasters increase, the scale of the damage becomes clearer. There have been many studies in this matter, demonstrating that the inevitable failure of hard engineering measures leads to magnified economic damage notwithstanding the loss of lives (Badola & Hussain, 2005). In this situation, there have been success stories for shifting the focus back onto the flexible; nature-based soft engineering approaches around the world (Mishra et al., 2021; Akber et al., 2018; Rizal et al., 2018). This presents an opportunity for the developing countries to integrate the Soft Engineering and Humanitarian Engineering approaches.

Traditionally the hard engineering structures are expensive to build. Structures are designed with a limit where the structure will fail. For example, a bank protection structure may fail if the water level rises to a certain point. These limiting points are being reached with unusual frequency due to the extreme weather events driven by climate change. On top of that, rigorous maintenance is required for ensuring the integrity of the structure throughout its life. This is one of the main reasons behind the fragile infrastructure in many developing countries. The constraint of resources in money and appropriately skilled manpower mean maintenance of infrastructure is highly complicated for an economically vulnerable country. If the maintenance is neglected, it creates a multidimensional problem. Often, the public has to rely on the infrastructure no matter what if it is their only option. In addition, that infrastructure can create a false sense of security. To use the example of the bank protection structure, if an embankment is not properly designed, constructed, and maintained, it cannot last to its expected structural life. Yet, the people living near the embankment may feel safe to live and invest in that area as they please. So naturally, if an extreme weather event causes the embankment to fail unexpectedly, then that causes damage to a much greater scale which might have been avoided with careful planning, quality design and construction, and proper maintenance. Moreover, for countries such as Bangladesh facing large floods virtually every year, the development efforts become paralyzed falling into a cycle of floods and recovery efforts in the vulnerable areas. The coastal areas of Bangladesh are particularly vulnerable to cyclones and storm surges. While the advancement of the early warning system has done wonders in saving lives and minimizing the economic losses to a degree, the dependency on hard engineering structures remains, with the construction of higher embankments every time an embankment fails.

Nature-based solutions and adaptation measures can offer a sustainable way forward in such situations for the affected countries. Nature-based solutions have demonstrated success in stabilizing riverbanks, reducing erosions, attenuating water levels and current speeds (Lee et al., 2021; Maza et al., 2019; Montgomery et al., 2019). Development of such techniques and increasing the efficiency of nature-based solutions should be prioritized for the resource-constrained countries because such measures require a less initial investment, and less maintenance compared to the concrete and steel-based structures. When natural solutions are used in conjunction with the hard engineering measures, they can reduce financial stress, create an efficient management structure, and provide enhanced resilience to the community in many aspects. Livelihood diversification and creation of alternative livelihood opportunities, as well as extraction and development of natural resources in a sustainable manner, come naturally with soft engineering measures.

This concept goes hand in hand with Humanitarian Engineering, as the benefits of the engineering projects equitably reach the community, and create socioeconomic security and resilience to natural disasters (Badola et al., 2012; Hussain & Badola, 2010). Adopting a strategy to integrate flexible engineering solutions for the emerging problems, and promoting necessary adaptation measures has the potential to induce sustainable development in the developing world.

Integration of Humanitarian and Soft Engineering

A brief example for the integration of humanitarian and soft engineering measures has been presented in this section based on a recent study conducted by the authors to assess the potential of soft engineering measures in mitigating cyclonic effects in coastal Bangladesh. In their study, the authors found that the existing mangroves and coastal forests – even when fragmented or sparse, play a significant role in attenuating storm surges and cyclonic winds compared to locations without coastal forests. The Normalized Difference Vegetation Index (NDVI) analysis was carried out on historical satellite images of coastal Bangladesh between 1990-2020 to assess coastal vegetation evolution over the years. This analysis showed the central coastal region to historically have a minimal buffer in the form of coastal vegetation compared to the western and eastern coastal regions. Figure 1 shows the result of the NDVI analysis for the coastal zone of Bangladesh using the satellite data from 1990.

The data from the following decades showed similar trends as illustrated in Figure 2. Figure 3 shows the NDVI analysis for 2020 where it can be seen that the vegetation extent and density have increased. This increase is mainly due to several afforestation initiatives of the government in the coastal area such as the "Community Based Adaptation to Climate Change through Coastal Afforestation" (CBACC-CF) project, following the National Adaptation Program of Action (NAPA) (Ahammad et al, 2013).

Subsequently, historical wind speed and storm surge data for individual cyclonic events were analyzed. The wind speed and water level stations were chosen based on the NDVI analysis. The study revealed that the wind speed and water level gauge stations around coastal mangrove cover consistently experienced less intense cyclonic effects compared to the stations that are virtually exposed to the coast. The reduced wind attack in the case of plantation along the embankment has also been validated in previous studies in coastal Bangladesh (Kader et al., 2019). Table 1 shows the data for wind speeds, where the Mongla station (near thick mangrove cover) generally experienced lower wind speeds than the other stations which had moderate or minimal covers – depending on the landfall locations and severity of the cyclones.

Similar results were observed for storm surge levels as well. Table 2 shows the data for the recorded water levels at different stations for a cyclone making landfall on 18 December 1990. The Water Level

Figure 1. NDVI of the coastal zone of Bangladesh for 1990

Stations SW 26, SW 29, SW 244, and SW 39 experienced lower storm surges as they were located around coastal forests. The remaining stations faced much higher water levels due to their more direct exposure to the Bay of Bengal. This scenario held for other cyclonic events in the following decades. The result of this study was also validated by primary data in the form of 40 semi-structured interviews with coastal residents, confirming the effectiveness of coastal buffers and soft engineering in disaster mitigation in the context of cyclones in coastal Bangladesh – even when the forests are not dense or continuous. The responses from the interviews were largely in agreement about the effectiveness of the coastal mangroves. Even the less dense and elongated strip-shaped mangroves were acknowledged to have a noticeable impact in retarding the cyclonic effects in the experience of the respondents. Some responses varied based on the regions of the coastal zone – for example, responses from the central coastal zone highlighted the need for more structural interventions due to being in direct interface with the Bay of Bengal, while the respondents from the eastern coastal zone highlighted the need to preserve the mangroves as well as the need to manage the land stress due to high density of population in the commercial hubs like the port city Chattogram. The economic significance of coastal forests in the livelihoods of local people was also highlighted. These responses underline the findings of the quantitative analysis of the study and reiterate the need to formulate and implement projects in a holistic way – with an emphasis on the

Figure 2. NDVI of the coastal zone of Bangladesh for 2015

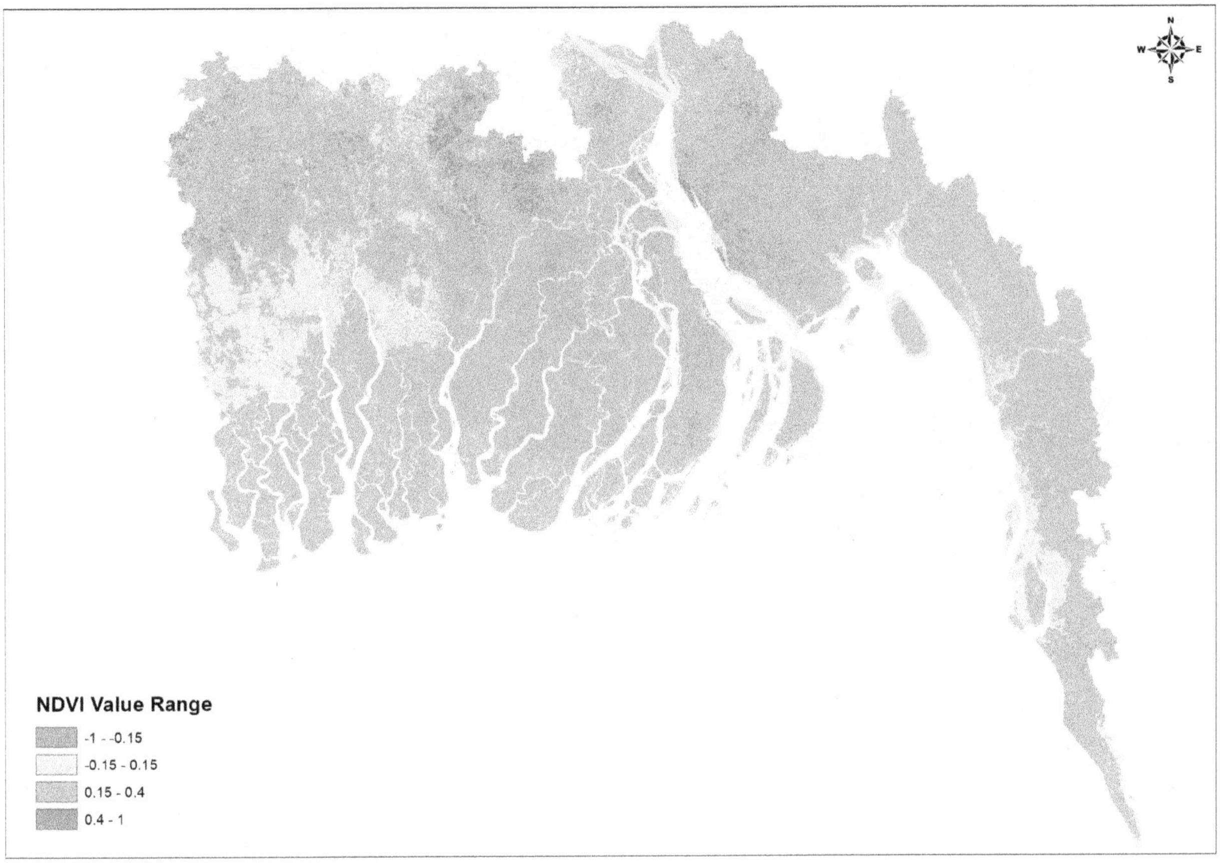

integration of soft and hard engineering measures. The detailed primary and secondary data analyses are available on request.

This provides a suitable basis for the integration of Humanitarian and Soft Engineering in disaster mitigation in coastal regions. Such integration can result in the rapid formulation and implementation of nature-based interventions, which contribute to combating climate change and its impacts as necessary for achieving the Sustainable Development Goal 13. The coastal population around the world is often vulnerable to socioeconomic and natural disaster issues (Rabby et al., 2019; Rudiarto et al., 2018). From literature review and experience, the major problems in coastal Bangladesh are socioeconomic security and natural disasters. The socioeconomic vulnerabilities arise from the livelihood practices of the region. Agriculture and fisheries are the predominant livelihood modes of coastal Bangladesh. However, due to the increasing salinity, sea-level rise, lack of freshwater availability, etc., the yield of agriculture is falling in many coastal districts (Hasan & Kumar, 2021), and the availability of fish is also reducing (Islam & Hoq, 2019). This leads to reduced income for the traditional farmers and fishers who often have to migrate or switch to aquaculture, day labor, etc. Aquaculture is often controlled by large businessmen and it is a politically influenced sector (Hoque et al., 2017) where the poorer farmers or fishers can be vulnerable to exploitation if they have no other alternatives. The tropical cyclones are

Figure 3. NDVI of the coastal zone of Bangladesh for 2020

almost annual events for coastal Bangladesh where storm surges cause havoc in the coastal area, which leaves the aquaculture vulnerable as well.

In recent years there have been significant improvements in this regard, with a number of initiatives taken by the local government authorities and other departments of the government. Non-structural

Table 1. Recorded maximum daily wind speeds at different stations in Bangladesh for different cyclones

Year and Date of Cyclonic Events		18 Dec 1990	25 Nov 1995	27 Sep 1997	21 May 2016	30 May 2017
Gauge Station Location	**Mangrove Cover**	**Wind Speed (knots) 1 knot = 1.858 km/hr**				
Mongla	Thick	12	20	15	12	5
Khepupara	Minimal	16	27	22	20	8
Hatiya	Minimal	20	18	18	25	20
Sandwip	Minimal	25	28	43	24	30
Chittagong	Moderate	15	25	47	26	45
Kutubdia	Moderate	8	17	62	29	18
Teknaf	Moderate	20	35	20	22	63

Table 2. Water Levels at different stations in Bangladesh for cyclone in 1990

Cyclonic Event	Water Level Station ID	Water Level (m)									
		17.12.90		18.12.90		19.12.90		20.12.90		21.12.90	
		High Tide	Low Tide	High Tide	Low Tide	High Tide	Low Tide	High Tide	Low Tide	High Tide	Low Tide
Cyclonic Storm (18.12.90) Landfall: Cox's Bazar	SW26	1.32	-1.70	1.54	-1.61	1.50	-1.64	1.36	-1.59	1.40	-1.56
	SW29	2.09	-0.57	2.17	-0.81	2.28	-0.63	2.19	-0.57	2.14	-0.65
	SW244	1.50	-0.80	1.52	-0.80	1.68	-0.90	1.70	-0.90	1.60	-0.85
	SW39	1.71	-0.58	1.79	-0.61	1.90	-0.60	1.81	-0.50	1.80	-0.20
	SW278	1.05	-0.48	1.37	-0.58	1.35	-0.52	1.42	-0.57	1.53	-0.53
	SW222	3.49	3.49	3.64	3.49	3.67	3.65	3.67	3.64	3.66	3.64
	SW121	1.90	-1.10	2.20	-1.00	2.30	-0.90	2.20	-1.00	2.10	-1.10
	SW250	2.16	-0.24	2.38	-0.08	2.16	-0.22	2.14	-0.48	2.10	-0.49
	SW200	2.86	-1.74	2.56	-1.64	2.36	-1.54	2.06	-1.44	1.76	-1.24

techniques in Bangladesh have enabled rapid adaptation to the dynamic river basins, socioeconomic conditions, and climatic scenarios, and are consistent with the purpose of ecologically sound and sustainable development (Shah, Rahman & Chowdhury, 2018). Bangladesh Water Development Board (BWDB) has constructed a network of hydrological stations linked to the Flood Forecasting & Warning Centre (FFWC) to develop and transmit flood forecast and warning information (Baten, González & Delgado, 2018). Coastal Embankment Rehabilitation Project (CERP), Coastal Embankment Improvement Project (CEIP), "Blue Gold" project, The Water Management Improvement Project (WMIP), etc. successfully combined participatory management with economic and market growth. Local communities were engaged in these projects, from identification through monitoring and assessment (Gain, Mondal & Rahman, 2017).

However, there is still a long way to go as community involvement is not yet embedded in many of the decision-making processes that come from a top-down approach. Often traditional knowledge is not taken seriously, while unplanned urbanization exacerbates emerging climate change related problems. In this situation, if flexible engineering solutions such as Dynamic Cobble Berm Revetment (Bayle et al., 2020), coastal green belt using mangroves, and structural measures such as sea walls and other structures are planned in an integrated way, then several benefits could be attained. Green belt and sea walls can act as two distinct lines of protection against storm surges. The use of mangroves can reduce the storm surge water levels and the wave attack speed. This would result in lower height requirements for the sea wall, decreasing the amount of investment required for the coastal defense system. At the same time, the mangroves and other afforested zones can contribute significantly to the ecosystem services as well as generate alternative income sources for the coastal population. A study in the Solomon Islands has revealed that the minimum annual value gained from mangroves ranges between USD 345 to 1501 per rural household (Warren-Rhodes et al., 2011). Such additional annual income can provide economically vulnerable coastal households a much-needed relief and assist in increasing their economic security. The increased economic security can create an environment for promoting education, creating an overall development in the societal situation, and equipping the coastal residents with the means to innovate, secure, and develop resilience against disasters – leading to equitable and sustainable development.

FORMULATION AND IMPLEMENTATION OF PROJECTS

Following on from the example presented in the previous section, it is undeniable that there is a multitude of other considerations to make while formulating an engineering project. This underlines the necessity of understanding the current situation at the project site, the necessity of the affected people, and the anticipatory foresight or long-term vision to ensure the project can be sustainable. Infrastructure development takes up a large chunk of the national budgets of the developing countries, and loans from different donor agencies are utilized for implementing various projects throughout the year. However, the taxpayers ultimately finance every project or initiative that is adopted in a country. As such, governance and accountability are of paramount importance when such large investments are made on behalf of the people of a resource-constrained country.

Objective and appropriate prioritization of projects is essential for sustainable development. The authorities responsible for planning thus play the most important role in this regard. Coordination between different ministries and departments is required to create a harmonized development plan that can truly be sustainable. The elements of Humanitarian Engineering have to be present at this formation stage, which can help in identifying the projects that need immediate attention and investment. Following this identification and prioritization, Humanitarian Engineering needs to be reflected in the project design phase as well.

However, all of this remains underpinned by the motivation of the authority behind the formulation and implementation of the projects, and the stakeholders whose inputs form the basis of the project design. If the motivation behind a project overlooks objectivity and sustainability in favor of an intentional or unintentional political or personal gain, then that project can hardly bring equitable benefit to the perceived project beneficiaries. Often projects get distorted and influenced by geopolitical considerations and the influence of local politics. Short-term political gains often get the priority instead of sustainable development, nevertheless, the gross investment costs still have to be borne by the general public. Such a situation in the project development phase must be avoided to maximize the achievable social, economic, and environmental benefits from Humanitarian and Soft Engineering.

Subsequently, the implementation of projects requires the involvement of different stakeholders throughout the project life. The developing countries face major challenges in the implementation of engineering projects due to the issues of quality control during the construction phase. The Humanitarian and Soft Engineering elements introduced during the project formulation also need to be properly implemented in this phase to reap the benefits. Mismanagement in the implementation phase leads to poor quality construction, the subpar performance of projects, and early failure to achieve the project objectives – representing a loss in the investment made for the project. In such scenarios, the vulnerable people are the biggest victims, being deprived of the anticipated equitable opportunities from the project. This highlights the importance of proper formulation and implementation of projects from a Humanitarian Engineering standpoint.

CHALLENGES

The major challenges for the developing countries in regards to Humanitarian and Soft Engineering can be identified as given in Table 3 based on the discussions presented in the previous sections.

Table 3. Challenges to humanitarian and soft engineering

Sl No.	Key Challenges	Descriptions
1	Transformation of Education System	Transformation of the traditional education system is particularly difficult for resource-constrained countries due to various technical and socioeconomic reasons. This change would require a change in the mindset of both the teachers and the students.
2	Skilled Professionals and Teachers	Building up sufficiently skilled teachers and trainers from the existing professionals is challenging in the absence of close cooperation between national and international stakeholders.
3	Awareness	Building awareness across society about these issues is highly important. Moreover, these awareness campaigns will need to have career-oriented messages to inform and motivate the students, to prompt them to think about the Humanitarian and Soft Engineering elements, as well as social justice, ethics, and sustainable development.
4	Good Governance	Ensuring good governance is crucial, and incredibly challenging for resource-constrained countries due to the prevalent socio-economic situations. This gets even more difficult if there is a presence of significant inequality in the economy and power structure of a country.
5	Accountability and Vision	Ensuring accountability for public affairs and clarity of political vision is challenging due to the power dynamics at the local, national, and international levels. This is particularly important to facilitate public trust in the system and get over the frustrations, helplessness, and exasperation of the public living around or below the poverty line.
6	Coherent Development Policies	Formulating a coherent nationwide development policy is a big challenge in the absence of close coordination between the different ministries and departments of a country. Improper coordination can often lead to conflicting visions of sustainable progress.
7	Transparency and Objectivity	Maintaining transparency and objectivity in the project formulation, design, and implementation phase is a huge challenge for developing countries. Keeping the feasibility studies of projects free from influence and interference is of high importance to ensure the efficient utilisation of public investments.
8	Science-informed Policy Decisions	Keeping the policy level decisions objective and science-informed is essential, yet challenging for developing countries due to short political cycles. Cooperation and transparency need to be maintained with all the stakeholders to create a harmonised vision for sustainable development and poverty alleviation in developing countries.
9	International Cooperation and Technology Transfer	Strengthening international cooperation and technology transfer should be prioritised in the context of Humanitarian and Soft Engineering – which can enable developing countries to make meaningful advances towards sustainable development. This is often challenging due to various geopolitical implications.

SOLUTIONS AND RECOMMENDATIONS

Based on the discussions in the preceding section and the challenges, the following solutions and recommendations have been presented.

1. The natural disasters in the coastal zone of Bangladesh need to be managed with the integration of soft and hard engineering measures. The existing natural and artificial mangroves need to be conserved in conjunction with strategic decisions of investments in science-informed, effective structural measures to attain significant disaster mitigation services from the combined coastal buffer.
2. Some of the recent initiatives with participatory project formulation and implementation have been a success, but this needs to be expanded to all the projects – particularly in the climate-sensitive coastal areas. Humanitarian Engineering elements need to be embedded in the related policy decisions.

3. Modernization of the education system is necessary to equip the professionals with the necessary hard and soft skills to address the requirements of Humanitarian and Soft Engineering. Students need to be familiarized with the practical scenarios and introduced to the emerging problems to help them prepare for meeting the challenges when they enter their professions.
4. Developing efficient internship programs attaching students with ongoing projects intensively for a while under the guidance of skilled professionals can be an option for resource-constrained countries in this regard.
5. International collaboration between universities and other institutions needs to be increased to share the knowledge and experiences of Humanitarian and Soft Engineering in the context of developing and developed countries.
6. Awareness building can rapidly create inherent social resilience, which in turn can facilitate the success of engineering projects in developing countries like Bangladesh.
7. Good governance needs to be ensured to rebuild the confidence in the authorities. Public participation and cooperation are essential for engineering projects, and transparency needs to be maintained.
8. Humanitarian Engineering as a concept should be integrated with every development project, expanding from aid and disaster prevention issues. Ingraining the elements of Humanitarian Engineering to all projects can ensure inherent disaster resilience and poverty alleviation. For counties like Bangladesh, this can be an effective tool to manage natural disasters in the coastal region.
9. Objectivity and science-informed decisions need to be made during the project formulation, design, and implementation phases. This will ensure the efficient use of the limited resources and can speed up the development process in a sustainable way, reducing the need for frequent massive reinvestments in the post-disaster recovery and rebuilding phase.
10. Clear and coordinated long-term visions should be in place, which will not be possible without developing a concentrated effort in coordination among different ministries and departments. These visions should be guided by the present situation, and the emerging problems, with a focus on the need of the public where Humanitarian Engineering can contribute significantly.

FUTURE RESEARCH DIRECTIONS

Future research directions should focus on the potentials of educational transformation issues in the socioeconomic and geopolitical context of resource-constrained countries. Any initiative of Humanitarian and Soft Engineering will find it difficult to utilize the expected benefits to the society and environment fully if a new generation of professionals with the suitable skillsets is not built up from the students. The governance and accountability issues should also be prioritized for research, in the light of the political landscape of the developing countries. Such research programs and policy level studies should involve the stakeholders from government and non-government sectors to explore the possibilities of Humanitarian and Soft Engineering in resource-constrained countries. The application of soft engineering projects in the coastal regions also needs to be investigated further from an economic point of view. In addition, land use related studies should be carried out to illustrate the impact of land use change on cyclones and other natural disasters.

CONCLUSION

This chapter has presented a comprehensive picture of the challenges faced by developing countries like Bangladesh. The resource-constrained countries are already going through climate change-induced crises in every sector. Food security, public health, and infrastructure are threatened by natural disasters and the most affected people are the socioeconomically vulnerable groups in developing countries. The engineering sector has to tackle these issues by incorporating the humanitarian and social aspects in engineering education and infrastructure development projects.

Coastal Bangladesh is an ideal example where Humanitarian and Soft Engineering can efficiently tackle the effects of natural disasters such as cyclones and storm surges. The authors showed the effectiveness of the sparse coastal vegetation in mitigating the cyclonic winds and water levels in coastal Bangladesh. The thick mangrove cover in the western coastal zone and the less dense coastal forests in the eastern coastal region consistently experienced significant attenuation in the cyclonic effects compared to the more exposed central coastal zone. The forests also provided more socioeconomic resilience to the local people who were very aware of the protective nature of their forests.

Climate change adaptation can only be possible through a united effort across society. These efforts have to include social awareness, good governance, political will, and accountability of public investments. Engineers and policymakers also have to adopt nature-based solutions where applicable. Flexible solutions for disaster mitigation in the coastal regions need to be studied and implemented with a long-term focus. Ministries and other organs of the government have to act in a coordinated way under clear policy guidance to make effective use of the limited resources. The private sectors also have to work towards this goal. The integration of Humanitarian and Soft Engineering has significant potential to create social resilience while performing engineering functions in resource-constrained countries, which is desperately needed for adapting to climate change.

REFERENCES

Ahammad, R., Nandy, P., & Husnain, P. (2013). Unlocking Ecosystem Based Adaptation Opportunities in Coastal Bangladesh. *Journal of Coastal Conservation, 17*(4), 833–840. doi:10.100711852-013-0284-x

Akber, M. A., Patwary, M. M., Islam, M. A., & Rahman, M. R. (2018). Storm Protection Service of the Sundarbans Mangrove Forest, Bangladesh. *Natural Hazards, 94*, 405–418. doi:10.100711069-018-3395-8

Ali, S. I., Hall, K. R., Aronson, K., & Philip, L. (2008). Humanitarian Engineering in Mylai Balaji Nagar: An integrated water, environment and public health project for slums in the Indian Subcontinent. *Desalination, 248*(1-3), 418–427. doi:10.1016/j.desal.2008.05.083

Azad, M. A. K., Uddin, M. S., Zaman, S., & Ashraf, M. A. (2020). Community-based Disaster Management and Its Salient Features: A Policy Approach to People-centred Risk Reduction in Bangladesh. *Asia-Pacific Journal of Rural Development, 29*(2), 135–160. Advance online publication. doi:10.1177/1018529119898036

Badola, R., Barthwal, S., & Hussain, S. A. (2012). Attitudes of Local Communities Towards Conservation of Mangrove Forests: A Case Study from the East Coast of India. *Estuarine, Coastal and Shelf Science, 96*, 188–196. doi:10.1016/j.ecss.2011.11.016

Badola, R., & Hussain, S. A. (2005). Valuing Ecosystem Functions: An Empirical Study on the Strom Protection Function of Bhitarkanika Mangrove Ecosystem, India. *Environmental Conservation*, *32*(1), 85–92. doi:10.1017/S0376892905001967

Baten, A., González, P. A., & Delgado, R. C. (2018). Natural Disasters and Management Systems of Bangladesh from 1972 to 2017: Special Focus on Flood. *OmniScience: A Multi-Disciplinary Journal*, *8*(3), 35–47.

Bayle, P. M., Blenkinsopp, C. E., Conley, D., Masselink, G., Beuzen, T., & Almar, R. (2020). Performance of a Dynamic Cobble Berm Revetment for Coastal Protection Under Increasing Water Level. *Coastal Engineering*, *159*, 103712. Advance online publication. doi:10.1016/j.coastaleng.2020.103712

Bhardwaj, S., & Gupta, A. K. (2021). Ecosystem-Based Approaches and Policy Perspective from India. In M. Mukherjee & R. Shaw (Eds.), *Ecosystem-Based Disaster and Climate Resilience*. Springer. doi:10.1007/978-981-16-4815-1_5

Bhatt, I. D., Negi, V. S., & Rawal, R. S. (2020). Promoting Nature-Based Solution (NbS) Through Restoration of Degraded Landscapes in the Indian Himalayan Region. In S. Dhyani, A. Gupta, & M. Karki (Eds.), *Nature-based Solutions for Resilient Ecosystems and Societies*. Springer. doi:10.1007/978-981-15-4712-6_12

Budny, D. (2018). Providing Solar Power to the Panama Rain Forest: A Senior Design Project. In *2018 IEEE World Engineering Education Conference (EDUNINE)* (pp. 1-8). IEEE. 10.1109/EDUNINE.2018.8450992

Chen, J., & Mueller, V. (2018). Coastal Climate Change, Soil Salinity and Human Migration in Bangladesh. *Nature Climate Change*, *8*(11), 981–985. doi:10.103841558-018-0313-8

Dodd, R. J., Chadwick, D. R., Harris, I. M., Hines, A., Hollis, D., Economou, T., Gwynn-Jones, D., Scullion, J., Robinson, D. A., & Jones, D. L. (2020). Spatial Co-localisation of Extreme Weather Events: A Clear and Present Danger. *Ecology Letters*, *24*(1), 60–72. doi:10.1111/ele.13620 PMID:33047444

Engineers Without Borders. (2021a). *Peru Project*. https://ewb.berkeley.edu/completed-projects/peru/

Engineers Without Borders. (2021b). *Panama Project*. https://ewb.berkeley.edu/completed-projects/panama-project/

Engineers Without Borders. (2021c). *Nicaragua Sanitation Project*. https://ewb.berkeley.edu/projects/nicaragua-sanitation-project/

Engineers Without Borders. (2021d). *Nicaragua Water Project*. https://ewb.berkeley.edu/projects/nicaragua-water-project/

Filho, W. L., Modesto, F., Nagy, G. J., Saroar, M., Toamukum, N. Y., & Ha'apio, M. (2017). Fostering coastal resilience to climate change vulnerability in Bangladesh, Brazil, Cameroon and Uruguay: A cross-country comparison. *Mitigation and Adaptation Strategies for Global Change*, *23*(4), 579–602. doi:10.100711027-017-9750-3

Frantzeskaki, N., McPhearson, T., Collier, M. J., Kendal, D., Bulkeley, H., Dumitru, A., Walsh, C., Noble, K., van Wyk, E., Ordóñez, C., Oke, C., & Pintér, L. (2019). Nature-Based Solutions for Urban Climate Change Adaptation: Linking Science, Policy, and Practice Communities for Evidence-Based Decision-Making. *Bioscience*, *69*(6), 455–466. doi:10.1093/biosci/biz042

Gain, A., Mondal, M., & Rahman, R. (2017). From Flood Control to Water Management: A Journey of Bangladesh Towards Integrated Water Resources Management. *Water (Basel)*, *9*(1), 55. doi:10.3390/w9010055

Ghafoori, N., & Dutta, S. (1995). Building and Nonpavement Applications of Non-Fines Concrete. *Journal of Materials in Civil Engineering*, *7*(4), 286–289. Advance online publication. doi:10.1061/(ASCE)0899-1561(1995)7:4(286)

Hasan, M. K., & Kumar, L. (2021). Yield Trends and Variabilities Explained by Climatic Change in Coastal and Non-coastal Areas of Bangladesh. *The Science of the Total Environment*, *795*, 148814. Advance online publication. doi:10.1016/j.scitotenv.2021.148814 PMID:34237533

Hoque, S. F., Quinn, C. H., & Sallu, S. M. (2017). Resilience, Political Ecology, and Well-being: An Interdisciplinary Approach to Understanding Social-ecological Change in Coastal Bangladesh. *Ecology and Society*, *22*(2), 45. doi:10.5751/ES-09422-220245

Hussain, S. A., & Badola, R. (2010). Valuing Mangrove Benefits: Contribution of Mangrove Forests to Local Livelihoods in Bhitarkanika Conservation Area, East Coast of India. *Wetlands Ecology and Management*, *18*(3), 321–331. doi:10.100711273-009-9173-3

Islam, M. S., & Hoq, M. E. (2019). Vulnerability of Aquaculture-Based Fish Production Systems to the Impacts of Climate Change: Insights from Inland Waters in Bangladesh. In M. N. Islam & A. van Amstel (Eds.), *Bangladesh I: Climate Change Impacts, Mitigation and Adaptation in Developing Countries* (pp. 67–97). Springer. doi:10.1007/978-3-319-26357-1_3

John, V. W. (2013). Water Supply to Communities and Challenges Facing the Lake Bogoria Basin Water Resource Users Association in the Upper Catchment of the Lake Bogoria Basin. *European International Journal of Science and Technology*, *2*(4).

Johnson, F., Higgins, P., & Stephens, C. (2021). Climate change and hydrological risk in the Pacific: A Humanitarian Engineering perspective. *Journal of Water and Climate Change*, *12*(3), 647–678. doi:10.2166/wcc.2021.277

Kader, M. A., Hossain, M. K., & Kabir, M. H. (2019). Wind Speed Reduction Efficiency of Potenga-Muhuri Irrigation Project Coastal Belt in Chittagong, Bangladesh. *Journal of Forest and Environmental Science*, *35*(2), 78–89. doi:10.7747/JFES.2019.35.2.78

Lee, W. K., Tay, S. H. X., Ooi, S. K., & Friess, D. A. (2021). Potential Short Wave Attenuation Function of Disturbed Mangroves. *Estuarine, Coastal and Shelf Science*, *248*, 106747. Advance online publication. doi:10.1016/j.ecss.2020.106747

Leydens, J. A., & Lucena, J. C. (2014). Social Justice: A Missing, Unelaborated Dimension in Humanitarian Engineering and Learning Through Service. *International Journal for Service Learning in Engineering*, *9*(2), 1–28. doi:10.24908/ijsle.v9i2.5447

Li, L., & Zha, Y. (2020). Population Exposure to Extreme Heat in China: Frequency, Intensity, Duration and Temporal Trends. *Sustainable Cities and Society*, *60*, 102282. Advance online publication. doi:10.1016/j.scs.2020.102282

Mamun, M. A., & Ullah, I. (2020). COVID-19 Suicides in Pakistan, Dying Off Not COVID-19 Fear but Poverty? – The Forthcoming Economic Challenges for a Developing Country. *Brain, Behavior, and Immunity*, *87*, 163–166. doi:10.1016/j.bbi.2020.05.028 PMID:32407859

Martin, D. A., Conlon, E., & Bowe, B. (2021). A Multi-level Review of Engineering Ethics Education: Towards a Socio-technical Orientation of Engineering Education for Ethics. *Science and Engineering Ethics*, *27*(5), 60. doi:10.100711948-021-00333-6 PMID:34427811

Maxwell, S. L., Butt, N., Maron, M., McAlpine, C. A., Chapman, S., Ullmann, A., Segan, D. B., & Watson, J. E. M. (2019). Conservation Implications of Ecological Responses to Extreme Weather and Climate Events. *Diversity & Distributions*, *25*(4), 613–625. doi:10.1111/ddi.12878

Maza, M., Lara, J. L., & Losada, I. J. (2019). Experimental Analysis of Wave Attenuation and Drag Forces in a Realistic Fringe Rhizophora Mangrove Forest. *Advances in Water Resources*, *131*, 103376. Advance online publication. doi:10.1016/j.advwatres.2019.07.006

Mazzurco, A., & Daniel, S. (2020). Socio-technical Thinking of Students and Practitioners in the Context of Humanitarian Engineering. *Journal of Engineering Education*, *109*(2), 243–261. doi:10.1002/jee.20307

Miller, L. (2017). *Source Protection And Drinking Water Quality in the Comarca Ngäbe-Buglé, Panama* [Master's Thesis]. Michigan Technological University. doi:10.37099/mtu.dc.etdr/365

Mishra, M., Acharyya, T., Santos, C. A. G., da Silva, R. M., Kar, D., Kamal, A. H. M., & Raulo, S. (2021). Geo-ecological Impact Assessment of Severe Cyclonic Storm Amphan on Sundarban Mangrove Forest Using Geospatial Technology. *Estuarine, Coastal and Shelf Science*, *260*, 107486. Advance online publication. doi:10.1016/j.ecss.2021.107486

Montgomery, J. M., Bryan, K. R., Mullarney, J. C., & Horstman, E. M. (2019). Attenuation of Storm Surges by Coastal Mangroves. *Geophysical Research Letters*, *46*(5), 2680–2689. doi:10.1029/2018GL081636

Montgomery, M., Knight, J., Heuckeroth, D., & Walker, W. (2020). Community-based Latrine Development with Engineering Without Borders – USA. In H. Tsao & L. K. Wang (Eds.), *Evolutionary Progress in Science, Technology, Engineering, Arts, and Mathematics (STEAM)*. Lennox Institute Press.

Pulhin, J. M., & Tapia, M. A. (2021). Climate Change Adaptation in the Philippines. In J. J. Pereira, M. K. Zain, & R. Shaw (Eds.), *Climate Change Adaptation in Southeast Asia*. Springer. doi:10.1007/978-981-16-6088-7_8

Rabby, Y. W., Hossain, M. B., & Hasan, M. U. (2019). Social Vulnerability in the Coastal Region of Bangladesh: An Investigation of Social Vulnerability Index and Scalar Change Effects. *International Journal of Disaster Risk Reduction*, *41*, 101329. Advance online publication. doi:10.1016/j.ijdrr.2019.101329

Rakib, M. A., Sasaki, J., Matsuda, H., & Fukunaga, M. (2019). Severe Salinity Contamination in Drinking Water and Associated Human Health Hazards Increase Migration Risk in the Southwestern Coastal Part of Bangladesh. *Journal of Environmental Management, 240*, 238–248. doi:10.1016/j.jenvman.2019.03.101 PMID:30952044

Rizal, A., Sahidin, A., & Herawati, H. (2018). Economic Value Estimation of Mangrove Ecosystems in Indonesia. *Biodiversity International Journal, 2*(3), 98–100. doi:10.15406/bij.2018.02.00051

Rudiarto, I., Handayani, W., & Setyono, J. S. (2018). A Regional Perspective on Urbanization and Climate-Related Disasters in the Northern Coastal Region on Central Java, Indonesia. *Land (Basel), 7*(1), 34. doi:10.3390/land7010034

Schismenos, S., Stevens, G., Emmanouloudis, D., Georgeou, N., Shrestha, S., Katopodes, N., & Wali, N. (2022, September). Humanitarian Engineering for Renewable Energy and Flood Early Warning in Remote Communities: A Scoping Review of Enabling Factors and Sustainability. *Journal of Sustainable Development of Energy, Water, and Environment Systems, 10*(3), 1090406. Advance online publication. doi:10.13044/j.sdewes.d9.0406

Schismenos, S., Stevens, G. J., Emmanouloudis, D., Georgeou, N., Shrestha, S., & Chalaris, M. (2021a). Humanitarian Engineering at the Sustainability-Development Nexus: Mapping Vulnerability and Capability Factors for Communities at Risk of Water-based Disasters. *Sustainability Science, 16*(4), 1185–1199. doi:10.100711625-020-00890-y PMID:33425037

Shah, M. A. R., Rahman, A., & Chowdhury, S. H. (2018). Challenges for Achieving Sustainable Flood Risk Management. *Journal of Flood Risk Management, 11*(S1), S352–S358. doi:10.1111/jfr3.12211

Smith, J., Anderson, B., Brown, N., Colley, A., Stoakley, A., & Turner, J. (2017). The Rise of Humanitarian Engineering Education in Australasia. In *Proceedings of the 28th Annual Conference of the Australasian Association for Engineering Education (AAEE 2017)* (pp. 312-320). Australasian Association for Engineering Education.

Tenzin, K., & Norbu, L. (2020). Leveraging Conservation Benefits through Ecosystem-Based Services Approach and Community Engagement in Wetland and Riparian Ecosystems: The Case of Conserving Black-Necked Crane and White-Bellied Heron in Bhutan. In S. Dhyani, A. Gupta, & M. Karki (Eds.), *Nature-based Solutions for Resilient Ecosystems and Societies*. Springer. doi:10.1007/978-981-15-4712-6_10

Volger, J. S., Thompson, P., Davis, D. W., Mayfield, B. E., Finley, P. M., & Yasseri, D. (2018). The Hard Work of Soft Skills: Augmenting the Project-based Learning Experience with Interdisciplinary Teamwork. *Instructional Science, 46*(3), 457–488. doi:10.100711251-017-9438-9

Walters, J. P., Greiner, B., O'Morrow, E., & Amadei, B. (2017). Fostering Systems Thinking Within Engineers Without Borders Student Teams Using Group Model Building. *International Journal of Engineering Education, 22*(1A), 247-260.

Warren-Rhodes, K., Schwarz, A. M., Boyle, L. N., Albert, J., Agalo, S. S., Warren, R., Bana, A., Paul, C., Kodosiku, R., Bosma, W., Yee, D., Ronnback, P., Crona, B., & Duke, N. (2011). Mangrove Ecosystem Services and the Potential for Carbon Revenue Programmes in Solomon Islands. *Environmental Conservation*, *38*(4), 485–496. doi:10.1017/S0376892911000373

White, D., Edmond, H., Eller, C., Lui, E., Sing, S., O'Toole, R., & Kremmyda, G. (2021). Design Thinking for Social Impact. In G. Kremmyda (Ed.), *Humanitarian Civil Engineering: Practical solutions for an interdisciplinary approach*. ICE Publishing.

Younger, J. S., Booth, D. J., Parry, D. E., & Kurniawan, K. (2018). Sustainable Humanitarian Engineering in Practice – the East Bali Poverty Project. *Proceedings of the Institution of Civil Engineers. Municipal Engineer*, *171*(2), 67–77. doi:10.1680/jmuen.16.00034

ADDITIONAL READING

Arfan, A., Sanusi, W., Rakib, M., Tuufieq, N. A. S., & Basram, N. F. (2021). Sustainable Management Modeling of Mangrove Ecosystem to Support the Local Economy in Small Islands, South Sulawesi Indonesia. *Journal of Environmental Treatment Techniques*, *9*(1), 296–304. doi:10.47277/JETT/9(1)304

Bayle, P. M., Kaminsky, G. M., Blenkinsopp, C. E., Weiner, H. M., & Cottrell, D. (2021). Behaviour and Performance of a Dynamic Cobble Berm Revetment During a Spring Tidal Cycle in North Cove, Washington State, USA. *Coastal Engineering*, *167*, 103898. Advance online publication. doi:10.1016/j.coastaleng.2021.103898

Catalano, G. D. (2020). Beyond Traditional Engineering: Green, Humanitarian, Social Justice, and Omnium Approaches. In D. P. Michelfelder & N. Doorn (Eds.), *The Routledge Handbook of the Philosophy of Engineering*. Routledge. doi:10.4324/9781315276502-61

Cools, C. (2020). Water Quality and Sustainability Assessment of Rural Water Systems in the Cornarca Ngäbe-Buglé, Panama [Thesis, University of South Florida]. *Graduate Theses and Dissertations*. https://scholarcommons.usf.edu/etd/8177

Hoque, S. (2021, November 2). The Karnaphuli Tunnel Fiasco and the Fault in Our Plans. *The Business Standard*. https://www.tbsnews.net/features/panorama/karnaphuli-tunnel-fiasco-and-fault-our-plans-324070

Kremmyda, G. (Ed.). (2021). *Humanitarian Civil Engineering: Practical solutions for an interdisciplinary approach*. ICE Publishing. doi:10.1680/hce.64683

Marlianingrum, P. R., Kusumastanto, T., Adrianto, L., & Fahrudin, A. (2021). Valuing Habitat Quality for Managing Mangrove Ecosystem Services in Coastal Tangerang District, Indonesia. *Marine Policy*, *133*, 104747. Advance online publication. doi:10.1016/j.marpol.2021.104747

Wang, L. K. (2021). Humanitarian Engineering Education of the Lenox Institute of Water Technology and Its New Potable Water Flotation Processes. In L. K. Wang, M. H. S. Wang, N. K. Shammas, & D. B. Aulenbach (Eds.), *Environmental Flotation Engineering*. Springer. doi:10.1007/978-3-030-54642-7_1

KEY TERMS AND DEFINITIONS

Dynamic Cobble Berm Revetment: Dynamic cobble berm revetments are designed to act similar to natural composite beaches. This is a flexible engineering approach for shoreline management that breaks away from the traditional rigid structures. In this approach, the revetment can reshape itself, and retreat as necessary in a self-sustainable way to respond to the sea level rise and other relevant parameters. Dynamic cobble berm revetments are also effective in reducing shoreline erosion, and drainage of accumulated runoffs.

Good Governance: Good governance mainly deals with the proper management of public and economic institutions to protect human rights. Efficient management of public resources falls within good governance.

Humanitarian Engineering: Humanitarian engineering is the development and application of engineering techniques to help people. It improves the socio-economic condition of the vulnerable communities in developing countries. The Humanitarian Engineering projects are usually focused on the marginalized and disadvantaged communities. These projects utilise extensive participatory and interdisciplinary engagement among various stakeholders to find simple solutions to basic needs.

Nature-Based Solutions: This term is used interchangeably with Soft Engineering. Nature-Based Solutions are engineering solutions that are flexible and incorporate natural elements.

Normalized Difference Vegetation Index (NDVI): NDVI is a remote sensing technique that uses different spectrum bands sensed by satellites to quantify vegetation status in an area. The value of this index varies between +1 to -1. A high positive value indicates a high density of vegetation. A value close to zero indicates built-up or urbanised areas. Negate values generally indicate the presence of water bodies.

Resource-Constrained Countries: Constrained resource is the limitation of any particular resource. Countries with such constraints are termed resource-constrained countries. Even though every country has some form of limitation in resources, in this chapter the developing countries with low and middle income are expressed as Resource-Constrained Countries.

Soft Engineering: Soft engineering is the use of flexible and nature-based solutions for engineering problems. This is usually applied around shorelines and riverbanks to reduce erosion and flooding. The natural elements are utilised to perform the engineering functions such as absorbing wave energy to reduce erosion or soil loss. The use of natural elements instead of artificial structures paves the way for a sustainable approach to managing the coastline and riverbanks.

Sustainable Development Goal 6: Sustainable Development Goal 6 (SDG 6) is one of 17 Sustainable Development Goals established by the United Nations General Assembly. SDG 6 aims to provide access to a safe water supply and sanitation for everyone. Humanitarian and Soft Engineering can play significant roles in attaining SDG 6.

Sustainable Development Goal 13: SDG 13 relates to taking urgent actions to combat climate change and its impacts. The targets of SDG 13 are to be achieved by 2030.

Compilation of References

Abduh, M. (2020). *Advancing the construction engineering and management in Indonesia.* Construction Plus Asia. https://www.constructionplusasia.com/id/advancing-the-construction-engineering-and-management-in-indonesia/

Abduh, M., & Putri, A. T. (2022). Capacity building in sustainable infrastructure through student's community engagement projects Indonesia. *Proceedings of the 9th Civil Engineering Conference in the Asian Region (CECAR).*

Abduh, M., Pribadi, K. S., Soemardi, B. W., & Chan, T.-K. (2022). Education, training, and certification. In T.-K. Chan & K. S. Pribadi (Eds.), *Construction in Indonesia: Looking back and moving forward* (p. 12). Routledge. https://doi.org/https://doi.org/10.1201/9781003149866

Abe, M., Ochiai, C., & Okazaki, K. (2018). Is post-disaster housing reconstruction with participatory method effective to increasing people's awareness for disaster prevention? *Procedia Engineering, 212*, 411–418. doi:10.1016/j.proeng.2018.01.053

Adrian, R. J. (1991). Particle imaging techniques for experimental fluid mechanics. *Annual Review of Fluid Mechanics, 23*, 261–304.

Afrasiabi, M., & Noroozian, N. (2015). Advantages and Limitations of Functional Magnetic Resonance Imaging (fMRI) of the Human Visual Brain. *Horizons in Neuroscience Research Series, 17.*

Ahammad, R., Nandy, P., & Husnain, P. (2013). Unlocking Ecosystem Based Adaptation Opportunities in Coastal Bangladesh. *Journal of Coastal Conservation, 17*(4), 833–840. doi:10.100711852-013-0284-x

Ahmed, A., Haque, M. M., Rahman, A., & Rahman, A. (2016). *Engineering Education in Bangladesh: Implementation of Online Delivery Approach. In Proceedings of International Conference on Engineering Education and Research.* Western Sydney University. http://handle.westernsydney.edu.au:8081/1959.7/uws:39681

Akber, M. A., Patwary, M. M., Islam, M. A., & Rahman, M. R. (2018). Storm Protection Service of the Sundarbans Mangrove Forest, Bangladesh. *Natural Hazards, 94*, 405–418. doi:10.100711069-018-3395-8

Akerblom, M. (1995). *Environmental Monitoring of Pesticide Residues.* Guidance for the SADC Region.

Akor, T. S., Bin Subari, K., Binti Jambari, H., Bin Noordin, M. K., & Onyilo, I. R. (2019). Engineering and related programs' teaching methods in Nigeria. *International Journal of Recent Technology and Engineering, 8*(2), 1279–1282. doi:10.35940/ijrte.B1915.078219

Al Bawaba. (n.d.). *BCG Releases "The Future of Education: How Governments Can Help Close the 21st Century Gap" Report.* Available at: https://www.albawaba.com/business/pr/bcg-releases-%E2%80%98-future-education-how-governments-can-help-close-21st-century
-gap%E2%80%99-report-

Al Madi, N. S., & Khan, J. I. (2016). Measuring learning performance and cognitive activity during multimodal comprehension. *7th International Conference on Information and Communication Systems (ICICS)*, 50-55. 10.1109/IACS.2016.7476085

Alam, F., Sarkar, R., La Brooy, R., & Chowdhury, H. (2016). Engineering Education in 21st Century. *AIP Conference Proceedings*, *1754*, 020002. doi:10.1063/1.4958344

Alam, G. M. (2020). Quality Assurance for Private Universities in Bangladesh: A Quest for Specialised Institutional Governance, Management and Regulatory Mechanism. *International Journal of Comparative Education and Development*, *22*(1), 1–15. doi:10.1108/IJCED-01-2019-0006

Alam, M. S., Sasaki, N., & Datta, A. (2017). Waterlogging, crop damage and adaptation interventions in the coastal region of Bangladesh: A perception analysis of local people. *Environmental Development*, *23*, 22–32. doi:10.1016/j.envdev.2017.02.009

Albashiry, N., Voogt, J., & Pieters, J. (2015). Improving curriculum development practices in a technical vocational community college: Examining effects of a professional development arrangement for middle managers. *Curriculum Journal*, *26*(3), 1–27. doi:10.1080/09585176.2015.1040041

Ali. (2021). Quality Assurance in Engineering Education: Accreditation and Its Global Influence. *Proceedings of the 11th Annual International Conference on Industrial Engineering and Operations Management.*

Ali, M. S., & Syfullah, K. (2017). Effect of sea level rise induced permanent inundation on the livelihood of polder enclosed beel communities in Bangladesh: People's perception. *Journal of Water and Climate Change*, *8*(2), 219–234. doi:10.2166/wcc.2016.236

Ali, S. I., Hall, K. R., Aronson, K., & Philip, L. (2008). Humanitarian Engineering in Mylai Balaji Nagar: An integrated water, environment and public health project for slums in the Indian Subcontinent. *Desalination*, *248*(1-3), 418–427. doi:10.1016/j.desal.2008.05.083

Alzen, J. L., Burkhardt, A., Diaz-Bilello, E., Elder, E., Sepulveda, A., Blankenheim, A., & Board, L. (2021). Academic Coaching and its Relationship to Student Performance, Retention, and Credit Completion. *Innovative Higher Education*, *46*(5), 539–563. Advance online publication. doi:10.100710755-021-09554-w

Ambraseys, N. (2010). On the long-term seismicity of the city of Athens. *Proceedings of the Academy of Athens*, (Vol A, pp. 81-136). Academy of Athens.

Amir, M. I., Khan, M. S. A., Khan, M. K., Rasul, M. G., & Akram, F. (2013). Tidal river sediment management–a case study in southwestern Bangladesh. *International Journal of Civil Science and Engineering*, *7*(3), 861–871.

Anastassiades, M., Lehotay, S. J., Stajnbaher, D., & Schenck, F. J. (2003). Fast and easy multiresidue method employing acetonitrile extraction/ partitioning and dispersive solid-phase extraction for the determination of pesticide residues in produce. *Journal of AOAC International*, *86*(2), 412–431. doi:10.1093/jaoac/86.2.412 PMID:12723926

Anene, A., Hosni, K., Chevalier, Y., Kalfat, R., & Hbaieb, S. (2016). Molecularly imprinted polymer for extraction of patulin in apple juice samples. *Food Control*, *70*, 90–95. doi:10.1016/j.foodcont.2016.05.042

Anmarkrud, Ø., Andresen, A., & Bråten, I. (2019). Cognitive Load and Working Memory in Multimedia Learning: Conceptual and Measurement Issues. *Educational Psychologist*, *54*(2), 1–23. doi:10.1080/00461520.2018.1554484

Arunkumar, S., Sasikala, S., & Kavitha, K. (2018). Towards enhancing engineering education through innovative practices in teaching learning. *International Journal of Engineering and Advanced Technology*, *8*(2), 153–159.

Arya, A. S., Boen, T., & Ishiyama, Y. (2013). *Guidelines for earthquake resistant non-engineered construction*. UNESCO.

ASCE. (2004). *Civil Engineering Body of Knowledge for the 21st Century*. American Society of Civil Engineers.

Ashby, F. G. (2015). An Introduction to fMRI. In B. U. Forstmann & E.-J. Wagenmakers (Eds.), *An Introduction to Model-Based Cognitive Neuroscience* (pp. 91–112). doi:10.1007/978-1-4939-2236-9_5

Asian Development Bank. (2015). *Innovations in Knowledge and Learning for Competitive Higher Education in Asia and the Pacific*. Author.

Atad, O. I., & Grant, A. M. (2021). Evidence-based coaching as a supplement to traditional lectures: Impact on undergraduates' goal attainment and measures of mental well-being. *International Journal of Mentoring and Coaching in Education*, *10*(3), 249–266. Advance online publication. doi:10.1108/IJMCE-05-2020-0024

Athanassas, C. D., Modis, K., Alcicek, M. C., & Theodorakopoulou, K. (2018). 'Contouring the Cataclysm: A Geographical Analysis of the Effects of the Minoan Eruption of the Santorini Volcano. *Environmental Archaeology*, *23*(2), 160–176.

Auerbach, L. W., Goodbred, S. L. Jr, Mondal, D. R., Wilson, C. A., Ahmed, K. R., Roy, K., Steckler, M. S., Small, C., Gilligan, J. M., & Ackerly, B. A. (2015). Reply to 'Tidal river management in Bangladesh'. *Nature Climate Change*, *5*(6), 492–493. doi:10.1038/nclimate2620

Awal, M. A. (2014). Water logging in south-western coastal region of Bangladesh: Local adaptation and policy options. *Science Postprint*, *1*(1), e00038. doi:10.14340pp.2014.12A0001

Azad, A. K., Jensen, K. R., & Lin, C. K. (2009). Coastal aquaculture development in Bangladesh: Unsustainable and sustainable experiences. *Environmental Management*, *44*(4), 800–809. doi:10.100700267-009-9356-y PMID:19701597

Azad, M. A. K., Uddin, M. S., Zaman, S., & Ashraf, M. A. (2020). Community-based Disaster Management and Its Salient Features: A Policy Approach to People-centred Risk Reduction in Bangladesh. *Asia-Pacific Journal of Rural Development*, *29*(2), 135–160. Advance online publication. doi:10.1177/1018529119898036

Aziz, M. A. (2005). Country reports Bangladesh. In *Proceedings Asia Regional Workshop: Implementation Monitoring and Observance International Code of Conduct on the Distribution and Use of Pesticide*. FAO.

Bachman, J., Howard, H., & Patterson. (1999). Photodecomposition of the Carbamate Pesticide Carbofuran: Kinetics and the Influence of Dissolved Organic Matter. *ACS Environ. Sci. Technol.*, *33,* 874-881. doi:10.1021/es9802652

Baddeley, A. (2012). Working memory: Theories, models, and controversies. *Annual Review of Psychology*, *63*(1), 1–29. doi:10.1146/annurev-psych-120710-100422 PMID:21961947

Badola, R., Barthwal, S., & Hussain, S. A. (2012). Attitudes of Local Communities Towards Conservation of Mangrove Forests: A Case Study from the East Coast of India. *Estuarine, Coastal and Shelf Science*, *96*, 188–196. doi:10.1016/j.ecss.2011.11.016

Badola, R., & Hussain, S. A. (2005). Valuing Ecosystem Functions: An Empirical Study on the Strom Protection Function of Bhitarkanika Mangrove Ecosystem, India. *Environmental Conservation*, *32*(1), 85–92. doi:10.1017/S0376892905001967

Bagheri, N., Khataee, A., Habibi, B., & Hassanzadeh, J. (2018). Mimetic Ag nanoparticle/Zn-based MOF nanocomposite capped with molecularly imprinted polymer for the selective detection of patulin. *Talanta*, *179*, 710–718. doi:10.1016/j.talanta.2017.12.009 PMID:29310298

Bair, C. R., & Haworth, J. G. (2006). Doctoral Student Attrition and Persistence: A Meta-Synthesis of Research. In Higher Education: Handbook of Theory and Research. doi:10.1007/1-4020-2456-8_11

Bai, Y. (2006). Organophosphorus pesticide residues in market foods in Shaanxi area. *Food Chemistry*, *98*, 240–242. doi:10.1016/j.foodchem.2005.05.070

Bakker, A. B., Boyd, C. M., Dollard, M., Gillespie, N., Winefield, A. H., & Stough, C. (2010). The role of personality in the job demands-resources model: A study of Australian academic staff. *Career Development International*, *15*(7), 622–636. Advance online publication. doi:10.1108/13620431011094050

Balamurugan, B., Mullai, M., Soundararajan, S., Selvakanmani, S., & Arun, D. (2020). Brain-computer interface for assessment of mental efforts in e-learning using the nonmarkovian queueing model. *Computer Applications in Engineering Education.*

Bangladesh Bereau of Statistics. (2011). *Agricultural Statistics Yearbook of Bangladesh*. BBS.

Bangladesh Bureau of Statistics. (2007). *Census of Agriculture 1983-84, 1996 and 2006*. BBS.

Bangladesh Bureau of Statistics. (2011). *Population Census 1981, 1991, 2001 and 2011*. BBS.

Bartlett, J. G. (2002). Antibiotic-Associated Diarrhea. *The New England Journal of Medicine*, *346*(5), 334–339. doi:10.1056/NEJMcp011603 PMID:11821511

Baten, A., González, P. A., & Delgado, R. C. (2018). Natural Disasters and Management Systems of Bangladesh from 1972 to 2017: Special Focus on Flood. *OmniScience: A Multi-Disciplinary Journal*, *8*(3), 35–47.

Bath, D., Smith, C., Stein, S., & Swann, R. (2004). Beyond mapping and embedding graduate attributes: Bringing together quality assurance and action learning to create a validated and living curriculum. *Higher Education Research & Development*, *23*(3), 313–328. doi:10.1080/0729436042000235427

Bayle, P. M., Blenkinsopp, C. E., Conley, D., Masselink, G., Beuzen, T., & Almar, R. (2020). Performance of a Dynamic Cobble Berm Revetment for Coastal Protection Under Increasing Water Level. *Coastal Engineering*, *159*, 103712. Advance online publication. doi:10.1016/j.coastaleng.2020.103712

BBS. (2018). *Fisheries Statistical Yearbook of Bangladesh 1983-84, 1990-91, 2000-01, 2010-11, 2016-17*. BBS.

Belsky, J., & Joshi, N. K. (2020). Effects of Fungicide and Herbicide Chemical Exposure on Apis and Non-Apis Bees in Agricultural Landscape. *Frontiers in Environmental Science*, *8*(81), 1–10. doi:10.3389/fenvs.2020.00081

Bernhardt, K. A., Poltavski, D., Petros, T., Ferraro, F. R., Jorgenson, T., Carlson, C., Drechsel, P., & Iseminger, C. (2019). The effects of dynamic workload and experience on commercially available EEG cognitive state metricsin a high-fidelity air traffic control environment. *Applied Ergonomics*, *77*, 83–91. doi:10.1016/j.apergo.2019.01.008 PMID:30832781

Bertolo, D., Dinet, J., & Vivian, R. (2014). Reducing cognitive workload during 3D geometry problem solving with an app on iPad. *Science and Information Conference*, 896–900. 10.1109/SAI.2014.6918292

Best, G. A., & Ruthven, D. (1995). Analytical Division & Royal Society of Chemistry. In *Pesticides: Developments, impacts, and controls*. Royal Society of Chemistry.

Bettinger, E. P., & Baker, R. B. (2014). The Effects of Student Coaching: An Evaluation of a Randomized Experiment in Student Advising. *Educational Evaluation and Policy Analysis*, *36*(1), 3–19. Advance online publication. doi:10.3102/0162373713500523

Bhardwaj, S., & Gupta, A. K. (2021). Ecosystem-Based Approaches and Policy Perspective from India. In M. Mukherjee & R. Shaw (Eds.), *Ecosystem-Based Disaster and Climate Resilience*. Springer. doi:10.1007/978-981-16-4815-1_5

Bhatt, I. D., Negi, V. S., & Rawal, R. S. (2020). Promoting Nature-Based Solution (NbS) Through Restoration of Degraded Landscapes in the Indian Himalayan Region. In S. Dhyani, A. Gupta, & M. Karki (Eds.), *Nature-based Solutions for Resilient Ecosystems and Societies*. Springer. doi:10.1007/978-981-15-4712-6_12

Bhuiyan, M. J. A. N., & Dutta, D. (2012). Assessing impacts of sea level rise on river salinity in the Gorai river network, Bangladesh. *Estuarine, Coastal and Shelf Science*, *96*, 219–227. doi:10.1016/j.ecss.2011.11.005

BNPB. (2021a). *Indonesia disaster risk index in 2020*. BNBP. http://mpbi.info/wp-content/uploads/2021/07/BUKU-IRBI-2020-KP.pdf

BNPB. (2021b). *Casualties and damage by disaster*. BNBP. https://dibi.bnpb.go.id/kbencana

BNPB. (2022). *Indonesian disaster information data*. BNBP. https://dibi.bnpb.go.id/xdibi

Boen, T. (2006a). *The minimum requirements for earthquake-resistant house: Masonry house with timber confinement*. https://teddyboen.com/Poster_Membangun_Rumah_Tahan_Gempa.html

Boen, T. (2006b). *The minimum requirements for earthquake-resistant house: One brick thick-walled masonry house*. https://teddyboen.com/Poster_Membangun_Rumah_Tahan_Gempa.html

Boen, T. (2006c). *The minimum requirements for earthquake-resistant house: One brick thick-walled masonry house with reinforced concrete*. https://teddyboen.com/Poster_Membangun_Rumah_Tahan_Gempa.html

Boen, T. (2006d). *The minimum requirements for earthquake-resistant house: Wooden house and several variations*. https://teddyboen.com/Poster_Membangun_Rumah_Tahan_Gempa.html

Boen, T. (2009a). *Constructing seismic-resistant masonry houses in Indonesia*. United Nations Centre for Regional Development. http://www.confinedmasonry.org/wp-content/uploads/2018/11/masonryhousesindonesiaguide.pdf

Boen, T. (2019). *Building earthquake-resistant masonry house with ferrocement layers*. https://teddyboen.com/Membangun_Rumah_Tembokan_Tahan_Gempa_dengan_Balutan_Lapisan_Ferosemen.html

Boen, T. (2021a). *Building earthquake-resistant masonry house with fibrocement (fibre reinforced cementitious/FRC) layers*. https://teddyboen.com/Membangun_Rumah_Tembokan_Tahan_Gempa_dengan_TRC_FRC.html

Boen, T. (2021b). *Building Earthquake-resistant Masonry House with Textile Fibre Composite/TRC Layers*. https://teddyboen.com/Membangun_Rumah_Tembokan_Tahan_Gempa_dengan_TRC_FRC.html

Boen, T., Suprobo, P., Sarwidi, S., Pribadi, K. S., Arkani, M. I., Satyarno, I., & Saputra, A. (2009b). *The minimum requirements for safer houses*. Directorate General of Housing and Settlement and Japan International Cooperation Agency. https://teddyboen.com/Persyaratan_Pokok_Rumah_yang_Lebih_Aman.html

Boen, T., Firmansjah, J., Ismail, F. A., Kusumastuti, D., Lase, Y., Pribadi, K. S., Sarwidi, S., Satyarno, I., & Widodo, L. (2010). *Ways to repair an earthquake-damaged simple structure*. Australia-Indonesia Facility for Disaster Reduction and Indonesian National Board for Disaster Management.

Boen, T., Saputra, A., Ismail, F. A., Pribadi, K. S., Satyarno, I., & Widodo, L. (2012). *The minimum requirements for safer houses*. Japan International Cooperation Agency.

Bonwell, C. C. (1996). Enhancing the lecture: Revitalizing a traditional format. In T. E. Sutherland & C. C. Bonwell (Eds.), *Using active learning in college classes: A range of options for faculty. New Directions for Teaching and Learning No. 67*. doi:10.1002/tl.37219966706

Botzen, W. J. W., Deschenes, O., & Sanders, M. (2019). The Economic Impacts of Natural Disasters: A Review of Models and Empirical Studies. *Review of Environmental Economics and Policy, 13*(2). doi:10.1093/reep/rez004

Bowden, J., Hart, G., King, B., Trigwell, K., & Watts, O. (2000). *Generic Capabilities of ATN University Graduates.* Final Report to DETYA, Australian Technologies Network Teaching and Learning Committee, Melbourne, Australia.

BPS. (2020). *Housing and Settlement Statistics 2019*. BPS. https://www.bps.go.id/publication/2020/08/31/6a9e70d6154fde75499239e6/statistik-perumahan-dan-permukiman-2019.html

BPS. (2021). *Percentage of Households by Province and Widest Non-Bamboo/Other Walls (Percent), 2019-2021*. BPS. https://www.bps.go.id/indicator/29/851/1/persentase-rumah-tangga-menurut-provinsi-dan-dinding-terluas-bukan-bambu-lainnya.html

Brammer, H. (2010). After the Bangladesh flood action plan: Looking to the future. *Environmental Hazards, 9*(1), 118–130. doi:10.3763/ehaz.2010.SI01

Brammer, H. (2014). Bangladesh's dynamic coastal regions and sea-level rise. *Climate Risk Management, 1*, 51–62. doi:10.1016/j.crm.2013.10.001

Bretveld, R.W., Thomas, C.M., Scheepers, P.T., Zielhuis, G. A., & Roeleveld, N. (2006). Pesticide exposure: the hormonal function of the female reproductive system disrupted? *Reprod Biol Endocrinol., 31*(4), 30. doi:10.1186/1477-7827-4-30

BRIN. (2019). *Bale Kohana, earthquake resistant residential technology innovation*. http://pusyantek.bppt.go.id/id/posts/artikel/bale-kohana-inovasi-teknologi-hunian-tahan-gempa

Britannica, The Editors of Encyclopaedia. (2019). Insecticide. *Encyclopedia Britannica.* https://www.britannica.com/technology/insecticide

Britannica, The Editors of Encyclopaedia. (2021). Rodenticide. *Encyclopedia Britannica.* https://www.britannica.com/technology/rodenticide

Britannica, The Editors of Encyclopaedia. (2021). Sulfa drug. *Encyclopedia Britannica.* https://www.britannica.com/science/sulfa-drug

Britannica, The Editors of Encyclopaedia. (2022). Fumigant. *Encyclopedia Britannica.* https://www.britannica.com/technology/fumigant

British Council. (2015). *Skills needed: Addressing South Asia's deficit of technical and soft skills Analysing the gap in.* Available at: https://www.britishcouncil.org/sites/default/files/skills_gap_jan2015_print.pdf

Brodsky, D., & Newman, L. (2010). Educational Perspectives: A Systematic Approach to Curriculum Development. *NeoReviews, 12*(1), e2–e7. doi:10.1542/neo.12-1-e2

Brody, J., & Rudel, R. (2003). Environmental pollutants and breast cancer. *Environmental Health Perspectives, 111*(8), 1007–1019. doi:10.1289/ehp.6310 PMID:12826474

Brown, G. B. (1939). The Evolution of Physics: The Growth of Ideas from the Early Concepts to Relativity and Quanta. By Albert Einstein and Leopold Infeld (Cambridge: at the University Press, 1938. Pp. x + 319. Price 8s. 6d.). *Philosophy (London, England), 14*(54), 242. Advance online publication. doi:10.1017/S0031819100011670

Brown, S., Nicholls, R. J., Hanson, S., Brundrit, G., Dearing, J. A., Dickson, M. E., & Jimenez, J. A. (2014). Shifting perspectives on coastal impacts and adaptation. *Nature Climate Change, 4*(9), 752–755. doi:10.1038/nclimate2344

Brzev, S., & Mitra, K. (2007). *Earthquake-resistant confined masonry construction.* National Information Center of Earthquake Engineering, Indian Institute of Technology Kanpur.

Budny, D. (2018). Providing Solar Power to the Panama Rain Forest: A Senior Design Project. In *2018 IEEE World Engineering Education Conference (EDUNINE)* (pp. 1-8). IEEE. 10.1109/EDUNINE.2018.8450992

Bui Thi Hong Dung. (2019). Improve the quality of higher education - Nâng cao chất lượng giáo dục đại học. *Figures and Event Journal.* https://consosukien.vn/nang-cao-chat-luong-giao-duc-dai-hoc.htmhọc

Burrill, G., Lappan, G., & Gonulates, F. (2015). Curriculum and the Role of Research. In S. J. Cho (Ed.), *The Proceedings of the 12th International Congress on Mathematical Education* (pp. 247–263). 10.1007/978-3-319-12688-3_17

Cao, X., Cheng, M., Xue, X., & Zhu, S. (2019). Effects of Lecture Video Types on Student Learning: An Analysis of Eye-Tracking and Electroencephalography Data. In F. Xhafa, S. Patnaik, & M. Tavana (Eds.), *Advances in Intelligent* (pp. 498–505). Interactive Systems, and Applications. doi:10.1007/978-3-030-02804-6_66

Çaparlar, C. Ö., & Dönmez, A. (2016). What is scientific research and how can it be done? Turk Anesteziyoloji ve Reanimasyon Dernegi Dergisi. doi:10.5152/TJAR.2016.34711

Capstick, M. K., Harrell-Williams, L. M., Cockrum, C. D., & West, S. L. (2019). Exploring the Effectiveness of Academic Coaching for Academically At-Risk College Students. *Innovative Higher Education, 44*(3), 219–231. Advance online publication. doi:10.100710755-019-9459-1

Carlevaro, N., Roux-Fouillet, G., & Schacher, T. (2018). *Guide book (for builders) for building earthquake-resistant houses in confined masonry.* Swiss Agency for Development and Cooperation and Earthquake Engineering Research Institute. https://www.shacc.ch/documents/files/library/Manual/1806-CM-masons-guide_eng_SMALL.pdf

Carson, R. (2002). Silent Spring. Houghton Mifflin Harcourt.

Carter, A. D., & Heather, A. I. J. (1995). Pesticides in groundwater. In G. A. Best & A. D. Ruthven (Eds.), *Pesticides-developments, impacts, and controls* (pp. 113–123). The Royals Society of Chemistry.

Cashin, W. E. (1985). *Improving lectures. Idea Paper No. 14.* Kansas State University, Center for Faculty Evaluation and Development.

Casida, J. E., & Durkin, K. A. (2013). Neuroactive insecticides: Targets, selectivity, resistance, and secondary effects. *Annual Review of Entomology, 58*(1), 99–117. doi:10.1146/annurev-ento-120811-153645 PMID:23317040

Castro-Meneses, L. J., Kruger, J.-L., & Doherty, S. (2020). Validating theta power as an objective measure of cognitive load in educational video. *Educational Technology Research and Development, 68*(1), 181–202. doi:10.100711423-019-09681-4

Centers For Disease Control and Prevention (CDC) Report 2019. https://www.cdc.gov/about/24-7/2019-Progress-Report.html

Centre for Research on the Epidemiology of Disasters. (2022). *Earthquake disaster data in Indonesia 1991–2020*. EM-DAT, The International Disaster Database. https://www.emdat.be/

Chang, H.-C., Hung, I.-C., Chew, S. W., & Chen, N.-S. (2016). Yet Another Objective Approach for Measuring Cognitive Load Using EEG-Based Workload. *IEEE 16th International Conference on Advanced Learning Technologies (ICALT)*, 501-502.

Chattopadhyay, M. K. (2014). Use of antibiotics as feed additives: A burning question. *Frontiers in Microbiology*, *5*. Advance online publication. doi:10.3389/fmicb.2014.00334 PMID:25071747

Chatzi, E. (2017). A Monitoring Approach to Smart Infrastructure Management. *Proceedings*, *1*(8), 755.

Chen, J., & Mueller, V. (2018). Coastal Climate Change, Soil Salinity and Human Migration in Bangladesh. *Nature Climate Change*, *8*(11), 981–985. doi:10.103841558-018-0313-8

Chen, J., Sun, B. M., Chen, D., Wu, X., Guo, L. Z., & Wang, G. (2014). Land use changes and their effects on the value of ecosystem services in the small Sanjiang plain in China. *TheScientificWorldJournal*, *2014*, 2014. doi:10.1155/2014/752846 PMID:24741356

Cheong, C. K., Hajeb, P., Jinap, S., & Ismail-Fitry, M. R. (2010). Sulfonamides determination in chicken meat products from Malaysia. *International Food Research Journal*, *17*, 885–892.

Chowdhury, H., Alam, F., Biswas, S. K., Islam, M. T., & Islam, A. K. M. S. (2013). Quality Assurance and Accreditation of Engineering Education in Bangladesh. *Procedia Engineering*, *56*, 864–869. doi:10.1016/j.proeng.2013.03.208

Clark, S. B., Turnipseed, S. B., Madson, M. R., Hurlbut, J. A., Kuck, L. R., & Sofos, J. N. (2005). Confirmation of Sulfamethazine, Sulfathiazole, and Sulfadimethoxine Residues in Condensed Milk and Soft-Cheese Products by Liquid Chromatography/Tandem Mass Spectrometry. *Journal of AOAC International*, *88*(3), 736–743. doi:10.1093/jaoac/88.3.736 PMID:16001847

Codex Alimentarius Commission no CX/MRL 2-2018, updated 2020. https://www.fao.org/fao-who-codexalimentarius/codex-texts/dbs/vetdrugs/en/

Codex Alimentarius Commission. (2013). *Codex Maximum Residue Limits for Pesticides*. https://www.who.int/entity/foodsafety/areas_work/chemicalrisks/IEDIcalculation0217clusterfinal.xlsm

Commission Regulation (European Union) no 37/2010, updated 2018. https://ec.europa.eu/health/sites/default/files/files/eudralex/vol-5/reg_2010_37/reg_2010_37_en.pdf

Confined Masonry Network. (2011). *Seismic design guide for low-rise confined masonry buildings*. Earthquake Engineering Research Institute.

Conrad, C., Bliemel, M. . (2016). Psychophysiological Measures of Cognitive Absorption and Cognitive Load in E-Learning Applications. *ICIS 2016 Proceedings*.

Council on Foreign Affairs. (2016). *Indonesia's Education Gap*. Retrieved from Asia Unbound website: https://www.cfr.org/blog/indonesias-education-gap

Cowan, N. (2001). The magical number 4 in short-term memory: A reconsideration of mental storage capacity. *Behavioral and Brain Sciences, 24*(1), 87–114. doi:10.1017/S0140525X01003922 PMID:11515286

Crk, I., & Kluthe, T. (2014). Toward using alpha and theta brain waves to quantify programmer expertise. *36th Annual International Conference of the IEEE Engineering in Medicine and Biology Society*, 5373–5376. 10.1109/EMBC.2014.6944840

Cropley, D. H., & Cropley, A. J. (2005). Engineering creativity: A systems concept of functional creativity. In J. C. Kaufman & J. Baer (Eds.), *Faces of the muse: How people think, work and act creatively in diverse domains* (pp. 169–185). Lawrence Erlbaum.

Cruz, L., & Rosemond, L. (2017). Coaching Academia: The Integration of Coaching, Educational Development, and the Culture of Higher Education. *Journal on Excellence in College Teaching.*

Culliney, T. W., Pimentel, D., & Pimentel, M. H. (1992). Pesticides and natural toxicants in foods. *Agriculture, Ecosystems & Environment, 41*(3-4), 297–320. doi:10.1016/0167-8809(92)90117-T

Dan, A., & Reiner, M. (2017). Real Time EEG Based Measurements of Cognitive Load Indicates Mental States During Learning. *Journal of Educational Data Mining, 9*(2), 31-44.

Dan, A., & Reiner, M. (2017a). EEG-based cognitive load of processing events in 3D virtual worlds is lower than processing events in 2D displays. *International Journal of Psychophysiology, 122*, 75–84. doi:10.1016/j.ijpsycho.2016.08.013 PMID:27592084

Dan, A., & Reiner, M. (2018). Reduced mental load in learning a motor visual task with virtual 3D method. *Journal of Computer Assisted Learning, 34*(1), 84–93. doi:10.1111/jcal.12216

Dasgupta, S., Hossain, M., Huq, M., & Wheeler, D. (2014). *Facing the hungry tide: climate change, livelihood threats, and household responses in coastal Bangladesh*. World Bank.

Dasgupta, S., Huq, M., & Wheeler, D. (2016). Drinking water salinity and infant mortality in coastal Bangladesh. *Water Economics and Policy, 2*(01), 1650003. doi:10.1142/S2382624X1650003X

Daun, M., Salmon, A., Weyer, T., Pohl, K., & Tenbergen, B. (2016). Project-Based Learning with Examples from Industry in University Courses: An Experience Report from an Undergraduate Requirements Engineering Course. *CSEET IEEE 29th International Conference*. 10.1109/CSEET.2016.15

Davies, D., Jindal-Snape, D., Collier, C., Digby, R., Hay, P., & Howe, A. (2013). Creative learning environments in education—A systematic literature review. *Thinking Skills and Creativity, 8*, 80–91. doi:10.1016/j.tsc.2012.07.004

De Groot, R. S., Wilson, M. A., & Boumans, R. M. (2002). A typology for the classification, description and valuation of ecosystem functions, goods and services. *Ecological Economics, 41*(3), 393–408. doi:10.1016/S0921-8009(02)00089-7

Delaplane, K. S. (1996). Pesticide Usage in the United States: History, Benefits, Risks, and Trends. Cooperative Extension Service, The University of Georgia College of Agricultural and Environmental Sciences.

Department of Public Works. (1993). *The Guideline of Constructing an Earthquake-Resistant Building - Decree of Directorate General of Housing and Settlement No. 111/KPTS/CK/1993*. Directorate General of Human Settlement.

Department of Public Works. (2006). *Technical Guideline for Earthquake-Resistant Houses and Building*. Directorate General of Housing and Settlement.

Devine, M., Meyers, R., & Houssemand, C. (2013). How can Coaching Make a Positive Impact Within Educational Settings? *Procedia: Social and Behavioral Sciences, 93*, 1382–1389. Advance online publication. doi:10.1016/j.sbspro.2013.10.048

Dewan, A. M., & Yamaguchi, Y. (2009). Land use and land cover change in Greater Dhaka, Bangladesh: Using remote sensing to promote sustainable urbanization. *Applied Geography (Sevenoaks, England)*, *29*(3), 390–401. doi:10.1016/j.apgeog.2008.12.005

Dewan, C., Mukherji, A., & Buisson, M. C. (2015). Evolution of water management in coastal Bangladesh: From temporary earthen embankments to depoliticized community-managed polders. *Water International*, *40*(3), 401–416. doi:10.1080/02508060.2015.1025196

DHI. (2017). *MIKE 11: A Modelling System for Rivers and Channels*. Danish Hydraulic Institute.

DHI. (2019). *MIKE 21 Flow Model FM*. Danish Hydraulic Institute.

Díaz, D., Ramírez, R., & Hernández-Leo, D. (2015). The Effect of Using a Talking Head in Academic Videos: An EEG Study. *IEEE 15th International Conference on Advanced Learning Technologies*, 367-369.

Dibner, J. J., & Richards, J. D. (2005). Antibiotic growth promoters in agriculture: History and mode of action. *Poultry Science*, *84*(4), 634–643. doi:10.1093/ps/84.4.634 PMID:15844822

Dodd, R. J., Chadwick, D. R., Harris, I. M., Hines, A., Hollis, D., Economou, T., Gwynn-Jones, D., Scullion, J., Robinson, D. A., & Jones, D. L. (2020). Spatial Co-localisation of Extreme Weather Events: A Clear and Present Danger. *Ecology Letters*, *24*(1), 60–72. doi:10.1111/ele.13620 PMID:33047444

Dowling, A., Dwyer, J. O., & Adley, C. C. (2017). *Antibiotics: Mode of action and mechanisms of resistance.* In A. Mendez-Vilas (Ed.), *Antimicrobial research: Novel bio knowledge and educational programs* (pp. 536–545). https://www.researchgate.net/publication/317381477_Antibiotics_Mode_of_action_and_mechanisms_of_resistance

Dubois, O. (2011). *The state of the world's land and water resources for food and agriculture: managing systems at risk*. Earthscan.

Dyguda-Kazimierowicz, E., Szczepan, R., & Andrzej Sokalski, W. (2014). *Alkaline Hydrolysis of Organophosphorus Pesticides: The Dependence of the Reaction Mechanism on the Incoming Group Conformation*. ACS J. Phys. Chem. B. doi:10.1021/jp503382j

Elassy, N. (2013). A Model of Student Involvement in the Quality Assurance System at Institutional Level. *Quality Assurance in Education*, *21*(2), 162–198. doi:10.1108/09684881311310692

Elassy, N. (2015). The Concepts of Quality, Quality Assurance and Quality Enhancement. *Quality Assurance in Education*, *23*(3), 250–261. doi:10.1108/QAE-11-2012-0046

Eldenfria, A., & Al-Samarraie, H. T. (2019). Towards an Online Continuous Adaptation Mechanism (OCAM) for Enhanced Engagement: An EEG Study. *International Journal of Human-Computer Interaction*, *35*(20), 1960–1974. doi:10.1080/10447318.2019.1595303

El-Sayed, M. G. A., El-Komy, A. A. A., El-barawy, A. M., & Mustafa, G. E. A. (2014). Pharmacokinetical Interactions of Amoxicillin and Amprolium in Broiler Chickens. *Journal of Physiology and Pharmacology Advances*, *4*(12), 515–524. doi:10.5455/jppa.20141208013007

Engineers Without Borders. (2021a). *Peru Project*. https://ewb.berkeley.edu/completed-projects/peru/

Engineers Without Borders. (2021b). *Panama Project*. https://ewb.berkeley.edu/completed-projects/panama-project/

Engineers Without Borders. (2021c). *Nicaragua Sanitation Project.* https://ewb.berkeley.edu/projects/nicaragua-sanitation-project/

Engineers Without Borders. (2021d). *Nicaragua Water Project.* https://ewb.berkeley.edu/projects/nicaragua-water-project/

EPA. (2001). *Dioxin Reassessment.* https://cfpub.epa.gov/ncea/cfm/dioxin doi:10.5772/57399

Eva, H.D., Brink, A. and Simonetti, D. (2006). *Monitoring land cover dynamics in sub-Saharan Africa.* Institute for Environmental and Sustainability, Tech. Rep. EUR, 22498.

Faizi, J., & Umar, M. S. (2021). A Conceptual Framework for Software Engineering Education: Project Based Learning Approach Integrated with Industrial Collaboration. *International Journal of Education and Management Engineering, 11*(5), 46–53. doi:10.5815/ijeme.2021.05.05

FAO. (2013). *International Code of Conduct on the Distribution and Use of Pesticides.* https://www.fao.org/3/bt565e/bt565e.pdf

Farida, R., Alba, A., Kurniawan, R., & Zainuddin, Z. (2019). The development of flipped learning model with Bloom's Taxonomy in teaching the Indonesian Political System course. *Kwangsan, 7*(2), 104–122.

Fernandez Rojas, R., Debie, E., Fidock, J., Barlow, M., Kasmarik, K., Anavatti, S., Garratt, M., & Abbass, H. (2020). Electroencephalographic Workload Indicators During Teleoperation of an Unmanned Aerial Vehicle Shepherding a Swarm of Unmanned Ground Vehicle. *Frontiers in Neuroscience, 14*, 40. doi:10.3389/fnins.2020.00040 PMID:32116498

Filho, W. L., Modesto, F., Nagy, G. J., Saroar, M., Toamukum, N. Y., & Ha'apio, M. (2017). Fostering coastal resilience to climate change vulnerability in Bangladesh, Brazil, Cameroon and Uruguay: A cross-country comparison. *Mitigation and Adaptation Strategies for Global Change, 23*(4), 579–602. doi:10.100711027-017-9750-3

Fisher, B., Turner, R. K., & Morling, P. (2009). Defining and classifying ecosystem services for decision making. *Ecological Economics, 68*(3), 643–653. doi:10.1016/j.ecolecon.2008.09.014

Food Safety Act, Bangladesh Food Safety Authority (BFSA), October 2013. https://puc.ac.bd/Content/PU-Journal/PU-Critical-Perspective-Article3.pdf

Food safety regulation, Bangladesh food safety Authority, 2017. https://www.fas.usda.gov/data/bangladesh-bangladesh-issues-regulation-food-safety-2017

Fraillon, J., Ainley, J., Schulz, W., Friedman, T., & Gebhardt, E. (2014). *Preparing for Life in a Digital Age* (1st ed.). doi:10.1007/978-3-319-14222-7

Frankl, V. E. (1997). Man's Search for Meaning. *Language.* doi:10.1037/a0019212

Frantzeskaki, N., McPhearson, T., Collier, M. J., Kendal, D., Bulkeley, H., Dumitru, A., Walsh, C., Noble, K., van Wyk, E., Ordóñez, C., Oke, C., & Pintér, L. (2019). Nature-Based Solutions for Urban Climate Change Adaptation: Linking Science, Policy, and Practice Communities for Evidence-Based Decision-Making. *Bioscience, 69*(6), 455–466. doi:10.1093/biosci/biz042

Fuoco, D. (2012). Classification Framework and Chemical Biology of Tetracycline-Structure-Based Drugs. *Antibiotics (Basel, Switzerland), 1*(1), 1–13. doi:10.3390/antibiotics1010001 PMID:27029415

Gain, A. K., Benson, D., Rahman, R., Datta, D. K., & Rouillard, J. J. (2017a). Tidal river management in the south west Ganges-Brahmaputra delta in Bangladesh: Moving towards a transdisciplinary approach? *Environmental Science & Policy*, *75*, 111–120. doi:10.1016/j.envsci.2017.05.020

Gain, A. K., Mondal, M. S., & Rahman, R. (2017b). From Flood Control to Water Management: A Journey of Bangladesh towards Integrated Water Resources Management. *Water (Basel)*, *9*(1), 55. doi:10.3390/w9010055

Gerlič, I., & Jaušovec, N. (2001). Differences in EEG power and coherence measures related to the type of presentation: Text versus multimedia. *Journal of Educational Computing Research*, *25*(2), 177–195. doi:10.2190/YDWY-U3FJ-4LY4-LYND

Ghafoori, N., & Dutta, S. (1995). Building and Nonpavement Applications of Non-Fines Concrete. *Journal of Materials in Civil Engineering*, *7*(4), 286–289. Advance online publication. doi:10.1061/(ASCE)0899-1561(1995)7:4(286)

Gill, J., Sharp, R., Mills, J., & Franzway, S. (2008). I still wanna be an engineer! Women, education and the engineering profession. *European Journal of Engineering Education*, *33*(4), 391–402. doi:10.1080/03043790802253459

Goodhew, P. (2010). *Teaching Engineering All you need to know about engineering education but were afraid to ask.* UK Centre for Materials Education.

Hadgraft, R. G., & Kolmos, A. (2020). Emerging Learning Environments in Engineering Education. *Australasian Journal of Engineering Education*, *25*(1), 3–16. doi:10.1080/22054952.2020.1713522

Hakro, A. N., & Mathew, P. (2020). Coaching and mentoring in higher education institutions: A case study in Oman. *International Journal of Mentoring and Coaching in Education*, *9*(3), 307–322. Advance online publication. doi:10.1108/IJMCE-05-2019-0060

Haque, K. N. H., Chowdhury, F. A., & Khatun, K. R. (2015). Participatory environmental governance and climate change adaptation: mainstreaming of tidal river management in south-west Bangladesh. In *Land and disaster management strategies in Asia* (pp. 189–208). Springer India. doi:10.1007/978-81-322-1976-7_13

Hasan, M. N., Hossain, M. S., Islam, M. R., Bari, M. A., Karim, D., & Rahman, M. Z. (2013). Trends in the availability of agricultural land in Bangladesh. Soil Resource Development Institute (SRDI), Ministry of Agriculture. http://www. nfpcsp.org/agridrupal/sites/default/files/Trends-in-the-availability-of agricultural-land-in-Bangladesh-SRDI-Supported-by-NFPCSP-FAO. pdf

Hasan, M. K., & Kumar, L. (2021). Yield Trends and Variabilities Explained by Climatic Change in Coastal and Non-coastal Areas of Bangladesh. *The Science of the Total Environment*, *795*, 148814. Advance online publication. doi:10.1016/j.scitotenv.2021.148814 PMID:34237533

Haskoning (1993) *Second Coastal Embankment Rehabilitation Project, Main Report.* Bangladesh Water Development Board.

Heierli, U., & Maithel, S. (2008). Brick by brick: The Herculean task of cleaning up the Asian brick industry. In P. Osborn (Ed.), *Swiss Agency for Development and Cooperation* (1st ed.). Swiss Agency for Development and Cooperation.

Herlina, N. (2020). *Direktorat Jenderal Pendidikan Tinggi Umumkan Klasterisasi Perguruan Tinggi Indonesia tahun 2020.* Retrieved September 20, 1BC, from Kemendikbud website: https://dikti.kemdikbud.go.id/kabar-dikti/kabar/direktorat-jenderal-pendidikan-tinggi-umumkan-klasterisasi-perguruan-tinggi-indonesia-tahun-2020/

Higgins, S. A., Overeem, I., Steckler, M. S., Syvitski, J. P., Seeber, L., & Akhter, S. H. (2014). InSAR measurements of compaction and subsidence in the Ganges-Brahmaputra Delta, Bangladesh. *Journal of Geophysical Research. Earth Surface*, *119*(8), 1768–1781. doi:10.1002/2014JF003117

Hinkel, J., Lincke, D., Vafeidis, A. T., Perrette, M., Nicholls, R. J., Tol, R. S., Marzeion, B., Fettweis, X., Ionescu, C., & Levermann, A. (2014). Coastal flood damage and adaptation costs under 21st century sea-level rise. *Proceedings of the National Academy of Sciences of the United States of America*, *111*(9), 3292–3297. doi:10.1073/pnas.1222469111 PMID:24596428

Holladay, A. (1977). Sparta's role in the First Peloponnesian War. *The Journal of Hellenic Studies*, *97*, 54–63.

Hoq, M. E. (2007). An analysis of fisheries exploitation and management practices in Sundarbans mangrove ecosystem, Bangladesh. *Ocean and Coastal Management*, *50*(5), 411–427. doi:10.1016/j.ocecoaman.2006.11.001

Hoque, R., Ahmed, S. M., Naher, N., Islam, M. A., Rousham, E. K., Islam, M. A., & Hassan, S. (2020). Tackling antimicrobial resistance in Bangladesh: A scoping review of policy and practice in human, animal and environment sectors. *PLoS One*, *15*(1), e0227947. doi:10.1371/journal.pone.0227947 PMID:31986167

Hoque, S. F., Quinn, C. H., & Sallu, S. M. (2017). Resilience, Political Ecology, and Well-being: An Interdisciplinary Approach to Understanding Social-ecological Change in Coastal Bangladesh. *Ecology and Society*, *22*(2), 45. doi:10.5751/ES-09422-220245

Hossain, M. L., Hossain, M. K., Salam, M. A., & Rubaiyat, A. (2012a). Seasonal variation of soil salinity in coastal areas of Bangladesh. International Journal of Environmental Science. *Management and Engineering Research*, *1*(4), 172–178.

Hossain, M. S., Das, N. G., Sarker, S., & Rahaman, M. Z. (2012b). Fish diversity and habitat relationship with environmental variables at Meghna river estuary, Bangladesh. *Egyptian Journal of Aquatic Research*, *38*(3), 213–226. doi:10.1016/j.ejar.2012.12.006

Hossain, M. S., Dearing, J. A., Rahman, M. M., & Salehin, M. (2016). Recent changes in ecosystem services and human well-being in the Bangladesh coastal zone. *Regional Environmental Change*, *16*(2), 429–443. doi:10.100710113-014-0748-z

Howlett, M. A., McWilliams, M. A., Rademacher, K., O'Neill, J. C., Maitland, T. L., Abels, K., Demetriou, C., & Panter, A. T. (2021). Investigating the Effects of Academic Coaching on College Students' Metacognition. *Innovative Higher Education*, *46*(2), 189–204. Advance online publication. doi:10.100710755-020-09533-7

Hua, J., & Relyea, R. (2019). The effect of a common pyrethroid insecticide on wetland communities. *Environ. Res. Commun.*, *1*(1), 015003. doi:10.1088/2515-7620/aaebb3

Hussain, S. A., & Badola, R. (2010). Valuing Mangrove Benefits: Contribution of Mangrove Forests to Local Livelihoods in Bhitarkanika Conservation Area, East Coast of India. *Wetlands Ecology and Management*, *18*(3), 321–331. doi:10.100711273-009-9173-3

Hwang, G.-J., & Chen, C.-H. (2016). Influences of an inquiry-based ubiquitous gaming design on students' learning achievements, motivation, behavioral patterns, and tendency towards critical thinking and problem solving: Inquiry-based ubiquitous game for students' learning. *British Journal of Educational Technology*, *48*(4), 950–971. Advance online publication. doi:10.1111/bjet.12464

Ibitayo, O. O. (2006). Egyptian farmers' attitudes and behaviours regarding agricultural pesticides: Implications for pesticide risk communication. *Risk Analysis*, *26*(4), 989–995. doi:10.1111/j.1539-6924.2006.00794.x PMID:16948691

ICF. (2021). *What is Coaching?* https://coachingfederation.org/about

ICH Q2. (2005). *Validation of Analytical Procedures: Text and Methodology*. https://www.ema.europa.eu/en/ich-q2-r1-validation-analytical-procedures-text-methodology

IFRC. (2010). *Disasters in Asia: The case for legal preparedness*. https://disasterlaw.ifrc.org/sites/default/files/media/disaster_law/2020-09/Disasters in Asia_EN 2010.pdf

Iftekhar, M. S. (2006, August). Conservation and management of the Bangladesh coastal ecosystem: Overview of an integrated approach. [). Blackwell Publishing Ltd.]. *Natural Resources Forum, 30*(3), 230–237. doi:10.1111/j.1477-8947.2006.00111.x

IPCS. (2002). *The WHO recommended classification of pesticides by hazard and guidelines to classification 2000-02*. https://www.who.int/ipcs/publications/en/pesticides_hazard.pdf

Irsyam, M., Asrurifak, M., Mikail, R., Sabaruddin, A., & Faisal, L. (2017). 2017 Indonesia earthquake source and map update. One Day Seminar on Disaster in Commemorating the 72nd Indonesian Independence Day, 28 August 2017, Jakarta, Indonesia.

Islam, G. T., Islam, A. S., Shopan, A. A., Rahman, M. M., Lázár, A. N., & Mukhopadhyay, A. (2015). Implications of agricultural land use change to ecosystem services in the Ganges delta. *Journal of Environmental Management, 161*, 443–452. doi:10.1016/j.jenvman.2014.11.018 PMID:25516384

Islam, M. A., Huq, S., Rahman, A., & Conway, G. R. (2000). *Consequences of increased pesticide use. In Environmental Aspects of Agricultural Development in Bangladesh*. The University Press.

Islam, M. S., & Hoq, M. E. (2019). Vulnerability of Aquaculture-Based Fish Production Systems to the Impacts of Climate Change: Insights from Inland Waters in Bangladesh. In M. N. Islam & A. van Amstel (Eds.), *Bangladesh I: Climate Change Impacts, Mitigation and Adaptation in Developing Countries* (pp. 67–97). Springer. doi:10.1007/978-3-319-26357-1_3

Islam, S. N., Reinstädtler, S., & Gnauck, A. (2017). Degraded Coastal Wetland Ecosystems in the Ganges-Brahmaputra Rivers Delta Region of Bangladesh. In *Coastal Wetlands: Alteration and Remediation* (pp. 187–213). Springer. doi:10.1007/978-3-319-56179-0_6

Ismatova, M. S. (2021). Development of coaching technologies in the educational space of the university. *Current Research Journal of Pedagogics*. doi:10.37547/pedagogics-crjp-02-05-25

IWM. (2005) *Monitoring of Hydrological and Hydraulic Parameters on Tidal River Management under KJDRP Areal, Final report*. Bangladesh Water Development Board.

Jakarya, M., Sikder, M. T., & Yousuf, A. I. (2016). Review of an Integrated Strategy of Climate Change Adaptation Using Tidal River Management. *Journal of Health and Environmental Research, 2*(1), 1–4. doi:10.11648/j.jher.20160201.11

Jaksztat, S., Neugebauer, M., & Brandt, G. (2021). Back out or hang on? An event history analysis of withdrawal from doctoral education in Germany. *Higher Education, 82*(5), 937–958. Advance online publication. doi:10.100710734-021-00683-x

John, V. W. (2013). Water Supply to Communities and Challenges Facing the Lake Bogoria Basin Water Resource Users Association in the Upper Catchment of the Lake Bogoria Basin. *European International Journal of Science and Technology, 2*(4).

Johnson, F., Higgins, P., & Stephens, C. (2021). Climate change and hydrological risk in the Pacific: A Humanitarian Engineering perspective. *Journal of Water and Climate Change*, *12*(3), 647–678. doi:10.2166/wcc.2021.277

Kabba, V. T. S., & Li, J. (2011). Analysis of land use and land cover changes, and their ecological implications in Wuhan, China. *Journal of Geography and Geology*, *3*(1), 104. doi:10.5539/jgg.v3n1p104

Kachuri, L., Harris, M. A., MacLeod, J. S., Tjepkema, M., Peters, P. A., & Demers, P. A. (2017). Cancer risks in a population-based study of 70,570 agricultural workers: Results from the Canadian census health and Environment cohort (CanCHEC). *BMC Cancer*, *17*(1), 343. doi:10.118612885-017-3346-x PMID:28525996

Kader, M. A., Hossain, M. K., & Kabir, M. H. (2019). Wind Speed Reduction Efficiency of Potenga-Muhuri Irrigation Project Coastal Belt in Chittagong, Bangladesh. *Journal of Forest and Environmental Science*, *35*(2), 78–89. doi:10.7747/JFES.2019.35.2.78

Kalogeras, I., Evangelidis, C., Melis, N., & Boukouras, K. (2012). *'The Athens' Acropolis Strong Motion Array', EGU General Assembly, Geophysical Research Abstracts* (Vol. 12). EGU.

Kapogianni, E. (2021). Geotechnical Structures in the Ancient World: The case of the Ziggurat of UR. *12th International Conference on Structural Analysis of Historical Constructions SAHC 2020*. P. Roca, L. Pelà, & C. Molins (Eds.)

Kapogianni, E., Sakellariou, M.G., & Laue, J. (2017). *Experimental investigation of reinforced soil slopes in a geotechnical centrifuge, with the use of optical fibre sensors, 35*(2), 585-605. Geotechnical and Geological Engineering, Springer International Publishing.

Kapogianni, E., Tsafou, S., & Sakellariou, M. G. (2020b). Image Analysis and Structural Monitoring of Physical Geotechnical Models Subjected to Seismic Loading, pp. 674-686. Springer Series in Geomechanics and Geoengineering.

Kapogianni, E., Kalogeras, I., Psarropoulos, P., Michalopoulou, D., Eleftheriou, V., & Sakellariou, M. (2019). 'Suitability of Optical Fibre Sensors and Accelerographs for the Multi-disciplinary Monitoring of a Historically Complex Site: The Case of the Acropolis Circuit Wall and Hill'. Geotechnical and Geological Engineering. *Springer International Publishing*, *37*(5), 4405.

Kapogianni, E., Psarropoulos, P. N., Kokoris, D., Kalogeras, I., Michalopoulou, D., Eleftheriou, V., & Sakellariou, M. G. (2020a). *Impact of Local Site Conditions on the Seismic Response of the Athenian Acropolis Hill. Geotechnical and Geological Engineering*. Springer International Publishing.

Kapogianni, E., Sakellariou, M., Laue, J., & Springman, S. (2016). Investigation of the mechanical behaviour of the interface between soil and reinforcement, via experimental and numerical modelling. *Procedia Engineering, Elsevier*, *143*, 419–426.

Karim, M. F., & Mimura, N. (2008). Impacts of climate change and sea-level rise on cyclonic storm surge floods in Bangladesh. *Global Environmental Change*, *18*(3), 490–500. doi:10.1016/j.gloenvcha.2008.05.002

Karim, R., & Mondal, R. (2017). Local Stakeholder Analysis of Tidal River Management (TRM) at BeelKapalia and the Implication of TRM as a Disaster Management Approach. *Hydrology*, *5*(1), 1. doi:10.11648/j.hyd.20170501.11

Kearns, H., Forbes, A., Gardiner, M., & Marshall, K. (2008). When a high distinction isn't good enough: A review of perfectionism and self-handicapping. *Australian Educational Researcher*.

Kearns, H., Gardiner, M., & Marshall, K. (2008). Innovation in PhD completion: The hardy shall succeed (and be happy!). *Higher Education Research & Development*, *27*(1), 77–89. Advance online publication. doi:10.1080/07294360701658781

Kebede, G., Zenebe, T., Disassa, H., & Tolos, T. (2014). Review on Detection of Antimicrobial Residues in Raw Bulk Milk in Dairy Farms. *African Journal of Basic & Applied Sciences*, *6*(4), 87–97. doi:10.5829/idosi.ajbas.2014.6.4.8642

Kelley, T. R., & Knowles, J. G. (2016). A conceptual framework for integrated STEM education. *International Journal of STEM Education*, *3*(1). Available at: https://stemeducationjournal.springeropen.com/articles/10.1186/s40594-016-0046-z

Khadim, F. K., Kar, K. K., Halder, P. K., Rahman, M. A., & Morshed, A. M. (2013). Integrated water resources management (IWRM) impacts in south west coastal zone of Bangladesh and fact-finding on tidal river management (TRM). *Journal of Water Resource and Protection*, *5*(10), 953–961. doi:10.4236/jwarp.2013.510098

Khan, A. E., Ireson, A., Kovats, S., Mojumder, S. K., Khusru, A., Rahman, A., & Vineis, P. (2011). Drinking water salinity and maternal health in coastal Bangladesh: Implications of climate change. *Environmental Health Perspectives*, *119*(9), 1328–1332. doi:10.1289/ehp.1002804 PMID:21486720

Khatri, R., Henderson, C., Cole, R., Froyd, J. E., Friedrichsen, D., & Stanford, C. (2017). Characteristics of well-propagated teaching innovations in undergraduate STEM. *International Journal of STEM Education*, *4*(1), 2. doi:10.118640594-017-0056-5

Khatun, R., Howlader, A. J., Ahmed, S., Islam, N., Alam, K., Haider, S., Mahmud, M. S., & Hasan, M. A. (2018). Validation of the Declared Withdrawal Periods of Antibiotics. *Universal Journal of Public Health*, *6*(1), 14–22. doi:10.13189/ujph.2018.060103

Kipper, H., & Ruutmann, T. (2012). Teaching for understanding in engineering education. *2012 15th International Conference on Interactive Collaborative Learning, 3*(1), 55–63. 10.1109/ICL.2012.6402148

Kiryakova, A., Olkhovaya, T., Melekesov, G., & Presnov, A. (2015). The Axiological Approach to the Analysis of the Problems of Modern University Education. *Mediterranean Journal of Social Sciences*, *6*. Advance online publication. doi:10.5901/mjss.2015.v6n2s3p22

Koerner, R. M. (2012). *Designing with Geosynthetics* (6th ed.). Xlibris Publishing Co.

Korbach, A., Brünken, R., & Park, B. (2017). Measurement of cognitive load in multimedia learning: A comparison of different objective measures. *Instructional Science*, *45*(4), 515–536. doi:10.100711251-017-9413-5

Krasnova, T., & Demeshko, M. (2015). Tutor-mediated Support in Blended Learning. *Procedia: Social and Behavioral Sciences*, *166*, 404–408. doi:10.1016/j.sbspro.2014.12.544

Kremmyda, G., Georgoulas, A., & Koumpouros, Y. (2020). Demystifying Humanitarian Engineering: A comparative study on perceptions in UK and Asia. *2020 IEEE Global Engineering Education Conference (EDUCON 2020) 27-30 April 2020*, 1009–1014.

Kremmyda, G., Georgoulas, A., & Koumpouros, Y. (2020). Demystifying humanitarian engineering: A comparative study on perceptions in UK and Asia. *2020 IEEE Global Engineering Education Conference (EDUCON 2020)*, 1009–1014.

Kremmyda, G., Georgoulas, A., Koumpouros, Y., & Mottram, J. T. (2020). Taxonomy of engineering attributes for tackling humanitarian challenges. *Proceedings of the 2020 IEEE Global Engineering Education Conference (EDUCON)*, 1003–1008.

Kruger, J.-L., Hefer, E., & Matthew, G. (2013). Measuring the impact of subtitles on cognitive load: Eye tracking and dynamic audiovisual texts. *Proceedings of the 2013 Conference on Eye Tracking South Africa*, 62-66. 10.1145/2509315.2509331

Kruger, J.-L., Hefer, E., & Matthew, G. (2014). Attention distribution and cognitive load in a subtitled academic lecture: L1 vs. L2. *Journal of Eye Movement Research*, *7*(5). Advance online publication. doi:10.16910/jemr.7.5.4

Kumar, P., & Yashiro, M. (2014). The Marginal Poor and Their Dependence on Ecosystem Services: Evidence from South Asia and Sub-Saharan Africa. In J. von Braun & F. Gatzweiler (Eds.), *Marginality*. Springer. doi:10.1007/978-94-007-7061-4_11

Kuswanto. (2007). *Report on reviewing findings of brick structure in Nglinguk Hamlet, Trowulan Village, Trowulan District, Mojokerto Regency*. BP3Jatim.

Kutz, F. W., Wood, P. H., & Bottimore, D. P. (1991). Organochlorine pesticides and polychlorinated biphenyls in human adipose tissue. *Reviews of Environmental Contamination and Toxicology*, *120*, 1. PMID:1899728

Lambrin, K., Areti, K., Kourkourikos, K., Iliadis, C., Plati, P., & Dimitriadou, A. (2018). History of Antibiotics. *Sumerianz Journal of Medical and Healthcare*, *1*(2), 51–54.

Landers, T. F., Cohen, B., Wittum, T. E., & Larson, E. L. (2012). A Review of Antibiotic Use in Food Animals: Perspective, Policy, and Potential. *Public Health Reports*, *127*(1), 4–22. doi:10.1177/003335491212700103 PMID:22298919

Larson, S. J., Capel, P. D., & Majewski, M. (2020). *Pesticides in surface waters: Distribution, trends, and governing factors (No. 3)*. CRC Press.

LeConte, L. J. (1905). Discussion on the paper, "Notes on the improvement of river and harbor outlets in the United States" by D. A. Watt, paper no. 1009. *Trans. ASCE*, *55*(December), 306–308.

Lee, H. (2014). Measuring cognitive load with electroencephalography and self-report: Focus on the effect of English-medium learning for Korean students. *Educational Psychology*, *34*(7), 838–848. doi:10.1080/01443410.2013.860217

Lee, W. K., Tay, S. H. X., Ooi, S. K., & Friess, D. A. (2021). Potential Short Wave Attenuation Function of Disturbed Mangroves. *Estuarine, Coastal and Shelf Science*, *248*, 106747. Advance online publication. doi:10.1016/j.ecss.2020.106747

Lehan, T. J., Hussey, H. D., & Shriner, M. (2018). The influence of academic coaching on persistence in online graduate students. *Mentoring & Tutoring*, *26*(3), 289–304. Advance online publication. doi:10.1080/13611267.2018.1511949

Leydens, J. A., & Lucena, J. C. (2014). Social Justice: A Missing, Unelaborated Dimension in Humanitarian Engineering and Learning Through Service. *International Journal for Service Learning in Engineering*, *9*(2), 1–28. doi:10.24908/ijsle.v9i2.5447

Li, V. C. (2008). Engineered cementitious composites (ECC) material, structural, and durability performance. In E. Nawy (Ed.), *Concrete construction engineering handbook* (2nd ed.). CRC Press.

Li, L., & Zha, Y. (2020). Population Exposure to Extreme Heat in China: Frequency, Intensity, Duration and Temporal Trends. *Sustainable Cities and Society*, *60*, 102282. Advance online publication. doi:10.1016/j.scs.2020.102282

Li, M., Chen, W., Zhang, Z., Zhang, Z., & Peng, B. (2018). Fermentative degradation of Patulin by Saccharomyces cerevisiae in aqueous solution. *Lebensmittel-Wissenschaft + Technologie*, *97*, 427–433. doi:10.1016/j.lwt.2018.07.040

Li, M., & Lei, Q. (2015). On Quality Assurance Mechanisms in Engineering Education: A Case Study of Purdue University. In *International Conference on Social Science, Education Management and Sports Education* (pp. 494-497). Atlantis Press. 10.2991semse-15.2015.126

Liu, C., Wang, R., Li, L., Ding, G., Yang, J., & Li, P. (2020). Effects of encoding modes on memory of naturalistic events. *Journal of Neurolinguistics*, *53*(100863), 100863. doi:10.1016/j.jneuroling.2019.100863

Liu, J., Liu, M., Tian, H., Zhuang, D., Zhang, Z., Zhang, W., Tang, X., & Deng, X. (2005). Spatial and temporal patterns of China's cropland during 1990–2000: An analysis based on Landsat TM data. *Remote Sensing of Environment*, *98*(4), 442–456. doi:10.1016/j.rse.2005.08.012

Liu, M., Wang, J., Yang, Q., Hu, N., Zhang, W., Zhu, W., Wang, R., Suo, Y., & Wang, J. (2019). Patulin removal from apple juice using a novel cysteine-functionalized metal-organic framework adsorbent. *Food Chemistry*, *270*, 1–9. doi:10.1016/j.foodchem.2018.07.072 PMID:30174021

Loftus, J. J., Jacobsen, M., & Wilson, T. D. (2017). Learning and assessment with images: A view of cognitive load through the lens of cerebral blood flow. *British Journal of Educational Technology*, *48*(4), 1030–1046. doi:10.1111/bjet.12474

Loftus, J. J., Jacobsen, M., & Wilson, T. D. (2018). The relationship between spatial ability, cerebral blood flow and learning with dynamic images: A transcranial Doppler ultrasonography study. *Medical Teacher*, *40*(2), 174–180. doi:10.1080/0142159X.2017.1395401 PMID:29113527

Mahato, D. K., Kamle, M., Sharma, B., Pandhi, S., Devi, S., Dhawan, K., Selvakumarf, R., Mishra, D., Kumar, A., Arora, S., Singh, N. A., & Kumar, P. (2021). Patulin in food: A mycotoxin concern for human health and its management strategies. *Toxicon*, *198*, 12–33. doi:10.1016/j.toxicon.2021.04.027 PMID:33933519

Mahmuduzzaman, M., Ahmed, Z. U., Nuruzzaman, A. K. M., & Ahmed, F. R. S. (2014). Causes of salinity intrusion in coastal belt of Bangladesh. *International Journal of Plant Research*, *4*(4A), 8–13.

Mahunu, G. K., Zhang, H., Apaliya, M. T., Yang, Q., Zhang, X., & Zhao, L. (2018). Bamboo leaf flavonoid enhances the control effect of Pichia caribbica against Penicillium expansum growth and patulin accumulation in apples. *Postharvest Biology and Technology*, *141*, 1–7. doi:10.1016/j.postharvbio.2018.03.005

Makransky, G., Terkildsen, T. S., & Mayer, R. E. (2019a). Adding immersive virtual reality to a science lab simulation causes more presence but less learning. *Learning and Instruction*, *60*, 225–236. doi:10.1016/j.learninstruc.2017.12.007

Makransky, G., Terkildsen, T. S., & Mayer, R. E. (2019b). Role of subjective and objective measures of cognitive processing during learning in explaining the spatial contiguity effect. *Learning and Instruction*, *61*, 23–34. doi:10.1016/j.learninstruc.2018.12.001

Mallick, D. L., Rahman, A., Alam, M., Juel, A. S. M., Ahmad, A. N., & Alam, S. S. (2005). Case study 3: Bangladesh floods in Bangladesh: A shift from disaster management towards disaster preparedness. *IDS Bulletin*, *36*(4), 53–70. doi:10.1111/j.1759-5436.2005.tb00234.x

Mamun, M. A., & Ullah, I. (2020). COVID-19 Suicides in Pakistan, Dying Off Not COVID-19 Fear but Poverty? – The Forthcoming Economic Challenges for a Developing Country. *Brain, Behavior, and Immunity*, *87*, 163–166. doi:10.1016/j.bbi.2020.05.028 PMID:32407859

Manathunga, C. (2005). Early warning signs in postgraduate research education: A different approach to ensuring timely completions. Teaching in Higher Education. doi:10.1080/1356251042000337963

Manathunga, C., & Goozée, J. (2007). Challenging the dual assumption of the "always/already" autonomous student and effective supervisor. *Teaching in Higher Education*, *12*(3), 309–322. Advance online publication. doi:10.1080/13562510701278658

Mandal, S., Rahman, A. M. H., Mamun, M. I. R., Shoeb, M., & Nahar, N. (2012). Method Development and validation for estimation of Commercially produced sex pheromones in Lure. *J. Bangladesh Chem. Soc.*, *25*(2), 180–185. doi:10.3329/jbcs.v25i2.15084

Manzoor, A., Aziz, H., Jahanzaib, M., Wasim, A., & Hussain, S. (2017). Transformational Model for Engineering Education from CoOntent-based to Outcome-based Education. *International Journal of Continuing Engineering Education and Life Long Learning*, *27*(4), 266-286.

Martin, D. A., Conlon, E., & Bowe, B. (2021). A Multi-level Review of Engineering Ethics Education: Towards a Socio-technical Orientation of Engineering Education for Ethics. *Science and Engineering Ethics*, *27*(5), 60. doi:10.100711948-021-00333-6 PMID:34427811

Maxwell, S. L., Butt, N., Maron, M., McAlpine, C. A., Chapman, S., Ullmann, A., Segan, D. B., & Watson, J. E. M. (2019). Conservation Implications of Ecological Responses to Extreme Weather and Climate Events. *Diversity & Distributions*, *25*(4), 613–625. doi:10.1111/ddi.12878

Mayer, R. (2014a). Cognitive Theory of Multimedia Learning. In R. Mayer (Ed.), *The Cambridge Handbook of Multimedia Learning* (2nd ed., pp. 43–72). doi:10.1017/CBO9781139547369.005

Mayer, R. (2014b). Introduction to Multimedia Learning. In R. Mayer (Ed.), *The Cambridge Handbook of Multimedia Learning* (2nd ed., pp. 1–24).

Mayer, R. E., & Moreno, R. (2003). Nine Ways to Reduce Cognitive Load in Multimedia Learning. *Educational Psychologist*, *38*(1), 43–52. doi:10.1207/S15326985EP3801_6

Maza, M., Lara, J. L., & Losada, I. J. (2019). Experimental Analysis of Wave Attenuation and Drag Forces in a Realistic Fringe Rhizophora Mangrove Forest. *Advances in Water Resources*, *131*, 103376. Advance online publication. doi:10.1016/j.advwatres.2019.07.006

Mazher, M., Aziz, A. A., Malik, A. S., Qayyum, A. (2015a). A statistical analysis on learning and nonlearning mental states using EEG. *IEEE Student Symposium in Biomedical Engineering Sciences (ISSBES)*, 36-40. 10.1109/ISSBES.2015.7435889

Mazher, M., Aziz, A. A., Malik, A. S., Qayyum, A. (2015b). A comparison of brain regions based on EEG during multimedia learning cognitive activity. *IEEE Student Symposium in Biomedical Engineering Sciences (ISSBES)*, 31-35. 10.1109/ISSBES.2015.7435888

Mazher, M., Abd Aziz, A., Malik, A. S., & Ullah Amin, H. (2017). An EEG-Based Cognitive Load Assessment in Multimedia Learning Using Feature Extraction and Partial Directed Coherence. *IEEE Access: Practical Innovations, Open Solutions*, *5*, 14819–14829. doi:10.1109/ACCESS.2017.2731784

Mazzurco, A., & Daniel, S. (2020). Socio-technical Thinking of Students and Practitioners in the Context of Humanitarian Engineering. *Journal of Engineering Education*, *109*(2), 243–261. doi:10.1002/jee.20307

Mazzurco, A., & Jesiek, B. (2017). Five Guiding Principles to Enhance Community Participation in Humanitarian Engineering Projects. *Journal of Humanitarian Engineering*, *5*(2). Advance online publication. doi:10.36479/jhe.v5i2.80

McDowell, L., White, S., & Davis, H. C. (2004). Changing assessment practice in engineering: How can understanding lecturer perspectives help? *European Journal of Engineering Education*, *29*(2), 173–181. doi:10.1080/030437903100 01633151

Mendoza-González, G., Martínez, M. L., Lithgow, D., Pérez-Maqueo, O., & Simonin, P. (2012). Land use change and its effects on the value of ecosystem services along the coast of the Gulf of Mexico. *Ecological Economics*, *82*, 23–32. doi:10.1016/j.ecolecon.2012.07.018

Menkem, Z. E., Ngagom, B. L., Shinwin, S., Tamunioh, A., & Boyom, F. (2018). Antibiotic residues in food animals: Public health concern. *Acta Ecologica Sinica*, *39*(5), 411–415. doi:10.1016/j.chnaes.2018.10.004

Millennium Ecosystem Assessment (MA). (2005). *Ecosystems and Human Well-Being: Synthesis "Archived copy" (PDF). Archived (PDF) from the original on 3 December 2013*. Island Press.

Miller, L. (2017). *Source Protection And Drinking Water Quality in the Comarca Ngäbe-Buglé, Panama* [Master's Thesis]. Michigan Technological University. doi:10.37099/mtu.dc.etdr/365

Miller, G. A. (1956). The magical number seven, plus or minus two: Some limits on our capacity for processing information. *Psychological Review, 63*(2), 81–97. doi:10.1037/h0043158 PMID:13310704

Milovanovic, J., Shealy, T., & Godwin, A. (2022). Senior Engineering Students in the USA Carry Misconceptions about Climate Change: Implication for Engineering Education. *Journal of Cleaner Production, 345*, 131129. doi:10.1016/j.jclepro.2022.131129

Minar, M. H., Hossain, B., & Shamsuddin, M. D. (2013). Climate change and coastal zone of Bangladesh: Vulnerability, resilience and adaptability. *Middle East Journal of Scientific Research, 13*(1), 114–120.

Ministry of Education and Culture. (2014). Minister Decree No. 49/ 2014 on National Standard of Higher Education, Indonesia.

Ministry of Education and Culture. (2022). *Muara Takus Temple*. https://petabudaya.belajar.kemdikbud.go.id/Repositorys/muara_takus/

Ministry of Manpower. (2018). *Indonesian national standard of work competence, category: Construction, Main group: Building construction, Field: RISHA building, Decree of Minister of Manpower No. 221/ 2018*. Ministry of Manpower.

Ministry of Public Works and Housing. (2017a). *Ministry of Public Works and Housing renovated 1,707 houses in Prabumulih*. https://pu.go.id/berita/kementerian-pupr-bedah-1-707-rumah-di-prabumulih

Ministry of Public Works and Housing. (2017b). *Practical guide on requirement of earthquake-resistant building for confined masonry building*. Ministry of Public Works and Housing.

Ministry of Public Works and Housing. (2017c). *Practical guide on requirement of earthquake-resistant building for RISHA*. Ministry of Public Works and Housing.

Ministry of Public Works and Housing. (2017d). *Practical guide on requirement of earthquake-resistant building for wooden building*. Ministry of Public Works and Housing.

Mishra, M., Acharyya, T., Santos, C. A. G., da Silva, R. M., Kar, D., Kamal, A. H. M., & Raulo, S. (2021). Geo-ecological Impact Assessment of Severe Cyclonic Storm Amphan on Sundarban Mangrove Forest Using Geospatial Technology. *Estuarine, Coastal and Shelf Science, 260*, 107486. Advance online publication. doi:10.1016/j.ecss.2021.107486

Mitcham, C., & Muñoz, D. (2010). *Humanitarian Engineering* (Vol. 5). Morgan & Claypool.

Mohamed, Z. (2006). *Managing Student Learning Time for Effective Learning*. Enhancing Human Capital Development Through Quality Technical Education.

Molyn, J. (2020). *The Role and Effectiveness of Coaching in Increasing Self-Efficacy and Employability Efforts of Higher Education Students*. doi:10.29007/294n

Mondal, M. K., Tuong, T. P., Ritu, S. P., Choudhury, M. H. K., Chasi, A. M., Majumder, P. K., & Adhikary, S. K. (2006). Coastal water resource use for higher productivity: participatory research for increasing cropping intensity in Bangladesh. *Environment and Livelihoods in Tropical Coastal Zones: Managing Agriculture-Fishery-Aquaculture Conflicts*, 72-84.

Mondal, M. S., Jalal, M. R., Khan, M., Kumar, U., Rahman, R., &Huq, H. (2013). Hydro-meteorological trends in southwest coastal Bangladesh: perspectives of climate change and human interventions.

Montgomery, D. (2016). Emperor Yu's Great Flood. *Science, 353*(6299), 538–539. PMID:27493166

Montgomery, J. M., Bryan, K. R., Mullarney, J. C., & Horstman, E. M. (2019). Attenuation of Storm Surges by Coastal Mangroves. *Geophysical Research Letters*, *46*(5), 2680–2689. doi:10.1029/2018GL081636

Montgomery, M., Knight, J., Heuckeroth, D., & Walker, W. (2020). Community-based Latrine Development with Engineering Without Borders – USA. In H. Tsao & L. K. Wang (Eds.), *Evolutionary Progress in Science, Technology, Engineering, Arts, and Mathematics (STEAM)*. Lennox Institute Press.

Moore, K., Jones, C., & Frazier, R. S. (2017). Engineering Education For Generation Z. *American Journal of Engineering Education*, *8*(2), 111–125. doi:10.19030/ajee.v8i2.10067

Mou, S. A., Islam, R., Shoeb, M., & Nahar, N. (2021). Determination of chloramphenicol in meat samples using liquid chromatography–tandem mass spectrometry. *Food Science & Nutrition*, *9*(10), 1–6. doi:10.1002/fsn3.2530 PMID:34646535

Musa, K.B. (2008). *Identifying Land Use Changes and Its Socio-economic Impacts: A Case Study of ChacoriaSundarban in Bangladesh*. BBS.

Nair, C. S., Patil, A., & Mertova, P. (2009). Re-engineering graduate skills - A case study. *European Journal of Engineering Education*, *34*(2), 131–139. doi:10.1080/03043790902829281

National Geographic Indonesia. (2015). *Tracing the site of the oldest observatory in Indonesia*. National Geographic. https://nationalgeographic.grid.id/read/13297532/melacak-tapak-observatorium-tertua-di-indonesia?page=all

Naturally:wood. (2022). *Cross-laminated timber (CLT)*. Naturally Wood. https://www.naturallywood.com/products/cross-laminated-timber/

Nelson, M.L. & Ismail, M Y. (2007). Comprehensive Medicinal Chemistry. *The Antibiotic and Nonantibiotic Tetracyclines,* 597–628. doi:10.1016/B0-08-045044-X/00221-2

Niethammer, K. R., White, D. H., Baskett, T. S., & Sayre, M. W. (1984). Presence and biomagnification of organochlorine chemical residues in oxbow lakes of northeastern Louisiana. *Archives of Environmental Contamination and Toxicology*, *13*(1), 63–74. doi:10.1007/BF01055647 PMID:6703785

Nilsson, L., Madon, T., & Sastry, S. S. (2014). Toward a new field of Development Engineering: Linking technology design to the demands of the poor. *Procedia Engineering*, *78*, 3–9. Advance online publication. doi:10.1016/j.proeng.2014.07.032

Nowreen, S., Jalal, M. R., & Khan, M. S. A. (2014). Historical analysis of rationalizing South West coastal polders of Bangladesh. *Water Policy*, *16*(2), 264–279. doi:10.2166/wp.2013.172

Nuruddin, M. F., Bayuaji, R., Masilamani, M. B., & Biyanto, T. R. (2010). Sidoarjo mud: A potential cement replacement material. *Civil Engineering Dimension*, *12*(1), 18–22.

O'Brien, M. P. (1931). Estuary and Tidal Prisms Related to Entrance Areas. *Civil Engineering (New York, N.Y.)*, *1*(8), 738–739.

O'Brien, M. P. (1969). Equilibrium flow areas of inlets on sandy coasts. *Journal of Waterway, Port, Coastal, and Ocean Engineering*, *95*(l), 43–52.

Oakley, A. (2012). Foreword. In An Introduction to Systematic Reviews. SAGE Publications.

Ohkawa, H., Miyagawa, H., & Lee, P. W. (2007). Pesticide Residues in Food and International Trade: Regulation and Safety Considerations. Pesticide Chemistry: Crop Protection, Public Health, Environmental Safety, 29-41.

Ortiz De Guinea, A., Titah, R., & Léger, P.-M. (2013). Measure for Measure: A two study multi-trait multi-method investigation of construct validity in IS research. *Computers in Human Behavior, 29*(3), 833–844. doi:10.1016/j.chb.2012.12.009

Örün, Ö., & Akbulut, Y. (2019). Effect of multitasking, physical environment and electroencephalography use on cognitive load and retention. *Computers in Human Behavior, 92*, 216–229. doi:10.1016/j.chb.2018.11.027

Ozel, P. B.-M. (2022). A Review for Neuroimaging Techniques in Multimedia Learning: Methodological Framework. In *AI-Enabled Smart Healthcare Using Biomedical Signals* (p. 26). IGI Global.

Ozel, P. B.-M., Mutlu-Bayraktar, D., Altan, T., Coskun, V., & Olamat, A. (2021). Neuroimaging tools in multimedia learning: A systematic review. *Interactive Learning Environments*, 1–18. doi:10.1080/10494820.2021.1984255

Paas, F., Renkl, A., & Sweller, J. C. (2003). Cognitive Load Theory and Instructional Design: Recent Developments. *Educational Psychologist, 38*(1), 1–4. doi:10.1207/S15326985EP3801_1

Park, J. J., Park, M., & Smith, J. (2021). *Engineering students' concepts of humanitarian engineering and their identity development as humanitarian engineers.* Sustainability. doi:10.3390u13168845

Parvin, G. A., Ali, M. H., Fujita, K., Abedin, M. A., Habiba, U., & Shaw, R. (2017). Land Use Change in Southwestern Coastal Bangladesh: Consequence to Food and Water Supply. In Land Use Management in Disaster Risk Reduction (pp. 381-401). Springer Japan.

Patel, S. J., Wellington, M., Shah, R. M., & Ferreira, M. J. (2020). Antibiotic Stewardship in Food-Producing Animals: Challenges, Progress, and Opportunities. *Clinical Therapeutics.* doi:10.1016/j.clinthera.2020.07.004

Patil, A., & Codner, G. (2007). Accreditation of engineering education: Review, observations and proposal for global accreditation. *European Journal of Engineering Education, 32*(6), 639–651. doi:10.1080/03043790701520594

Patrick, G. L. (2013). *An Introduction to Medicinal Chemistry* (5th ed.). Oxford University Press.

PCA. (2019). *Autoclaved Aerated Concrete.* PCA. https://www.cement.org/cement-concrete/paving/buildings-structures/concrete-homes/building-systems-for-every-need/autocl
aved-aerated-concrete

Pearson, M., Cowan, A., & Liston, A. (2009). PhD education in science. In Changing practices of doctoral education (pp. 100–112). Academic Press.

Pereira, E., Queiroz, C., Pereira, H. M., & Vicente, L. (2005). Ecosystem services and human well-being: A participatory study in a mountain community in Portugal. *Ecology and Society, 10*(2), art14. doi:10.5751/ES-01353-100214

Perez, N., Gutierrez, R., Nao, M., Diaz, G., Luna, H., Escobar, I., & Muniev, Z. (2002). Liquid chromatographic determination of multiple sulfonamides, nitrofurans, and chloramphenicol residues in pasteurized milk. *Journal of AOAC International, 85*(1), 20–24. doi:10.1093/jaoac/85.1.20 PMID:11878601

Pesticides and Human Health. (n.d.). *Californians for Pesticide Reforms.* http://www.Pesticidereform.org/section.php?id=38

Pethick, J., & Orford, J. D. (2013). Rapid rise in effective sea-level in southwest Bangladesh: Its causes and contemporary rates. *Global and Planetary Change, 111*, 237–245. doi:10.1016/j.gloplacha.2013.09.019

Pham Thi Hang. (2021). University and enterprise cooperation in the current context - Hợp tác đại học và doanh nghiệp trong bối cảnh hiện nay. *State Management Review.*

Pinhome. (2021). *RISHA (Rumah Instan Sehat Sederhana).* https://www.pinhome.id/kamus-istilah-properti/risha/

Potkonjak, V., Gardner, M., Callaghan, V., Mattila, P., Guetl, C., Petrović, V. M., & Jovanović, K. (2016). Virtual laboratories for education in science, technology, and engineering: A review. *Computers & Education*, *95*, 309–327. doi:10.1016/j.compedu.2016.02.002

Pradoto, R., Oktavianus, A., Pribadi, K. S., Rasmawan, I. M. A. B., & Wulandari, L. D. (2022). Palu housing reconstruction process: Reviewing and learning after the 2018 earthquake. Sustainable Infrastructure and Built Environment (SIBE-2022), Bandung, Indonesia.

Prakasam, C. (2010). Land use and land cover change detection through remote sensing approach: A case study of Kodaikanaltaluk, Tamil Nadu. *International Journal of Geomatics and Geosciences*, *1*(2), 150.

Pribadi, K. S., Kusumastuti, D., & Rildova. (2008). *Learning from recent Indonesian earthquakes: An overview to improve structural performance.* The 14th World Conference on Earthquake Engineering, Beijing, China.

Pribadi, K. S., Pradoto, R., Hanafi, E., & Rasmawan, I. M. (2020). Lombok earthquake, one year later: Housing sector recovery. *E3S Web of Conferences, 156*, 1012. https://doi.org/ doi:10.1051/e3sconf/202015601012

Pribadi, K. S., Abduh, M., Wirahadikusumah, R. D., Hanifa, N. R., Irsyam, M., Kusumaningrum, P., & Puri, E. (2021). Learning from past earthquake disasters: The need for knowledge management system to enhance infrastructure resilience in Indonesia. *International Journal of Disaster Risk Reduction*, *64*, 102424. https://doi.org/https://doi.org/10.1016/j.ijdrr.2021.102424

Pribadi, K. S., & Chan, T.-K. (2022). Construction workers and the informal sector. In T.-K. Chan & K. S. Pribadi (Eds.), *Construction in Indonesia: Looking back and moving forward* (p. 18). https://doi.org/https://doi.org/10.1201/9781003149866

Pribadi, K. S., Hidayat, B., Triyadi, S., & Harapan, A. (2009). Indigenous knowledge on house building system in West Sumatra and Southern-West Jawa, Indonesia. In R. Shaw, A. Sharma, & Y. Takeuchi (Eds.), *Indigenous knowledge and disaster risk reduction*. Nova Science.

Psarropoulos, P., & Gkantona, A. (2020). *Information Technology and a New Theory for the Verification of the Location of the Ancient City of Helike in Achaia, Greece*. Springer Series in Geomechanics and Geoengineering.

Psarropoulos, P., Kapogianni, E., Kalogeras, I., Michalopoulou, D., Eleftheriou, V., Dimopoulos, G., & Sakellariou, M. (2018). *Seismic response of the Circuit Wall of the Acropolis of Athens: Recordings versus numerical simulations*. Journal Soil Dynamics and Earthquake Engineering.

Pulhin, J. M., & Tapia, M. A. (2021). Climate Change Adaptation in the Philippines. In J. J. Pereira, M. K. Zain, & R. Shaw (Eds.), *Climate Change Adaptation in Southeast Asia*. Springer. doi:10.1007/978-981-16-6088-7_8

Qayyum, A., Faye, I., Malik, A. S., & Mazher, M. (2018). Assessment of Cognitive Load using Multimedia Learning and Resting States with Deep Learning Perspective. *IEEE-EMBS Conference on Biomedical Engineering and Sciences (IECBES)*, 600-605. 10.1109/IECBES.2018.8626702

Qayyum, A., Khan, M. K. A. A., Mazher, M., & Suresh, M. (2018). Classification of EEG Learning and Resting States using 1D-Convolutional Neural Network for Cognitive Load Assessment. *IEEE Student Conference on Research and Development (SCOReD)*, 1-5.

QS Asia University Rankings. (2021). https://www.topuniversities.com/university-rankings/asian-university-rankings/2021

Rabby, Y. W., Hossain, M. B., & Hasan, M. U. (2019). Social Vulnerability in the Coastal Region of Bangladesh: An Investigation of Social Vulnerability Index and Scalar Change Effects. *International Journal of Disaster Risk Reduction*, *41*, 101329. Advance online publication. doi:10.1016/j.ijdrr.2019.101329

Radt, S. (2011). *Strabons Geographika* (C. Apparatus, Trans.). Vandenhoeck & Ruprecht.

Rahman, M. H., & Alam, M. J. B. (1997). Risk assessment of pesticides in Bangladesh. *J. Civil Engineering.*, *25*(1), 97–106.

Rahman, M. R., & Asaduzzaman, M. (2013). Ecology of Sundarban, Bangladesh. *Journal of Science Foundation*, *8*(1-2), 35–47. doi:10.3329/jsf.v8i1-2.14618

Rahman, R., & Salehin, M. (2013). Flood risks and reduction approaches in Bangladesh. In *Disaster risk reduction approaches in Bangladesh* (pp. 65–90). Springer Japan. doi:10.1007/978-4-431-54252-0_4

Rahman, S. (2003). Farm-level pesticide use in Bangladesh: Determinants and awareness. *Agriculture, Ecosystems & Environment*, *95*(1), 241–252. doi:10.1016/S0167-8809(02)00089-0

Rahman, S. (2010). Six decades of agricultural land use change in Bangladesh: Effects on crop diversity, productivity, food availability and the environment, 1948–2006. *Singapore Journal of Tropical Geography*, *31*(2), 254–269. doi:10.1111/j.1467-9493.2010.00394.x

Rahnuma, N. (2020). The Bangladeshi Higher Education Quality Assurance Framework: A Pathway for Transformation. *Quality in Higher Education*, *26*(1), 14–31. doi:10.1080/13538322.2020.1729309

Rakib, M. A., Sasaki, J., Matsuda, H., & Fukunaga, M. (2019). Severe Salinity Contamination in Drinking Water and Associated Human Health Hazards Increase Migration Risk in the Southwestern Coastal Part of Bangladesh. *Journal of Environmental Management*, *240*, 238–248. doi:10.1016/j.jenvman.2019.03.101 PMID:30952044

Rasul, G., & Chowdhury, A. M. J. U. (2010). *Equity and social justice in water resource management in Bangladesh.* Bangladesh Water Development Board.

Ratchet, G. (1983). *Dictionnaire L' Archéologie* (S. A. Robert Laffont, Ed.).

Reinerman-Jones, L., Barber, D. J., Szalma, J. L., & Hancock, P. A. (2017). Human interaction with robotic systems: Performance and workload evaluations. *Ergonomics*, *60*(10), 1351–1368. doi:10.1080/00140139.2016.1254282 PMID:28745552

Reinhardt, O., & Oldroyd, D. R. (1983). 'Kant's theory of earthquakes and volcanic action'. *Annals of Science*, *40*(3), 247–272.

Research and Development Agency PUPR. (2017). 2017 *Indonesia earthquake source and hazard map*. Ministry of Public Works and Housing.

Reynolds, P., Hurley, S. E., Gunier, R. B., Yerbati, S., Quach, T., & Hertz, A. (2005). Residential proximity to agricultural pesticide use and incidence of breast cancer in california. *Environmental Health Perspectives*, *113*(8), 993–1000. doi:10.1289/ehp.7765 PMID:16079069

Ridha, R., Rahmawaty, A. A., & Santoso, H. (2021). Strategy to accelerate post-earthquake rehabilitation and reconstruction through earthquake resistant house zoning in North Lombok Regency. *Planoearth National Seminar*, *2*, 33–41.

Rizal, A., Sahidin, A., & Herawati, H. (2018). Economic Value Estimation of Mangrove Ecosystems in Indonesia. *Biodiversity International Journal*, *2*(3), 98–100. doi:10.15406/bij.2018.02.00051

Roberts, T., & Hutson, D. (1999). *Metabolic Pathways of Agrochemicals. Part 2. Insecticides and Fungicides*. The Royal Society of Chemistry.

Roebeling, P. C., Costa, L., Magalhães-Filho, L., & Tekken, V. (2013). Ecosystem service value losses from coastal erosion in Europe: Historical trends and future projections. *Journal of Coastal Conservation*, *17*(3), 389–395. doi:10.100711852-013-0235-6

Rose, L., & Bhaskaran, P. K. (2016). The role of environmental forcing on tidal dynamics along complex near-shore waters off Bangladesh. *Ocean Engineering*, *116*, 68–81. doi:10.1016/j.oceaneng.2016.02.031

Rouillard, J. J., Benson, D., & Gain, A. K. (2014). Evaluating IWRM implementation success: Are water policies in Bangladesh enhancing adaptive capacity to climate change impacts? *International Journal of Water Resources Development*, *30*(3), 515–527. doi:10.1080/07900627.2014.910756

Roy, A. K. D., & Roy, A. K. D. (2017a). An investigation into the major environment and climate change policy issues in southwest coastal Bangladesh. *International Journal of Climate Change Strategies and Management*, *9*(1), 123–136. doi:10.1108/IJCCSM-03-2016-0028

Roy, K., Gain, A. K., Mallick, B., & Vogt, J. (2017b). Social, hydro-ecological and climatic change in the southwest coastal region of Bangladesh. *Regional Environmental Change*, *17*(7), 1–12. doi:10.100710113-017-1158-9

Rudiarto, I., Handayani, W., & Setyono, J. S. (2018). A Regional Perspective on Urbanization and Climate-Related Disasters in the Northern Coastal Region on Central Java, Indonesia. *Land (Basel)*, *7*(1), 34. doi:10.3390/land7010034

Sadiqi, Z., Trigunarsyah, B., & Coffey, V. (2017). A framework for community participation in post-disaster housing reconstruction projects: A case of Afghanistan. *International Journal of Project Management*, *35*(5), 900–912. https://doi.org/https://doi.org/10.1016/j.ijproman.2016.11.008

Safe, S. (2004). Endocrine disruptors and human health: Is there a problem. *Toxicology*, *205*(1-2), 3–10. doi:10.1016/j.tox.2004.06.032 PMID:15458784

Samuels, P., Klijn, F., & Dijkman, J. (2006). An analysis of the current practice of policies on river flood risk management in different countries. *Irrigation and Drainage*, *55*(S1), S141–S150. doi:10.1002/ird.257

Sarwidi, S. (2020). *Barrataga*. http://barrataga.com/

Savic, S. (2018). Introductory Chapter: Antibiotic Use in Animals Today. In Antibiotic Use in Animals. doi:10.5772/intechopen.73076

Sayed, M. B., & Haruyama, S. (2015). Dynamics of Land Use/Cover Change in Manikganj District, Bangladesh: A Case Study of Manikganj Sadar Upazila. *Journal of Geographic Information System*, *7*(06), 579–587. doi:10.4236/jgis.2015.76046

Schismenos, S., Stevens, G. J., Emmanouloudis, D., Georgeou, N., Shrestha, S., & Chalaris, M. (2021a). Humanitarian Engineering at the Sustainability-Development Nexus: Mapping Vulnerability and Capability Factors for Communities at Risk of Water-based Disasters. *Sustainability Science*, *16*(4), 1185–1199. doi:10.100711625-020-00890-y PMID:33425037

Schismenos, S., Stevens, G., Emmanouloudis, D., Georgeou, N., Shrestha, S., Katopodes, N., & Wali, N. (2022, September). Humanitarian Engineering for Renewable Energy and Flood Early Warning in Remote Communities: A Scoping Review of Enabling Factors and Sustainability. *Journal of Sustainable Development of Energy, Water, and Environment Systems*, *10*(3), 1090406. Advance online publication. doi:10.13044/j.sdewes.d9.0406

Sepulveda, A. (2021). Coaching college students to thrive: Exploring coaching practices in higher education. *Dissertation Abstracts International. B, The Sciences and Engineering*.

Shah, M. A. R., Rahman, A., & Chowdhury, S. H. (2018). Challenges for Achieving Sustainable Flood Risk Management. *Journal of Flood Risk Management*, *11*(S1), S352–S358. doi:10.1111/jfr3.12211

Shampa, M. I. M. P., & Pramanik, I. M. (2012). Tidal river management (TRM) for selected coastal area of Bangladesh to mitigate drainage congestion. *International Journal of Scientific & Technology Research*, *1*(5), 1–6.

Shekar, A., & Tunnicliffe, M. (2018). An introductory course with a humanitarian engineering context. *Proceedings of the 13th International CDIO Conference*, 137–148. Retrieved from https://prism.ucalgary.ca/handle/1880/52101

Shekar, A., & Drain, A. (2016). Community engineering: Raising awareness, skills and knowledge to contribute towards sustainable development. *International Journal of Mechanical Engineering Education*, *44*(4), 272–283. Advance online publication. doi:10.1177/0306419016668966

Siagian, T. H., Purhadi, P., Suhartono, S., & Ritonga, H. (2014). Social vulnerability to natural hazards in Indonesia: Driving factors and policy implications. *Natural Hazards*, *70*(2), 1603–1617. doi:10.100711069-013-0888-3

Singh, S., Sharma, R., & Parween, T. (2018). Pesticide Contamination and Human Health Risk Factor. *Modern Age Environmental Problems and Their Remediation*, 49-68. doi:10.1007/978-3-319-64501-8_3

Singh, G. S. (2020). Carbohydrate-based antibiotics: Opportunities and Challenges. In *Carbohydrates in Drug Discovery and Development* (pp. 523–559). Synthesis and Application. doi:10.1016/B978-0-12-816675-8.00013-0

Skakkebaek, N., Rajpert-De, M., & Main, K. (2001). Testicular dysgenesis syndrome: An increasingly common developmental disorder with environmental aspects. *Human Reproduction (Oxford, England)*, *16*(5), 972–978. doi:10.1093/humrep/16.5.972 PMID:11331648

SMEC. (1997). *Khulna-Jessore Drainage Rehabilitation Project, Feasibility Study Report for Overall Drainage Plan.* Bangladesh Water Development Board.

SMEC. (2002). *Khulna Jessore Drainage Rehabilitation Project, Project Completion Report (Final).* Bangladesh Water Development Board.

Smith, J. (2019). Impacts of a humanitarian engineering education pathway on student learning and graduate outcomes. *International Journal for Service Learning in Engineering, Humanitarian Engineering and Social Entrepreneurship.* doi:10.24908/ijsle.v14i1.12555

Smith, J., Anderson, B., Brown, N., Colley, A., Stoakley, A., & Turner, J. (2017). The Rise of Humanitarian Engineering Education in Australasia. In *Proceedings of the 28th Annual Conference of the Australasian Association for Engineering Education (AAEE 2017)* (pp. 312-320). Australasian Association for Engineering Education.

Smith, J., Tran, A. L. H., & Compston, P. (2020). Review of humanitarian action and development engineering education programmes. *European Journal of Engineering Education*, *45*(2), 249–272. doi:10.1080/03043797.2019.1623179

Sodhi, G. S. (2002). *Fungicides and Herbicides*. Fundamental Concepts of Environmental Chemistry.

Sönmez, S. (2018). *"11 steps" process as a research method.* Universal Journal of Educational Research. doi:10.13189/ujer.2018.061125

Sood, A., & Mathukumalli, B. K. P. (2011). Managing international river basins: Reviewing India–Bangladesh transboundary water issues. *Intl. J. River Basin Management*, *9*(1), 43–52. doi:10.1080/15715124.2011.553832

Souza, M. R. A., Veado, L., Moreira, R. T., Figueiredo, E., & Costa, H. (2018). A systematic mapping study on game-related methods for software engineering education. *Information and Software Technology*, *95*, 201–218. doi:10.1016/j.infsof.2017.09.014

Spector, J., Merrill, M.D., Elen, J., & Bishop, M.J. (2014). Handbook of Research on Educational Communications and Technology: Fourth edition. *Technology, Knowledge and Learning, 20*(1), 123–128. doi:10.1007/s10758-014-9231-7

Spence, G. B., & Grant, A. M. (2005). Individual and group life coaching: Initial findings from a randomised, controlled trial. In Evidence-based coaching, Vol 1: Theory, research and practice from the behavioural sciences. Academic Press.

Spencer, D. (2021). Understanding the Coaching Experiences of Non-traditional Students in Higher Education in the UK. *International Journal of Evidence Based Coaching and Mentoring*. Advance online publication. doi:10.24384/00v3-nm61

Springman, S. M., Laue, J., Herzog, R., & El-Hamalawi, A. (2014). The evolution of a physical modelling course over two decades. *Proceedings of the 8th International Conference on Physical Modelling in Geotechnics*, Perth, Australia.

Sridhara, S., Turbett, S. E., Harrisd, J. B., & Larocque, R. C. (2021). Antimicrobial-resistant bacteria in international travelers. *Infectious Diseases*, *34*(5), 423–431. doi:10.1097/QCO.0000000000000751 PMID:34267046

Stanley, S. L., Jr. (2010). Antiparasitic agents. Infectious Diseases, 2, 1490–1507. doi:10.1016/B978-0-323-04579-7.00150-7

Sternberg, R. J. (2007). Creativity as a habit. In A.-G. Tan (Ed.), *Creativity: A handbook for teachers* (pp. 3–25). World Scientific. doi:10.1142/9789812770868_0001

Stocker, T. F., Qin, D., Plattner, G. K., Tignor, M., Allen, S. K., Boschung, J., & Midgley, P. M. (2013). *Climate Change 2013: The Physical Science Basis. Working Group 1 (WG1) Contribution to the Intergovernmental Panel on Climate Change (IPCC) 5th Assessment Report (AR5).* Cambridge, United Kingdom and New York, NY. Cambridge, UK and New York, New York, USA.

Subheesh, N. P., & Sethy, S. S. (2018). Assessment and evaluation practices in engineering education: A global perspective. *3rd International Conference of the Portuguese Society for Engineering Education, CISPEE 2018*, 1–5. 10.1109/CISPEE.2018.8593451

Sun, D., Lee, V., & Lu, Y. (2016). An intelligent data fusion framework for structural health monitoring. *IEEE 11th Conference on Industrial Electronics and Applications*. IEEE. 10.1109/ICIEA.2016.7603550

Suryaman, M. (2020). Orientasi Pengembangan Kurikulum Merdeka Belajar. *Seminar Nasional Pendidikan Bahasa Dan Sastra, 1*(1).

Suwantara, I. K., & Rusli, R. (2013). Experimental study of shear-friction connection of lontar pillars (Borassus flabellifer) with stone in Ammu Hawu traditional house (NTT). *Jurnal Teknik Sipil*, *20*(2), 79. doi:10.5614/jts.2013.20.2.1

Sweller, J. (2010). Element Interactivity and Intrinsic, Extraneous, and Germane Cognitive Load. *Educational Psychology Review*, *22*(2), 123–138. doi:10.100710648-010-9128-5

Sweller, J. (2020). Cognitive load theory and educational technology. *Educational Technology Research and Development*, *68*(1), 1–16. doi:10.100711423-019-09701-3

Sweller, J., van Merrienboer, J. J. G., & Paas, F. (1998). Cognitive architecture and instructional design. *Educational Psychology Review*, *10*(3), 251–296. doi:10.1023/A:1022193728205

Sweller, J., van Merriënboer, J. J. G., & Paas, F. (2019). Cognitive Architecture and Instructional Design: 20 Years Later. *Educational Psychology Review*, *31*(2), 261–292. doi:10.100710648-019-09465-5

Syvitski, J. P., Kettner, A. J., Overeem, I., Hutton, E. W., Hannon, M. T., Brakenridge, G. R., Day, J., Vörösmarty, C., Saito, Y., Giosan, L., & Nicholls, R. J. (2009). Sinking deltas due to human activities. *Nature Geoscience*, *2*(10), 681–686. doi:10.1038/ngeo629

Talchabhadel, R., Nakagawa, H., & Kawaike, K. (2016). Tidal River Management (TRM) and Tidal Basin Management (TBM): A case study on Bangladesh. In E3S Web of Conferences (Vol. 7, p. 12009). EDP Sciences.

Teilmann, G., Juul, A., Skakkebaek, N., & Toppari, J. (2002). Putative effects of endocrine disrupters on pubertal development in the human. *Best Practice & Research. Clinical Endocrinology & Metabolism*, *16*(1), 105–121. doi:10.1053/beem.2002.0184 PMID:11987902

Tejedor, G., Segalàs, J., & Rosas-Casals, M. (2018). Transdisciplinarity in higher education for sustainability: How discourses are approached in engineering education. *Journal of Cleaner Production*, *175*, 29–37. doi:10.1016/j.jclepro.2017.11.085

Tenzin, K., & Norbu, L. (2020). Leveraging Conservation Benefits through Ecosystem-Based Services Approach and Community Engagement in Wetland and Riparian Ecosystems: The Case of Conserving Black-Necked Crane and White-Bellied Heron in Bhutan. In S. Dhyani, A. Gupta, & M. Karki (Eds.), *Nature-based Solutions for Resilient Ecosystems and Societies*. Springer. doi:10.1007/978-981-15-4712-6_10

The Environment Court Act, 2000. extwprlegs1.fao.org/docs/pdf/bgd42277.pdf

The Environment Court Act, 2000. National Integrated Pest Management Policy, B. Distribution and Use of Pesticides in Bangladesh. http://www.doebd.org/ pop/pdf/Report

The Ministry of Education and Culture. *Peraturan Menteri Pendidikan dan Kebudayaan (The Regulation of Ministry of Education and Culture) No. 3 of 2020.*, Pub. L. No. Permen No 3/2020 (2020).

The Republic of Indonesia. (2012). *Undang-Undang Nomor 12 Tahun 2012 tentang Pendidikan Tinggi (Higher Education Act)*. Indonesian Act.

Thomas, E., Salvinelli, C., Harper, J., MacDonald, L., Klees, R., Platais, G., Javernick-Will, A., & Linden, K. (2021). A Body of Knowledge and Pedagogy for Global Engineering. *International Journal for Service Learning in Engineering, Humanitarian Engineering and Social Entrepreneurship*. doi:10.24908/ijsle.v16i1.14483

Thomas, K. A. (2017). The Ganges water treaty: 20 years of csooperation, on India's terms. *Water Policy*, *19*(4), wp2017109. doi:10.2166/wp.2017.109

Thompson, P. M., & Sultana, P. (1996). Distributional and social impacts of flood control in Bangladesh. *The Geographical Journal*, *162*(1), 1–13. doi:10.2307/3060212

Tingsanchali, T., & Karim, M. F. (2005). Flood hazard and risk analysis in the southwest region of Bangladesh. *Hydrological Processes*, *19*(10), 2055–2069. doi:10.1002/hyp.5666

Tirmazi, S. A. A. S., Qadir, M. A., Ahmed, M., Imran, M., Hussain, R., Sharif, M., & Muddassar, M. (2021). Levofloxacin and sulfa drugs linked via Schiff bases: Exploring their urease inhibition, enzyme kinetics and in silico studies. *Journal of Molecular Structure*, *1235*, 130226. doi:10.1016/j.molstruc.2021.130226

Toft, G., Axmon, A., Giwercman, A., Thulstrun, A. M., Rignell-Hydborn, A., Pedersem, H. S., Luwicki, J. K., Zyyezday, V., Zinchuk, A., Spano, M., Manicardi, G. C., Bonefeld-Jorgensen, E. C., Hagmar, L., Bonde, J. P., & INUENDO. (2005). Fertility in four regions spanning large contrasts in serum levels of widespread persistent organochlorines: a cross sectional study. *Environmental Health: A Global Access Science Source, 4*, 26-37. doi:10.1186/1476-069X-4-26

Topcuoglu, M. A. (2012). Transcranial Doppler ultrasound in neurovascular diseases: Diagnostic and therapeutic aspects. *Journal of Neurochemistry*, *123*(S2), 39–51. doi:10.1111/j.1471-4159.2012.07942.x PMID:23050641

Travaglianti, F., Babic, A., & Hansez, I. (2018). Relationships between employment quality and intention to quit: Focus on PhD candidates as traditional workers. *Studies in Continuing Education*, *40*(1), 115–131. Advance online publication. doi:10.1080/0158037X.2017.1396448

Trikkalinos, I.K. (1977). *Remarks on the published study on the geology of the Acropolis*. Proc Acad Athens, Athens.

Uekane, T. M., Aquino Neto, F. R., & Gomes, L. N. F. (2011). Development and validation of a method for the analysis of tetracyclines in chicken-muscle by liquid chromatography-electrospray-mass spectrometry in tandem (LC-ESI-MS/MS). *Quimica Nova*, *34*(1), 43–48. doi:10.1590/S0100-40422011000100009

UFO Dome House Indonesia. (2022). *Dome house*. https://www.domehouse.id/

UNEP. (2019). *Stockholm Convention, Protecting human health and the environment from the persistent organic pollutants*. www.pops.int

UNESCO. (2014). *Education Systems in ASEAN+6 Countries: A Comparative Analysis of Selected Educational Issues*. Education Policy Research Series Discussion Document No. 5, United Nations Educational, Scientific and Cultural Organization.

Unger, M., & Polt, W. (2017). The Knowledge Triangle between Research, Education and Innovation – A Conceptual Discussion. *Foresight and STI Governance, 11*(2), 10–26. doi:10.17323/2500-2597.2017.2.10.26

United Nations General Assembly. (2015). *Transforming our world: the 2030 Agenda for Sustainable Development*. A/RES/70/1.

United Nations. (2005). *Hyogo Framework for Action 2005-2015: Building the Resilience of Nations and Communities to Disasters*. A/CONF.206/6.

United Nations. (2015). *Sendai framework for disaster risk reduction 2015–2030*. United Nations Office for Disaster Risk Reduction.

Unsworth, J., (2010). *History of pesticide use, International union of pure and applied chemistry*. Academic Press.

Uysal, M. P. (2016). Evaluation of learning environments for object-oriented programming: Measuring cognitive load with a novel measurement technique. *Interactive Learning Environments, 24*(7), 1590–1609. doi:10.1080/10494820.2015.1041400

van Staveren, M. F., Warner, J. F., & Khan, M. S. A. (2017). Bringing in the tides. From closing down to opening up delta polders via Tidal River Management in the southwest delta of Bangladesh. *Water Policy, 19*(1), 147–164. doi:10.2166/wp.2016.029

van Woezik, T., Koksma, J., Reuzel, R., Jaarsma, D., & Jan van der Wilt, G. (2020). How to encourage a lifelong learner? The complex relation between learning strategies and assessment in a medical curriculum. *Assessment & Evaluation in Higher Education, 45*(4), 513–526. doi:10.1080/02602938.2019.1667954

Ventola, C. L. (2015). *The Antibiotic Resistance Crisis. Part 1: Causes and Threats*. https://pubmed.ncbi.nlm.nih.gov/25859123/

Vihervaara, P., Marjokorpi, A., Kumpula, T., Walls, M., & Kamppinen, M. (2012). Ecosystem services of fast-growing tree plantations: A case study on integrating social valuations with land-use changes in Uruguay. *Forest Policy and Economics, 14*(1), 58–68. doi:10.1016/j.forpol.2011.08.008

Vitae. (2011). *Coaching for research in UK higher education institutions: A review*. https://www.vitae.ac.uk/vitae-publications/reports/coaching-report-2012-vitae.pdf

Vogler, C. (1999). The Stages of the Hero's Journey. *Myth and the Movies: Discovering the Mythic Structure of 50 Unforgettable Films,* 300.

Volger, J. S., Thompson, P., Davis, D. W., Mayfield, B. E., Finley, P. M., & Yasseri, D. (2018). The Hard Work of Soft Skills: Augmenting the Project-based Learning Experience with Interdisciplinary Teamwork. *Instructional Science, 46*(3), 457–488. doi:10.100711251-017-9438-9

Volodymyr, K. (2018). Indonesias Higher Education: Context, Policy, And Perspective. *Asian Journal of Contemporary Education, 2*(2), 159–172. doi:10.18488/journal.137.2018.22.159.172

Vuran, B., Ulusoy, H. I., Sarp, G., Yilmaz, E., Morgül, U., Kabir, A., Tartagla, A., Locatelli, M., & Soylak, M. (2021). Determination of chloramphenicol and tetracycline residues in milk samples by means of nanofiber coated magnetic particles prior to high-performance liquid chromatography-diode array detection. *Talanta, 230*, 122307. doi:10.1016/j.talanta.2021.122307 PMID:33934773

Walter, C., Schmidt, S., Rosenstiel, W., Gerjets, P., & Bogdan, M. (2013). Using Cross-Task Classification for Classifying Workload Levels in Complex Learning Tasks. *Humaine Association Conference on Affective Computing and Intelligent Interaction*, 876–881. 10.1109/ACII.2013.164

Walters, J. P., Greiner, B., O'Morrow, E., & Amadei, B. (2017). Fostering Systems Thinking Within Engineers Without Borders Student Teams Using Group Model Building. *International Journal of Engineering Education, 22*(1A), 247-260.

Waluyo, D. T. (2021). *RISHA instant technology builds disaster victim's house*. https://infopublik.id/kategori/sorot-ekonomi-bisnis/537996/risha-teknologi-instan-membangun-rumah-korban-bencana?show=

Wang, J., Antonenko, P., Keil, A., & Dawson, K. (2020). Converging Subjective and Psychophysiological Measures of Cognitive Load to Study the Effects of Instructor-Present Video. *Mind, Brain and Education: the Official Journal of the International Mind, Brain, and Education Society, 14*(3), 279–291. doi:10.1111/mbe.12239

Wang, S., Zhang, H. Y., Wang, L., Duan, Z. J., & Kennedy, I. (2006). Analysis of sulphonamide residues in edible animal products: A review. *Food Additives and Contaminants, 23*(4), 362–384. doi:10.1080/02652030500499359 PMID:16546883

Wang, Y., Yu, X., He, K., Song, S., Jia, G., & Huang, Z. (2012). Ecosystem services value based on land use change in Huairou reservoir watershed. *Nongye Gongcheng Xuebao (Beijing), 28*(5), 246–251.

Warren-Rhodes, K., Schwarz, A. M., Boyle, L. N., Albert, J., Agalo, S. S., Warren, R., Bana, A., Paul, C., Kodosiku, R., Bosma, W., Yee, D., Ronnback, P., Crona, B., & Duke, N. (2011). Mangrove Ecosystem Services and the Potential for Carbon Revenue Programmes in Solomon Islands. *Environmental Conservation, 38*(4), 485–496. doi:10.1017/S0376892911000373

Wesselink, A., Warner, J., Syed, M. A., Chan, F., Tran, D. D., Huq, H., & Wester, P. (2015). Trends in flood risk management in deltas around the world: Are we going'soft'? *International Journal of Water Governance, 3*(4), 25–46. doi:10.7564/15-IJWG90

White, D., Edmond, H., Eller, C., Lui, E., Sing, S., O'Toole, R., & Kremmyda, G. (2021). Design Thinking for Social Impact. In G. Kremmyda (Ed.), *Humanitarian Civil Engineering: Practical solutions for an interdisciplinary approach*. ICE Publishing.

WHO. (2020). *Antibiotics Resistance*. https://www.who.int/news-room/fact-sheets/detail/antibiotic-resistance

WHO. (2021). *Antimicrobial Resistance*. https://www.who.int/news-room/fact-sheets/detail/antimicrobial-resistance

Wikipedia.org. (n.d.). *Patulin*. https://en.wikipedia.org/wiki/Patulin

Williams, C. (2019). The Hero's Journey: A Mudmap for Change. *Journal of Humanistic Psychology, 59*(4), 522–539. Advance online publication. doi:10.1177/0022167817705499

Williams, C. A. (1919). *History of the Rivers in the Gangetic Delta (1750-1918). Netherlands Engineering Consultants (NEDECO)*. Bengal Secretariat Press.

Winarno, S., Griffith, A., & Stephenson, P. (2010). Reducing earthquake risk to non-engineered buildings in Indonesia. *Construction Innovation*, *10*(4), 421–434. doi:10.1108/14714171011083588

Wong, F., Alegria, H. A., Jantunen, L. M., Bidleman, T. F., Salvador-Figueroa, M., Gold-Bouchot, G., Ceja-Moreno, V., Waliszewski, S. M., & Infanzon, R. (2008). Organochlorine pesticides in soils and air of southern Mexico: Chemical profiles and potential for soil emissions. *Atmospheric Environment*, *42*(33), 7737–7745. doi:10.1016/j.atmosenv.2008.05.028

Yadav, K., Garg, L., & Kaur, L. (2020). Pesticides Usage: Effect on Human Health. *International Journal of Current Microbiology and Applied Sciences*, *9*(8), 2605–2615. doi:10.20546/ijcmas.2020.908.298

Yang, G., Bai, Y., Wu, X., Sun, M., Liu, X., Yao, X., Zhang, C., Chu, Q., Jiang, L., & Wang, S. (2018). Patulin induced ROS-dependent autophagic cell death in Human Hepatoma G2 cells. *Chemico-Biological Interactions*, 24–31. Advance online publication. doi:10.1016/j.cbi.2018.03.018 PMID:29604266

Yang, X., Wang, F., Meng, L., Zhang, W., Fan, L., Geissen, V., & Ritsema, C. J. (2014). Farmer and retailer knowledge and awareness of the risks from pesticide use: A case study in the Wei River catchment, China. *The Science of the Total Environment*, *497–498*, 172–179. doi:10.1016/j.scitotenv.2014.07.118 PMID:25128887

Younger, J. S., Booth, D. J., Parry, D. E., & Kurniawan, K. (2018). Sustainable Humanitarian Engineering in Practice – the East Bali Poverty Project. *Proceedings of the Institution of Civil Engineers. Municipal Engineer*, *171*(2), 67–77. doi:10.1680/jmuen.16.00034

Young, S. N., Vanwye, W. R., Schafer, M. A., Robertson, T. A., & Poore, A. V. (2019). Factors Affecting PhD Student Success. *International Journal of Exercise Science*. PMID:30761191

Yu, H., Tao, Y., Chen, D., Wang, Y., Huang, L., Peng, D., Dai, M., Liu, Z., Wang, X., & Yuan, Z. (2011). Development of a high-performance liquid chromatography method and a liquid chromatography–tandem mass spectrometry method with the pressurized liquid extraction for the quantification and confirmation of sulfonamides in the foods of animal origin. *Journal of Chromatography. B, Analytical Technologies in the Biomedical and Life Sciences*, *879*(25), 2653–2662. doi:10.1016/j.jchromb.2011.07.032 PMID:21840270

Zahabi, M., Abdul Razak, A. M., Shortz, A. E., Mehta, R. K., & Manser, M. (2020). Evaluating advanced driver-assistance system trainings using driver performance, attention allocation, and neural efficiency measures. *Applied Ergonomics*, *84*, 84. doi:10.1016/j.apergo.2019.103036 PMID:31987518

Zainuddin, Z., & Keumala, C. M. (2018). Blended Learning Method within Indonesian Higher Education Institutions. *Jurnal Pendidikan Humaniora*, *6*, 69–77.

Zhang, X., & Liu, S. (2017). Understanding Reading Comprehension in Multi-display Presenting System: Visual Distribution and Cognitive Effect. In C. Stephanidis (Ed.), *HCI International 2017 – Posters' Extended Abstracts* (pp. 207–214). doi:10.1007/978-3-319-58753-0_32

Zimdahl, R. L. (2015). Antibiotics. *Six Chemicals That Changed Agriculture*, 165–182. doi:10.1016/b978-0-12-800561-3

About the Contributors

Yiannis Koumpouros (BSc, MSc, MBA, PhD) is Ass. Professor in the University of West Attica, Department of Public and Community Health. Specialized in health informatics, e-health, m-health, telemedicine, assistive technologies, rehabilitation, users satisfaction assessment, and strategic management. Expert for the European Commission in the e-health and m-health sectors, and evaluator of numerous R&D projects. Responsible for the WHODAS 2.0 scale cultural adaptation for the World Health Organization in Greece. Top-level managerial experience in the private and public healthcare sector (chairman of hospitals, member of the board of directors, business development and strategic management director, etc.). Chairman of several committees in the Ministry of Development. Regular member of the Consultative Technical Council (Ministry of Internal Public Administration and Decentralisation), etc. He has published more than 70 papers in reputed journals and conferences and has been serving as an editorial board member of repute. He has participated in more than 50 R&D funded projects in the field of healthcare and ICTs.

Angelos Georgoulas is Adjunct Assistant Professor and Research Associate at the University of West Attica, teaching a number of undergraduate courses both in the Faculty of Engineering and the Faculty of Health and Caring Professions. He has completed his undergraduate, postgraduate taught and doctoral studies at the School of Electrical and Computer Engineering of the National Technical University of Athens, Greece. His related areas of expertise include mobile applications development, digital security, medical informatics, e-health and m-health Applications. Dr. Georgoulas has published several scientific papers in international journals and conferences. He is also registered expert for the European Commission, participating in several peer review panels for various programmes, proposals and projects in the field of healthcare and ICTs (e.g. FP7, HORIZON 2020, etc.). He is a Chartered Engineer by the Technical Chamber of Greece (1999) and member of the IEEE, the Greek Society of Biomedical Technology, the Hellenic Association of Computer & Communication Engineers and the Hellenic Association of Electrical & Mechanical Engineers.

Georgia Kremmyda is Professor and Head of Teaching at the School of Engineering, University of Warwick. Georgia completed her undergraduate, postgraduate taught and doctoral studies at the National Technical University of Athens, Greece. Before joining Warwick in 2015 and for a period of 15 years, Georgia shared her time between carrying leading engineering roles in industry and delivering teaching and research in academia. Her teaching and research interests are oriented, but not limited, to the design of earthquake resilient structures, sustainable cities, infrastructure for emergencies and the delivery of interdisciplinary, problem-based education, especially in resource-constrained environments. Within her academic role, Georgia is leading initiatives related to the development of innovative programmes which are promoting equitable access to education, in Science and Engineering. Georgia leads the Warwick Humanitarian Engineering programme and Co-leads the Warwick Global Research Priority in Sustainable Cities, both aligned to the UN Sustainable Development Agenda 2030. On international level, Georgia is elected Vice President (Conferences) of the International Network of Women Engineers and Scientists for 2021-2023. Within Warwick, she is Deputy Chair of the Institutional Athena Swan Self-Assessment Team, leading the subgroup working on 'Organisation and culture for inclusivity, diversity and equality' and 'Support to trans people'. She is also Chair of the STEM Education and Student Experience within the framework of the Warwick STEM Grand Challenge programme.

* * *

Mohammad Al-Amin has been working as an Assistant Engineer at the Local Government Engineering Department (LGED) of Bangladesh since 2018. He graduated in Civil Engineering from Dhaka University of Engineering and Technology (DUET) in 2013 and did his Master's in Water Resources Development from the Institute of Water and Flood Management (IWFM), BUET in 2018.

Tugba Altan, Ph.D., is a research assistant at the Department of Educational Sciences in the Faculty of Education at Kahramanmaras Sutcu Imam University, Turkey. Her research focuses on designing and evaluating multimedia learning environments such as textbooks and mobile apps in K12 and teacher education, integrating massively multiplayer online role-playing games into K12 education, and using mobile technologies in teacher education.

Sujit Kumar Bala has been working as Professor under the Institute of Water and Flood Management (IWFM) since 2011, (BUET). He has joined the IWFM, BUET as lecturer in 1997. Before joining the BUET, he worked as consultant in many national consultancy projects. He served in Surface Water Modelling Centre now Institute of Water Modelling. He was involved in the construction of the Jamuna Multipurpose Bridge. Dr. Bala had his M.Sc. in Irrigation Engineering from Tashkent Institute of Irrigation and Land Reclamation, Uzbekistan (ex-USSR) in 1982. He received Ph.D. from the same institute in 1986. His field of interests include bridge hydraulics, irrigation, gender and water, water control structures, etc. He had several international journal publications and was involved in international and national research projects funded by EU, JICA, DFID, IFAD, etc. He participated in many national and international conferences. He served as the Director, IWFM from 2017 to 2019.

Biswa Bhattacharya is an Associate Professor of Hydroinformatics in the Hydroinformatics Chair Group of the Integrated Water Systems and Governance at IHE Delft Institute for Water Education. Biswa Bhattacharya has a PhD degree in Hydroinformatics from Delft University of Technology and IHE Delft Institute for Water Education. He also holds an MSc degree in Hydroinfomratics (with distinction) from IHE Delft and a Bachelor of Engineering (Civil) from Indian Institute of Engineering Science and Technology, India. Prior to his PhD he was engaged in India in numerical modelling of river systems and in hydraulic and hydrological data collection, analysis, and management, and in designing and monitoring of engineering protection. His professional interests include flood risk management, hydrological modeling and forecasting, flood mapping and coastal flooding. He also has long experience in coordinating and managing research, educational and capacity-building projects of national and international standards. He is connected with networks of hydrologists (e.g., International Association of Hydrological Sciences) and international organizations (e.g., WMO and UNESCO) and with their activities. He wrote several publications, journal papers, conference proceedings and chapters in books. Since 2010 he is coordinating a pan-European Masters programme on flood risk management.

Veysel Coskun received his Ph.D. degree in Computer Education and Instructional Technology from Marmara University in January 2018. He obtained his master's degree in Computer Education and Instructional Technology from Sakarya University in June 2009. His research interests include human–computer interaction, multimedia learning, computer science education, and instructional design.

Dong Doan, Ph.D. Science and Technology Application for Sustainable Development Research Group, Ho Chi Minh City University of Transport, Ho Chi Minh City, Vietnam. Research interest: Content-Centric Networking; Internet of Things; Ad-hoc Networks; Image Processing; Sensing; Machine Learning; Programing, Humanitarian engineering.

Khai Ha is currently a lecturer at Faculty of Environment & Natural Resources (FENR) – Ho Chi Minh City University of Technology (HCMUT). He obtained his Bachelor degree in Hydrogeology and Geological engineering, Ha Noi University of Mining and Geology in 2007, and received Master in Water Resources Engineering, Chulalongkorn University in 2015, and PhD degrees in Environmental Engineering at Kunsan National University, South Korea in 2019. He started his academic career at HCMUT since 2021. Dr. Ha's professional experience in teaching, research and consultancy spans over 14 years in Vietnam. He teaches several undergraduate and graduate subjects, including Water quality management, Environmental Monitoring. Dr. Ha have worked as groundwater planner and researchers in Division for Water Resources Planning and Investigation for the south of Vietnam and Groundwater and soil Pollution laboratory of Kunsan National University. In these institutes, he implemented several project on groundwater quantity and quality management, modelling, and investigation.

Budi Hasiholan is a lecturer in Civil Engineering, particularly Construction Engineering and Management, at Institut Teknologi Bandung (ITB), Indonesia. He has a bachelor's in Civil Engineering from the ITB (Bandung, Indonesia), an M.Eng. in Civil Engineering from Chulalongkorn University (Bangkok, Thailand), and a Ph.D. in Civil Engineering from Hokkaido University (Sapporo, Japan). His research interests include Automation in Construction, Lean Construction, Productivity Improvement, and Construction e-Procurement. Since December 2018, he has been involved in "“Enabling Humanitarian Attributes for Nurturing Community-based Engineering/ ENHANCE" Project at ITB, funded by Erasmus+.

Md. Saiful Hossain is Superintending Engineer of Processing and Flood Forecasting Circle (PFFC) of Bangladesh Water Development Board (BWDB).

A. K. M. Saiful Islam is a Professor at the Institute of Water and Flood management of Bangladesh University of Engineering and Technology (BUET). He has received a Bachelor of Science in Civil Engineering (1989) from BUET, a Master of Science in Water Resources Engineering (1999) from BUET, and a Ph.D. in Civil Engineering (2004) from Drexel University, USA. His research interests aim at climate change impact on hydrology, water resources management, urban and coastal flood management, remote sensing for disaster risks reduction. His research adopted regional climate modelling, basin-scale hydrological modelling, river hydrodynamic modelling, urban flood modelling, coastal and storm surge modelling, weather forecast modelling, satellite remote sensing for environmental monitoring, hydro-informatics for disaster risks reduction and climate change adaption. He has published over 67 peer-reviewed journal articles, more than 100 conference papers, and contributed to 9 book chapters. Over the past decade, Dr. Islam has led several major international scientific climate change assessments and the reports by the Intergovernmental Panel on Climate Change (IPCC). He contributed to the IPCC's Fifth Assessment Report and IPCC Special Report on Global Warming of 1.5°C as an Expert Reviewer. He is currently a Lead Author on the IPCC's Sixth Assessment Report. Dr. Islam has given more than 100 invited talks and seminars throughout the world, speaking on diverse topics focusing on climate change impacts on water resources, challenges of weather and flood forecasting, urban drainage management, estimation of riverbank erosion using remote sensing, cyclonic storm surges under sea-level rise, changes in meteorological droughts, environmental impact assessment & mitigation, and climate vulnerability & risks assessment. He is teaching about 10 graduate courses and organized many workshops locally and overseas. He also currently serves as the Coordinator of the Climate Change Study Cell at IWFM, BUET.

G. M. Tarekul Islam is a Professor at the Institute of Water and Flood Management (IWFM) of the Bangladesh University of Engineering and Technology (BUET). He graduated in Civil Engineering in 1994. He did his PhD from the University of Tokyo in 2000. He was a recipient of the prestigious JSPS fellowship to pursue post-doctoral research at the Department of Civil and Environmental Engineering of Hiroshima University, Japan during 2006-2008. His research areas include hydro-morphology of rivers and floodplains, applications of remote sensing and GIS in water and environmental monitoring, disaster management and humanitarian engineering. He was awarded by the University Grants Commission (UGC) of Bangladesh as the best researcher in Engineering and Technology category in 2003. He has served as a member of different committees of national and international organizations. He has rendered many consultancy and advisory services to nationally important projects of different government and non-government organizations. He has published over 145 papers in international journals and conference proceedings.

Rafiza Islam obtained her B.Sc. and M.Sc. degree from Department of Chemistry, National University, Bangladesh. Now she is continuing her PhD on "Investigation of antibiotic residues in meat, milk and eggs samples" at the Department of Chemistry, University of Dhaka from 2017. She got a fellowship from Bangabandhu Science and Technology Fellowship Trust, Ministry of Science and Technology, Government of Bangladesh. She attended a workshop on "Gas Chromatography Mass Spectrometry (GC-MS)" at National University Laos, Vientiane, Lao PDR, organized by ANFEC and another workshop on "Food Analysis Workshop: Proficiency Testing and Reference Materials Development" at Asia Pacific Food Analysis Network (APFAN), Bangkok, Thailand. She has four publications including an international publication entitled "Determination of chloramphenicol in meat samples using liquid chromatography–tandem mass spectrometry" in the Journal of Food Science & Nutrition. She has a particular experience in Liquid Chromatography Mass Spectrometer (LC-MS/MS), Gas Chromatography Mass Spectrometer (GC-MS), Gas Chromatography (GC-ECD), Gas Chromatography (GC-FID), High Performance Liquid Chromatography (HPLC), UV-Visible (UV-Vis) Spectrophotometer and Total Organic Carbon Analyzer (TOC).

Syed Labib Ul Islam completed his BSc in Civil Engineering from Rajshahi University of Engineering & Technology (RUET) in 2017. He completed his MSc in Water Resources Development from the Institute of Water and Flood Management (IWFM), Bangladesh University of Engineering and Technology (BUET) in 2021. At present he is working towards a PhD in Civil Engineering from the University of Birmingham. Professionally he is a civil engineer who has worked on a number of engineering megaprojects in Bangladesh - including several projects of Bangladesh Railway as a Railway Track Engineer. He has also worked as a Research Associate with the IWFM. He is passionate about a diverse range of topics including long-term sustainability of infrastructure, resource management, climate change adaptation, soft engineering, and coastal zone management.

Abu Khan graduated in Civil Engineering from BUET in 1980 and later completed his MSc. Engg in Water Resources from the same university with a specialization in the hydro-morphological impact assessment using mathematical models. Mr. Khan has a long professional career of 30 years in the planning and management of water resources. He served in the Bangladesh Water Development Board in various capacities in the planning and design of major flood control, irrigation and drainage projects. Since 1990 he has been engaged in the use of Mathematical Hydraulic Models for improved planning of water and water-related environmental studies.

Irwanda Laory is an Associate Professor at the University of Warwick, United Kinfdom. He graduated from Bandung Institute of Technology, Indonesia, with a BSc in Civil Engineering and then went on to Bauhaus Universiteit Weimar, Germany, where he undertook a master program (Natural Hazard and Mitigation in Structural Engineering) and obtained his MSc in civil engineering. This was followed by his PhD at Swiss Federal Institute of Technology Lausanne (EPFL), Switzerland. His research is focused on development of methodologies for monitoring and maintenance of civil infrastructure.

Trung Le is currently an Associate Professor and Head of Department of Resources and Environmental Informatics, Faculty of Environment & Natural Resources (FENR) – Ho Chi Minh City University of Technology (HCMUT). He obtained his Bachelor degree in Geomatics, University of Technology, Vietnam in 1981, and received Master and PhD degrees in Civil Engineering at the University of Tokyo, Japan in 1994 and 1997, respectively. He started his academic career at HCMUT since September 1981. Dr. Trung was a Research Fellow at University of Hannover – Germany in 1998; University of Hosei – Japan in 2008; Duke University, North Carolina – USA in 2011 and Maryland University - USA in 2016. He was a key member of the Union Association of Science and Technology, HCM City and Vice President of the HCM City Association of Geomatics. Dr. Trung's professional experience in teaching, research, and consultancy spans over 40 years in Vietnam. He teaches several undergraduate subjects, including Surveying, Remote Sensing, Land use planning and graduate courses on Neural Network, Remote Sensing, GIS, Land Evaluation. Since 2005, he became a Director of Geomatics Center - Information Technology Park (ITP) Viet Nam National University of Ho Chi Minh City (VNU-HCM).

Hiep Luu obtained his Bachelor Engineer in Telecommunications in 1999 and Master's degrees in GIS in 2003 in Ho Chi Minh City University of Technology (HCMUT). He is currently a senior lecturer at Faculty of Environment and Natural Resources, Ho Chi Minh City University of Technology (HCMUT) - VNU HCM. He used to be an active member of Annual Conference Organisation Board of GISnet Conference in Vietnam from 2000 to 2005, and was member of Academy Board GIScience of HCMUT from 2000 to 2006. Mr. Luu's professional experience in teaching, technology transfer and consultancy spans over 18 years in Vietnam. He is currently a Director of Centre for IT And GIS (DITAGIS).

J. Toby Mottram, BSc, PhD, DSc, CEng, FIStructE, SFHEA, is affiliated to The School of Engineering at The University of Warwick, UK. He's a member of the Civil and Environmental Engineering stream and his specialist field is structural engineering. He is internationally known for his research contributions towards us reliably and safely designing structures of fibre-polymer composite. Toby delivers an undergraduate module on Forensic Engineering for how professional engineers can learn from failures. As a partner in the ENHANCE Erasmus+ project he has been involved in supporting pedagogy projects that are for nurturing community-based engineering in Asian and Sub-Sahara countries.

Duygu Mutlu-Bayraktar received her Ph.D. in Computer Education and Instructional Technology Department from Marmara University in July 2014. She obtained her master's degree in Computer Education and Instructional Technology Department from Hacettepe University in June 2010. Her research interests include multimedia learning, instructional design, eye-tracking, cognitive science, and human-computer interaction.

Nilufar Nahar is a former Chairperson, Department of Chemistry, University of Dhaka. She graduated (BSc and MSc) from the University of Dhaka and did Ph.D. from the Swedish University of Agricultural Sciences, Uppsala, Sweden. She started her academic career at the Department of Chemistry as a lecturer and retired as a full Professor. She did Post-Doctoral Research at Arrhenius Laboratory, Stockholm University, Sweden, she was a Visiting Scientist at NCI, USA, Shaman Pharmaceutical Company, USA, Mahidol University, Bangkok, Thailand. Her main research areas are Natural Products Chemistry, Food and Environmental Chemistry and Carbohydrate Chemistry. Prof Nahar supervised 19 PhD and 10 MPhil students, more than 200 undergraduate students received their degree under her supervision, she published 171 papers in peer reviewed international and national journals and presented more than 165 papers on her research results in many different countries of the Asia, Europe, America and Latin America. She is co-author of a book on Co-author of a book "Use of medicinal plants for management of Diabetes" and co-author of two Chapters of two different books. Prof Nahar received research grant from ISP, IFS, OPCW and UNESCO besides national grant from University Grants Commission of Bangladesh. She has arranged many conference/seminar/symposia in Bangladesh including farmers´ awareness programs regarding toxicity of pesticides and environmental issues. She established national, regional and international research collaboration with distinguished scientists during her tenure at the University. She is one of the founding members three Swedish supported networks; ANRAP, NITUB and ANFEC. She worked as an Executive Editor of Journal of Bangladesh Chemical Society, she was national representative IUPAC and member of Division VII (Chemistry and Human Health). She was also a Titular member of IUPAC Division VII. Professor Nahar received three distinguished Scientists awards and one gold medal for her devotion and contribution to science and chemistry.

Ali Olamat received his bachelor's degree in Biomedical Engineering from Hashemite University, Zarqa-Jordan, in 2007. He received his master's degree in Biomedical Engineering from the University of Luebeck, Germany, in 2010. He received his Ph.D. in Biomedical Engineering from Istanbul University, Istanbul-Turkey, in 2019. His research interests include Signal and Systems, Digital Signal Processing, Imaging and Image processing, Numerical methods and Mathematics for Engineering, Machine-Learning and Deep-Learning.

Pınar Özel received her BS degree in Electrical-Electronics Engineering from Erciyes University, Kayseri, in 2007, and MS degree in Biomedical Engineering from Bogazici University, İstanbul in 2011. She obtained her Ph.D. degree in Biomedical Engineering from Istanbul University in 2019. As a professional experience, she worked as a product manager. Her research areas are signal processing and machine learning techniques to biomedical signals.

Nargis Parvin obtained Bachelor of Science (BSc) and Master of Science (MSc) degrees from Department of Chemistry, National University. At present, she is doing her PhD on "Studies of Antibiotic Residues in Beef and Chicken Meat Samples" at the Department of Chemistry (Organic), University of Dhaka which was started from 2018. She is experienced in Organic, Analytical, and Environmental Chemistry. She has particularly expertise in UV-Visible (UV-Vis) Spectrophotometer, High Performance Liquid Chromatography (HPLC), Liquid Chromatography Mass Spectrometer (LC-MS/MS), Gas Chromatography (GC-ECD), Gas Chromatography Mass Spectrometer (GC-MS) and Total Organic Carbon Analyzer (TOC). She participated in many training programs, conferences relevant to her research and attended seminars for expanding scientific knowledges for her research field and she can contribute in the field of Chemistry. She participated in a workshop on "Gas Chromatography Mass Spectrometry (GC-MS)" at The Royal University of Cambodia, Phnom Penh, Cambodia. She has two publications in Bangladesh Pharmaceutical Journal and Asian Journal of Pharmacognosy. She got a fellowship from BAN:04, ISP, Uppsala, Sweden and University of Grant Commissions of Bangladesh.

Indra Perdana completed his undergraduate study in chemical engineering from Universitas Gadjah Mada. in 1997 and his master in 2002 from the same university. He received his PhD in chemical reaction engineering from Chalmers University of Technology, Sweden in 2008. Currently he is working as a university lecturer and researcher at the Department of Chemical Engineering, Universitas Gadjah Mada. His research area is of circular economy in mineral processing, and he currently leads a research group in lithium-ion battery (LIB) recycling supported by Indonesian government and industrial sectors. Related to humanitarian engineering, He has been consistently involved in student community services as a supervisor for many years. Since December 2018, together with partners from Greece, Bangladesh, Vietnam and Indonesia, Indra Perdana has also been active within Enhance Project under coordination of University of Warwick in UK funded by Erasmus+ EU Program which addresses engineering attributes for humanitarian issues.

Krishna Suryanto Pribadi is currently a Professor in Disaster Mitigation and Chair of the Construction Engineering and Management Research Group and Vice Head of the Disaster Mitigation Research Centre at the Bandung Institute of Technology. He served as the Vice Chair for Research and Development, Training and International Collaboration in the National Construction Service Development Board (CSDB) during the period of 2016 to 2020. From 2009 to 2011 he also chaired the National

Platform for Disaster Risk Reduction/PLANAS and during the period of 2008-2015 he also chaired the National University Forum for Disaster Risk Reduction. Dr. Pribadi obtained a Bachelor degree in Civil Engineering from Bandung Institute of Technology in 1977 and a Docteur d'Ingenieur en Genie Civil degree from Institute National des Sciences Appliquees de Lyon, France, in 1985. His research interest includes disaster mitigation, business continuity planning, post-disaster recovery, construction industry and human resource development. He is also a holder of project management expert certificate from Indonesian Association of Project Management, as well as mediator certificate from the Indonesian Mediation Centre at Gajah Mada University, Yogyakarta.

Mohammad Shoeb is a Professor of Chemistry at the Department of Chemistry, University of Dhaka. He obtained B.Sc and M.Sc from Department of Chemistry, University of Dhaka, and M.Sc from University of Aberdeen, UK. He pursued PhD in Natural Product Chemistry from the Robert Gordon, University, UK. He is leading the research group consisting of staffs, Post Doc, Postgraduate students (PhD, MPhil, MS) and technical staffs on "Analytical, Environmental and Natural Product Chemistry". He received research grants from International Science Programme (ISP), Uppsala University; International Foundation for Science (IFS), Stockholm; OPCW, Ministry of Science and Technology, Government of Bangladesh; University Grant Commissions of Bangladesh and Bose Centre. He supervised 7 PhD, 3 M.Phil. and quite a large number of MS and BS (project) students in natural products, analytical and environmental Chemistry and currently 6 PhD, 2 M.Phil. and MS and BS project students are working under his supervision. He has more than 105 research publications in peer reviewed journals with current citation: 2104, h-index: 24, i10-index: 46 and has attended 90 National and International conferences, and delivered lectures. He has collaborations with different universities of Australia, UK, Sweden, Belgium, Cambodia, Laos, Malaysia, Iran, Tanzania, Indonesia, India and Turkey.

Intan Supraba is the Faculty Member at the Department of Civil and Environmental Engineering, Faculty of Engineering, Universitas Gadjah Mada (UGM), Indonesia. In the year of 2004, she obtained her bachelor's degree in civil engineering from UGM. Then in the year of 2006, she completed her master's degree in environmental science and engineering from Nanyang Technological University, Singapore in a partnership with Stanford University, United States of America. In the year of 2015, she earned her doctoral degree in the Field of Engineering for the Environment from Hokkaido University, Japan. Before joining UGM in 2011, she worked in Singapore from 2006 to 2010 as a contractor in the Plant Engineering Construction Ltd., and as a structural engineering consultant in the Tritech Group Limited. She was the Secretary of University Network for Indonesia Infrastructure Development that is a consortium of 33 universities in Indonesia from 2016 to 2019. She is the current Head of Sanitation and Environmental Engineering Laboratory for period 2022-2025. On December 2022 she was elected as Executive Committee Member of International Association for Hydro Environment Engineering and Research, Asia Pacific Division (IAHR-APD). Her main research interest area is in water conservation, especially related to rainwater harvesting and algae bloom mitigation using constructed wetlands and bioswales.

Surjono is a professor in the urban and regional planning Department, Faculty of Engineering Universitas Brawijaya. He was graduated from Architecture Department at Universitas Brawijaya and completed his master's degree (Town Planning) and doctoral degree (Urban and Regional Planning) in Australia. His research is in the planning of poverty reduction, urban planning and design and development planning. His role in the project is to synchronize the new curriculum development with the standards at the University and national levels. Several courses that he teaches match with core issues of humanitarian engineering, such as poverty and social resilience, resilient cities, and urban management. The scope and perspective proposed by the Enhance Project have inspired him to write reference books that were: 1) Sustainable World, 2) Settlement Planning, 3) City and Community Resilience, and 4) Livable city and resilience community, all in the Indonesian language. He hopes that humanitarian engineering can contribute to the development of better higher education curriculum in Indonesia.

Van Tran Thi is currently an Associate Professor of the Faculty of Environment & Natural Resources (FENR) – Ho Chi Minh City University of Technology (HCMUT). She obtained her Bachelor degree in Meteorology at Russian State Hydrometeorological University (former, Leningrad Hydrometeorological Institute, USSR). She then went on to get a master degree in Geomatics from Ho Chi Minh City University of Technology (VNUHCM). She received a PhD degree in Using and Conserving Environmental Resources at Vietnam National University Ho Chi Minh City (VNUHCM) in 2012. She started her academic career as a Researcher at Geoscience Center, Vietnam National Academy of Sciences, in 1989. In 2005 she was elected Lecturer in Institute for Environment and Resources (VNUHCM). From 2014 up to now, she has worked for HCMUT. Dr. Van's professional experience in research and teaching over 20 years in Vietnam. She has implemented several projects and published many papers in refereed journals in the field of Geomatics, Meteo-Hydrology, Climate Change and Environment.

Phu Le Vo is currently an Associate Professor and Dean of the Faculty of Environment & Natural Resources (FENR) – Ho Chi Minh City University of Technology (HCMUT). He obtained his Bachelor degree in Biochemistry, Hue University, Vietnam in 1994, and received Master and PhD degrees in Environmental Studies at the University of Adelaide, South Australia in 2000 and 2008, respectively. He started his academic career at HCMUT since September 2000. Dr. Phu was a Research Fellow at Sanford School of Public Policy and Nicholas School of Environmental Science (NSOE), Duke University, North Carolina – USA in 2010 and 2011. He was also a Visiting Professor at Faculty of Life and Environmental Sciences, Tsukuba University, Japan in 2012. Dr. Phu was a Visiting Scholar at Faculty of Environmental Sciences, Wageningen UR and UNESCO-IHE (Delft), the Netherlands in 2013, and was a key member of the Consultancy Board for developing a Master curriculum program on Water Management and Climate Change Adaptation which was run by the Center for Water Management and Climate Change – VNU HCM. Dr. Phu's professional experience in teaching, research and consultancy spans over 21 years in Vietnam. He teaches several undergraduate subjects, including Water Resources Management, Coastal Zone Management, Clean Technology and graduate courses on Water Governance, Integrated Urban Water Management, Climate Change and Adaptation Measures. His field of expertise is Water Resources Management and Policy, IWRM, Groundwater Chemistry, Climate Change and Water Management. Since 2015, he became a national Short-term Consultant (STC) on Environmental Safeguards for various World Bank-financed projects in Vietnam in the Water and Transport Sectors.

Index

G

H

I

L

M

N

O

P

Q

R

S

T

U

V

W

CPSIA information can be obtained
at www.ICGtesting.com
Printed in the USA
LVHW022303220323
742312LV00006B/243

9 781668 456194